W9-BRT-664

THE CLIMATE OF THE EARTH

THE CLIMATE OF THE EARTH

Paul E. Lydolph
Professor of Geography
University of Wisconsin–Milwaukee

Don and Denise Temple,
Cartographers

ROWMAN & LITTLEFIELD
PUBLISHERS, INC.

Rowman & Littlefield Publishers, Inc.

Published in the United States of America in 1985
by Rowman & Allanheld, Publishers
(a division of Littlefield, Adams & Company)

Library of Congress Cataloging in Publication Data
Lydolph, Paul E.
The climate of the earth.
Includes bibliographical references and index.
1. Climatology. I. Title.
QC981.L93 1985 551.5 84-18081
ISBN 0-86598-119-1

Reprinted in 1989 by Rowman & Littlefield Publishers, Inc.
Printed in the United States of America

This book is dedicated to
my wife, Mary,
who typed the manuscript again and again.

CONTENTS

TABLES AND FIGURES

Tables

Figures

1 | INTRODUCTION

The Significance and Scope of Climatology

Climatology is the study of the exchanges of energy and mass at the interface between the atmosphere and the surface of the earth. The temperature, moisture, and motion characteristics of the atmosphere are generally the initiating factors at this interface for much of the activity in other natural phenomena, such as plant and animal growth, soil development, weathering and erosion processes, and the evolution of the landscape. These atmospheric characteristics are all-pervading in human affairs, so much so that they become taken for granted and one is apt to forget that the climate of a place and the weather events comprising that climate overwhelmingly determine the rhythm of life. The alternation of day and night—the receipt or not of sunlight—sets the basic cycle of life.

At breakfast each morning, one turns to the radio, television, or newspaper for the weather report to help plan what to wear, what precautions to take in traveling to work, and so forth. The economic fortunes of farmers completely depend upon heat and moisture conditions during the growing season. The well-being of humans around the globe to a great extent depends upon the availability of food and fiber and the avoidance of natural hazards such as damaging winds and floods, which are caused by the vagaries of weather and time variations of climate.

Subfields of Climatology

Curiosity about and dependence on the state of the atmosphere have prompted the development of a broad field of climatology that contains within it subfields specifically conceived to serve the needs of certain facets of life and economic activity. Within the general scope of climatology, the subfields of physical and synoptic climatology seek primarily to understand the distribution of climate over the earth by developing physical laws that relate such things as radiation and other forms of heat transfer, temperature, moisture, and atmospheric pressure and motion. Set in instantaneous time frames, these relations reveal synoptic patterns of areal distributions across the face of the earth. Such studies are invaluable to weather forecasters, who must know the causes behind weather events and the areal relationships among them. The subfield of regional climatology develops naturally from an understanding of the physical and synoptic aspects of climate. It serves to understand the climate of the earth more precisely by sectioning the earth's surface into regions that are small enough and homogeneous enough to be comprehended as areal units that can be compared with other areal units around the world.

A subfield of paleoclimatology has evolved that deals with past climate and attempts to ascertain causes and effects of long-term climatic changes outside the timespan of normal climatic variation. This subfield draws upon evidence compiled by such specialists as geologists, biologists, archeologists, and historians to piece together long-term climatic changes that have affected the earth through its long history of at least 4.5 billion years. The study of drastic climatic change in the past stems primarily from intellectual curiosity. But there are some pragmatic motives, for an identification of the causes of natural changes in the past may illuminate ways that man might induce changes in the future for his own benefit. It may also shed additional light on the course of human history. Closely related to climatic change are purposeful and inadvertent modifications of climate as a result of human activities. A study of past climatic change may identify key processes that, taken to some critical point, could set off series of chain reactions that could culminate in such dramatic events as the ice ages.

Pragmatic motives have stimulated the development of the broad field of applied climatology, which approaches climatology primarily from its effects on other phenomena, generally those that profoundly influence human lives and their economic well-being. Originally, most investigations of an applied nature related to agriculture, since intimate linkages exist between climate and the agricultural part of economic activity. In a broader scope it related to the entire field of biology, which became known as bioclimatology. Such studies quickly revealed the inadequacy of climatic data that had been gathered primarily in weather shelters about 1.5 meters above ground level. It soon became apparent that most crops are affected primarily by the air next to the ground surface or even within the soil itself, and this led to the (rapidly developing) field of microclimatology, the study of the air next to the earth's surface. As a result, new measuring instruments have been invented to collect data on temperature, humidity, and air motion within the first two meters of air.

Agroclimatology has continued to develop primarily as a subfield of microclimatology, but it has developed indices of its own that identify those elements of climate that are solely or primarily in control of specific phases of plant growth. More recently, as the world's population has become increasingly urbanized, the field of urban climatology has developed as a subfield of applied climatology. This again has resulted in the evolution of distinct sets of observations and climatic indices specifically designed to suit urban planning and construction considerations.

The Scope and Organization of This Book

Obviously, one book of reasonable length cannot fully cover all these diverse subfields of climatology. In a single semester's time, one can only establish a solid basis for comprehending the contemporary distribution of climate over the earth and understanding the reasons for it. Such knowledge will lay the groundwork for further studies on topics such as climatic change and the various applications of climatology to other phenomena. Therefore, this book will be limited primarily to such subfields as physical, synoptic, and regional climatology. Its two objectives are: (a) to describe the climate of the earth and (b) to develop principles that will make the distribution of climate meaningful. In practice, the second objective must be accomplished first. To attach significance to regional description, one must master the atmospheric processes relating heat, moisture, and motion. Although the focus is on properties and processes at the bottom of the atmosphere at the interface between the atmosphere and the earth's surface, linkages exist that cause upper portions of the atmosphere to affect what is going on below. Therefore, to understand actions at the surface one must know something about the entire atmosphere.

To grasp the complete makeup of the climate of a place, one must know both the statistical indices of individual climatic elements and the day-to-day weather events that create these statistics. In other words, one must have a compendium of statistical means and deviations in one hand and a set of daily weather maps in the other. As will be elaborated in Chapter 15, places in different parts of the earth having essentially the same statistical means of, for example, temperature and precipitation may experience entirely different types of weather. Thus, it is not enough to deal only in statistical averages and variations.

This book will consist of essentially three interwoven parts: (a) presentation of the basic principles governing atmospheric processes, (b) world distributions of statistical means and variations of individual elements and climatic types, and (c) analyses of weather complexes from which these statistics have been derived. To accomplish this the book will be organized into three parts. Chapters 2 through 11 will deal with the processes taking place in the atmosphere and between the atmosphere and the earth's surface. Chapters 12 through 14 will describe the resultant spatial and temporal distributions of the three primary resultant climatic elements: condensation forms and precipitation, temperature, and wind. Chapters 15 through 21 will describe and explain the climate of the earth on a regional basis.

Although it is impossible to talk about heat and moisture without mentioning temperature and precipitation, the complete discussions of elements per se have been delayed until all the atmospheric processes have been presented, since all the processes have bearings on the final, measured, climatic elements. More than heat affects temperature, and more than moisture affects precipitation. In fact, it is difficult to consider even one process at a time; so many feedback mechanisms exist in the atmosphere that almost every action causes a series of reactions. Thus, a complete consideration of temperature, probably the most commonly observed and all-pervading element of climate, is delayed until Chapter 14, after everything that affects it has been discussed.

The description of the climate of the earth in Chapters 16–21 will proceed on a continent to continent basis rather than on a climatic-type basis. The author believes it is better to describe climate and its causes and influences in terms of weather events, which are real happenings in real areas, rather than in terms of somewhat arbitrary climatic categories. A climatic type is an abstraction, but weather events actually happen. A consideration of them will not only impart a better understanding of the weather makeup of the climate of an area, but will also give the student some knowledge of world geography, real locations, and earth features.

To provide a worldwide framework, within which to fit the climate of each part of the earth and allow comparisons with other parts, Chapter 15 will briefly present a classification scheme and a distribution map of climatic types based on temperature and precipitation data. It must be realized, however, that the categories utilized in such classifications are very broad, with much variation within any given type from one continent to another. Students should remember that the types are not immutable nor identical from region to region and that their boundaries are not real. They are only transition zones across which sweep synoptic situations that transgress climatic-type boundaries.

Finally, this book is written for students. Generalizations have been presented without the clutter of finer qualifications that would satisfy more seasoned specialists. Since undoubtedly the largest group of users will come from North America, the regional discussion in Chapters 16–21 will start with that continent. Within this chapter certain principles of regional discussion can be established which can then be used in discussion of the other continents: first Eurasia, another primarily middle-latitude Northern Hemisphere landmass, then Africa because of its proximity to Eurasia, and finally South America, Australia–New Zealand, and the Polar areas.

2 CHEMICAL AND PHYSICAL STRUCTURE OF THE ATMOSPHERE

Atmospheric Composition

The earth's atmosphere is a mixture of gases in portions essentially as are shown in Table 2.1. Nitrogen makes up almost four-fifths the volume of the entire atmosphere, and oxygen more than one-fifth. All the other gases combined make up less than the remaining 1 percent. These ratios remain essentially the same to the height of more than 100 kilometers (more than 60 miles). Even so, molecular structures tend to change above 50 kilometers since exposure to direct bombardment of high-energy particles from the sun splits many molecules into their component atoms, often ionizing the air. For instance, O_2 typically splits into O_1, some of which may recombine with O_2 to form O_3, ozone. Thus, although oxygen is found at all levels, the molecule O_2 that is breathed by humans at the earth's surface is not necessarily present higher in the atmosphere.

Although nitrogen and oxygen are by far the most abundant gases, and oxygen is necessary to all life on earth, neither gas

Table 2.1 Average Composition of the Lower Atmosphere

Gas	Molecular composition	Percent by volume	Approximate molecular mass
Nitrogen	N_2	78.08	28
Oxygen	O_2	20.94	32
Argon	Ar	0.93	40
Carbon dioxide	CO_2	0.03(variable)	44
Neon	Ne	0.0018	20
Helium	He	0.0005	4
Ozone	O_3	0.00006	48
Hydrogen	H_2	0.00005	2
Krypton	Kr	Trace	
Xenon	Xe	Trace	
Methane	CH_4	Trace	
Water vapor	H_2O	Variable	18
Average			29

figures prominently in the study of climate. More important to weather processes are the heat absorption and radiation properties of gases, such as carbon dioxide and water vapor, and the consequent hydrologic cycles.

The amount of water vapor varies greatly from one place to another, according to its availability for evaporation at the surface and the holding capacity of the air above. Since all water vapor is derived from the earth's surface, it decreases rapidly with height. Unlike the other gases, water changes state within the range of temperatures commonly observed within the atmosphere; therefore, water vapor derived from the earth's surface does not accumulate in the air and mix upward thoroughly, but recondenses into liquid water and precipitates back to earth before it can rise more than a few kilometers into the atmosphere.

For these reasons, throughout much of the atmosphere there is practically no water vapor, and the average for the entire atmosphere is a small fraction of 1 percent. But near the earth's surface water vapor may account for as much as 4 percent of the atmosphere by volume in very warm, humid air. Since the water molecule is made up of two atoms of hydrogen and one atom of oxygen, its molecular mass is 18, which is considerably lighter than the molecular mass for most of the gases listed in Table 2.1. Thus, it makes up less of the atmosphere by mass than by volume, and humid air masses will be lighter than dry air masses with identical temperatures. The significance of this fact will become evident later.

Vertical Structure of the Atmosphere

Density and Pressure

Since air is a compressible mixture of gases, it is not the same density throughout. The upper air, pulled by the force of gravity toward the earth's surface, weighs down on the lower air, compresses it, and makes it more dense. Therefore, most of the mass of the atmosphere is in the bottom few kilometers. Upward from the earth's surface, the air thins very rapidly at first and then at slower and slower rates. So even though most of the atmosphere is near the earth's surface, it cannot be defined as having an outer edge. Gas molecules as high as 60,000 kilometers (37,000 miles) above the earth's surface move with the earth, attracted to it by magnetic and gravitational fields, and therefore can be considered as part of the earth's atmosphere. Of course, the air is very rare at this altitude. Half the mass of the atmosphere lies below approximately 5–6 kilometers (17,000–19,500 feet), depending on the temperature. Practically all weather phenomena with which this book will be dealing occur within the bottom 8–16 kilometers, which contain about 80 percent of the atmosphere by weight, and most of the considerations in this book will be limited to only the lower portion of that.

Since the thickness of the entire atmosphere, for practical purposes, is 100–200 kilometers, the atmosphere is only a thin shell compared to the earth, whose radius averages about 6360 kilometers. It can be seen that we are living at the bottom of a relatively shallow ocean of air, which we will consider briefly in its entirety before descending into its bottom layer, where we will be spending the rest of our time. Processes in the upper atmosphere may influence those in the lower atmosphere, although linkages are not clearly known at present.

Vertical Temperature Distribution

Evidence gathered from natural phenomena, such as the transmission of sound waves, and from direct observations by radiosondes, balloons, rockets, and satellites, have been patched together to construct a composite model of the vertical structure of the atmosphere (see Figure 2.1, Standard Atmosphere). On the average for

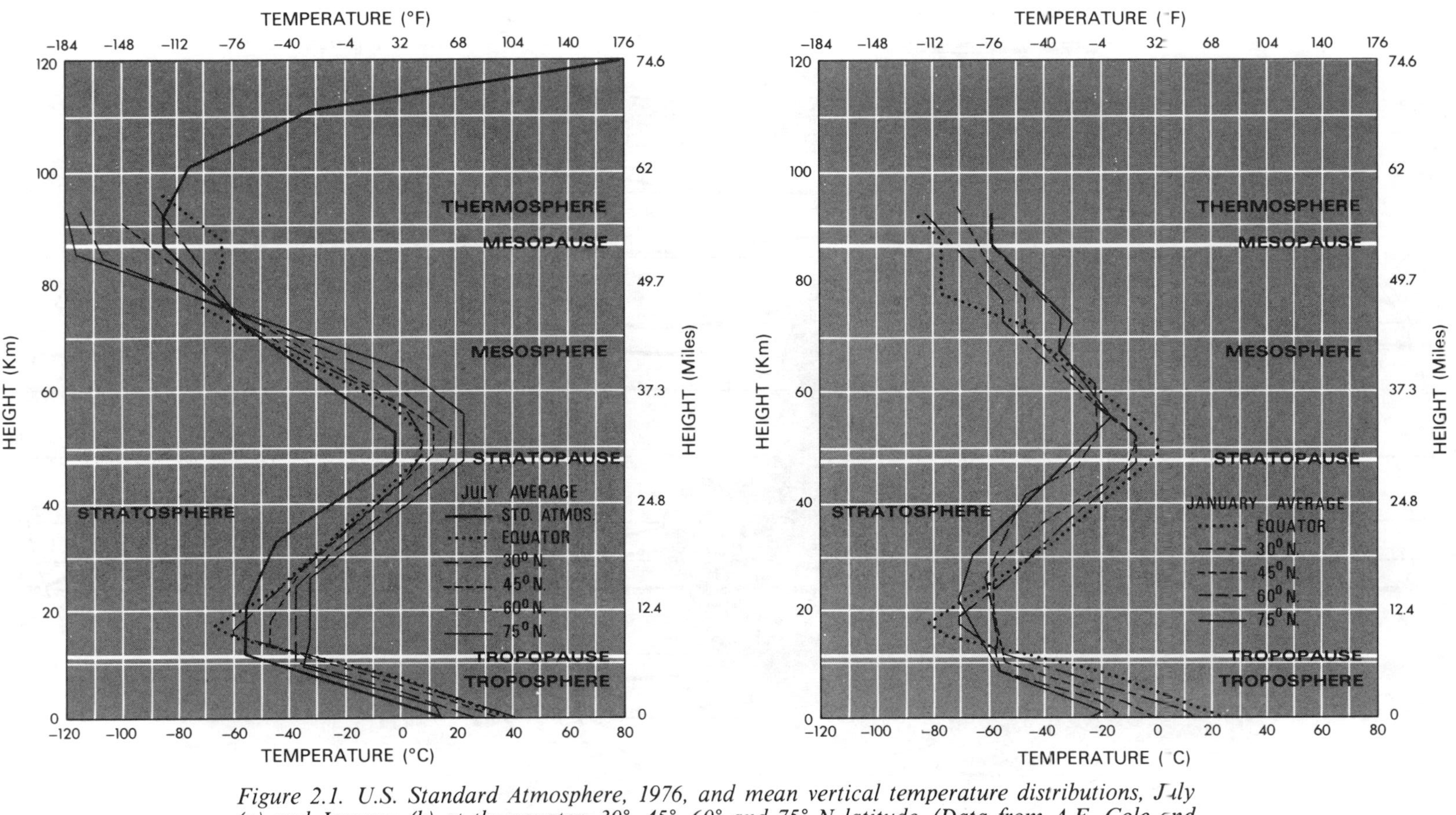

Figure 2.1. U.S. Standard Atmosphere, 1976, and mean vertical temperature distributions, July (a) and January (b) at the equator, 30°, 45°, 60° and 75° N latitude. (Data from A.E. Cole and A.J. Kantor, Air Force Reference Atmosphere, *Air Force Surveys in Geophysics No. 382.)*

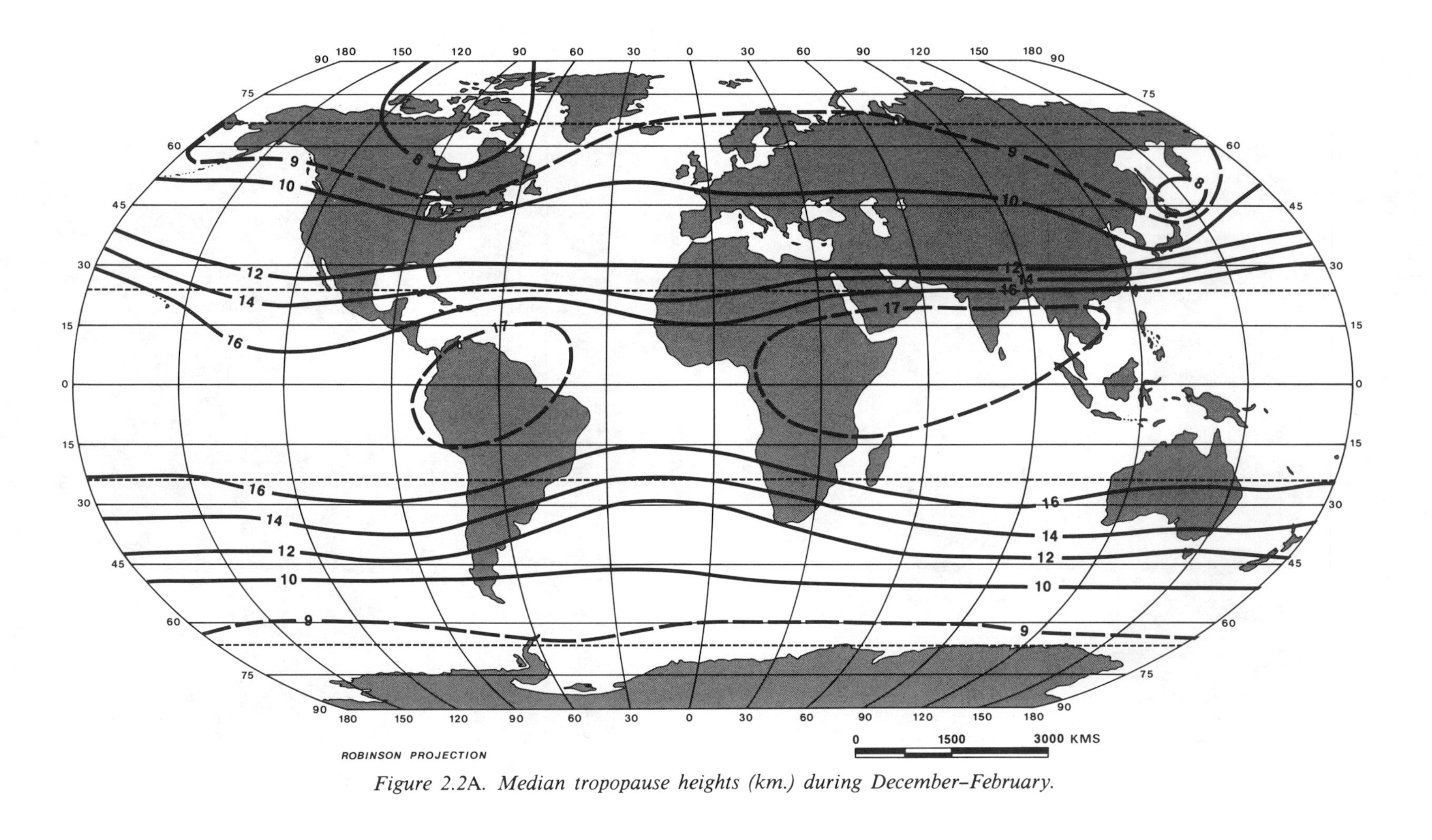

Figure 2.2A. Median tropopause heights (km.) during December–February.

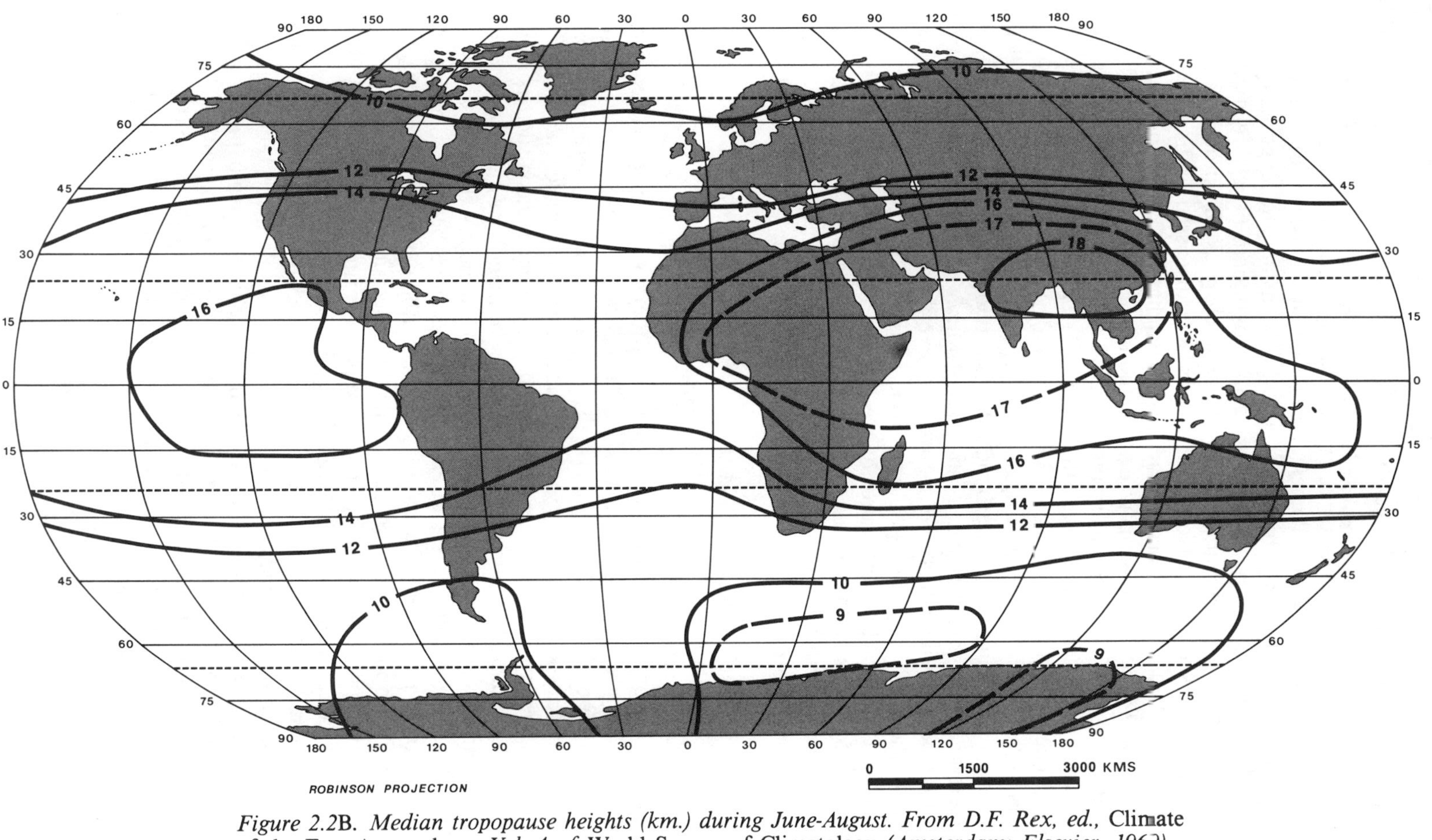

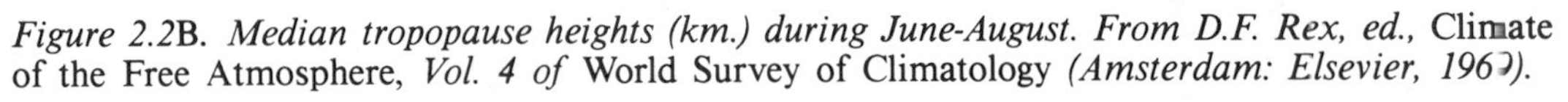

Figure 2.2B. Median tropopause heights (km.) during June-August. From D.F. Rex, ed., Climate of the Free Atmosphere, *Vol. 4 of* World Survey of Climatology *(Amsterdam: Elsevier, 196[illegible]).*

the entire earth, temperature at the earth's surface is around 15°C (59°F), and it decreases upward through the atmosphere at an average rate of about 6.5°C per kilometer (3.5°F per thousand feet) to an altitude of about 11 kilometers (6.8 miles), at which point the temperature is about −57°C. Above 11 kilometers it remains constant (isothermal) to about 20 kilometers and then increases. A maximum temperature of approximately −2°C is reached at around 47 kilometers, and the temperature remains at approximately −2°C up to about 52 kilometers, after which it decreases again to a minimum of about −86°C at around 86 kilometers. Above that it remains isothermal for about 5 kilometers and then increases to the outer edge of the atmosphere where it may reach temperatures of as much as 725°C or more. At this altitude the air is so rarified that lay concepts of temperature break down, as will be explained later.

Thus, the earth's atmosphere is divided into rather discrete layers with only minimal exchanges of air between them because of stability conditions, which will be explained later. The layer next to the earth's surface with a lapse (decrease) of temperature with altitude is known as the *troposphere.* This is the layer in which we live and where most weather phenomena take place. It is topped by the *tropopause* at the point where the temperature ceases to decrease and becomes isothermal. The nature and height of the tropopause has profound effects on weather phenomena in the troposphere below, and therefore a great deal will be said about it later. Above the tropopause the temperature increases in a very stable layer of air known as the *stratosphere.* The top of the stratosphere is the *stratopause,* and above that the temperature decreases through the *mesosphere* to a minimum temperature at the *mesopause,* above which the temperature increases again throughout the *thermosphere.*

Much of the vertical temperature structure of the atmosphere can be explained by considerations of absorption and transmission of heat energy originating from the sun and by temperature changes resulting from expansion and contraction of vertically moving air. (These relationships will become clear as the processes controlling them are discussed individually in succeeding chapters.) Perhaps the most-outstanding feature of the atmosphere's vertical temperature structure is the relatively warm layer in the middle, centering on about 50 kilometers height. This appears to be due to the absorption of much of the ultraviolet light from the sun by ozone (O_3), which is concentrated at and below that level. As explained earlier, ozone is one of the results of the splitting of molecules by high-energy particles in the sunlight, which causes the atmospheric molecular makeup to be somewhat different in the upper layers than it is in the troposphere at the bottom.

Other Layers

In addition to the layers named according to temperature lapse rates, certain other layers in the atmosphere are known by names that reflect molecular structures and their related phenomena such as ionization. The *ozonosphere,* for instance, extends from 15 to 50 kilometers. The *ionosphere,* where molecules have been split into their constituent atoms, with resultant negative and positive electrical charges, exists in several layers extending from about 50 kilometers to about 600 kilometers. In the far reaches of the rarefied atmosphere, between 6000 and 60,000 kilometers above the earth's surface, charged atmospheric particles, whose movement apparently is largely controlled by the magnetic field of the earth, form the Van Allen radiation belts in what is known as the *magnetosphere.* Although much remains to be learned about the connections between these upper layers of the atmosphere and the climate at the earth's

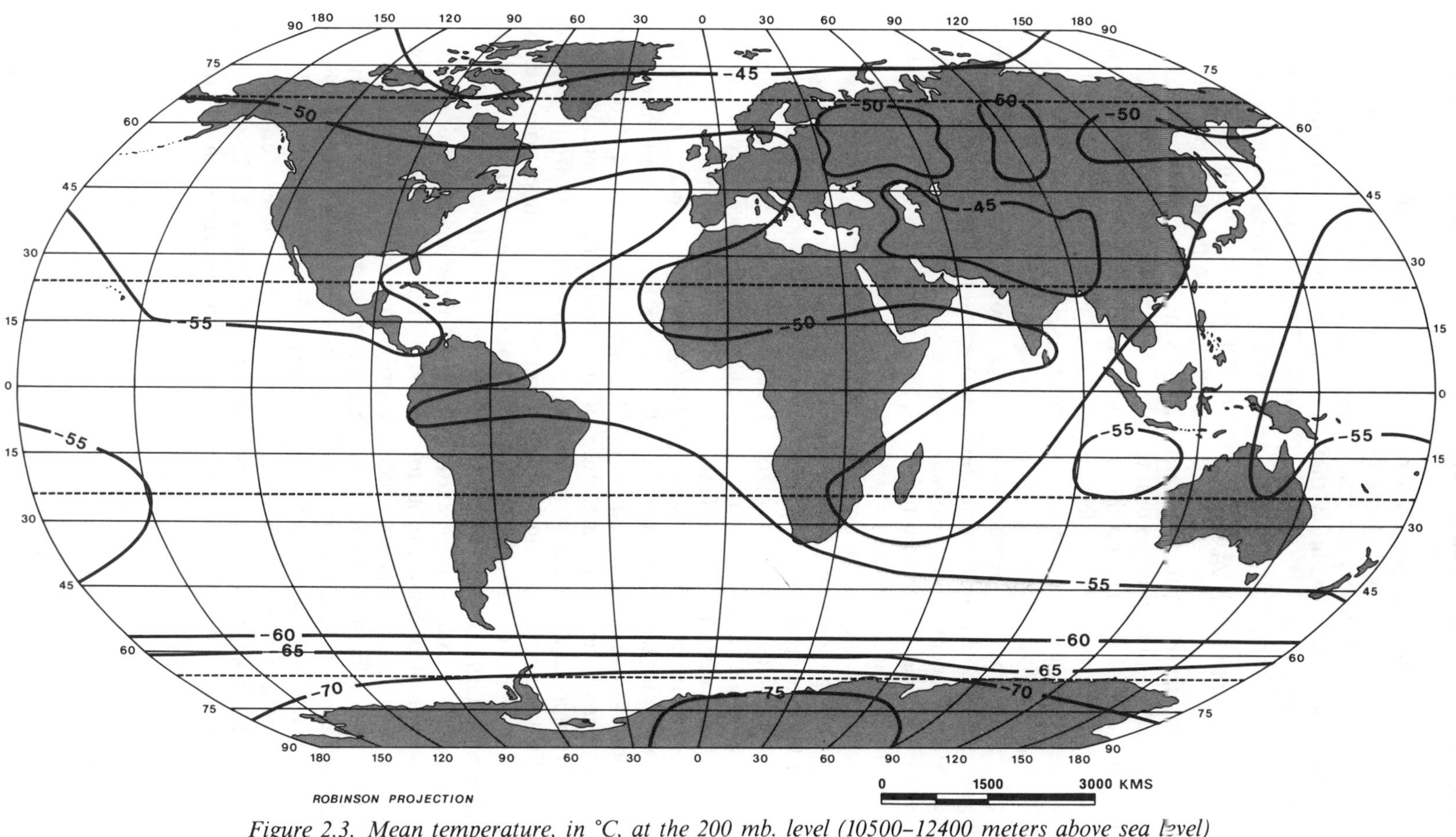

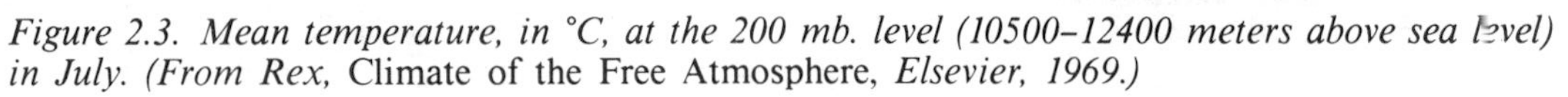

Figure 2.3. Mean temperature, in °C, at the 200 mb. level (10500–12400 meters above sea level) in July. (From Rex, Climate of the Free Atmosphere, *Elsevier, 1969.)*

surface, it is obvious that the ozonosphere is very important to life at the bottom of the atmosphere because it filters out much of the harmful short wave radiation that could destroy living tissue.

Latitudinal and Seasonal Variations

As can be seen in Figure 2.1, the standard atmosphere varies significantly according to latitude and season. In the tropical atmosphere, extending about 25 to 30 degrees on either side of the equator, the troposphere is thick, the tropopause is high and consequently cold, and the temperature generally increases immediately above the tropopause rather than remaining isothermal for a while as it does at higher latitudes. The higher the latitude, the lower and warmer the tropopause and the thicker the isothermal layer in the lower stratosphere. The height of the tropopause is controlled primarily by the average temperature of the troposphere, which expands when it is warm and contracts when it is cold. Thus, outside the tropics, the tropopause is usually higher in summer than in winter.

Figure 2.2 shows these general zonal and seasonal variations of the tropopause, but it also reveals important deviations from simple latitudinal patterns, particularly in lower latitudes. The most noticeable irregularity is the great height of the tropopause during June, July, and August over southeast Asia; this is probably due to the topography of the high Tibetan Plateau, which acts as a heat source for the middle troposphere during summer. Although temperatures on the plateau surface are cool because of its elevation, they are warmer than they would be at that altitude in the free atmosphere, were not the plateau surface present to absorb sunlight.

Since the troposphere extends to greater heights in the equatorial region than in the polar areas, the normal decrease of temperature with height continues through a thicker layer in the equatorial area and eventually reaches lower temperatures at the tropopause level than is true above the poles. The tropopause is typically about 25°C colder above the equator than above the poles. Since the temperature eventually increases above the tropopause in the stratosphere, the lower stratosphere becomes as much as 30°C warmer over the poles than over the equator at the same altitude. This brings about a reversal in air flow from east to west throughout much of the stratosphere, while below, in the upper troposphere, the general circulation is from west to east.

During winter the temperature increases upward through the stratosphere faster above the equator than above the poles, so that in the middle and upper stratosphere the temperature once more becomes warmer over the equator than over the poles. But during summer the equator remains colder than the poles throughout the stratosphere. During the solstice periods throughout much of the stratosphere, there is a consistent temperature gradient from pole to pole, warm over the summer pole and cold over the winter pole, with intermediate temperatures over the equator (Figure 2.3). Thus, at these heights the equator ceases to be a climatic divide with symmetrical distributions of temperature on either side. Such a temperature distribution results in easterly winds in the summer hemisphere stratosphere and, simultaneously, westerly winds in the winter hemisphere stratosphere (see Figure 6.18). (The reason for this will be explained in Chapters 5 and 6.)

3 | GLOBAL HEAT EXCHANGE

Heat is the energy input that initiates all atmospheric processes. Its unequal distribution over the earth causes density differences in the atmosphere that bring about atmospheric motion. Heat is the energy that evaporates water from the earth's surface and causes it to mix upward into the atmosphere. Heat is intimately exchanged with other forms of energy, such as molecular motion, kinetic energy of atmospheric flow, the latent energy of the change of state of water, and the potential energy of position above the earth's surface.

For practical purposes all the heat energy that the earth's atmosphere receives originates from the sun. What happens to the sunlight as it penetrates the earth-atmosphere system depends on the characteristics of the atmosphere and the underlying surface. For the earth as a whole the entirety of climate can be explained by the amount of sunlight received and the character of the surface receiving it. For any portion of the earth-atmosphere system, however, the climate is profoundly influenced by atmospheric motion, which itself is a product of heat receipt and nature of the earth's surface. Since the heat input from the sun is the energy that initiates motion, it is important that it be considered first in considerable detail. Before that can be done, some definitions are in order.

Heat and Temperature

Heat and temperature are not the same thing, although a change in one often reflects a change in the other. Heat is a form of energy that exists in quanta that can be added or subtracted. Temperature is a measure of energy level, an expression of the mean square speed of the molecular motion of a substance. Although temperature will be dealt with in a later chapter, it is necessary to define its units of measure here, since the unit of measure for heat is based on the unit of measure for temperature.

In 1714 Gabriel Fahrenheit, of Holland, adapted the Fahrenheit temperature scale to the mercurial thermometer. He based his zero point on the lowest temperature he could obtain by mixing ice and salt. A second point on his scale was fixed by what he thought was the human temperature, which he called 96°. (This subsequently proved to be slightly inaccurate.) At any rate, on this scale the freezing point of water turned out to be 32° and the boiling point at sea level, 212°. Much of the English-speaking world adopted this scale, which has turned out to be a rather awkward one. Now the United States stands almost alone in its use, and there are plans to convert to the Celsius scale in the near future.

In 1742 Anders Celsius, a Swedish astronomer, introduced the Celsius (centigrade) scale, which has its zero degree based on the freezing point of water, and 100 degrees based on the boiling point of water at normal sea-level pressure. This has proved to be a more logical scale, particularly since the size of the degree is the same as that of the absolute (Kelvin) scale invented by Lord Kelvin (William Thompson) in 1848.

The zero point on this scale is based on the cessation of molecular motion. Since temperature is the measure of the amount of molecular motion in a substance, it is logical that with no molecular motion the temperature is zero.

Only the absolute scale has a true zero point on which to base temperature measurements that will allow additions, subtractions, multiplications, and divisions of temperatures as if they were quanta. Therefore, this scale has to be used in all scientific formulas. Since weather records around the world are all recorded in either Celsius or Fahrenheit degrees, these have to be converted first to absolute or Kelvin degrees before they can be used in formulas. Lord Kelvin chose his degree size to equal that of the Celsius degree, and therefore it is easy to convert between Celsius and Kelvin temperatures. He found that the absolute zero point was 273°C below the freezing point of water. Therefore, 0° on the Celsius scale is equal to 273° on the absolute scale, and to convert from Celsius to Kelvin all one need do is add 273 (Figure 3.1).

It is not quite so simple to convert from Fahrenheit to Celsius, since the two scales not only have different zero points, but also different degree sizes. Since it is 100°C between the boiling and freezing points of water and 180°F between the two, the Celsius degree is 1.8 times, or nine-fifths, as large as the Fahrenheit degree. To convert from Celsius to Fahrenheit, one must multiply the Celsius reading by 9/5 and then add 32. To convert from Fahrenheit to Celsius, one must first subtract 32 and then multiply by 5/9. Thus,

$$°F = 9/5°C + 32$$

and

$$°C = 5/9\ (°F - 32).$$

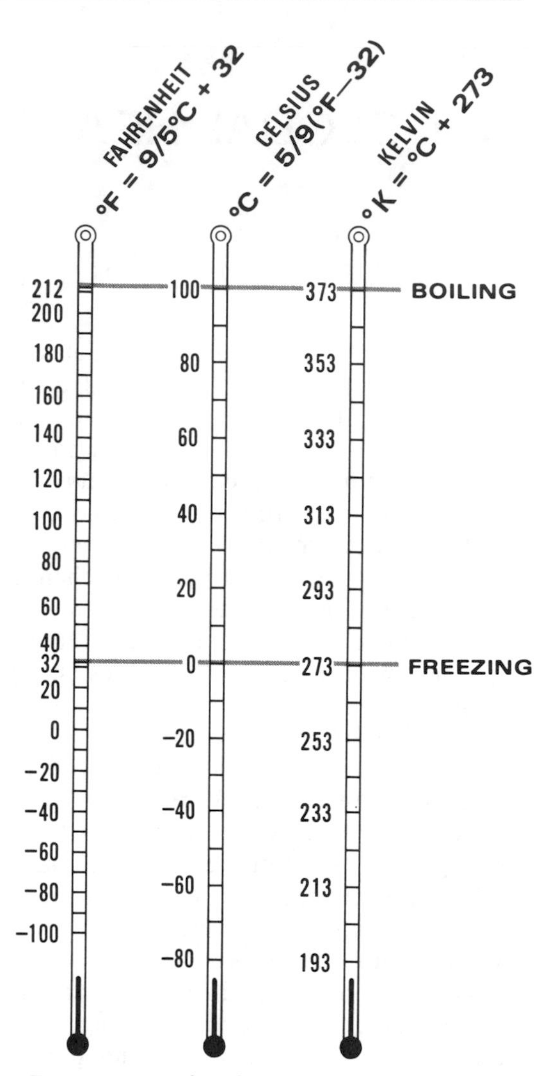

Figure 3.1. Fahrenheit, Celsius, and Kelvin temperature scales.

Heat energy is expressed in a number of ways, but for the purposes of this book it will be expressed in calories.* One calorie is the amount of heat required to raise the temperature of 1 gram of water 1°C (from 14.5°C to 15.5°C). This is not the same calorie as the food calorie, which is 1000

*The international system of units (SI) recently adopted by the World Meteorological Organization (WMO) specifies the joule (J) as the official unit of energy. One calorie equals 4.186 joules or 4.186×10^7 ergs. The official unit of power (rate of expenditure of energy) is the watt (W). One watt equals one joule per second. The rate at which energy passes through or falls on a unit area is called radiant flux density. This is usually expressed in langleys per minute, where one langley equals one calorie per square centimeter (ly = cal/cm^2), or in watts per square meter. One ly/min = 697.5 W/m^2.

times as large. To distinguish between the two, some books use the term "gram calorie" to indicate the small unit of heat that raises the temperature of 1 gram of water 1°C. Since this book always alludes to the heat unit, the simple term "calorie" will be used.

Insolation and Terrestrial Radiation

The heat received from incoming sunlight is known as *insolation*. This has been measured approximately at many different points on the earth over a great number of years and, taking into account various states of the atmosphere, it has been estimated that at the top of the atmosphere slightly less than two calories of heat are received per minute on a square centimeter of surface oriented perpendicular to the sunlight. Within the range of error of calculations, this appears to be a constant, known as the solar constant.*

The sunlight is the result of radiation of heat outward from the sun which on the average is about 150 million kilometers (93 million miles) away from the earth. Since the sun is essentially a sphere, and heat is radiated perpendicular to its surface in all directions, sun rays diverge away from the sun's surface so that the heat flux passing through any unit area decreases rapidly with distance from the sun. The rule is that the heat flux through a unit area diminishes at the rate of the distance squared that it has traveled (Figure 3.2). Thus, knowing the amount of heat received at the earth, and knowing (through astronomical measurement) the distance between the earth and the sun, one can compute the emission power of the sun. This turns out to be approximately 10^{23} kilowatts, an enormous magnitude that exceeds by many millions of times all the electrical generating capacity on earth. If

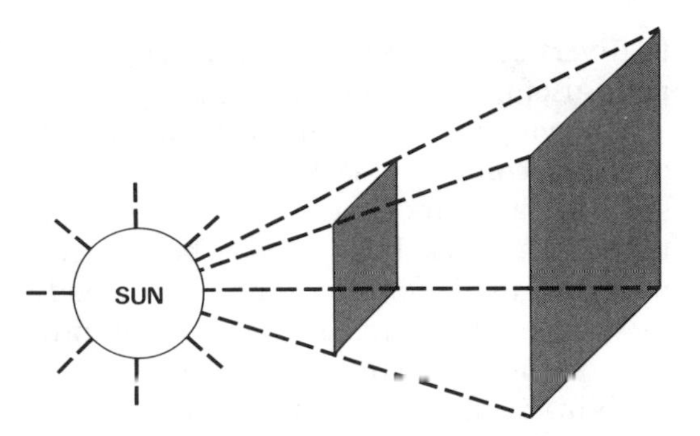

Figure 3.2. The heat radiated from the sun spreads over a larger area the farther from the sun. Thus the heat flux per unit area decreases with the square of the distance from the sun.

this amount of energy were being produced by the combustion of coal, the sun would have burned up in 5000 years, even though it is a large body with a volume approximately one million times that of the earth. The sun's energy is not being produced by the combustion of coal, however, but by the fusion of hydrogen to helium. This process will consume the sun's mass in about 14×10^{12} years, which is about 3000 times longer than the earth's estimated age at present.

By comparison to the sun and the distance between the sun and the earth, the earth is a small speck in space that intercepts only a minute portion of the total heat emission from the sun. The entire earth intercepts only about 1/2 billionth of the sun's energy output, but this still amounts to 1.8×10^{14} kilowatts, which is more than 300,000 times all the electrical generating capacity in the United States. Obviously, then, any efforts by man to add to or subtract from the heat budget of the earth by burning fossil fuels, or even by producing nuclear energy, will be puny compared to the amount of heat the earth receives from the sun every instant. Thus, man cannot hope to alter the earth's climate over broad areas by the artificial addition of heat. But man *can* hope to find ways

*There appears to be a little variation in solar emission, perhaps of a cyclic nature. Recent measurements from satellites have yielded an average value for the "solar constant" of 1370 W/m² (1.963 ly/min).

to divert and convert the sun's energy into more-useful forms and processes.

Knowing the emission power of the sun, one can compute the temperature of the sun's surface, since the sun radiates essentially as a black body. It has been determined in the laboratory that a black body (perfect radiator) radiates energy in proportion to the fourth power of the absolute (Kelvin) surface temperature of the body (Stefan-Boltzmann Law). Knowing the energy emission, one can solve for the temperature, which turns out to be approximately 5784°K. This, of course, is much hotter than the earth's surface, which averages about 59°F or 15°C or 288°K. Therefore, the sun emits much more heat energy than the earth does.

But the earth's temperature is a long way from absolute zero, so the earth does radiate a significant amount of heat. Most of the heat radiated to space by the earth-atmosphere system is radiated by the atmosphere at heights that are considerably colder than the earth's surface. Terrestrial (earth) radiation measured by satellites indicates that the earth-atmosphere system has an average temperature of approximately 250°K. As a result, the sun emits energy at a rate about 160,000 times that of the earth-atmosphere system.

Not only do the two bodies emit heat at very different rates, but they also emit it at very different wavelengths. All bodies not at absolute zero temperature radiate energy over a considerable spectrum of wavelengths, but this spectrum is arranged around a wavelength at which the maximum amount of energy is radiated, which has been determined to be inversely proportional to the temperature of the radiating body (Wien's Law). Therefore, the hot sun radiates its energy at much shorter wavelengths than the cool earth-atmosphere system does (Figure 3.3).

The wavelength at which the sun emits the most heat is about 0.47 micrometers (μm = 1 millionth of a meter), which is in the yellow-green portion of the visible spectrum of light. Most of the sun's energy is emitted in a spectrum from 0.15μm to about 4μm. Forty-one percent of it is visible, 9 percent is ultraviolet (shorter than visible), and 50 percent is infrared (longer than visible). Terrestrial radiation, on the other hand, stretches from about 4μm to about 100μm, with maximum energy falling at about 10.1μm.

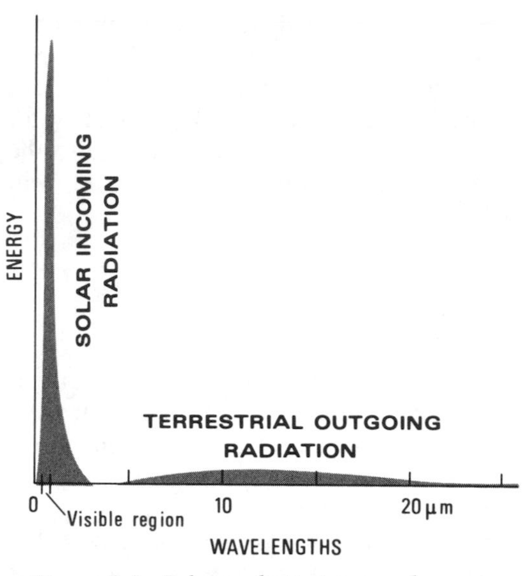

Figure 3.3. Solar radiation is much greater in magnitude and at shorter wavelengths than is earth radiation.

Disposition of Insolation Within the Earth-Atmosphere System

The so-called "solar constant," as mentioned before, at the outer limits of the atmosphere is approximately two calories per square centimeter per minute. But this is on a theoretical stationary surface perpendicular to the sunlight, and since the earth is a rotating sphere, sunlight hits the earth perpendicularly only at one point at any instant. Half the earth's surface is always in shadow; therefore, the average amount of heat being received on the earth's surface is considerably less than two calories per square centimeter per minute. The entire surface of the spherical earth intercepts only the same amount of sunlight as a disc the same diameter oriented perpendicular to the sunlight. Since the surface area of a sphere is four times that of a circle with the same diameter, the average amount of energy being received on the earth's surface is approximately 0.5 calories per square centimeter per minute. This, then, is the quantity of energy that one works with when considering the heat budget of the entire earth-atmosphere system. Many things happen to this energy within the atmosphere and at the earth-atmosphere interface before it is finally lost back to space again.

Scattering and absorption attenuate the sunlight by approximately half on its way down through the atmosphere. Air molecules and suspended particles scatter part of the sun's radiation in all directions. About half of this scattered light is lost back to space immediately and has no heating effect on the earth. The rest finds its way to the surface, bouncing back and forth in all directions, and provides the indirect lighting that comes through the windows of buildings and other shaded places during the day. It also lights up the daytime sky. Molecules, being very small, scatter primarily short-wave radiation and therefore cause the sky to appear blue or violet. With larger particles in the air, such as fog or smog, longer wavelengths are scattered, and the sky takes on a more whitish hue. With heavy smog, some of the light is absorbed, and the sky becomes yellowish-brown. Absorption is quite different from scattering. With absorption the air molecules take up the radiant energy and convert it to internal energy, whereby the motion of the molecules is increased or their atomic makeup is changed. This process is manifested as temperature change.

Unlike solids and liquids, which radiate and absorb heat much as a black body across a continuous spectrum of wavelengths, gases radiate and absorb heat at selected wavelengths and are transparent to others. Nitrogen (N_2), which makes up

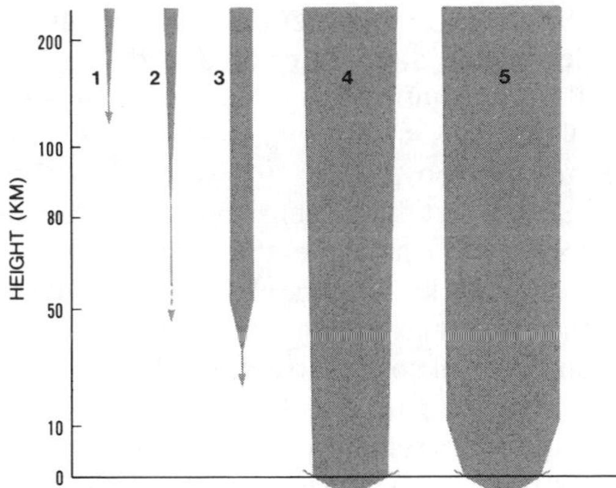

Figure 3.4. Absorption of solar radiation. Arrow 1 represents the quantity of ultraviolet radiation at wavelengths less than 0.12μm absorbed by N_2 and O_2 in the high atmosphere above 125 km. above sea level. Arrow 2 represents quantity of ultraviolet at wavelengths between 0.12 and 0.21μm absorbed by O_2, mostly above 80 km. Arrow 3 represents quantity of ultraviolet at wavelengths between 0.21 and 0.34μm absorbed by O_3 (ozone) between 20 km. and 60 km. Arrow 4 represents quantity of near ultraviolet and visible light at wavelengths between 0.34 and 0.7μm transmitted to the earth's surface with some diminution due to scattering. Arrow 5 represents quantity of near infrared at wavelengths between 0.7 and 4.0μm transmitted to the earth's surface, absorbed slightly by O_2 and CO_2 in the upper atmosphere and more strongly by H_2O in the troposphere. (From Understanding Our Atmospheric Environment, *2nd ed., by Neiburger, M., Edinger, T.G., and Bonner, W.D., Copyright © 1982 by W.H. Freeman and Company. All rights reserved.)*

more than three-fourths of the atmosphere, absorbs only very short wavelengths, generally less than 0.12μm. Only a little of this far ultraviolet radiation is in the sunlight, and it is absorbed as soon as it enters the atmosphere by the nitrogen at high altitudes, generally above 125 kilometers. Here it dissociates molecular nitrogen, ionizes the resultant atoms, and heats the rarified air to form the so-called "thermosphere" (Figure 3.4).

Oxygen (O_2) absorbs over a somewhat

wider range of wavelengths in the ultraviolet and is also dissociated and ionized in the upper atmosphere. By the time the sunlight passes downward to about 100 kilometers above the earth's surface, all the far ultraviolet has been absorbed. Below 100 kilometers a layer extending downward to below 80 kilometers absorbs very little sunlight, because none of the radiation that is strongly absorbed reaches that low an altitude, and the density of the air is insufficient for weakly absorbed radiation to be important. The lack of absorption of much radiation centered on 80-kilometers height causes the minimum of temperature there, the mesopause (see Figure 2.1).

Below 80 kilometers a little longer-wave ultraviolet, around 0.2μm, is absorbed by oxygen molecules (O_2), some of which then disintegrate into atomic oxygen (O) and mix downward to join with diatomic oxygen to produce ozone (O_3). The ozone absorbs ultraviolet radiation of wavelengths up to 0.3μm, which heats up the atmosphere and produces the temperature maximum, the stratopause, at approximately 50-kilometers height. Practically all the ultraviolet that is absorbed by ozone is used up by the time it descends to 20 kilometers above the earth's surface. This is fortunate, since this short-wave radiation, if it reached the earth, would deteriorate much living tissue and perhaps make all life impossible. This is the cause of the concern about the destruction of the ozonosphere concentrated around 30 kilometers above the earth's surface by such things as supersonic transport planes that would fly in the stratosphere, and by certain chemicals in aerosol sprays that eventually mix upward to that level.

The rest of the solar radiation, between 0.3 and 4.0μm, is not absorbed much by the upper atmosphere. There is some absorption of visible and infrared radiation at lower levels of the troposphere where water vapor content is significant.

The left side of Figure 3.5 diagramatically indicates the relative percentages of the different dispositions of incoming solar radiation, taking as 100 percent the 0.5 calories per square centimeter per minute that, on the average, reaches the outer limits of the atmosphere. All together the various constituents of the atmosphere just described absorb, on the average, 25 percent of the incoming solar radiation. An average of 35 percent of the incoming radiation is intercepted by clouds, which reflect about 22 percent back to space and allow 13 percent to reach the earth's surface as diffuse radiation. Another 13 percent is intercepted by atmospheric molecules that scatter approximately 5 percent of the radiation back to space and allow 8 percent to reach the earth's surface as diffuse radiation. About 27 percent of the solar radiation penetrates directly to the earth's surface, where about 24 percent is absorbed and 3 percent is reflected back to space. Hence, of the 100 percent of incoming sunlight, about 45 percent is absorbed at the earth's surface, 25 percent is absorbed in the atmosphere, and 30 percent is reflected and scattered back to space by clouds, air molecules, and the earth's surface.

The 30 percent of sunlight that is reflected and scattered immediately back to space has no effect on the earth-atmosphere system and is lost energy, except for the fact that it causes the earth to shine. This is known as the albedo (reflectivity) of the earth. Since most of the albedo is caused by reflection from cloud tops and scattering by atmospheric molecules, a planetary body without an atmosphere, such as the earth's moon, would have a much lower albedo than the earth. The moon has an average albedo of only about 7 percent. Therefore, the earth is a much brighter object in the heavens, when viewed from the moon, than the moon is, when viewed from the earth. The astronauts on their way to the moon commented on this difference in brightness of the two bodies as they looked back at the earth and exclaimed about its beauty as an object in space. On the other hand, some of the largely gaseous planets have

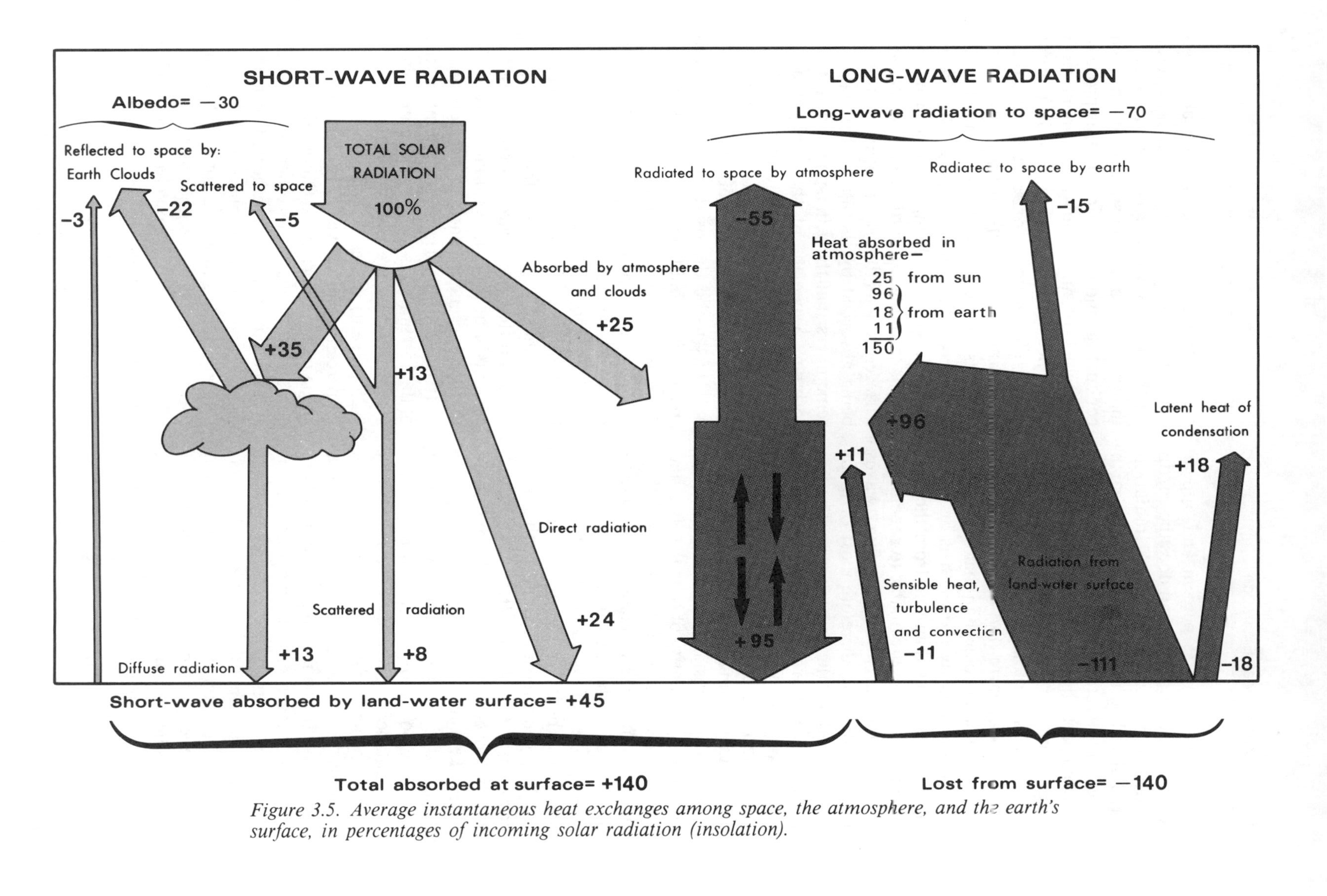

Figure 3.5. Average instantaneous heat exchanges among space, the atmosphere, and the earth's surface, in percentages of incoming solar radiation (insolation).

albedos exceeding 75 percent and appear much brighter in space than does the earth.

The 45 percent of the sun's radiation that is absorbed at the surface of the earth is used mostly to heat the surface of the earth and partially to evaporate water, which requires much heat. As the earth's temperature rises, it radiates more heat upward into the atmosphere. But since the earth's surface is much cooler than that of the sun, the heat radiated by the earth is at much longer wavelengths than that from the sun. These longer waves are unable to peneterate the atmosphere as well as the short waves from the sun, and therefore most of them are absorbed in the atmosphere. The atmosphere then warms up and radiates heat both downward and upward. The downward radiation of heat from the atmosphere adds to the heating of the earth's surface, so that at any instant the earth's surface is receiving a quantity of heat that is equal to about 140 percent of the sun's radiation. Thus the atmosphere acts as an insulating coverlet over the earth—a reservoir of heat—that is not lost immediately to space, but is absorbed and reradiated back and forth between atmosphere and earth a number of times as it slowly filters out to space. This holdover of heat in the atmosphere causes the earth's surface to be approximately 40°C warmer than it would be if there were no atmosphere and if there were an instantaneous exchange of heat between the surface and space. This entire process is known as the greenhouse effect, and it is the result of the fact that the atmosphere is not as transparent to long waves as it is to short waves.

At any instant, the earth's surface radiates upward, on the average, 111 percent as much energy as comes in at the top of the atmosphere from the sun (see right side, Figure 3.5). Most of this, 96 percent, is absorbed by the atmosphere, and only 15 percent goes directly out to space. Heat also moves upward from the earth's surface into the atmosphere through the process of evaporation of moisture (latent heat of condensation). About 18 percent is accounted for by this process. Another 11 percent moves upward as sensible heat through conduction and convection. In this process, molecules of air come into contact with molecules of earth and receive heat from them. The heated air, being lighter than colder air, moves upward in convective currents and carries the heat into the atmosphere. Thus, on the average, 140 percent as much heat as is being received from the sun is moving upward from the earth's surface, and all but 15 percent of this is being absorbed by the atmosphere. If one remembers that the atmosphere is also absorbing 25 percent of the sunlight as it moves down through the atmosphere, the total absorption by the atmosphere at any instant is $125 + 25 = 150$ percent. This heat has to be disposed of by the atmosphere, and about 95 percent, on the average, is radiated back to the earth's surface, and about 55 percent is radiated out to space. These percentages may be diminished by about 1 percent, which is transformed into the kinetic energy of motion of the atmosphere, a small fraction of which is transmitted downward to drive the ocean currents. But friction eventually reconverts this kinetic energy back to heat energy.

It is very important to remember that of the 150 units absorbed by the atmosphere at any instant, the overwhelming amount (125 percent) comes from below, and only 25 percent from above. This is perhaps the single most important fact about the earth's climate. If it were the other way around and the atmosphere were heated primarily from above, the atmosphere would be absolutely stable and would not circulate, and there would be no weather as we know it, with cloud formation, precipitation, and so forth. But as it is, the pot is constantly being boiled from below, and the atmosphere is a turbulent one.

Over the long run, there must be balances of incoming and outgoing heat at the earth's surface, at the outer limit of the atmosphere, and within the atmosphere; otherwise there would be heating or cooling trends at these levels. Figure 3.5 shows that these balances do in fact exist. At the top of the atmosphere, 100 percent radiation is received as short-wave radiation from the sun. Of this short-wave radiation, 22 percent is immediately reflected back to space by clouds, 3 percent is reflected back from the earth's surface, and 5 percent is scattered to space by atmospheric molecules. In addition to this 30 percent short-wave loss, 55 percent long waves are radiated to space from the atmosphere and 15 percent long waves from the earth's surface, for a total of 100 percent. Within the atmosphere are absorbed 25 percent of the sun's short-wave radiation, 96 percent of the earth's long-wave radiation, 18 percent latent heat of evaporation, and 11 percent sensible heat (temperature rise) due to conduction and convection between the earth's surface and the atmosphere, for a total of 150 percent. This is lost as follows: 95 percent radiated downward from the atmosphere to the earth's surface and 55 percent radiated upward to space. At the earth's surface, 45 percent of the short-wave radiation from the sun is absorbed and 95 percent of the long wave radiation from the atmosphere, for a total input of 140 percent. This is balanced by 111 percent movement upward as long-wave radiation, 18 percent as latent heat of evaporation, and 11 percent as turbulent heat through conduction and convection, for a total of 140 percent.

All the numbers in Figure 3.5 refer to radiated heat, except that being carried upward from the earth's surface by turbulent air currents, 18 percent as latent heat of evaporation and 11 percent as sensible heat. This 29 percent is referred to as the *net radiation balance* at the earth's surface, and in essence is the budget of heat energy available on the earth's surface for natural and human processes, such as the growing of plants and the weathering of soil. In many cases, from the viewpoint of humans, it is desirable to try to increase this usable amount of heat energy. In some ways it can be altered significantly. In most cases the greatest shifts in forms of heat can probably be effected by changes in albedo. As can be seen in Figure 3.6, different types of surfaces have very different reflectivities. Fresh snow and thick clouds reflect more than 75 percent of the incident radiation, whereas wet earth may reflect as little as 5 percent. Water has the lowest reflectivity if the radiation is incident at a perpendicular angle, but at low angles water becomes quite reflective. Albedos of many types of surfaces vary significantly according to both angle and wavelength of the incident radiation. Also, cloud cover may affect the albedo of the underlying surface.

A classic example, often cited to illustrate the profound results that might be expected by changing the albedo, involves the intriguing notion of melting the Arctic ice cap. In its present state, the Arctic is a frozen sea that presents a highly reflective ice-and-snow surface to the small amount of incident sunlight reaching it during the year. Therefore, most of the small amount of heat that does reach it is immediately lost back to space through reflection. If the reflectivity of the ice-and-snow surface could be reduced by, say, covering it with coal dust or some other substance, more of the incident sunlight would be absorbed and available for melting the ice. Over a few years' time, perhaps as little as seven or eight, the Arctic ice cap, which averages only about three meters thick, might melt. Once melted, the sea would remain unfrozen because the open water surface, having a significantly lower albedo than ice and snow, would absorb enough incident sunlight to maintain its surface temperature above freezing.

This is an irreversible process, according to present technology and economic feasibility, so before one undertook such a project one should be completely sure of

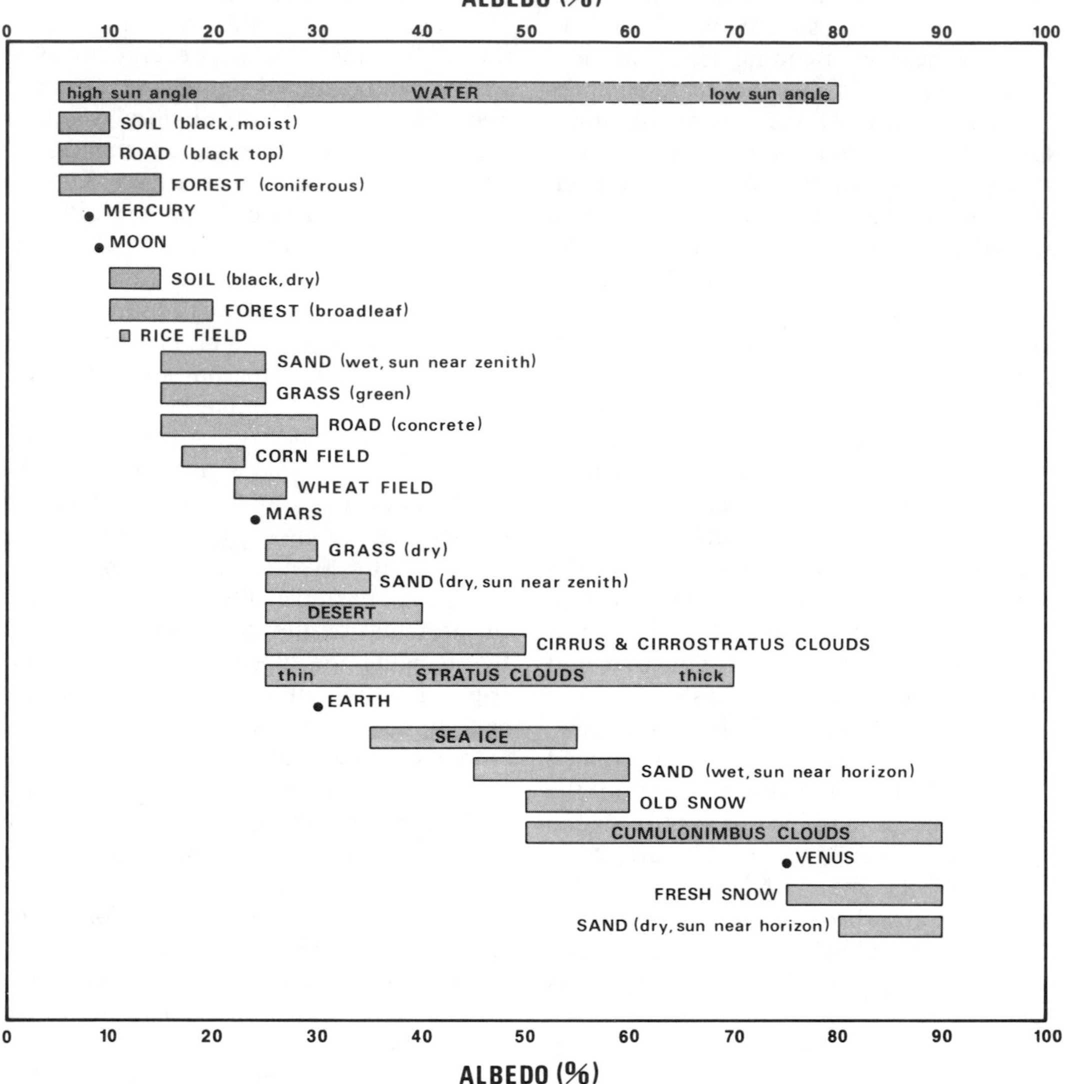

Figure 3.6. Albedos of various kinds of surfaces.

the consequences. As will be explained in Chapter 6, such a change in the Arctic might effect a circulation change in the atmosphere of the entire Northern Hemisphere, or even the whole earth, and therefore the consequences are not as simple as might at first be assumed. Not only would there be an expansion of thermal belts northward, but areas that are now humid might be turned into deserts, and so forth. Also, such a change in the Arctic would probably bring on at least partial melting of ice caps on Greenland, Antarctica, and high mountains, which would cause a catastrophic rise in sea level. Therefore, from lack of complete knowledge about the consequences of such action, no one has yet attempted such a grandiose change on the face of the earth, even though it is physically, and perhaps even economically, possible.

Of much more subtle nature, and less

risky, are heat transfers effected by irrigation, which all seem to be beneficial. Irrigation is an age-old practice used to add water to dry land. But recently it has been discovered that, in addition, irrigation normally brings about a number of beneficial and significant shifts in the heat budget at the earth's surface. By adding water to dry soil the albedo is reduced, thereby allowing for greater absorption of incoming heat energy. At the same time the temperature of the surface of the earth is lowered, because less heat goes into sensible heat and is instead transformed into latent heat of evaporation, which then is unavailable for heating the atmosphere. Since the earth is cooler, it radiates less heat, and therefore the energy budget is enhanced by a smaller loss of heat through terrestrial radiation.

If one goes to the expense of irrigating land, it is probably for growing crops on it. This will likely further reduce its reflectivity. The crops will transpire much moisture, which will increase the latent heat component of the heat transfer from earth to atmosphere. It will also transfer rather useless evaporation from a soil surface into moisture transfer through transpiration, which will effect plant growth. The crops will shade the earth, thereby further reducing the earth's surface temperature and terrestrial radiation, and will also slow down air movement so that evaporation rates are reduced. The reduction of surface temperature and motion of the air will necessitate less transpiration from each plant, so that moisture is utilized more efficiently by each plant.

Generally, plant growth is not a function of transpiration rate. Transpiration is primarily a cooling process and, if excessive, can place extreme stress on a plant because the plant is trying to evaporate water from its leaves faster than it is taking it up through its roots. The plant must then resort to a defense mechanism such as wilting, which over a prolonged period is detrimental to its growth. Of course, some transpiration of moisture through the plant is necessary to circulate nutrients from the soil upward into the plant, but this requires only a very small circulation of moisture. Thus, in most cases reduced transpiration rates are beneficial to plant growth.

In summary, then, irrigation not only adds the necessary water for plant growth, but also increases the net radiation balance at the earth's surface by cooling the surface of the earth, thereby transferring heat energy from sensible processes to latent heat processes. And much of the latent heat is transferred from simple evaporation processes to more useful transpiration processes.

The foregoing estimates of the components of the earth-atmosphere heat budget are just that and nothing more—estimates of average conditions over the entire earth. Not all the components are measurable with present technology: some have been derived from empirically determined relationships, and others have been realized only as residuals in equations. The latter might encompass considerable error if errors were inherent in measurements of the other variables, since the residuals would absorb the sum of the errors. Thus, magnitudes will be found to vary from one textbook to another, but the nature of the variables themselves are the same, and the quantitative estimates are better than mere guesses.

4 SPATIAL AND TEMPORAL HEAT BUDGET VARIATION

The heat budget components vary greatly from place to place and from time to time at any one place. Insolation varies by latitude because the earth is a sphere. It varies seasonally at any one place because the earth is rotating on an axis, revolving about the sun, and the axis of rotation is inclined with respect to the orbit around the sun. By the same token, the insolation varies with time of day because the sun comes up in the east, crosses the zenith at noon, and sets in the west, thereby causing the angle of incidence at any point on the earth to vary constantly during the day.

Undulations in the earth's surface, of course, vary the amounts of heat received because of different angles of slope. A north slope obviously receives less heat than a south slope in the middle and high latitudes of the Northern Hemisphere. In the morning an east slope will receive more heat, and in the afternoon a west slope.

In addition to all these geometrical relationships, the irregularities of atmospheric conditions, particularly clouds and humidity, as well as the characteristics of the receiving surface, complicate the pattern immeasurably. Since the earth's surface is approximately three-fourths water, the widely varying albedo of water with angle of incident sunlight is an important variable.

Sun-Earth Relations

First, consider the regularly varying spatial characteristics brought about by the spherical shape of the earth and its rotation and revolution about the sun. The earth revolves around the sun approximately once per year in a slightly elliptical orbit with the sun at one focus of the ellipse. The distance between the earth and the sun averages about 150 million kilometers (93 million miles). This distance varies by about 1.7 percent. The greatest distance, *aphelion,* occurs on 5 July, and the shortest distance, *perihelion,* occurs on 4 January. This variation in distance causes a plus or minus 3.4 percent variation in heat receipt by surfaces perpendicular to the sun's rays at the outer limit of the atmosphere. At the surface of the earth, however, local factors generally mask this variation.

Angle of Incidence

Much more important than distance in determining the amount of heat received is the curvature of the earth's surface, which varies the angle at which the sunlight falls on the earth. Since the earth is so far away from the sun and such a small speck in space compared to the sun, the small amount of sun's energy that is incident upon the earth's surface has been radiated from only a very small part of the sun's surface, which for practical purposes can be considered as a flat surface, and the rays then can be considered to be parallel to each other. Therefore, the angle at which the sun's rays strike the earth's surface depends essentially on the latitude of the earth and is a complement to the angle of latitude. That is, if the sun is directly overhead at the equator, which is 0° latitude, the angle of incidence will be 90°. At 30° latitude the angle of incidence will be 60°, at 45° latitude, 45°, and so forth. At the pole, 90° latitude, the angle of incidence will be zero.

A direct ray yields more heat per unit area than an oblique ray does, for three reasons. First, the vertical ray will spread over a smaller area when it hits the earth's surface than a slanting ray will. Second, the slanting ray will pass through a greater thickness of atmosphere than the vertical ray and therefore become more attenuated on its way through the atmosphere. When the sun is only 5° above the horizon, its rays must travel through almost eleven times as much atmosphere as when the rays are perpendicular. The radiation intensity on a horizontal surface, then, is little more than 1 percent what it is when the rays are perpendicular (Table 4.1). And third, the albedo over many surfaces is greater for slanting rays than for direct rays.

In addition to the facts that the earth is spherical and is revolving around the sun, it is also spinning on an axis once a day, and this axis of rotation is tilted at an angle approximately 23.5° away from the perpendicular to the orbit of revolution around the sun. This tilt of the axis remains essentially parallel to itself as the earth revolves about the sun, so that at different points in the orbit of revolution different portions of the earth's surface are exposed to sunlight (Figure 4.1). This combination of circumstances produces the seasons.

On approximately 21 June the north pole is tilted most directly toward the sun, and the sun's last tangent rays pass across the pole to 66.5°N latitude on the other side (Figure 4.2). At this time the direct rays of the sun are overhead at 23.5°N latitude. On 21 December the situation is reversed: the south pole is tilted toward the sun, the north pole away from it, and the last tangent rays of the sun in the Northern Hemisphere reach only to 66.5° latitude. The direct rays of the sun are now overhead at 23.5°S latitude. These extreme positions of the direct and tangent rays of the sun set the positions of the Tropic of Cancer at 23.5°N latitude, the Tropic of Capricorn at 23.5°S latitude, the Arctic Circle at 66.5°N latitude, and the Antarctic Circle at 66.5°S latitude.

Table 4.1 Atmospheric Thickness and Radiation Intensity on a Horizontal Surface for Varying Sun Altitude above the Horizon, Assuming a Transmission Coefficient of 78 Percent

Sun's altitude	Number of atmospheres rays must pass through	Radiation intensity (in percent)
90°	1.00	78
80°	1.02	76
70°	1.06	72
60°	1.15	65
50°	1.31	55
40°	1.56	44
30°	2.00	31
20°	2.92	17
10°	5.70	5
5°	10.80	1
0°	45.00	0

Daylight Period

Not only does the angle vary at which the sun's rays are incident upon the earth's surface, but also the length of daylight varies with latitude because of the tilt of the axis. As can be seen in Figure 4.2, the circle of illumination, which divides the earth into a light half and a dark half, includes the entire North Polar area southward to the Arctic Circle within the light half on 21 June (summer solstice) and within the dark half on 21 December (winter solstice). This means that on 21 June the pole and all the area around it south-

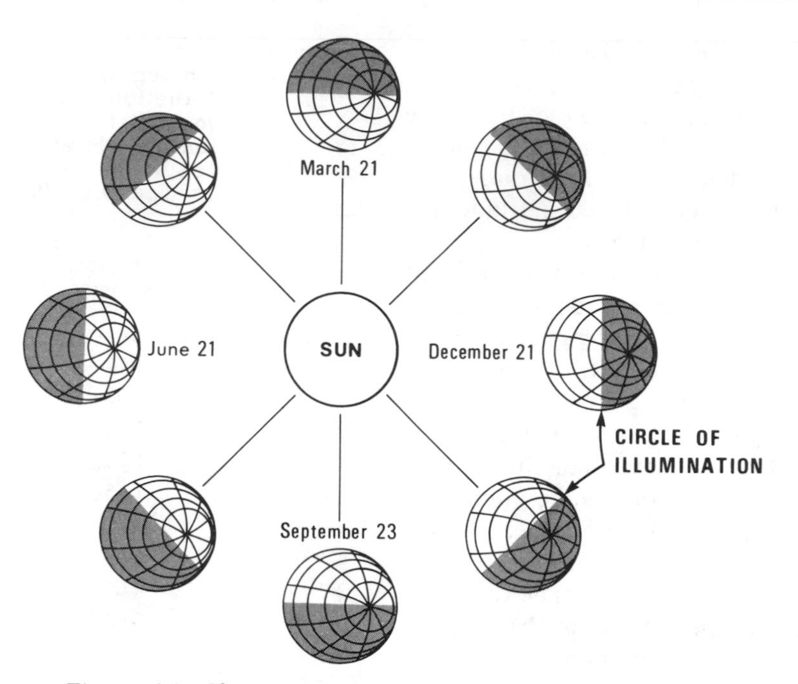

Figure 4.1. Changes in earth's position relative to the sun during the course of a year.

ward to the Arctic Circle receives sunlight all 24 hours of the day, while on 21 December this area receives no direct sunlight at all. On 21 March and 23 September the circle of illumination passes through the poles, and therefore all points on the earth's surface have 12 hours of daylight and 12 hours of darkness on these dates, which are called the vernal (spring) and autumnal (fall) equinoxes.

The length of daylight period at any place at any date can be judged by the portion of the latitudinal parallel that falls within the lighted half of the earth relative to the portion that falls within the darkened half (Table 4.2). At the equator there are 12 hours of daylight and 12 hours of darkness every day of the year. This varies more widely the farther one goes toward the poles, and at the poles themselves there is a six-months' daylight period followed by a six-months' darkness period. At intermediate positions between the equator and poles, the length of daylight period increases rapidly during spring, then levels off as the summer solstice is approached, falls rapidly during autumn, and tapers off again as winter is approached (see Table 4.2). For instance, at 40° latitude, the daylight period lengthens from 9 hours 37 minutes on 15 January to 10 hours 42 minutes on 15 February, a change of 1 hour 5 minutes during the month. It lengthens from 11 hours 53 minutes on 15 March to 13 hours 14 minutes on 15 April, a change of 1 hour 21 minutes during that month. From 15 May to 15 June it lengthens from 14 hours 22 minutes to 15 hours, a change of only 38 minutes.

Latitudinal Effects on Insolation Receipts

Latitude, then, exercises two variables on insolation receipt, angle of incidence and length of daylight period. These sometimes vary with latitude in the same direction and sometimes in opposite directions to produce a composite latitudinal heat distribution as shown in Figure 4.3. In the

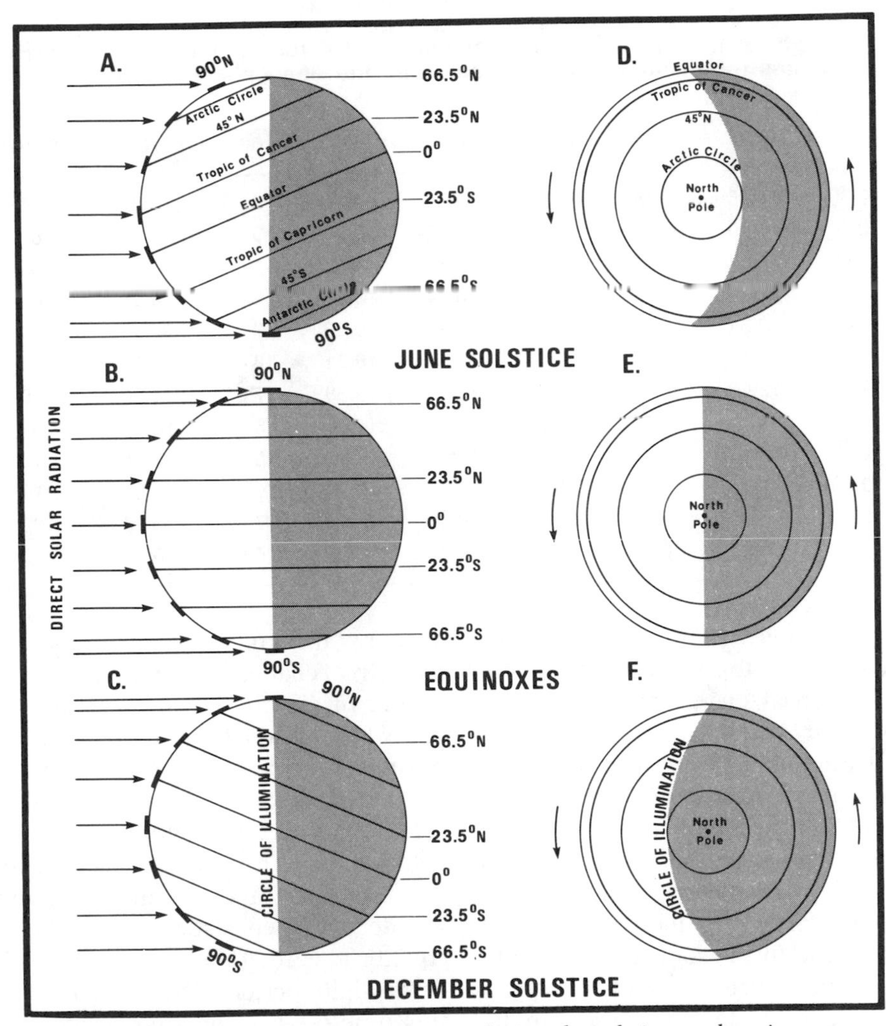

Figure 4.2. The circle of illumination during the solstices and equinoxes. (Modified from John B. Leighly.)

Northern Hemisphere on 21 June (s) the sun at noon is directly overhead at 23.5°N latitude. The elevation above the horizon (H) decreases in both directions from 90° at the Tropic of Cancer to 66.5° at the equator and 23.5° at the pole. The length of day (D) varies with latitude from 12 hours at the equator, slowly at first and then more rapidly, until it reaches 24 hours at 66.5°, at which it remains all the way to the pole. These two curves, marked H_s and D_s in the diagram, add up graphically to I_s, the amount of insolation received at the outer limit of the atmosphere. Since from the equator to the Tropic of Cancer, both the angle of incidence and the daylight period increase initially, the amount of insolation received increases with latitude. But shortly poleward of the Tropic of Cancer, the insolation begins to decrease slightly because the angle of incidence is decreasing faster than the daylight period is increasing. Poleward of about 50° latitude, however, the daylight period increases more rapidly than the angle of incidence decreases, and the insolation receipt increases again all

Table 4.2 Length of Daylight Period by Latitude on the 15th of Each Month, in Hours and Minutes (not including light refracted by the atmosphere)

	0°	10°	20°	30°	40°	50°	60°	70°	80°	90°
Jan.	12:00	11:35	11:02	10:24	9:37	8:30	6:38	0:00	0:00	0:00
Feb.	12:00	11:49	11:21	11:10	10:42	10:07	9:11	7:20	0:00	0:00
Mar.	12:00	12:04	12:00	11:57	11:53	11:48	11:41	11:28	10:52	0:00
Apr.	12:00	12:21	12:36	12:53	13:14	13:44	14:31	16:06	24:00	24:00
May	12:00	12:34	13:04	13:38	14:22	15:22	17:04	22:13	24:00	24:00
June	12:00	12:42	13:20	14:04	15:00	16:21	18:49	24:00	24:00	24:00
July	12:00	12:40	13:16	13:56	14:49	15:38	17:31	24:00	24:00	24:00
Aug.	12:00	12:28	12:50	13:16	13:48	14:33	15:46	18:26	24:00	24:00
Sept.	12:00	12:12	12:17	12:23	12:31	12:42	13:00	13:34	15:16	24:00
Oct.	12:00	11:55	11:42	11:28	11:10	10:47	10:11	9:03	5:10	0:00
Nov.	12:00	11:40	11:12	10:40	10:01	9:06	7:37	3:06	0:00	0:00
Dec.	12:00	11:32	10:56	10:14	9:20	8:05	5:54	0:00	0:00	0:00

the way to the pole. The pole at this time of the year at the top of the atmosphere receives more heat energy over a 24-hour period than any other latitude.

On 21 December (W), on the other hand, the sun is directly over the Tropic of Capricorn, 23.5°S latitude, and the equator again experiences a sun that at noon stands only 66.5° above the horizon. The length of the daylight period at the equator is 12 hours, as is true throughout the year. Both the angle of incidence, H_w, and the length of daylight period, D_w, decrease rapidly from the equator to 66.5°N latitude, where they both become zero and remain at that level all the way to the pole. Thus, in the Northern Hemisphere the insolation receipt, I_w, decreases steadily from a maximum over the equator to zero at the Arctic Circle and beyond. Insolation over the equator at this time of year is slightly greater than it was on 21 June, because the earth is now closer to the sun.

It is evident that in the middle latitudes, the latitudinal changes of daylight period and angle of incidence are in the same direction during winter but in opposite directions during summer. Thus, latitudinal differences in insolation are much greater during winter than during summer. Hence, latitudinal changes in temperature are much greater in the middle latitudes during winter, which leads to a more vigorous atmospheric circulation during winter.

At the equinoxes (E), every latitude receives 12 hours of daylight and 12 hours of darkness. The sun is directly overhead at the equator, and the angle of incidence, H_E, decreases consistently from 90° at the equator to zero at the pole. Since the length of daylight period, D_E, does not vary at this time of year, the amount of insolation received, I_E, varies directly with the angle of incidence of the sunlight on the earth.

The longest and shortest days of the year are known as the summer and winter solstices, respectively. In the Northern Hemisphere these fall approximately on 21 June and 21 December. In the Southern Hemisphere they are just the reverse. Since the earth does not revolve around the sun in exactly 365 days, every fourth year is a leap year (has an extra day). Thus, the dates of the solstices and equinoxes vary one or two days from year to year. Also, the earth's axis itself rotates slowly during periods of about 26,000 years. Therefore, the seasons precess slowly through time. In 13,000 years,

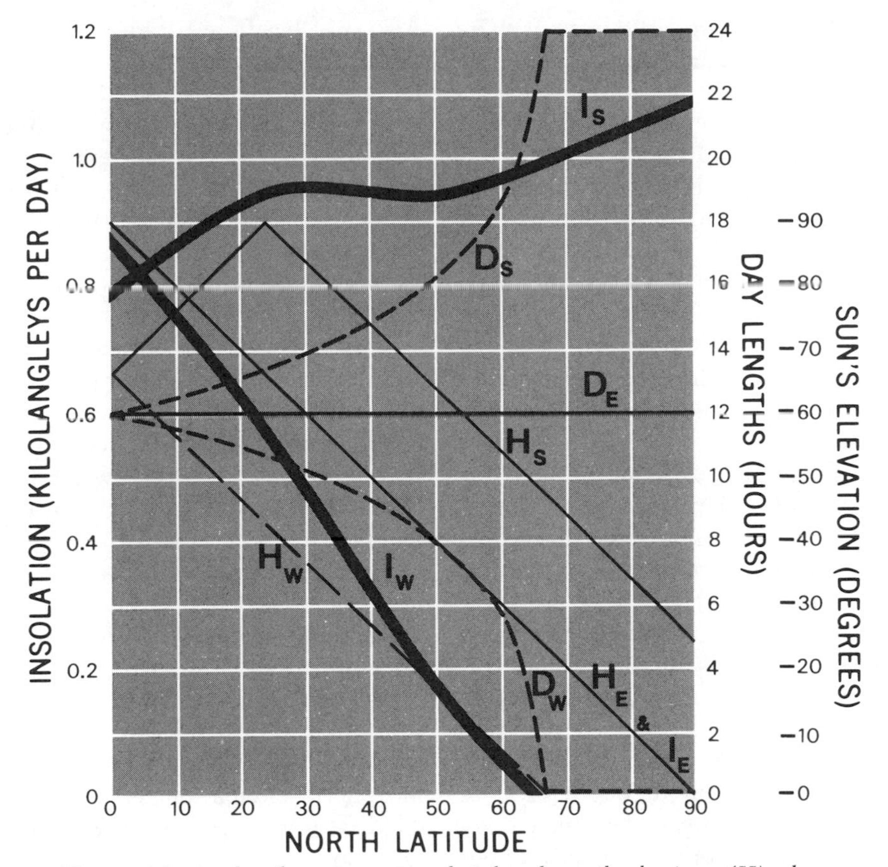

Figure 4.3. Angle of sun's rays or height above the horizon (H), day length (D), and insolation receipt (I) at the top of the atmosphere, by latitude, during the summer (s) and winter (w) solstices and the equinoxes (E).

the seasons will be just the reverse of what they are now: the Northern Hemisphere winter solstice will fall in June and the summer solstice in December.

It happens that at present the summer solstice of the Northern Hemisphere nearly coincides with the position of the earth at aphelion (farthest point from the sun), and the winter solstice nearly coincides with the timing of the earth in its position at perihelion (closest to the sun). The angle of tilt of the pole toward the sun and the nearest and farthest distances of the earth in its orbit from the sun are not one and the same thing, and there is no reason for them to coincide in time. As the seasons precess through the years, this coincidence will become less marked.

The Southern Hemisphere, of course, is closest to the sun during its summer season and farthest from the sun during its winter season. Knowing this one fact, one might conclude that the Southern Hemisphere would have greater seasonal extremes than the Northern Hemisphere does. This, in fact, is true in the case of insolation receipt. As far as resultant surface temperatures are concerned, however, the situation is the reverse. The greater preponderance of water in the Southern Hemisphere tempers seasonal extremes there.

Figure 4.4 shows the distribution of insolation at the outer limit of the atmosphere by month and latitude. It can be seen that the greatest values on the entire earth are realized over the poles during their short summer seasons, and the South Pole' receives slightly more insolation during its summer than the North Pole does during its summer, because of the ellipticity of the orbit. The equator varies the least through the year and receives its maximum insolation during the equinoxes. Over the course of the year the equator receives considerably more heat than any other latitude.

If the effects of a clear atmosphere are added, assuming 70 percent transmission through the normal thickness of the atmosphere, the distribution of insolation changes significantly (Figure 4.5). Since the slanting rays in the high latitudes must pass through more atmosphere, the insolation is depleted more at high latitudes than at low latitudes. Heat receipt is much less everywhere on the earth's surface with the atmosphere than without it, but it reduces from more than 44 megajoules/square meter (1000 langleys) per day to only about 18 MJ/m^2 (450 langleys) per day at the pole, while it declines only from about 40 MJ/m^2 (925 langleys) per day to about 25 MJ/m^2 (575 langleys) per day at 30° latitude during the summer solstice. The result is that the maximum heat receipt at the earth's surface during the solstice periods occurs in the middle latitudes centered on about 30°–35° latitude. Maximum heating at the earth's surface thus shifts during the year over a span of latitude averaging about 65°, more than the 47°-latitude shift of the direct rays of the sun.

Atmospheric Effects

This is still a very hypothetical picture, since it does not take into account the state of the atmosphere, primarily clouds and moisture content. The assumed transmissivity of 70 percent is frequently exceeded. Also, sky (scattered) radiation is added to direct radiation, so that total daily global radiation in excess of 30 MJ/m^2 (720 langleys) is not rare even at low altitudes, and mountain stations may experience even higher values. On cloudy days, of course, transmissivity is reduced. Figure 4.6 depicts the pattern of insolation receipt at the earth's surface as actually measured. Here the pattern is far from a purely zonal (latitudinal) one. The greatest amounts are received in such desert areas as the Sahara of northern Africa, the Kalahari of southern Africa, central Australia, and southwestern United States and northwestern Mexico. Here the skies are generally clear and the air relatively free of moisture. The equator, on the other hand, which receives the most heat during the year at the outer limits of the atmosphere, has considerable heat attenuation due to great amounts of cloudiness and high humidity content. (The reasons for such distributions of clouds and humidity will become clear later when atmospheric circulation and resultant cloud conditions are discussed.) Seasonal patterns of insolation are similar to the annual pattern, with some latitudinal shifts corresponding to shifts in the angle of incidence of the sun and the length of daylight period.

Heat Exchange at the Earth's Surface

Radiation Balance

The same dry atmospheric conditions that allow maximum insolation over dry land areas also allow maximum terrestrial radiation to escape to space, and therefore these areas utilize the heat they receive very inefficiently and show up as minimum areas of radiation balance at the earth's surface (Figure 4.7). The world pattern of radiation balance, then, is much different from that of radiation receipt. Whereas land areas, with their lower atmospheric mois-

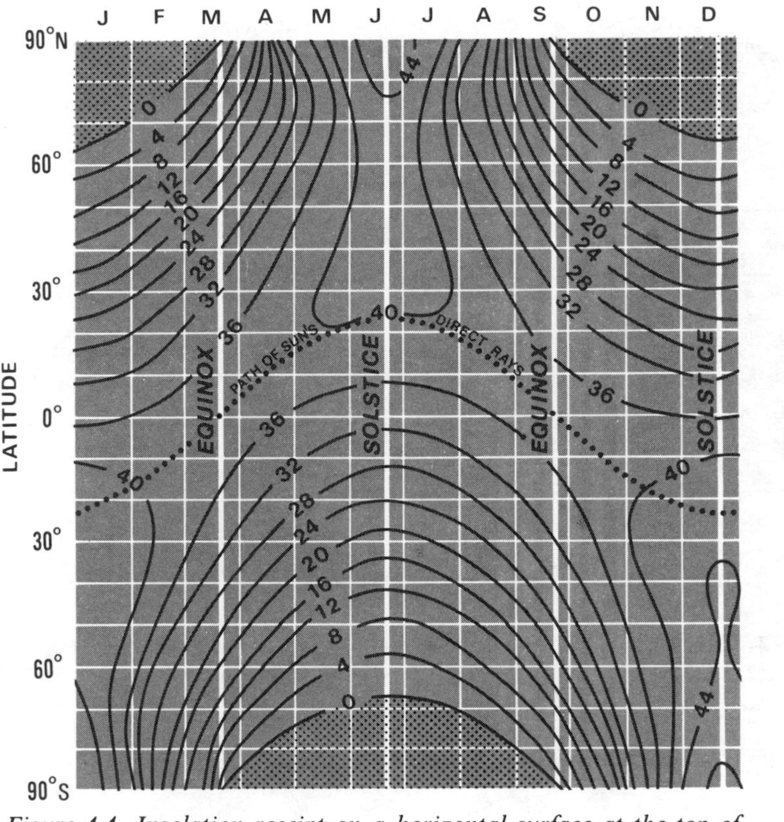

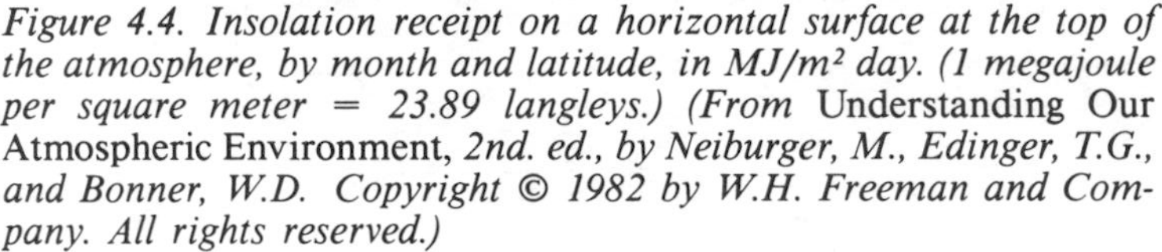
Figure 4.4. Insolation receipt on a horizontal surface at the top of the atmosphere, by month and latitude, in MJ/m² day. (1 megajoule per square meter = 23.89 langleys.) (From Understanding Our Atmospheric Environment, *2nd. ed., by Neiburger, M., Edinger, T.G., and Bonner, W.D. Copyright © 1982 by W.H. Freeman and Company. All rights reserved.)*

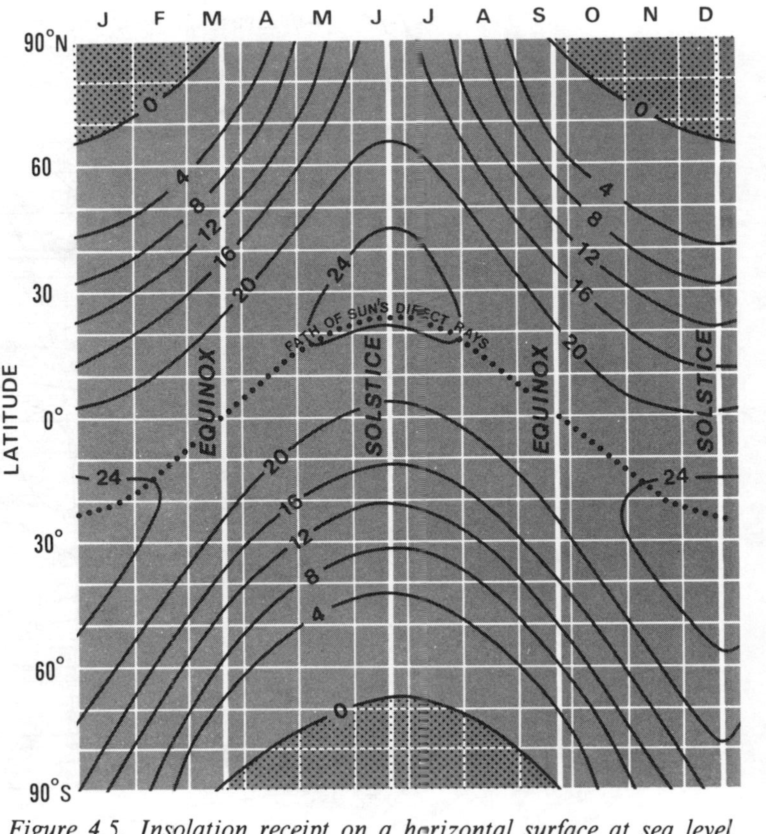

Figure 4.5. Insolation receipt on a horizontal surface at sea level, assuming 70 percent transmission through a normal atmosphere, by month and latitude, in MJ/m² day. (From Understanding Our Atmospheric Environment, *2nd ed., by Neiburger, M., Edinger, T.G., and Bonner, W.D. Copyright © 1982 by W.H. Freeman and Company. All rights reserved.)*

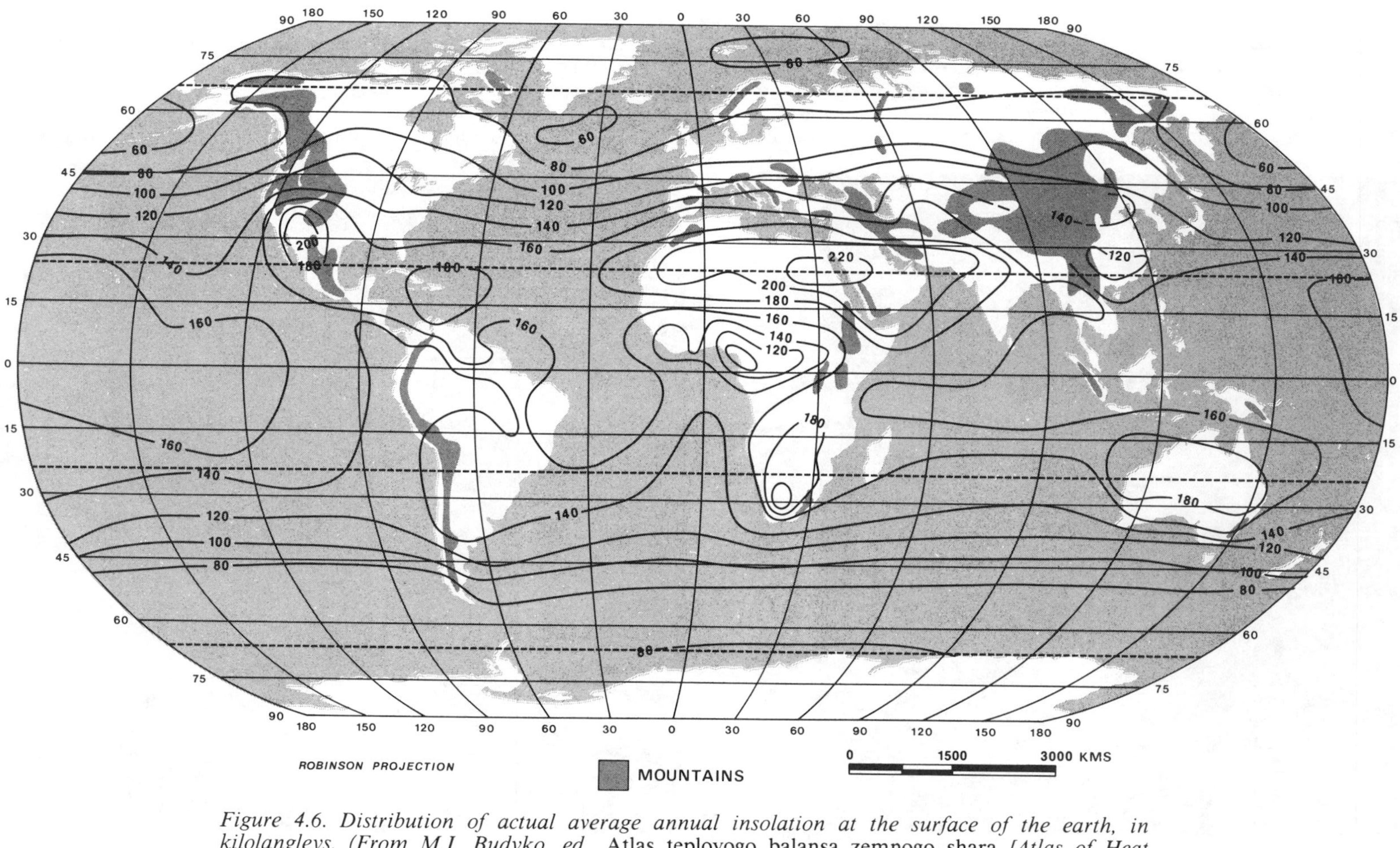

Figure 4.6. Distribution of actual average annual insolation at the surface of the earth, in kilolangleys. (From M.I. Budyko, ed., Atlas teplovogo balansa zemnogo shara *[Atlas of Heat Balance of the Earth] [Moscow, 1963].)*

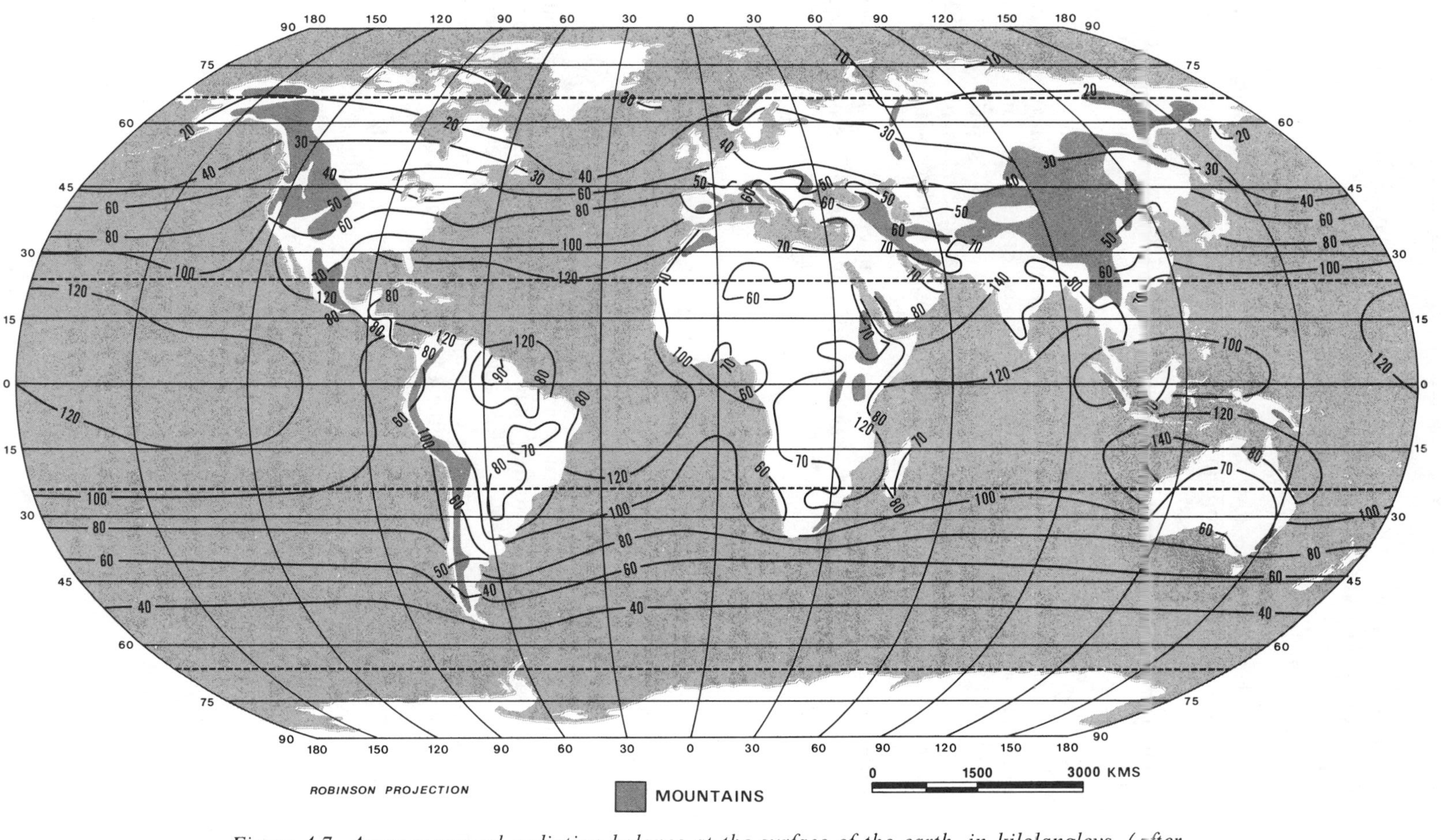

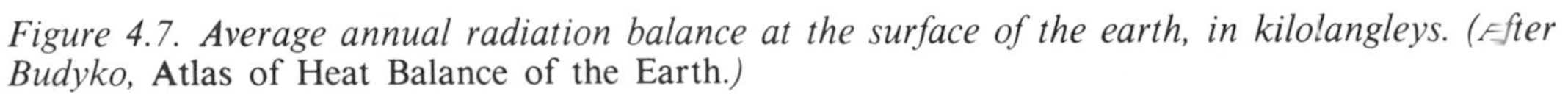

Figure 4.7. Average annual radiation balance at the surface of the earth, in kilolangleys. (After Budyko, Atlas of Heat Balance of the Earth.*)*

ture contents, generally receive more insolation than adjacent water areas and the highest values of all are in the large deserts of the world, the radiation balance pattern generally shows higher values over water than over adjacent land, and the lowest values of all occur over the great subtropical deserts, such as the Sahara and Australia, or in the earth's high-mountain regions, where the air is thin and low in moisture. The deserts and high mountains of the earth have often been called the "windows of the world," because both areas are associated with atmosphere that allows ready passage of both incoming and outgoing radiation.

In general, oceans absorb much more of the incoming solar radiation than land surfaces, and radiate much less back to space because of their consistently moderate temperatures. The oceans thus show a high radiation balance and serve as a huge reservoir for heat storage on the earth.

It is interesting to note that over the entire course of the year, the radiation balance at the earth's surface almost everywhere is positive. The pattern of radiation balance, of course, changes drastically with seasons. From May through August all the earth's surface, except for the Southern Hemisphere south of about 42°S latitude and perhaps parts of the Arctic, shows a positive radiation balance. By September a negative balance begins to show up on the northern fringes of the Canadian Archipelago and the Arctic islands off Siberia. This spreads rapidly southward in succeeding months and reaches a latitude of about 40°N (37°N over Japan) at its maximum extent during the Northern Hemisphere winter solstice in December. The area of negative radiation balance begins to recede northward again in January, and by April it has practically disappeared off the Northern Hemisphere continents except for the northern islands in the Canadian Archipelago and Spitsbergen, Franz Joseph Land, and Severnaya Zemlya in the Arctic north of Eurasia.

Latent Heat Exchange

As might be expected, much of the radiation balance over the oceans is used as latent heat to evaporate water, particularly in subtropical latitudes where the atmosphere is usually clear of clouds and relatively dry (Figure 4.8). Values are even higher in certain ocean areas adjacent to middle latitude continents, such as off the east coast of the United States and portions of the western North Pacific next to Japan. In these middle-latitude areas the excessive evaporation takes place primarily during winter. (The reasons for this will be discussed in Chapter 7.) The land masses, where moisture generally is not as readily available to be evaporated, show much lower values, although some of the equatorial land areas, such as the Amazon Basin, the Congo, and the Indonesian Islands, still show relatively high values.

Sensible Heat Exchange

The sensible heat exchange map is almost the exact complement of the evaporation map (Figure 4.9). Heat is moved upward through conduction and convection (as an air temperature increase) most rapidly where the earth's surface is hottest—the earth's subtropical deserts, such as the Sahara and Kalahari in Africa and the large Australian desert. The oceans, with their relatively cool summer temperatures, engage in very little sensible heat exchange with the atmosphere above them. A few oceanic areas, such as the North Atlantic and western North Pacific, where large, warm ocean currents carry anomalously warm waters into middle and higher latitudes, show high values of sensible heat exchange, particularly in winter. The Barents Sea north of Scandinavia is outstanding in this respect.

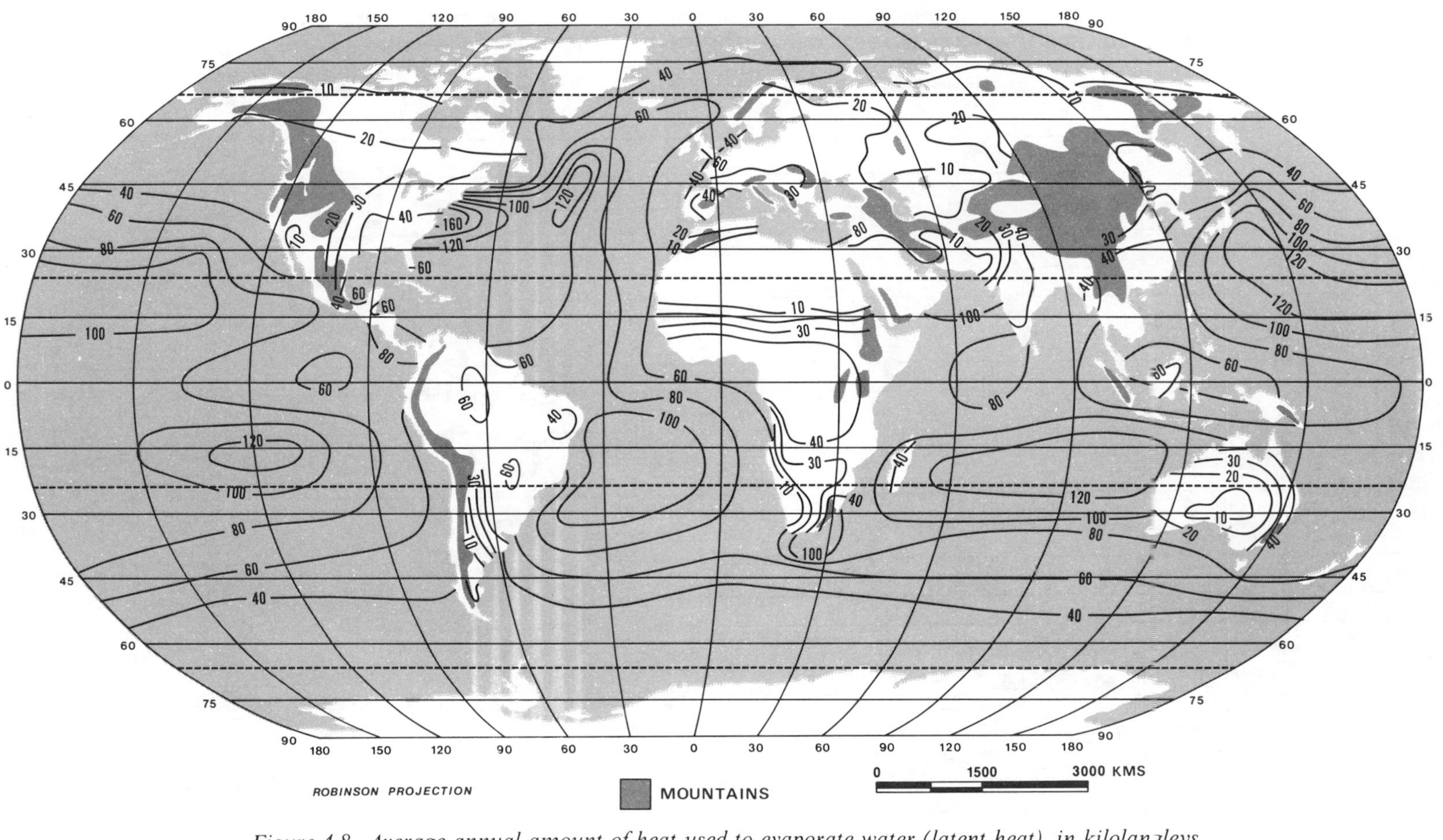

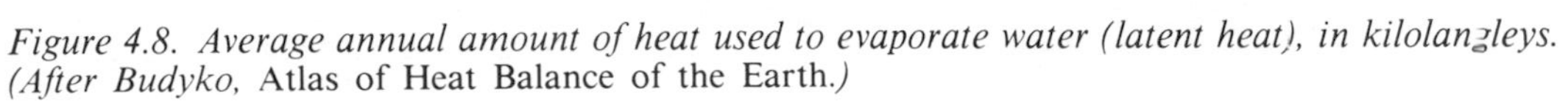

Figure 4.8. Average annual amount of heat used to evaporate water (latent heat), in kilolangleys. (After Budyko, Atlas of Heat Balance of the Earth.*)*

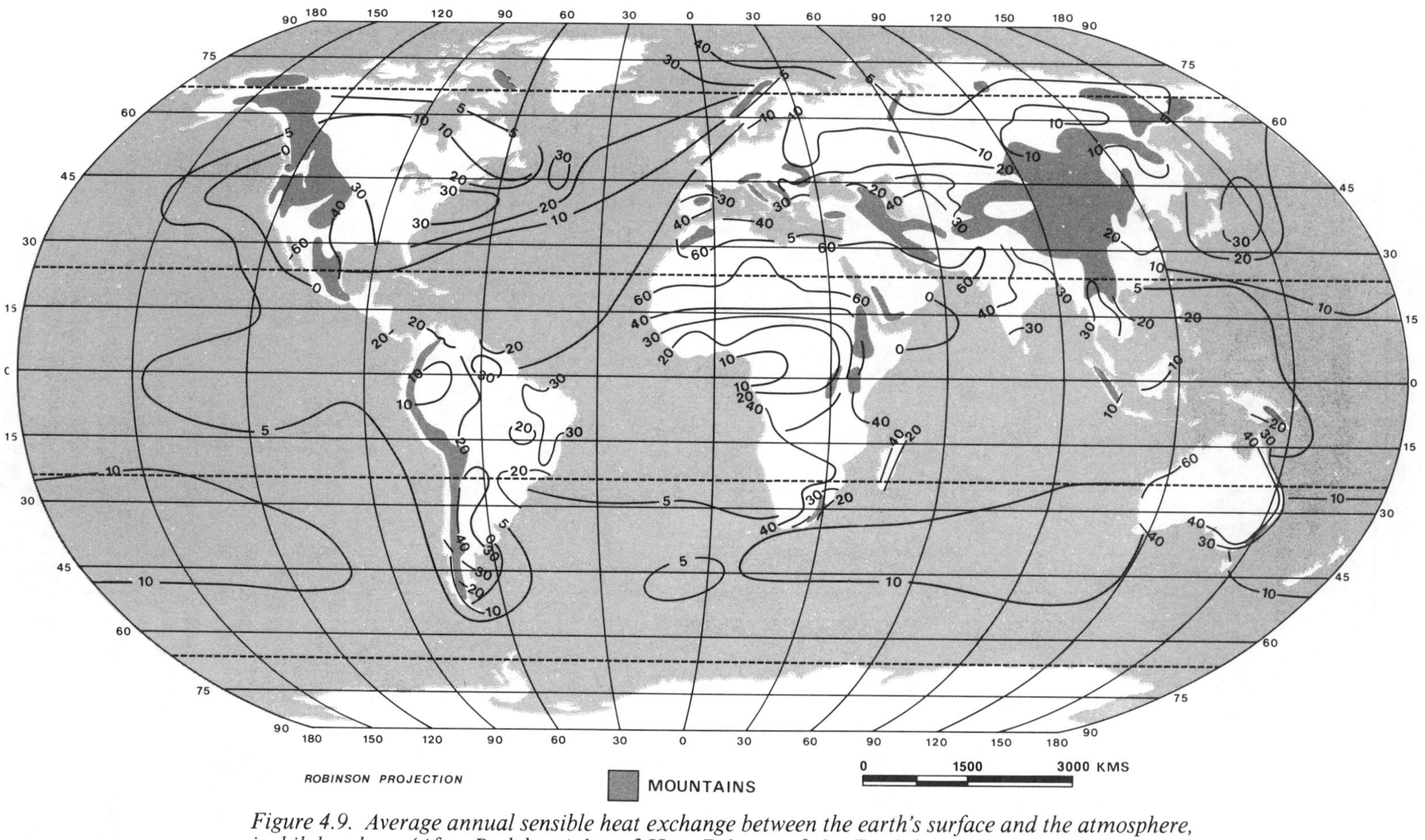

Figure 4.9. Average annual sensible heat exchange between the earth's surface and the atmosphere, in kilolangleys. (After Budyko, Atlas of Heat Balance of the Earth.*)*

On the other hand, some ocean areas with cool surface currents experience a sensible heat exchange that is directed downward from the atmosphere to the water.

Heat Balance of the Earth-Atmosphere System

Over the course of a year, the net radiation balance at the earth's surface is positive over most of the earth. Therefore, the net radiation balance of the atmosphere must be negative for the radiation balance of the earth-atmosphere system to be zero so there is no net heating or cooling over a long period of time. Most of the heat lost to space is radiated from the atmosphere, particularly from cloud tops. The atmosphere thus has a radiation balance of approximately −80 kilolangleys per year at every latitude. It varies slightly by latitude, but not very much. The atmosphere, of course, makes up for this heat deficit by receipt from the surface of the earth.

Horizontal Transfers of Heat

The earth-atmosphere system has a net gain of heat from approximately 32°N latitude to 28°S latitude, and a net loss poleward from those latitudes. Figure 4.10 shows these relationships for the Northern Hemisphere. The latitudinal scale is a cosine one in order to represent equivalent areas of the earth's surface properly. Since the parallels of latitude become shorter from equator to pole, there is less area between equal latitudinal intervals near the poles than near the equator. Half the earth's surface falls between 30°S latitude and 30°N latitude. In Figure 4.10 it can be seen that the surplus in the low latitudes balances the deficit in the middle and high latitudes.

For the low latitudes to get rid of surplus heat and for the high latitudes to radiate more heat to space than they receive from the sun, there must be a horizontal transfer of heat from low latitudes to high latitudes.

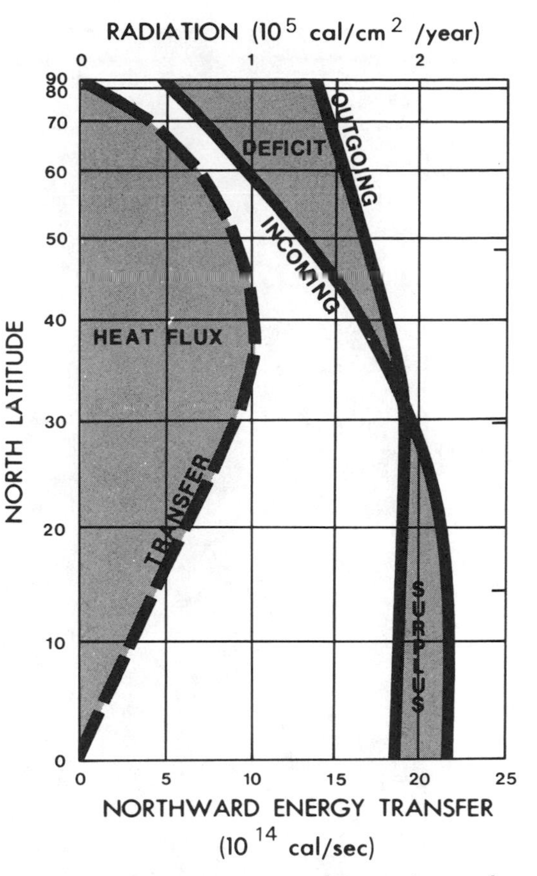

Figure 4.10. Average annual incoming and outgoing radiation for the earth-atmosphere system and the poleward transfer of heat.

The dashed line in Figure 4.10 indicates roughly the magnitude of the heat flux by latitude. It is obvious that the greatest heat flux is in the middle latitudes around 30°–40°, for all the heat that moves from low latitudes to high latitudes must pass through the middle latitudes. This necessitates a maximum of air movement in the middle latitudes to effect the heat transfer. In Chapters 6 and 12 it will become evident that the middle latitudes, on the average, have the highest wind speeds and are the stormiest and in general, climatically most vigorous of the earth's zones.

The transfer of heat from low latitudes to high latitudes is accomplished by three

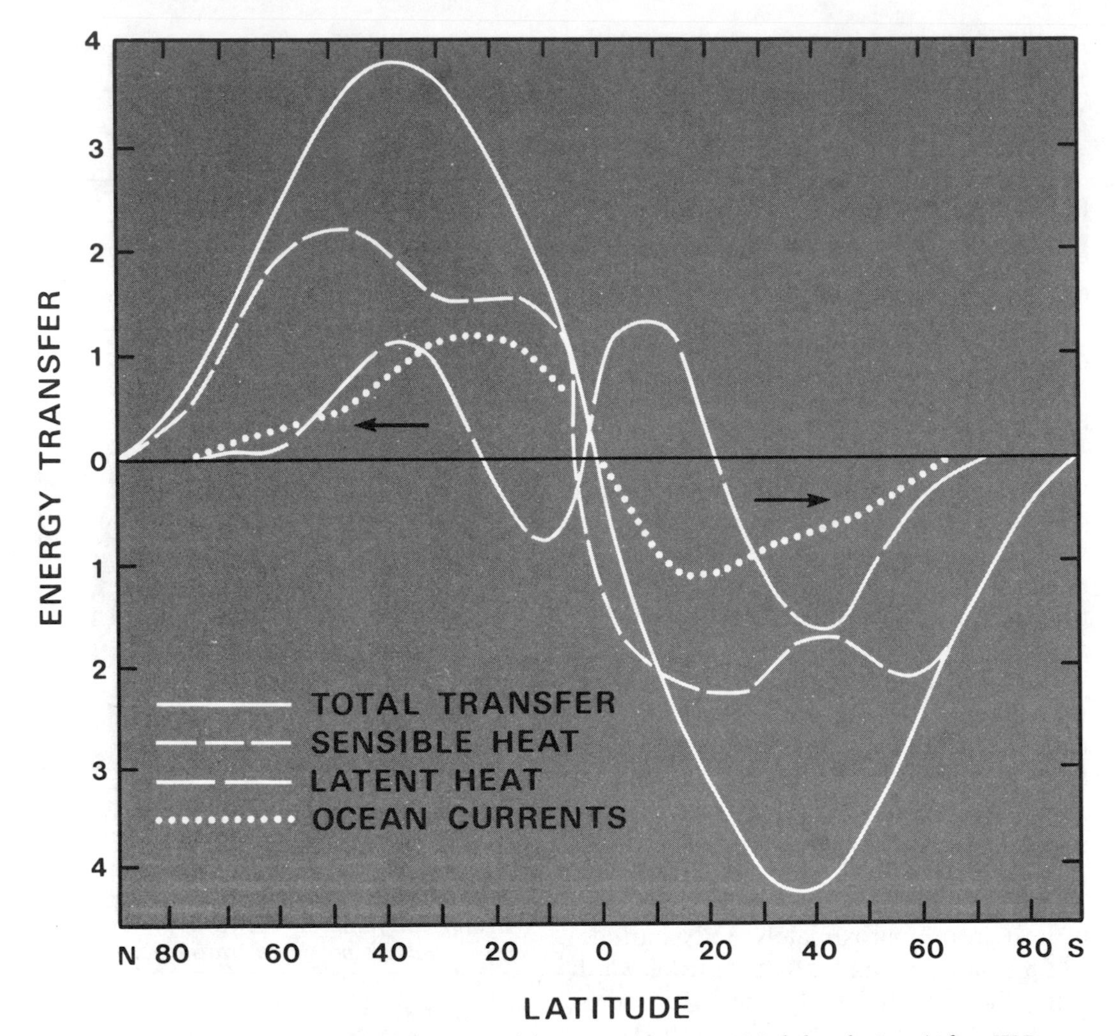

Figure 4.11. Average annual horizontal heat transfers, in 10^{19} kilocalories. (After W.D. Sellers, Physical Climatology *[Chicago: University of Chicago Press, 1965]. © 1965 by the University of Chicago. All rights reserved.)*

means (see Figure 4.11). The largest component is sensible heat transfer in the atmosphere; that is, the movement of warmer air into higher latitudes. In each hemisphere there is a fairly simple poleward movement, with maximum flux taking place in the middle latitudes. Ocean currents tend to show the same pattern of heat transfer, at lesser magnitudes. The movement of latent heat via water vapor, however, shows a different pattern. Since most of the evaporation on earth takes place over the subtropical oceans, latent heat moves both equatorward and poleward from the subtropical areas on either side of the equator. Since the magnitude of latent heat movement is considerably smaller than that of sensible heat movement in the atmosphere and oceans combined, this more complicated pattern of latent heat transfer does not perceptibly affect the overall pattern of heat transfer.

5 ATMOSPHERIC PRESSURE AND MOTION

Since different amounts of heat energy are received at different points on the earth's surface, and since these inputs change from time to time at any one place, differences in density are set up in the earth's atmosphere that cause atmospheric motion. Heat energy is the driving force that initiates the motion, but once in motion, the atmosphere is acted upon by such forces as that due to the rotation of the earth, and a circulation evolves that tends to maintain itself and take on a character of its own, which in turn influences air pressure and temperature. Thus, on any portion of the earth's surface, atmospheric motion and the heat budget must be considered as major controls on climatic elements. In the middle latitudes, particularly, the surface air temperature on any given day is probably due more to movements of air over the region than to the radiation balance at that point on the earth's surface. The same is true of atmospheric humidity, cloud formation, precipitation, and so forth. Therefore, a knowledge of the movements in the atmosphere and their consequences is all-important. Since this motion is a result of atmospheric pressure differences, it is necessary to define atmospheric pressure.

Air Pressure

The mixture of gases that make up the atmosphere has a certain mass and, although not very dense, adds up to considerable weight when one considers the entire thickness of the atmosphere attracted by the force of gravity to the earth's surface. On the average at sea level, the atmosphere weighs approximately one kilogram per square centimeter (14.7 pounds per square inch). This is quite a great weight if one considers how many square centimeters are on, say, the surface of the human body or the roof of a building. For instance, the ceiling of a room 3 × 4 meters square would contain 120,000 square centimeters, which would be under an atmospheric weight of 120 metric tons. Of course, in a gas, pressure is exerted equally in all directions, and as long as there is as much air pressure inside a room as outside there is no net force on the structure. But if the air were suddenly evacuated from the room, it is easy to see that the structure would collapse under the great weight of the atmosphere.

Measurement and Expression

The most accurate instrument for measuring air pressure is the mercurial barometer, which consists of an evacuated glass tube about one meter long, mounted vertically, with a surface of mercury at the base. The weight of the air pressing down on the mercury surface balances a column of mercury in the tube equal in weight to the weight of the atmosphere. Thus, air pressure is often given in terms of the height of the column of mercury balanced by the air. Normal sea-level pressure in these terms is 76 centimeters or 29.92 inches.

Another kind of instrument is the aneroid barometer, which consists of an evacuated metal wafer that expands and contracts with air pressure and moves a pointer across a scale. Since this is entirely mechanical, contains no liquids, and can be moved easily and held in any position, it is much more convenient to use. It is not as accurate as the mercurial barometer, however. Since, for most purposes, not the absolute pressure but the relative pressure from one place to another or the change at a given point over time is important, an instrument called a barograph was invented to show a continuous trace of air pressure over time. This uses an aneroid barometer to move a pen arm on a graph paper wrapped around a slowly rotating cylinder, thereby tracing a continual record of the changes of air pressure.

Neither pounds per square inch nor inches or centimeters of mercury are very convenient for plugging into scientific equations, and therefore most of the world has gone to the expression of air pressure in millibars (mb). Since one millibar equals 1000 dynes per square centimeter, and one dyne equals one gram-centimeter per second per second, the millibar is based on the c.g.s. system and is convenient to use in scientific equations. Average sea-level pressure in millibars is 1013.2. Almost all atmospheric pressures recorded in the United States since 1940 have been expressed in millibars, as is true in most of the rest of the world.* Nevertheless, airports still report altimeter settings in inches of mercury, so it is important to be able to convert from one to the other. One inch of mercury equals approximately 34 millibars.

Variation

Since the atmosphere is usually involved in some motion, and different amounts of heat energy are being expended in different ways, atmospheric pressure undergoes constant variation at any given point on the earth's surface and varies from place to place at any given time. In a horizontal direction, the variations are not great. At sea level the variation is usually no more than 5 percent on either side of the normal, which is probably not enough for the human body to detect consciously. Air pressure varies much more rapidly with height, however; as was pointed out before, the air pressure decreases by half about every 5000–6000 meters. Therefore, roughly half the mass of the atmosphere lies below 5500 meters (18,000 feet). Anyone who has climbed a high mountain or flown in an unpressurized airplane knows that changes in air pressure of this magnitude can cause extreme reactions in the human body, ranging from accelerated breathing and general fatigue in milder cases, to severe headaches, nosebleeds, nausea, dizziness, and fainting, in more severe cases.

Although the horizontal variations in air pressure might not be of physiological significance, they are all-important to atmospheric motion. And since atmospheric motion is such an important control on all other weather phenomena, the distribution of air pressure over the earth's surface at any given time is probably the single most important tool for predicting a whole range of weather elements. Because the motion itself, once initiated by heat differences, becomes a control on the air pressure pattern, there is a feedback mechanism be-

*The recently adopted SI system specifies the kilopascal (kPa) as the official unit of air pressure. One kPa equals 10 mb. One pascal (Pa) equals one newton (N) per square meter, where one newton is the force that, acting on a mass of one kilogram, produces an acceleration of one meter per second per second.

tween the two that requires air pressure and motion to be considered jointly. First to be considered are broad-scale patterns of pressure and motion over the earth, and then smaller circulation systems. So we may grasp the workings of the general circulation of the atmosphere, a simple case will be considered first, and then complications will be added one at a time until reality is approached.

Atmospheric Circulation

Wind-Force Relations

Let us begin with a homogeneous, stationary earth receiving the sun's energy on its spherical surface. Under these conditions, the equatorial region would receive more heat than the polar areas, and an air pressure gradient (decrease) would be set up at the earth's surface directed from the cold, dense air over the pole to the warm, light air over the equator. The resultant circulation theoretically would be a single vertical cell between equator and pole (Figure 5.1). The warm, light air at the equator would rise, and the cold, dense air at the pole would sink. This would induce a pole-to-equator movement of air at the surface and an equator-to-pole movement of air aloft.

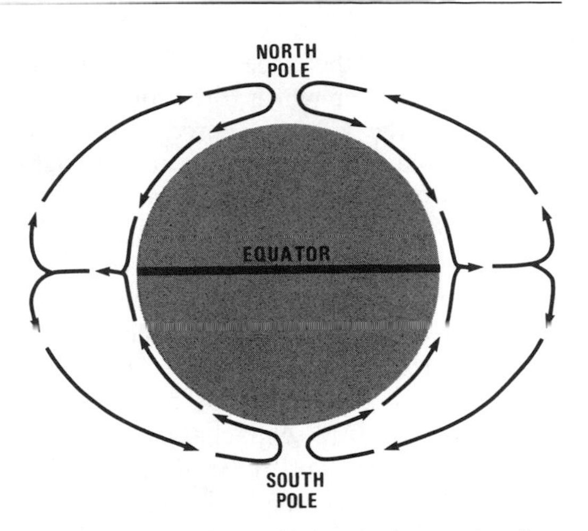

Figure 5.1. Single-cell hemispheric circulations on a homogeneous, stationary earth.

Coriolis Force

The earth is not just sitting in space, however, but is rotating on an axis, and therefore the north-south-moving air streams are flowing above a rotating plane of reference from which the motion is being judged. An observer rotating with the earth's surface would perceive the air streams to curve with respect to this frame of reference even though they might appear to an observer in space to be straight lines (Figure 5.2). Since the earth rotates from west to east, it is rotating counterclockwise as viewed from above the North Pole. Therefore, in the Northern Hemisphere this apparent deflection of an object moving across the face of the earth appears to be clockwise, or to the right of the motion, viewed downstream with the motion. As viewed from the South Pole, the earth appears to be rotating clockwise, so in the Southern Hemisphere the deflection appears counterclockwise, or to the left of the motion, as viewed downstream. This apparent "force" is known as the Coriolis Force after G.G. de Coriolis (1792–1843), a French mathematician who first treated the phenomenon quantitatively.

It can be shown by such devices as Foucault pendulums (heavy balls on long, thin wires) in museums that because of the earth's rotation about its axis, any point on the surface of the earth acts as a center of rotation for its own horizon about an axis perpendicular to the earth's surface at that point. The speed of rotation of the horizon depends upon the latitude, increasing from zero at the equator to a maximum at the poles. Thus, the Coriolis Force also increases from zero at the equator to a maximum at the poles. In addition, since the Coriolis Force is a relative motion between a moving object and the rotating surface of the earth, the Coriolis Force also increases with the speed of the moving object. No matter which direction the motion takes, the Coriolis Force is directed perpendicular to the motion, to the right in the Northern Hemisphere and to the left in the Southern Hemisphere.

Thus there are two basic forces acting upon the air, the pressure gradient force (pressure decrease from one place to another) and the Coriolis Force. As long as these two forces are out of balance with each other, the air will change its speed and/or direction, and by so doing quickly bring into balance the two forces so that the air will move with a constant velocity (speed and direction).

Figure 5.3 illustrates in slow motion what is essentially an instantaneous process of adjustment. Assume that over a small portion of the earth's surface in the Northern Hemisphere the upper air contours (lines of equal altitude on a surface of constant pressure) are straight and equally spaced, as shown in Figure 5.3. The pressure gradient force, P, is directed from high pressure to low pressure. The air begins to move down this pressure gradient, but as soon as motion begins the Coriolis Force, C, deflects the air to the right. As long as the air is moving partially in the direction of the pressure gradient, it will accelerate, and the faster it goes, the stronger the Coriolis

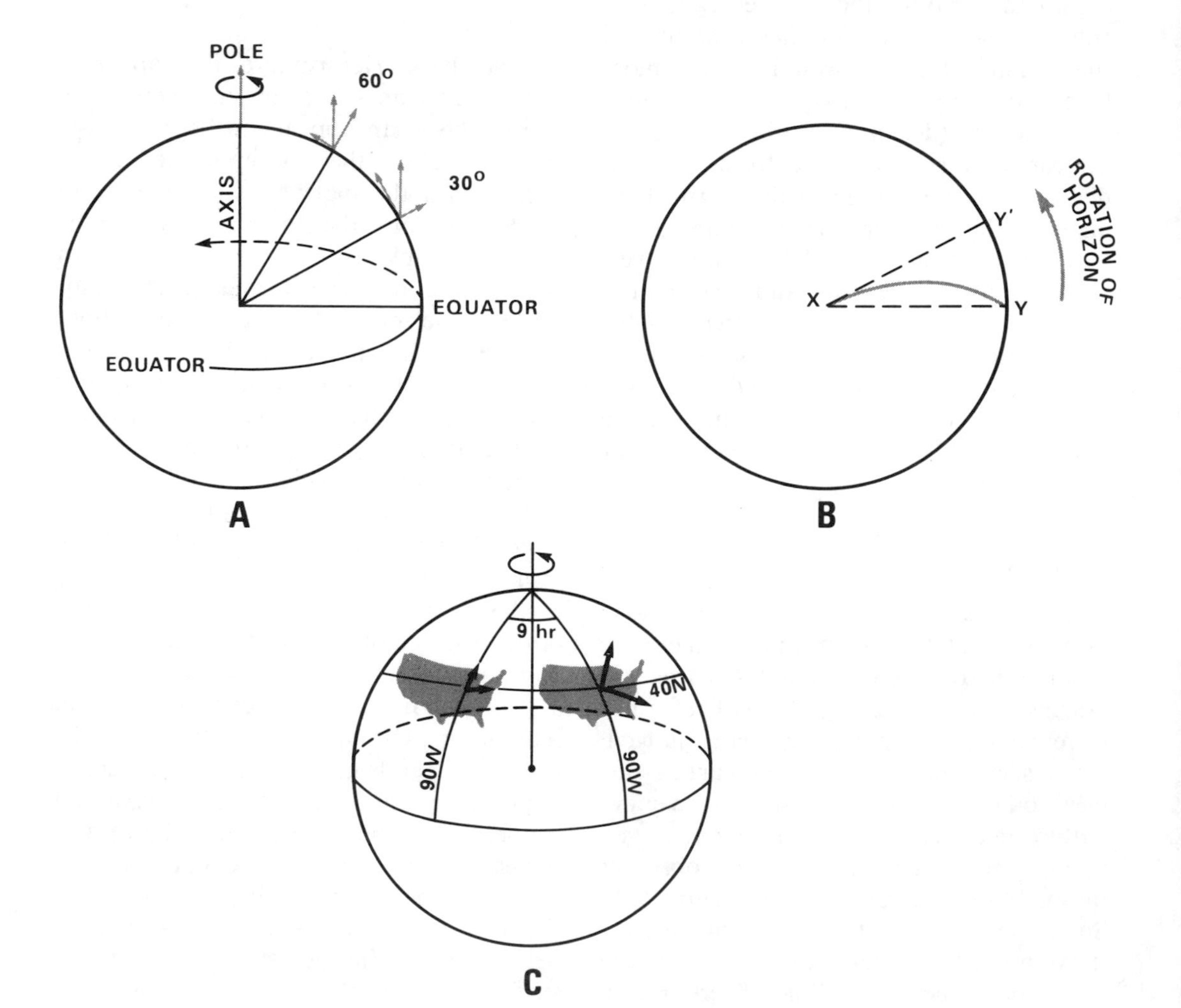

Force. Therefore, the deflection to the right increases, and the Coriolis Force swings around with the changing direction of the wind, remaining perpendicular to it at all points. Ultimately, the air moves with the Coriolis Force directed opposite to the pressure gradient. The wind then has a terminal velocity fast enough to cause the magnitude of the Coriolis Force to equal that of the pressure gradient force, and from then on there is no net force. Therefore, the wind neither accelerates nor decelerates, but moves with a constant speed and direction that is parallel to the contours, perpendicular to the pressure gradient. Thus, if the pressure gradient and Coriolis Force are the only two forces acting on the air, the wind does not move from high to low pressure but along lines of equal pressure, and thereby tends to perpetuate the original pressure pattern. This straight-line motion of air between the balanced pressure gradient and Coriolis Force is known as the *geostrophic wind.* It can be computed if one knows the pressure gradient and the latitude. The closer together the contours (the stronger the pressure gradient), the stronger the wind; and the lower the latitude, with the same pressure gradient, the stronger the wind.

Returning to the consideration of the atmospheric motion between equator and poles, because of the Coriolis Force neither the surface air, which tends to move from pole to equator nor the upper air, which tends to move from equator to pole, moves in a north-south direction but ends up moving in a zonal direction (parallel to the parallels of latitude). The pole-to-equator movement of surface air is deflected to the right in the Northern Hemisphere and the left in the Southern Hemisphere, which causes east-to-west winds at the surface in both hemispheres. The equator-to-pole-

Figure 5.2 (opposite page). *The Coriolis Force. Part A illustrates that at any point on the earth's surface an object rotates about an imaginary axis parallel to the earth's axis of rotation. This rotation can be broken into two component parts, one that is parallel to the earth's surface and one that is perpendicular to it at that point. At the pole the entire rotation is parallel to the earth's surface around a vertical axis. Here the earth's grid rotates completely around once per day. At successively lower latitudes less and less of the rotation is parallel to the earth's surface around a vertical axis, and more and more is perpendicular to the earth's surface around a horizontal axis. At 30° latitude the horizontal rotation around the vertical axis is only one-half that at the pole. At the equator there is no horizontal rotation at all; an object moving across the surface would experience one complete rotation per day in a plane perpendicular to the earth's surface and would not change its direction with respect to the earth's grid. Part B illustrates the apparent deflection of a projectile shot from point X on the earth's surface in the Northern Hemisphere toward a target at point Y. During the flight of the projectile this portion of the earth's surface rotates counterclockwise around the observer at point X so that the target moves from point Y to point Y′. The projectile moves to point Y and thus misses the target. To an observer in space it appears that the projectile moved along a straight line from X to Y, but to the gunner at X, who is rotating with the earth, it appears that the projectile moved from X to Y along the curved solid line. Part C shows the rotation of the position of the United States during a nine-hour period. Assuming frictionless airflow and constant pressure gradient, air that initially was blowing from west or south at 40°N, 90°W will end up blowing from northwest or southwest because of the rotation of the continent under it.*

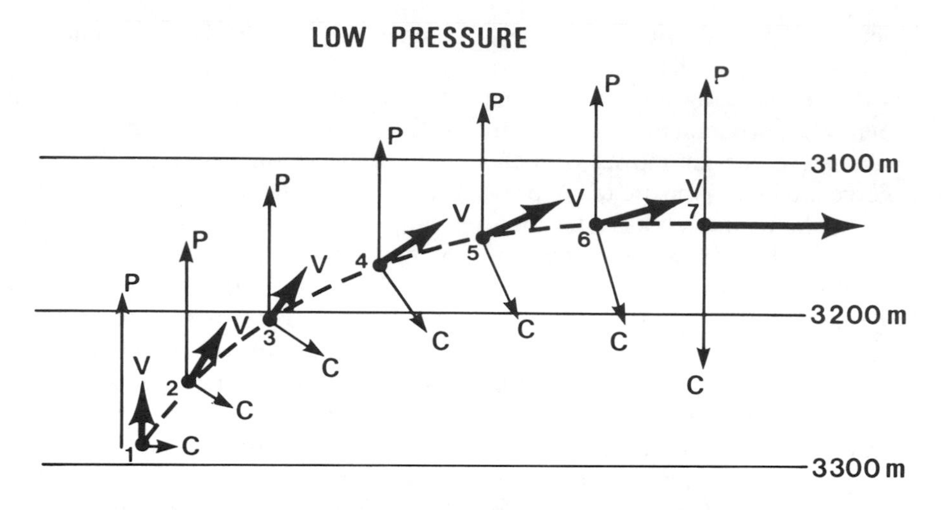

Figure 5.3. Frictionless airflow at 3100–3300 meters altitude quickly adjusts to the pressure pattern, so the wind blows with constant velocity parallel to the contours with the pressure gradient and Coriolis forces balanced against each other.

moving upper air ends up moving from west-to-east in both hemispheres.

Friction

All the foregoing considerations assume frictionless flow, and this is essentially true above the influences of the earth's surface, perhaps 300–600 meters in the air. But at the surface of the earth, friction between the moving air and the surface below becomes a significant force that must be taken into consideration. This changes the wind-force diagram to that shown in Figure 5.4. Friction (F) always acts opposite to the wind (W). This slows down the motion, which reduces the Coriolis Force, which allows the pressure gradient force (P) to overbalance the Coriolis Force (C) and pull the wind toward the pressure gradient. The Coriolis Force swings with the wind, remaining perpendicular to it at all times, and the friction swings with the wind, opposing it at all times. The motion soon adjusts to the three forces, so that the resultant of the Coriolis Force and friction is equal and opposite to the pressure gradient. Thus, a balance of forces is again achieved, and the wind moves with constant speed and direction, crossing the isobars at some acute angle, commonly about 15°–30° over land and less than that over smoother water.

Hadley Cell

Thus, the surface air, which on a differentially heated earth might be expected to move from the cold poles toward the hot equator, ends up moving neither north-south nor east-west, but in some diagonal direction, from east-northeast to west-southwest in the Northern Hemisphere. There are now both zonal (latitudinal) and meridional (longitudinal) components. The upper air flow, which is essentially free of friction, nevertheless also takes on a meridional component from equator to pole in order to achieve mass continuity with the pole-to-equatorward component of the surface air. Although the zonal component is usually much larger than the meridional one, the meridional movement effects the latitudinal heat and momentum transfers

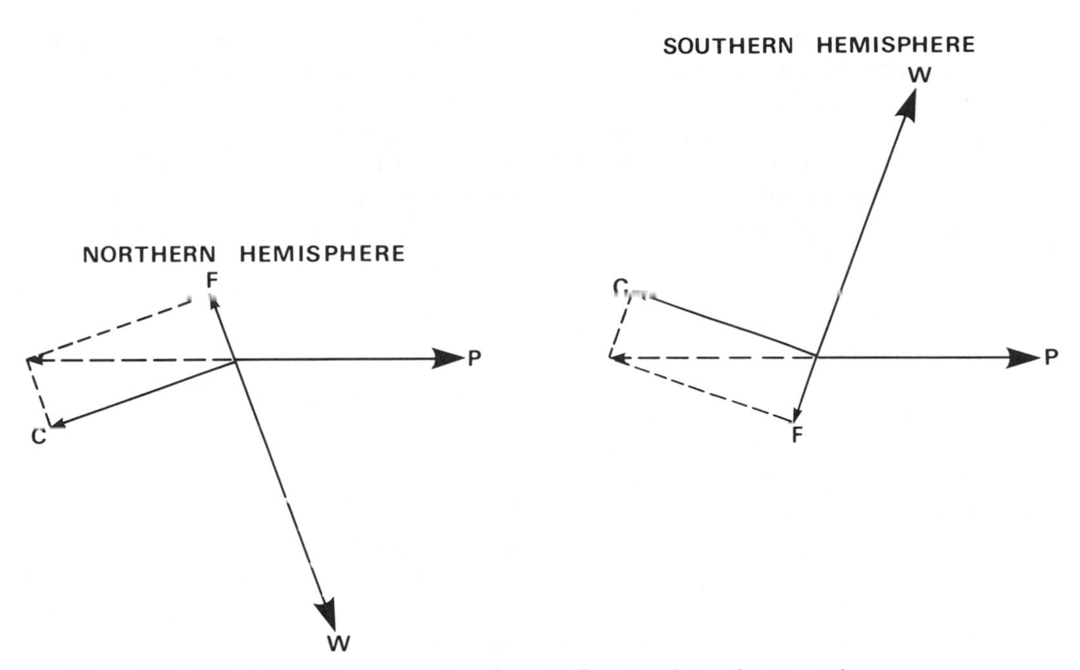

Figure 5.4. Wind-force diagrams of surface air flow involving friction. The same pressure gradient force (P) produces different wind directions (W) in the two hemispheres because of the reversal across the equator of the direction of the Coriolis Force (C).

necessary for the maintenance of the general circulation of the atmosphere and the heat balance of the earth. This meridional component comprises a vertical circulation cell that is known as the "Hadley cell," after the Englishman George Hadley who, in the eighteenth century, postulated these movements qualitatively to explain the trade winds (which will be explained in Chapter 12).

Momentum

Observation has indicated, and theory has demonstrated, that a simple Hadley cell cannot exist over an entire hemisphere. If the surface winds everywhere were easterly (from east to west), friction between the air and the earth's surface would soon slow the winds to no wind at all. In order to keep the trade winds blowing in low latitudes, there must be a mechanism for transferring the westerly (west to east) momentum of the earth's rotation away from the easterly winds in low latitudes to the air at higher latitudes, and this transport of westerly momentum will then feed westerly winds at extratropical latitudes. Friction with the earth's surface would stall the winds whether they were all easterly or all westerly. There must be a balance between belts of alternating easterly and westerly surface winds, and there must be mechanisms for transferring westerly momentum from easterly wind belts to westerly wind belts.

Also, a simple meridional component of flow all the way from equator to pole in the middle troposphere is not feasible. This air, rising over the equator, is essentially stationary relative to the earth, and since the earth is rotating once a day, it has an average linear speed of more than 450 meters per second (m/s) (1000 miles per hour) from west to east. As the air moves poleward in essentially frictionless flow at upper levels, it comes into latitudes with constantly decreasing radii about the earth's axis of rotation. In order to maintain constant westerly momentum, the west-east velocity must increase correspondingly. By the time the air has moved half way to the pole, at 45° latitude, the radius of rotation has reduced to approximately 0.7 what it was at the equator, and the west-east absolute motion of the air must have speeded up to about 635 m/s (1400 mph). At the same time, the west-east speed of the earth's surface has decreased to a ratio of 0.7 what it was at the equator, and therefore the west-east air speed relative to the earth must be 330 m/s (737 mph)! Thus, the great west-east momentum imparted to the air by the rotation of the earth would cause inconceivably high wind speeds relative to the earth's surface at increasing latitudes. Such flow would become extremely turbulent and unstable, and zonal flow would break down into large whirls or eddies that would dissipate much of the energy.

Only at low latitudes, then, does a semblance of a Hadley cell exist. At higher latitudes momentum considerations become paramount, and the circulation of the atmosphere shows little or no direct connection with temperature patterns. The indirect "Ferrel" cell results in the middle latitudes from the transfer of westerly momentum from lower latitudes. It is related only indirectly to latitudinal temperature differences through momentum transfers effected by the Hadley cell of the lower latitudes, which is directly driven by heat differentials.

Resonance

The result is that, with the given energy inputs (latitudinal heat differentials and earth's rotation) and the character of the atmosphere (mass, vertical-temperature structure, and hydrostatic balance between gravity and pressure decrease with height), the earth's atmosphere resonates in a given mode, much as a tuned instrument, that sets up circulatory systems that tend to be persistent in time and location. The air masses within circulating cells between resonating nodes tend to be somewhat self-contained, but there are slow exchanges of flow between them, effected largely by transitory circulatory systems on smaller spatial scales, that through their movements facilitate transfers of heat and momentum necessary to keep the general circulation going.

Cellular Flow

Most of the time the air does not flow along straight lines, but along curved lines around cells and vortices. For the wind to blow along curved isobars, another apparent force must be introduced. For example, suppose there is a circular cell of frictionless wind flow around a center of high pressure (Figure 5.5). A pressure gradient force (P) is directed outward in all directions from the center. Air will begin to move down the pressure gradients from the center toward the periphery of the cell. If this is in the Northern Hemisphere, the Coriolis Force (C) will act to the right and eventually cause the wind to flow clockwise parallel to the isobars at that point, with C equal and opposite to P. If the wind continues in a constant direction, however, it will go off on a tangent to the circular isobars and leave the circulation system. An additional force is necessary, directed toward the center of the high pressure cell, to cause a continual change in direction of the wind to keep it moving parallel to the curved isobars. In other words, the Coriolis Force must slightly overbalance the pressure gradient force. For this to happen, the wind must accelerate. Therefore, the wind flow around the curved isobars of a high pressure cell must be slightly stronger than the geostrophic wind with the same pressure gra-

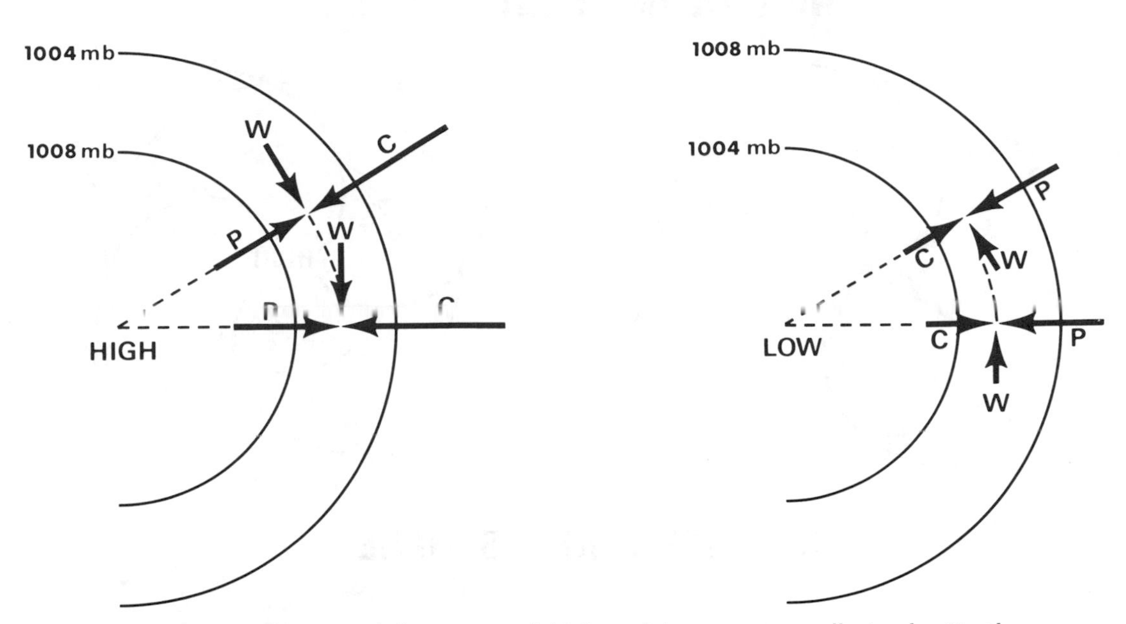

Figure 5.5. Gradient wind flows around high and low pressure cells in the Northern Hemisphere.

dient along straight isobars. The additional force thus produced is known as centripetal force, directed toward the center of the cell.

In the case of a low pressure cell, the pressure gradient is everywhere directed from the periphery toward the center (see Figure 5.5). The air begins to move from the periphery toward the center of the cell, but in the Northern Hemisphere the Coriolis Force acts to the right and makes the wind flow counterclockwise, parallel to the isobars. The centripetal force is again directed from the periphery toward the center, so that in the case of a low pressure cell the centripetal force and the pressure gradient force oppose the Coriolis Force. Therefore, around a low pressure cell the Coriolis Force is less than the pressure gradient force, and consequently the velocity of the wind must be less than it is around the high pressure cell or even the geostrophic wind along straight isobars.

Thus, for a given pressure gradient, the resultant wind will be fastest around a high pressure cell, slowest around a low pressure cell, and intermediate along straight isobars. Nonetheless, in nature the pressure gradient around low pressure cells is usually significantly greater than it is around high pressure cells, and it can be shown mathematically that the pressure gradient must weaken toward the center of a high pressure cell so that, ultimately, the wind becomes essentially calm at the center of a high pressure cell. In a low pressure cell, on the other hand, the pressure gradient normally increases toward the center of the cell, and high wind speeds may be achieved.

The balanced, frictionless flow just described around curved isobars is known as the *gradient wind.* Like the geostrophic wind, the gradient wind may be computed from the pressure gradient and the latitude. In the Southern Hemisphere, since the Coriolis Force is directed to the left of the motion, the air circulates around high and low pressure cells in opposite directions to those described for the Northern Hemisphere, counterclockwise around highs and clockwise around lows.

NORTHERN HEMISPHERE

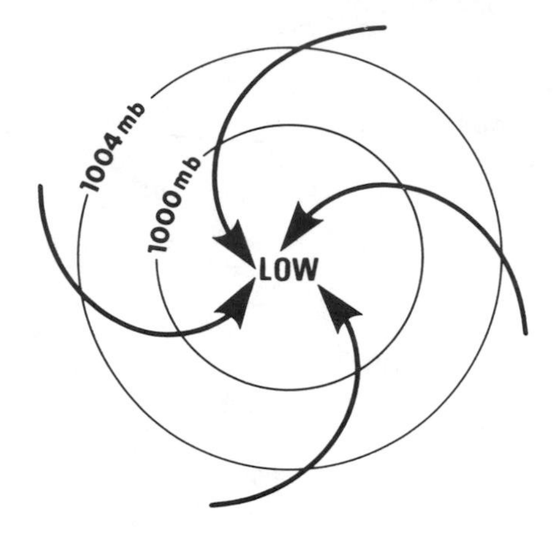

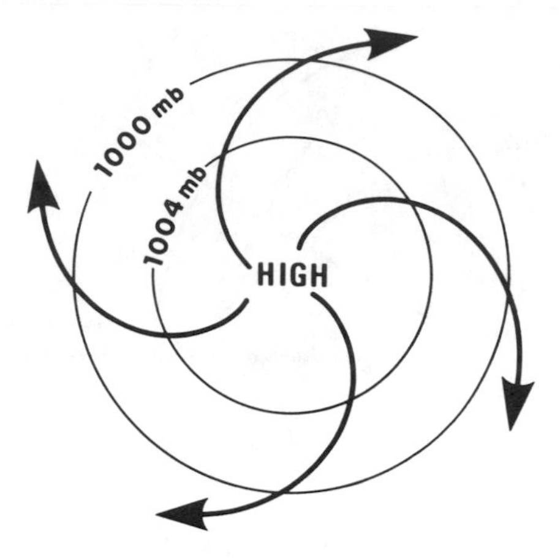

SOUTHERN HEMISPHERE

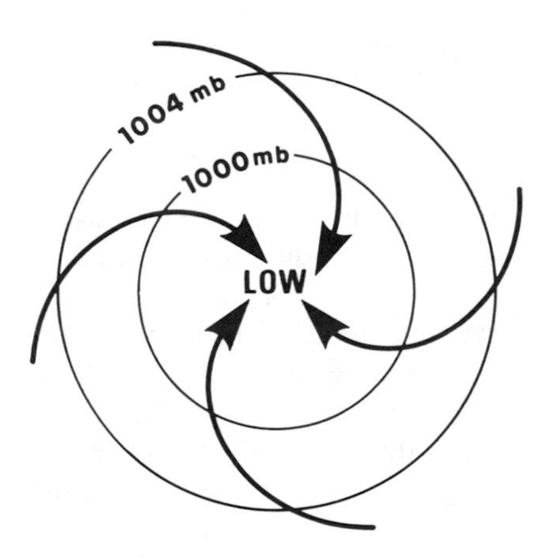

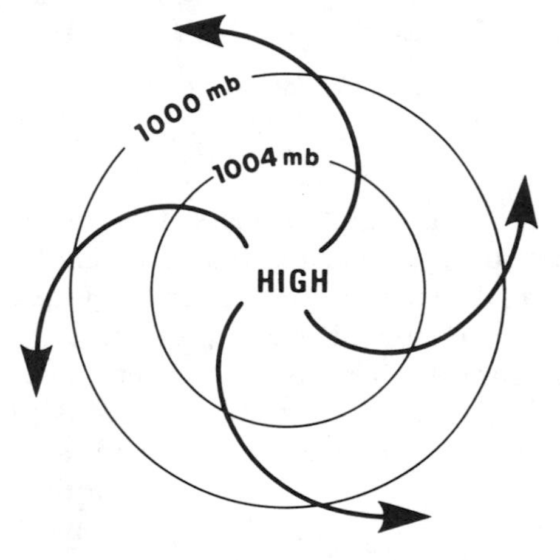

EITHER HEMISPHERE

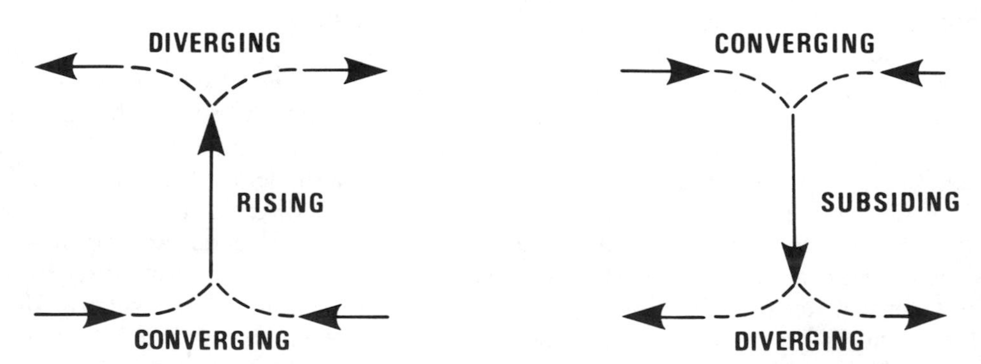

Figure 5.6. Horizontal and vertical air flows in high and low pressure cells in the Northern and Southern Hemispheres.

Near the surface of the earth, friction must be taken into account. This, again, will act opposite to the motion. It will reduce the velocity of the wind, thereby reducing the Coriolis Force and allowing the air to move gradually toward lower pressure. Therefore, in a high pressure cell, the air will gradually diverge from the central point in a spiraling motion outward toward the periphery in all directions. In a low pressure cell the air will spiral inward toward the center, thereby converging on the center from all points around the periphery (Figure 5.6).

In a high pressure cell, the constantly diverging air will tend to leave a partial vacuum at the center, and in a low pressure cell the converging air will cause an excess of air at the center. These imbalances must be compensated for by some additional air flow, and since these circulation systems are assumed to be at the surface of the earth, the only air flow that can compensate for deficits or excesses of air created by horizontal circulations must come from above. Therefore, in a high pressure cell a constant subsiding (downward) motion must take place in the middle of the high in order to replenish the air that is diverging from the center at the surface. In a low the excess air at the center must rise and diverge aloft. Therefore, the horizontal motion around highs and lows induces vertical motion, and although the magnitude of this vertical motion generally is much less than that of the horizontal motion, the vertical movement is the significant component that influences development of clouds and precipitation.

In summary, in the Northern Hemisphere atmospheric motion associated with high pressure cells is characterized by clockwise circulation, divergence, and subsidence. That associated with low pressure cells is characterized by counterclockwise circulation, convergence, and uplift. In the Southern Hemisphere the circulation directions are reversed, but the other motions remain the same: a high pressure cell is characterized by counterclockwise circulation, divergence, and subsidence, while a low pressure cell has clockwise circulation, convergence, and uplift (see Figure 5.6).

6 OBSERVED ATMOSPHERIC AND OCEANIC CIRCULATION PATTERNS

Surface Atmospheric Flow

Schematic

The general features of the atmospheric circulation near the surface are essentially as shown in Figure 6.1. There are three pressure belts and three wind belts in each hemisphere. At or near the equator a continuous zone of low pressure relates to the greatest heat input over the course of the year, with resultant high average temperatures and reduced atmospheric density. Around 30° latitude in either hemisphere, a high pressure belt marks one of the resonating nodes of the atmosphere where excess air circulates into the zone in the upper troposphere and undergoes steady subsidence (settling motion). Around 60° latitude in both hemispheres another resonating node produces low pressure, and in the polar regions, generally low temperatures tend to produce dense air and shallow high pressure cells.

The surface winds flow between these pressure zones, responding to pressure gradients between high belts and low belts but being deflected by the earth's rotation, to the right in the Northern Hemisphere and to the left in the Southern Hemisphere. The broadest and most consistent of these wind belts are the trades, from the northeast in the Northern Hemisphere and from the southeast in the Southern Hemisphere, which blow between the subtropical highs and the equatorial low (or doldrums). The subtropical highs and equatorial low are the most consistent pressure belts on earth, so the trades are the most consistent wind belts. Hence their names, which relate to the commercial activities of the early sailing vessels that made use of these reliable winds, particularly between Europe and the New World. Rather than set out directly west from Europe to reach America, sailors traditionally moved parallel to the west coast of Europe and North Africa until they reached the trade winds and then headed west across the Atlantic. In so doing they had to cross the subtropical high belt (horse latitudes) with generally weak or even calm surface winds, where ships often became becalmed for days. Here was where the ship was becalmed in "*The Rime of the Ancient Mariner.*"

Between the subtropical highs and the subpolar lows the prevailing westerlies blow at the surface. In the Northern Hemisphere these are very inconstant because of the great amount of land at this latitude, which causes great land-sea differences in circulation, and the inconstancy of the subpolar lows, which generally are well formed only over the northern oceans during winter. Traveling circulation systems on a smaller scale cause the surface winds to shift frequently. In the Southern Hemisphere the westerlies are more constant, since there is practically no land at this latitude and the essentially circular, high plateau-like ice

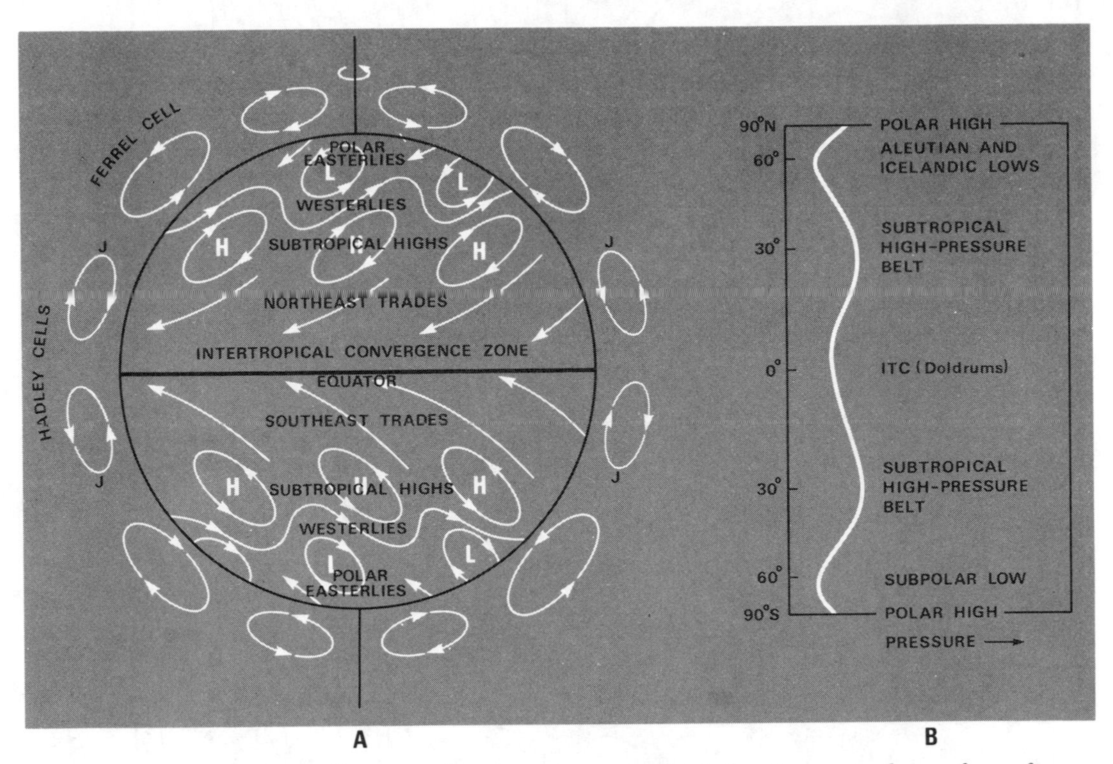

Figure 6.1. Schematic diagram of major pressure and wind patterns on the surface of the earth. Right-hand diagram has a cosine scale for latitude to represent correctly equivalent surface areas on the earth. This dramatically illustrates the dominance of the doldrums and subtropical high pressure belts in the global atmospheric circulation. And the left hand diagram correctly illustrates the great areal influence of the consequent trade winds and intertropical convergence zone. Idealized vertical circulations, including the Hadley and Ferrel cells, illustrate three cellular circulations in each hemisphere. The J's represent the subtropical jet streams at the tropopause level. (From Understanding Our Atmospheric Environment, *2nd ed., by Neiburger, M., Edinger, T.G., and Bonner, W.D., Copyright © 1982 by W.H. Freeman and Company. All rights reserved.)*

surface of Antarctica forms a good topographic barrier for the air to flow around. But even here, traveling circulation systems tend to break up the zonal circulation to some extent. Poleward of the subpolar lows the surface winds are commonly easterly, but this varies a great deal, particularly in the Northern Hemisphere where the Arctic Ocean is often not a region of consistently high pressure.

Mean Observed Patterns

Figures 6.2 and 6.3, which show observed sea-level pressures and prevailing surface winds averaged over many years for January and July, reflect the major features of the schematic presented in Figure 6.1, plus additional complications produced by land-sea differences and seasonal changes in latitudinal positions and intensities. Here it can be seen that a low pressure belt in the vicinity of the equator forms a fairly continuous zone of convergence between the trade winds on either side at all times of the year, although it is distorted by monsoonal flows in Africa and Asia during summer. The subtropical highs are conspicuous throughout much of the year, particularly in the Southern Hemisphere where

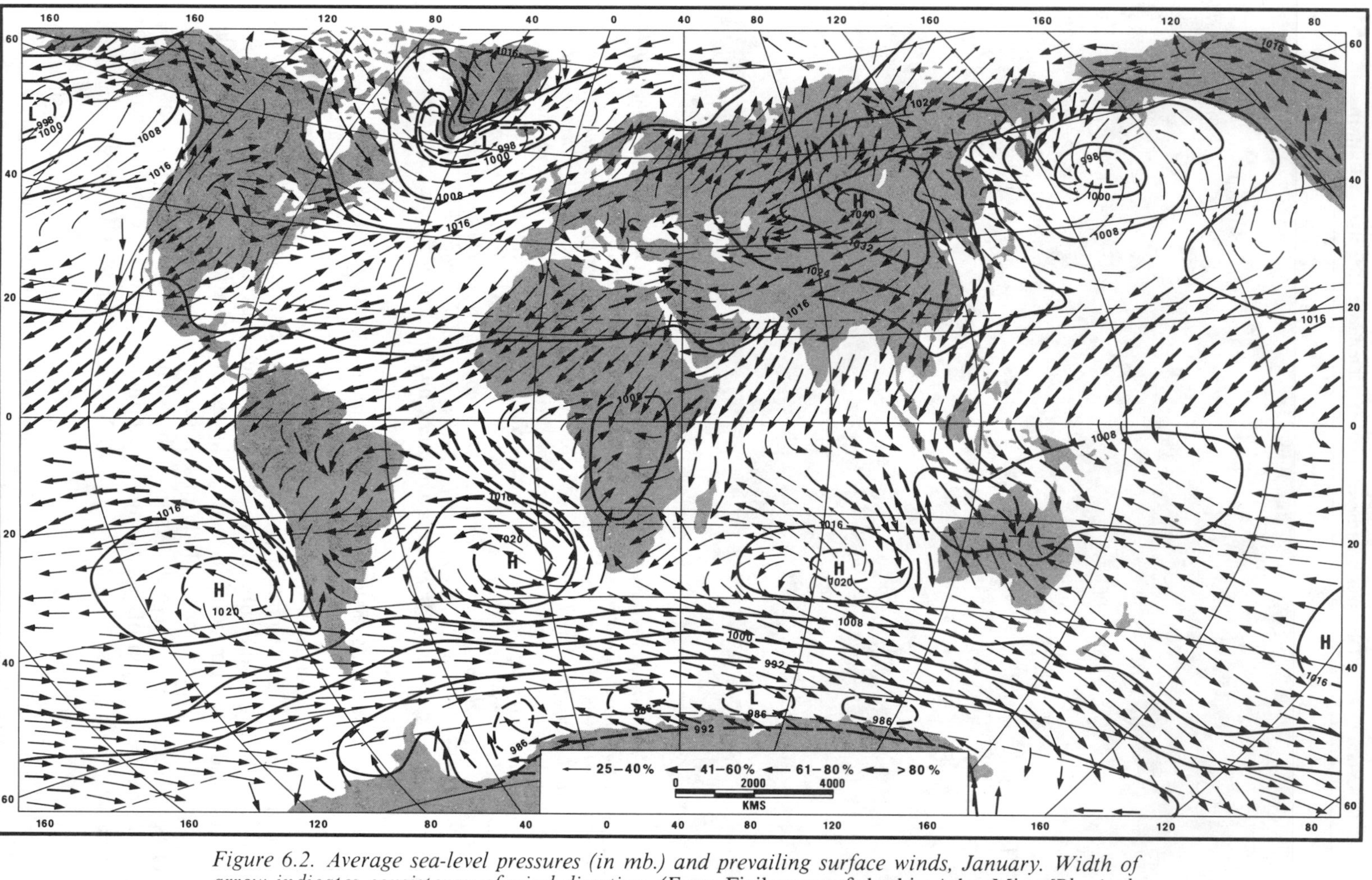

Figure 6.2. Average sea-level pressures (in mb.) and prevailing surface winds, January. Width of arrow indicates consistency of wind direction. (From Fizikogeograficheskiy Atlas Mira *[Physical-Geographical Atlas of the World] [Moscow, 1964], p. 40.)*

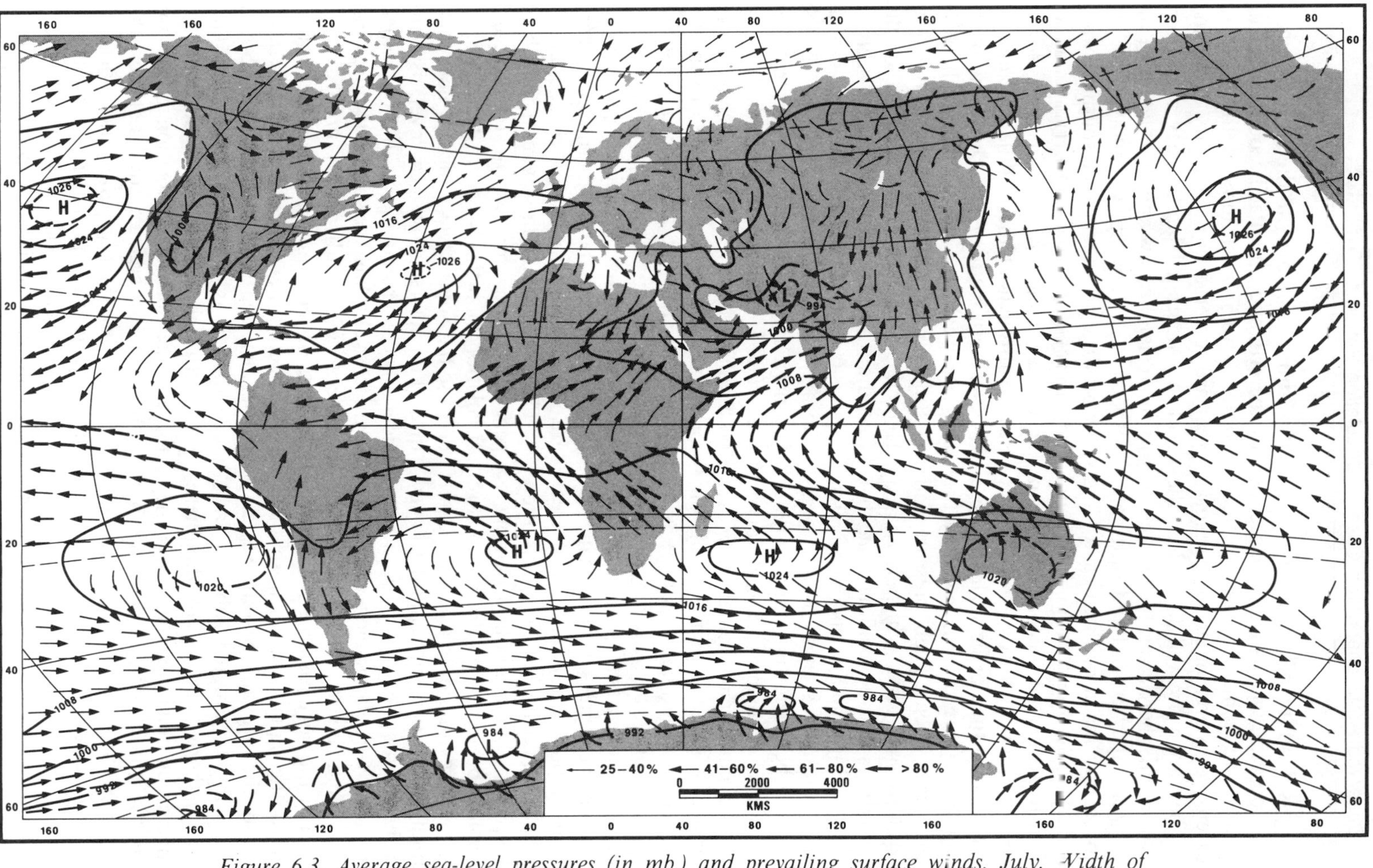

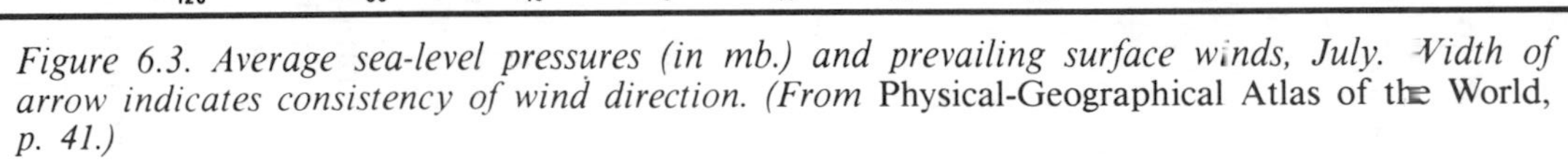

Figure 6.3. Average sea-level pressures (in mb.) and prevailing surface winds, July. Width of arrow indicates consistency of wind direction. (From Physical-Geographical Atlas of the World, *p. 41.)*

there is less land to break up the zone. They are most prevalent over the oceans and usually are somewhat stronger and more extensive in area in summer than in winter. They are very well formed in the Northern Hemisphere over the oceans in summer where they expand and shift poleward to occupy much of the areas of the North Pacific and North Atlantic oceans. In winter, however, they become shrunken and weakened in intensity as they shift equatorward over the Northern Hemisphere oceans; and they are overshadowed by the apparent huge, intense high pressure cell that develops over Asia, which is related to the coldness of the landmass at that time of year. Both the Eurasian and North American landmasses tend to show pressure reversals between winter and summer because of the changing temperatures by seasons, the land being colder in winter and warmer in summer than the adjacent oceans. Since the Eurasian landmass is by far the larger one, this tendency is most pronounced over Eurasia.

The subtropical highs are dominant features in the troposphere, typically with great areal extent and great depth. The cells often extend upward to the middle troposphere and beyond, because they are warm-core highs whose pressures decrease with height less rapidly than their surroundings. They control the wind systems over broad areas on either side. Circulating clockwise in the Northern Hemisphere, they lead to the formation of the northeasterly trades blowing toward the equator on their southern flanks and the prevailing southwesterly winds blowing toward the subpolar lows on their northern sides. In the Southern Hemisphere, rotating counterclockwise, they originate the southeasterly trades blowing northwestward toward the equator and the northwesterly winds blowing southeastward toward the subpolar low.

The subpolar low is well formed in the Southern Hemisphere around the fringes of Antarctica. In the Northern Hemisphere it is limited primarily to the northern parts of the Atlantic and Pacific oceans during winter. Two intense lows at that time of year are centered in the vicinity of Iceland and the Aleutian Islands. They are known as the Icelandic Low and the Aleutian Low respectively. They spawn many traveling low pressure cells that sweep southeastward into Europe and North America to influence the weather over broad areas downstream throughout the middle latitudes of the Northern Hemisphere. During summer they are weak and diffuse, although there is still some semblance of the Low over Iceland and the northeastern part of the Canadian Archipelago. As can be seen in Figures 6.4 to 6.7, the postulated polar high appears to exist over Antarctica in the Southern Hemisphere but is illusory over the Arctic in the Northern Hemisphere. In both polar areas there is a greater tendency for high pressure over the pole in summer than in winter, but a rather nondescript pressure pattern occurs over the Arctic during all seasons. Therefore, the so-called polar easterlies in the Northern Hemisphere are not very discernable on most maps.

The Southern Hemisphere troposphere generally has a stronger zonal (west-east) circulation than the Northern Hemisphere, because the South Pole, which is more than three kilometers above sea level, is considerably colder than the North Pole, which is at sea level. Thus, the surface temperature gradient, which ultimately controls the strength of the atmospheric circulation in its respective hemisphere, is considerably greater between the equator and the South Pole than it is between the equator and the North Pole. This probably accounts for the average northward position of the intertropical convergence zone, between the trade winds, which generally lies north of the geographical equator (Figures 6.8 and 6.9). The Southern Hemisphere circulation, being stronger, tends to slop over the geographical equator so that the meteorological equator, on the average, lies to the north

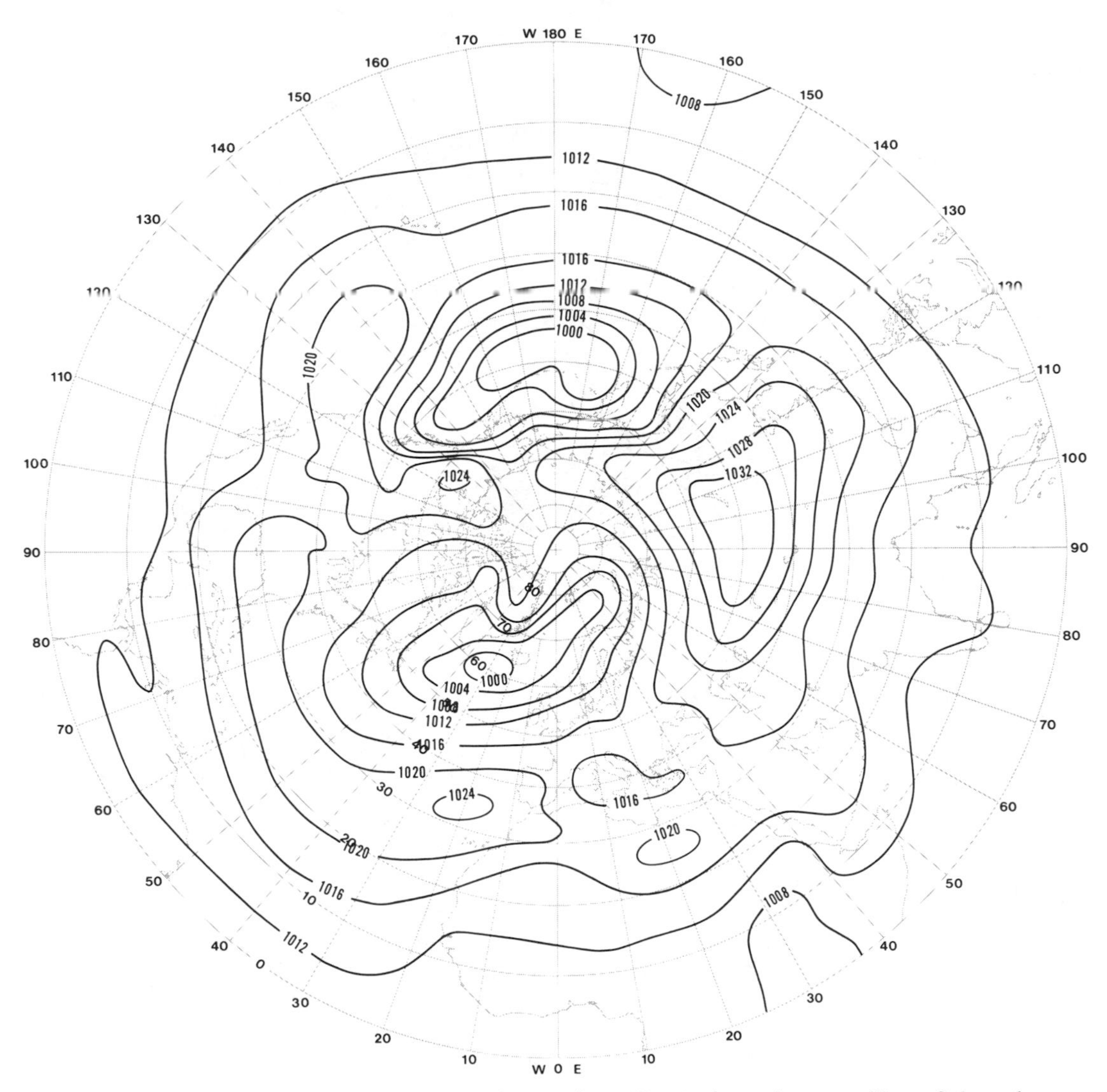

Figure 6.4. Mean sea-level pressure in mb., Northern Hemisphere, January. (From Selected Level Heights, Temperatures, and Dew Points for the Northern Hemisphere, *NAVAIR 50-1C-52.)*

of the geographical equator. During the Northern Hemisphere summer, when the Northern Hemisphere circulation is at its weakest, the convergence zone is pushed northward beyond the Tropic of Cancer in India, and in north Africa it is pushed northward to about 20° latitude (Figure 6.9). During January, on the other hand, the convergence zone lies only about 8°–10° south of the equator in the Indian Ocean (see Figure 6.8). In South Africa it tends to orient north-south through the center of the continent. Over the Atlantic and Pacific oceans its position does not fluctuate as much, but on the average the convergence zone lies slightly north of the equator.

In both hemispheres the atmospheric circulation is stronger in middle latitudes and weaker in high latitudes during winter than during summer, because of seasonal

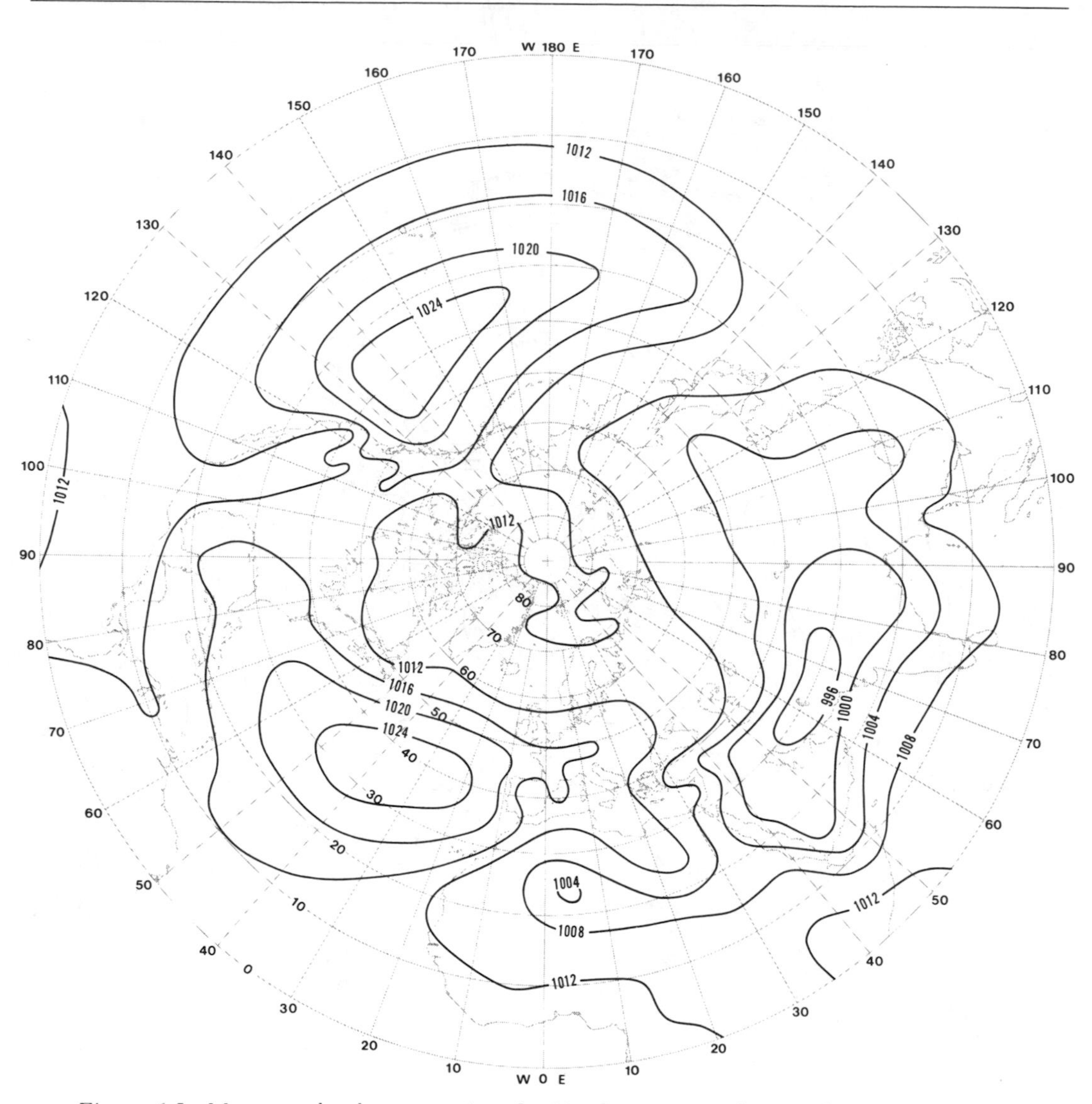

Figure 6.5. Mean sea-level pressure in mb., Northern Hemisphere, July. (From NAVAIR 50-1C-52.)

differences in latitudinal distributions of insolation (refer back to Figure 4.3). During summer the subtropical highs expand and move poleward at the expense of the westerlies, while during winter the westerlies expand their influences equatorward. The entire system of pressure and wind belts tends to shift poleward during summer and equatorward during winter, but not as much as the latitudinal shifts of insolation itself.

Causes of the Observed Pattern

Questions have naturally arisen as to why the earth's atmosphere near the surface circulates with three belts of winds in each hemisphere, rather than some other number. Experiments have been conducted in various ways (using both laboratory equipment and computer technology) to try to simulate the circulation of the atmosphere

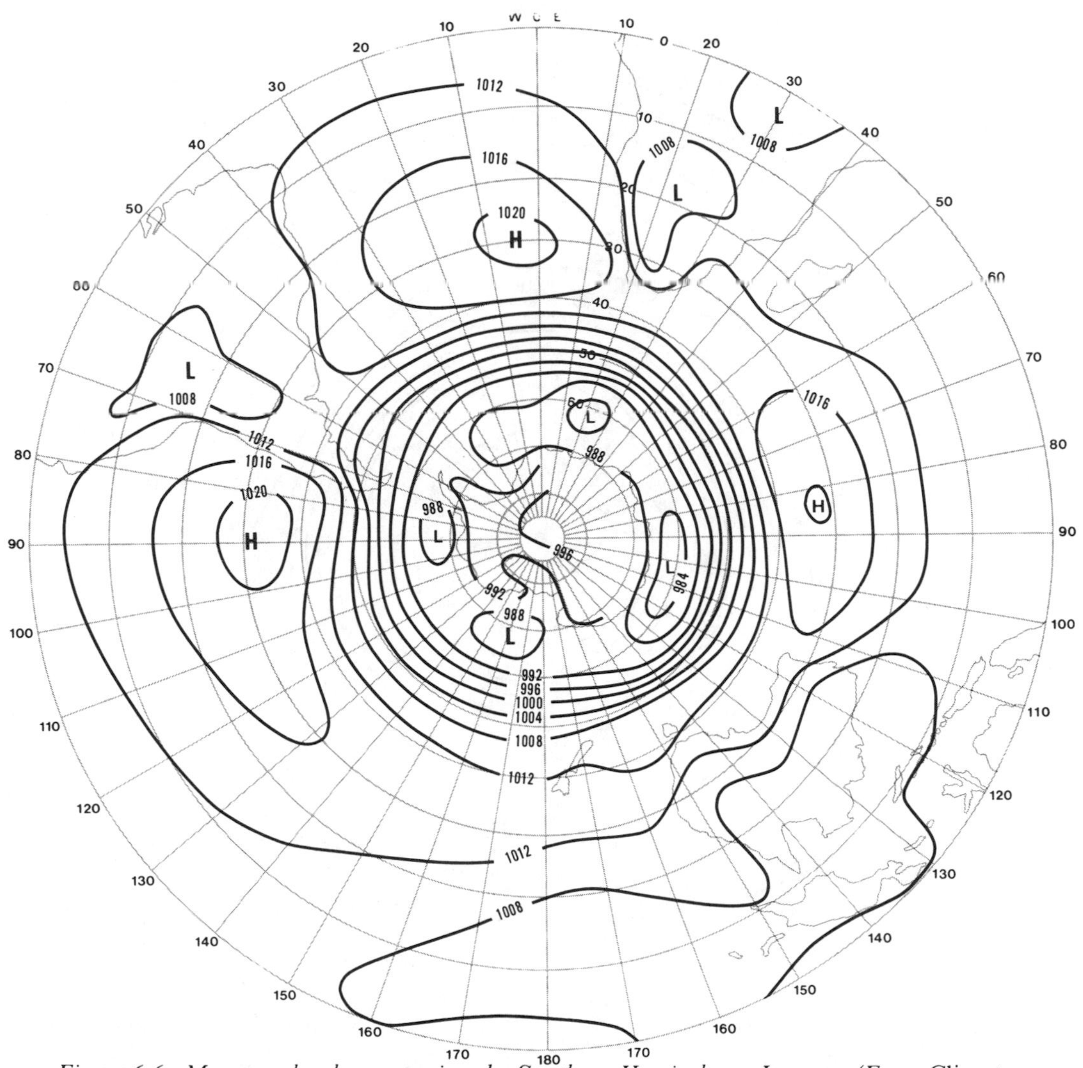

Figure 6.6. Mean sea-level pressure in mb., Southern Hemisphere, January. (From Climate of the Upper Air, Part 1—Southern Hemisphere, *NAVAIR 50-1C-55.)*

and thereby ascertain its operative controls. One of the famous experiments of this kind was the so-called "dishpan experiment" carried out at the University of Chicago by David Fultz, who poured liquid into a dishpan-like vessel mounted on a rotating platform, which he heated at the rim and cooled at the center to simulate the equator and the pole. A substance was added to the liquid to make the circulation cells visible. Fultz found that different nodal patterns could be produced by varying the speed of rotation and/or the differential heating between the rim and the center of the pan. The faster the rotation or the less the heating differential, the greater the number of circulation nodes, and the slower the rotation or the greater the heating differential, the simpler the flow.

Thus, it appears that the earth's atmosphere circulates as it does because of the rate at which it rotates and the difference

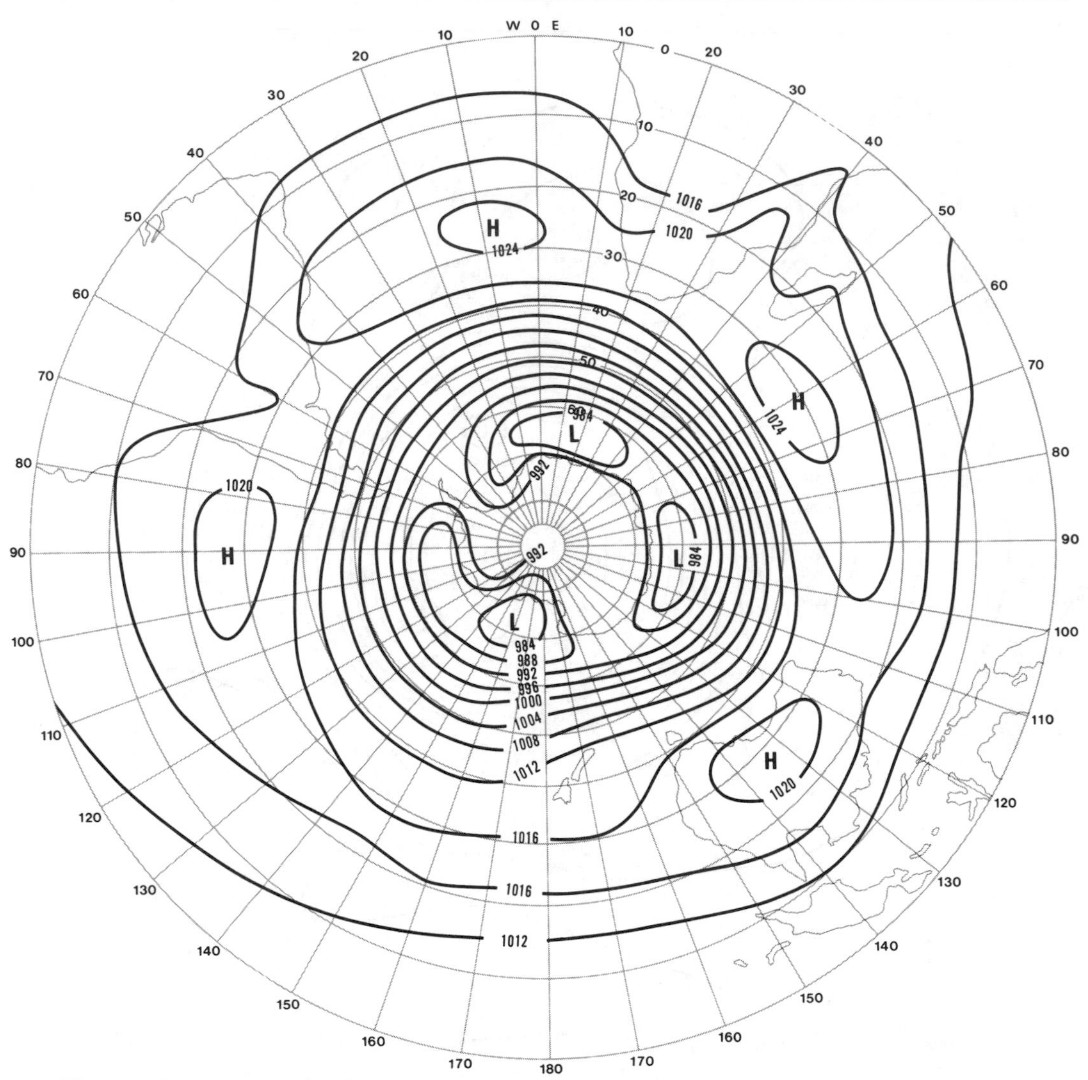

Figure 6.7. Mean sea-level pressure in mb., Southern Hemisphere, July. (From NAVAIR 50-1C-55.)

in heating between the equator and poles. These, of course, are simply astronomical accidents, and if either of these inputs were changed the atmosphere might circulate quite differently. This is the primary reason why no one yet has dared to tamper with the Arctic ice cap. If it were melted, the temperature difference between the North Pole and the equator would be much reduced, and this might change the entire atmospheric circulation of the Northern Hemisphere, or perhaps of the entire earth. It probably would not simply cause a shift in temperature belts with improved conditions along the Arctic coasts of America and Eurasia. It might even induce desert conditions or other detrimental effects over large sections of the earth's surface that now have tolerable climates for plant growth and human occupancy.

Ocean Currents

The same planetary controls that determine the general circulation of the atmosphere act on the water of the oceans, so it is logical to assume that oceanic circulations, at the surface at least, will mirror surface atmospheric circulations. To a great extent this is true (Figures 6.10 and 6.11). In the North Pacific and North Atlantic the general circulations are two large clockwise cells that correspond to the subtropical high pressure cells in the atmosphere above these regions. In the Southern Hemispheric oceans, large counterclockwise cells correspond to the subtropical highs above those regions. Near land masses, configurations of coastlines may set up localized swirls tangential to the main oceanic circulations; in higher latitudes, particularly around the fringes of the Arctic, currents may be induced by rapid changes in water temperatures.

Other factors enter into oceanic circulations beside density differences set up by temperature differences and Coriolis Force due to rotation of the earth. Currents may be caused by density differences due to salinity differences or turbidity (sediment content) differences along coasts. In addition, the atmospheric flow above the ocean and the surface of the ocean itself set up a frictional linkage. If the atmospheric flow is consistent, the water surface will take on a similar movement displaced somewhat to the right in the Northern Hemisphere and to the left in the Southern Hemisphere. Since the subtropical high pressure cells are consistent features of the atmosphere, under these cells the water movement is most reflective of the air movement.

Water is much denser than air and much more confined to basins, so it is not as free to react to density differences and the Coriolis Force as is the air. Nevertheless, surface waters, in particular, are free enough to move so that their patterns of circulation do resemble those of the atmosphere. Deeper currents may be quite different from the surface drifts, but they are of little climatic consequence, so this book will be concerned only with the surface currents of the oceans, primarily as they affect the climate along adjacent coasts.

Undoubtedly, the climatically most significant ocean current on earth is the Gulf Stream in the North Atlantic, which flows northeastward between Cuba and the southern tip of Florida and then parallels the eastern coast of North America almost to Newfoundland. It continues as the North Atlantic Drift northeastward past Britain and Norway, around the northern tip of Scandinavia, and into the Barents Sea of the Arctic north of European USSR. This large, warm current has a profound effect on much of Europe since there are few terrane barriers at these latitudes to stop the westerly winds that carry the marine air eastward. In the Norwegian Sea in winter, the Gulf Stream causes surface air temperatures to be as much as 26°C above the normal for the latitude. Much of Britain, the southern parts of which lie north of the Canadian border, experiences winters as mild as northern Florida, Georgia, and South Carolina, 15°–20° latitude farther south.

The counterpart to the Gulf Stream in the North Pacific is the Japan (Kuroshio) Current, which branches off the North Equatorial Current in the southwestern part of the North Pacific near the Philippine Islands and moves northward past Taiwan and Japan and then eastward across the Pacific as the North Pacific Drift. The Japan Current does not have as profound climatic effects on the eastern side of the Pacific as the Gulf Stream does on the eastern side of the Atlantic, because high mountains paralleling the west coast of North America do not allow much penetration of fresh Pacific marine air across the continent. Also, the North Pacific is separated from the Arctic during the winter by the narrow and frozen-over Bering Straits, which block any movement of surface water from the North Pacific into the Arctic during middle and late winter. This is not the case through the broad Norwegian Sea between the North Atlantic and the Arctic. Therefore, the climatic influences of the North Pacific Drift are not carried nearly as far poleward as

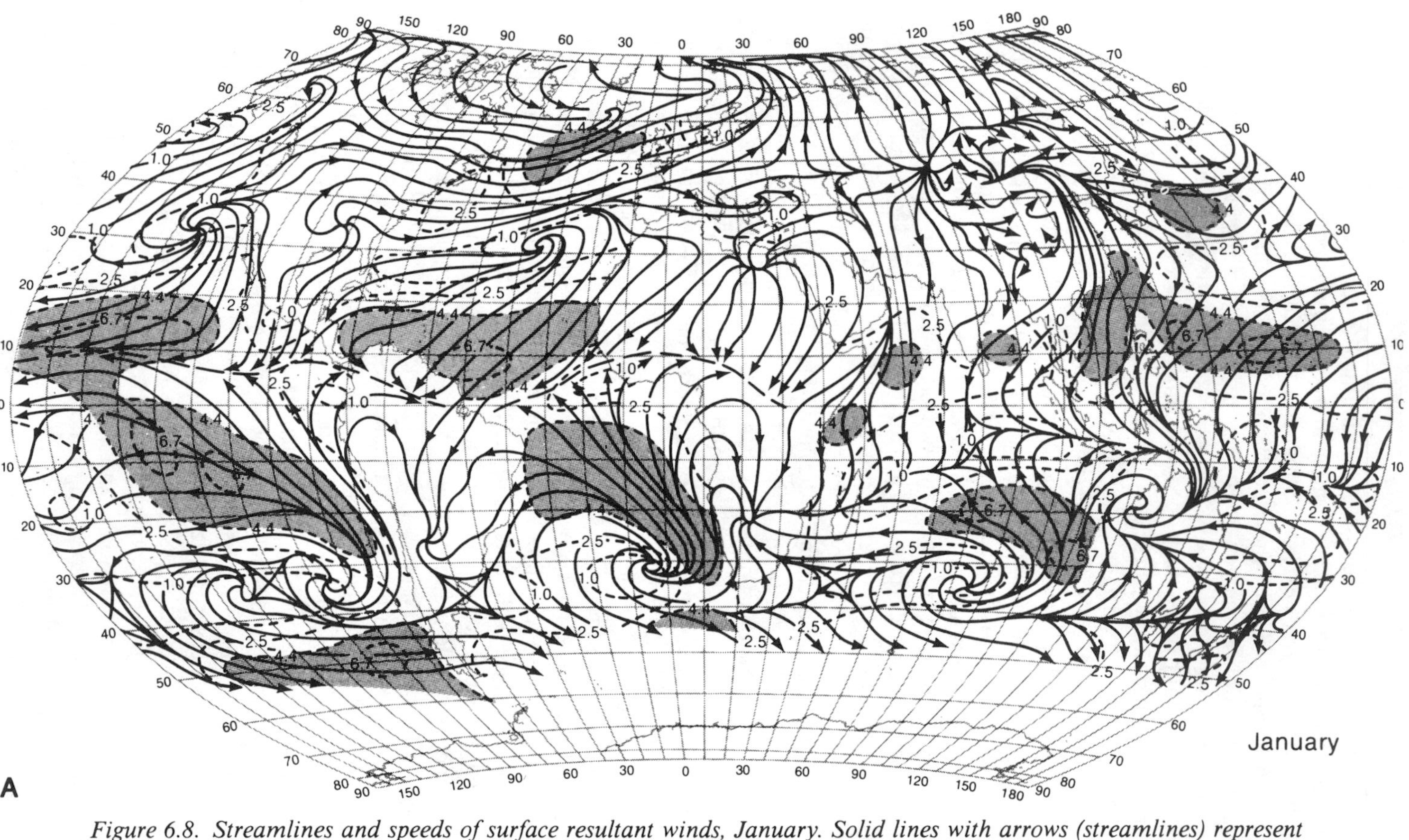

Figure 6.8. Streamlines and speeds of surface resultant winds, January. Solid lines with arrows (streamlines) represent net movement of air during January, and dashed lines (isotachs) represent resultant wind speeds in m/s (one m/s=2.237 mph). Speeds greater than 4.4 m/s (10 mph) are shaded. (Northern Hemisphere streamlines after W.M. Wendland and R.A. Bryson, "Northern Hemisphere Airstream Regions," Monthly Weather Review *109, no. 2 [Feb. 1981]. Isotachs and Southern Hemisphere streamlines from Y. Mintz and G. Dean,* The Observed Mean Field of Motion of the Atmosphere, *Geophysical Research Papers No. 17, Geophysics Research Directorate [Air Force Cambridge Research Center, 1952]. As shown in* Understanding Our Atmospheric Environment, *2nd ed., by Neiburger, M., Edinger, T.G., and Bonner, W.D. Copyright © 1982 by W.H. Freeman and Company. All rights reserved.)*

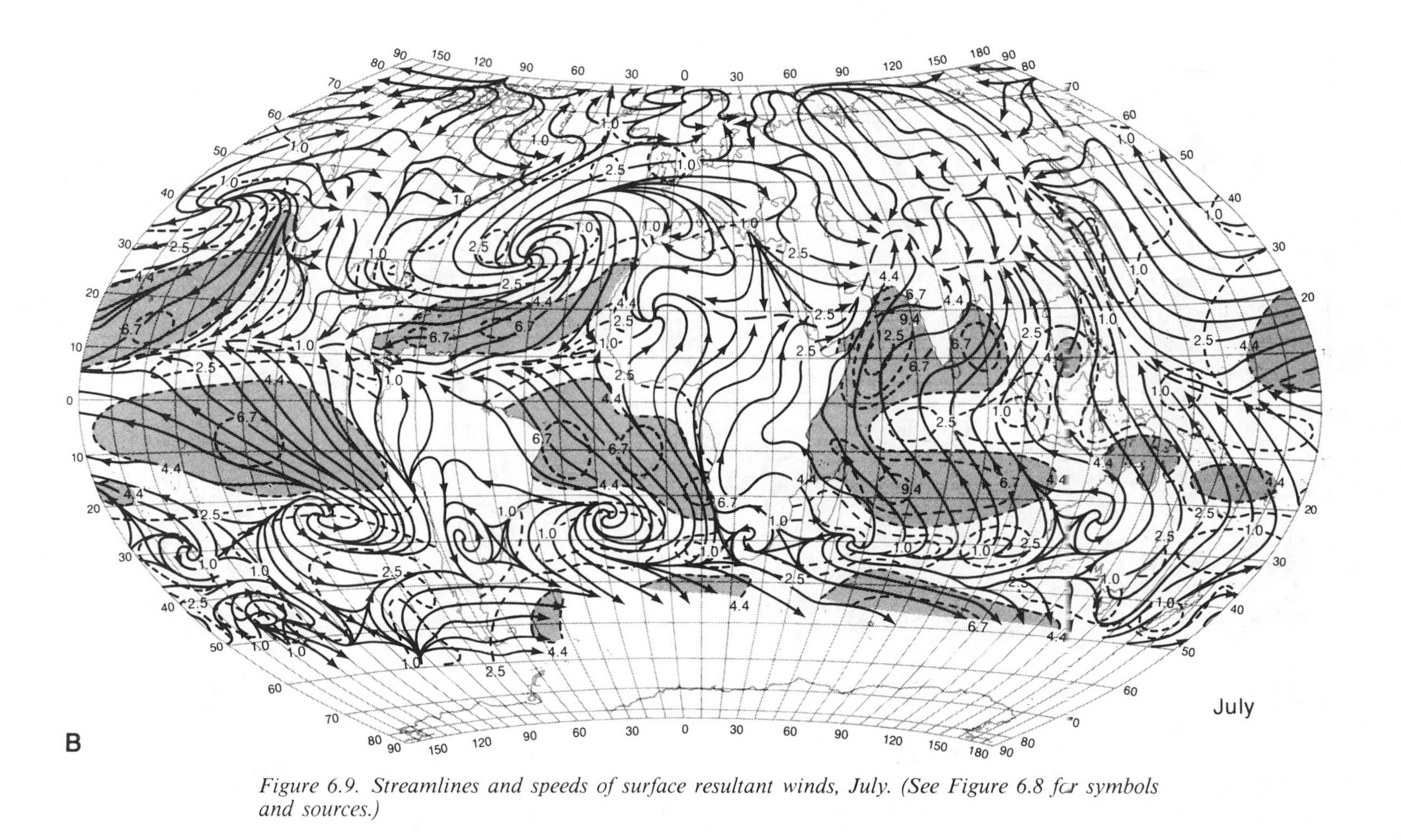

Figure 6.9. Streamlines and speeds of surface resultant winds, July. (See Figure 6.8 for symbols and sources.)

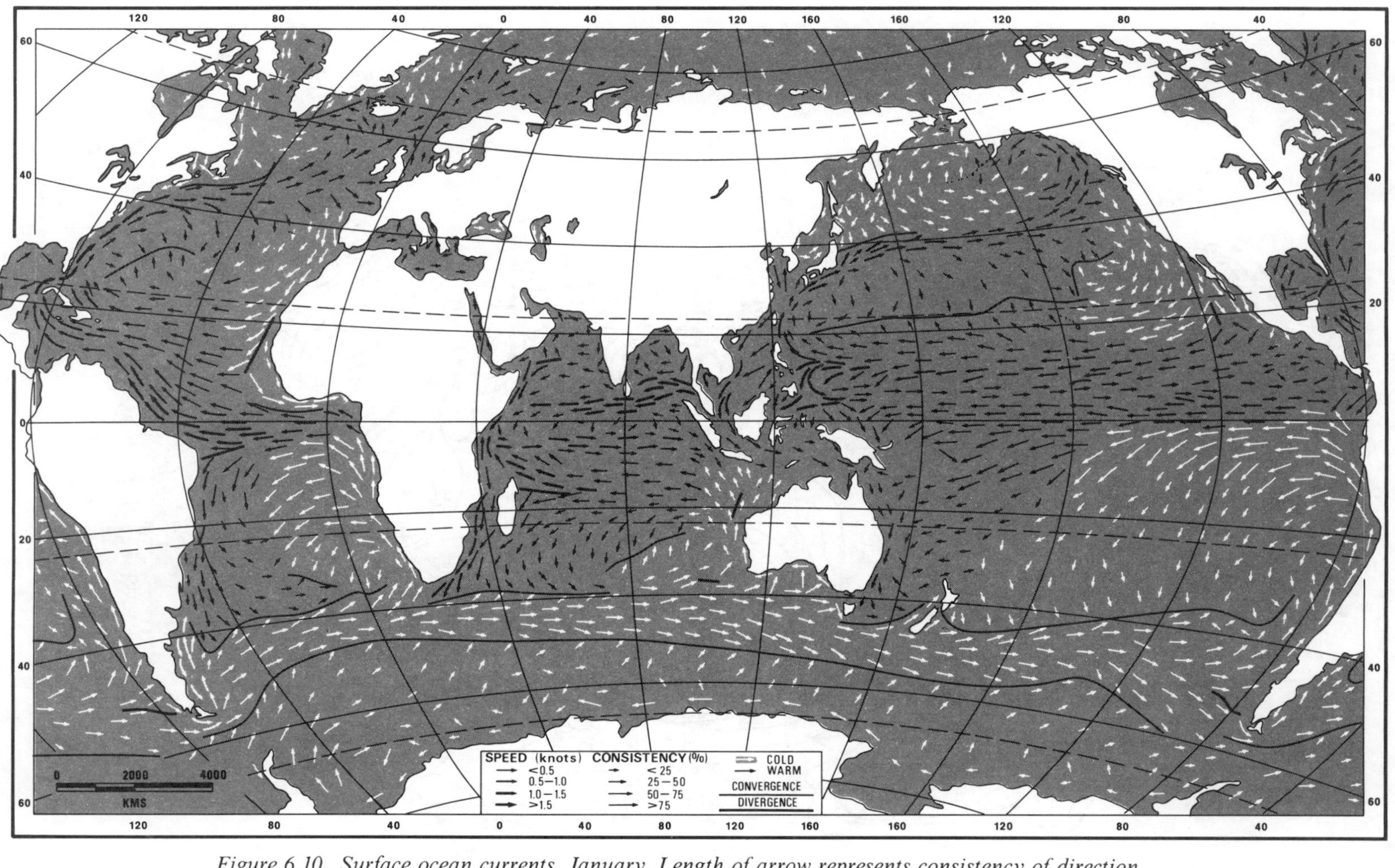

Figure 6.10. Surface ocean currents, January. Length of arrow represents consistency of direction, and width of arrow represents speed in knots (1.15 statute mph). Black arrows represent warm currents; white arrows, cold currents. Significant zones of convergence and divergence are indicated. (From Physical-Geographical Atlas of the World.*)*

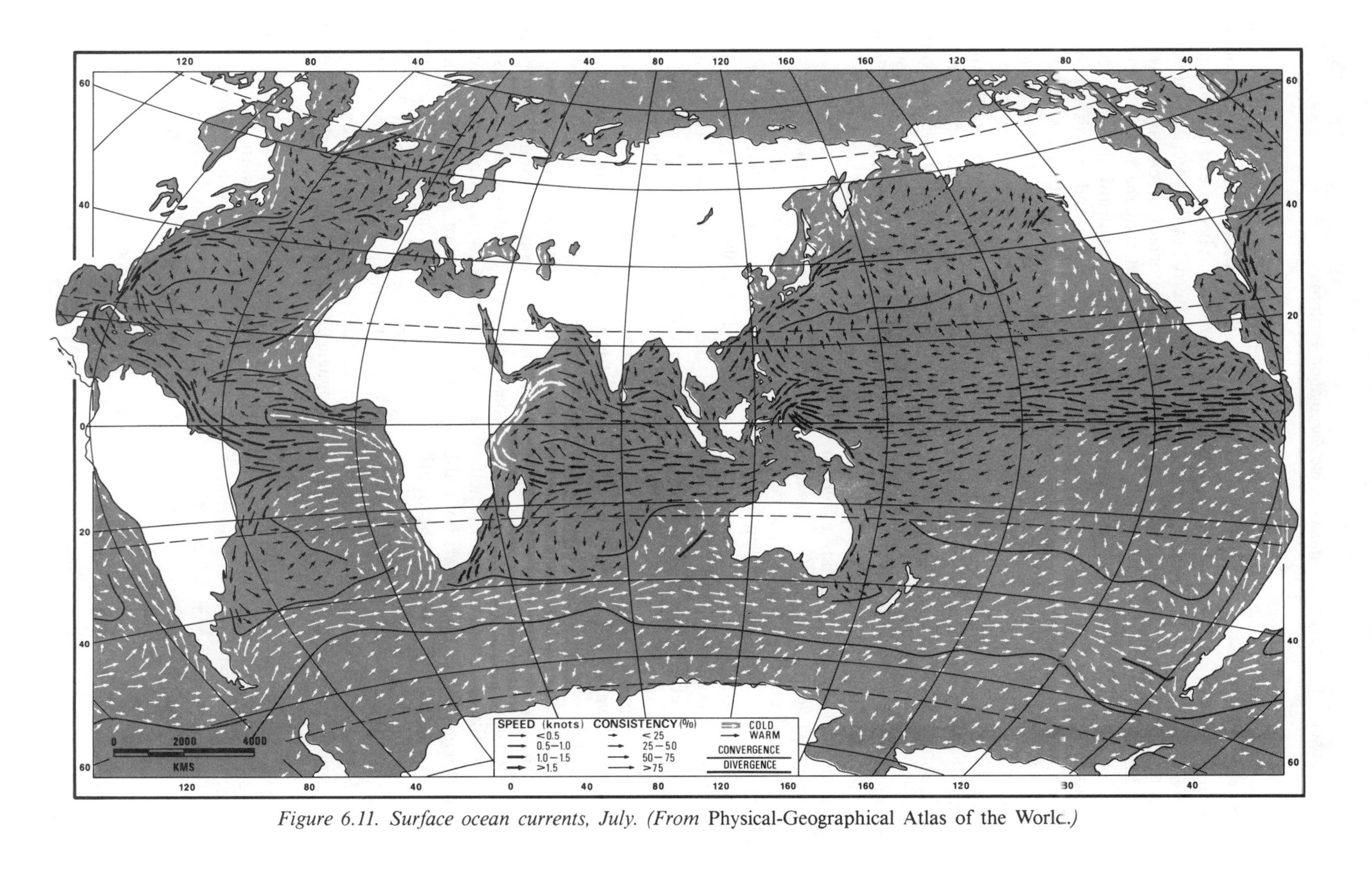

Figure 6.11. Surface ocean currents, July. (From Physical-Geographical Atlas of the World.*)*

are the influences of the North Atlantic Drift.

Both the North Atlantic and North Pacific warm currents have branches moving southward along the eastern edges of their basins to complete the clockwise swirls of general oceanic circulations. These currents are cold relative to the latitudes into which they are flowing. (In all cases the designations "warm" or "cold" are relative terms.) Obviously, in most cases the California Current, flowing southward along the west coast of California and Baja California, has warmer temperatures, particularly in winter, than does the so-called "warm" Alaskan Current flowing northward into the Gulf of Alaska. But the California Current has a cooling effect on its region, while the Alaskan Current has a warming effect on its region. The counterpart of the California Current in the eastern North Atlantic is the Canaries Current which flows southwestward along the coasts of the Iberian Peninsula and northwestern Africa past the Canary Islands.

These equatorward-flowing cold currents continue westward across their respective oceans as relatively warm equatorial currents to complete the general clockwise circulation in each ocean, but these equatorial extensions have little climatic effect. It is interesting to note that in the Pacific Ocean during most of the year, both the North Equatorial Current and the South Equatorial Current lie to the north of the geographical equator, a reflection of an earlier statement regarding the more vigorous atmospheric circulation of the Southern Hemisphere. In between the two equatorial currents in the Pacific is an eastward flowing gravity current, the so-called "counter current," which serves to maintain mass continuity in the area, so the two equatorial currents do not pile up the water higher on the west side of the Pacific than on the east.

Warm currents are poorly developed in the Southern Hemisphere. Perhaps the most consistent one is the Brazil Current, which flows southwestward along the southeast coast of Brazil. But because of the triangular shape of eastern South America, much of the south Equatorial Current in the Atlantic, which should feed the Brazil Current, is shunted northwestward along the coast of Cape São Roque and moves north of the equator to join the current feeding the Gulf Stream. This leaves the Brazilian Current with much less flow than might be the case if the coastal shape of South America were different. Currents are localized and inconstant in the Indian Ocean, where they are greatly affected by the seasonal reversal of the monsoons during the year. In the South Pacific only a weak, diffuse warm current moves southwestward along the eastern coast of Australia.

The two currents with by far the most pronounced climatic influences in the Southern Hemisphere are large cold currents, the largest cold currents on earth. The most voluminous and climatically significant is the Peru (Humboldt) Current that flows northward along the west coast of South America from about 50°S latitude to almost the equator. In fact, it crosses the equator in the eastern Pacific shortly after diverging from the South American coast near the Gulf of Guayaquil. This has a profound effect on the climate along much of the west coast of South America, although its influences are limited to a narrow coastal strip because of the high Andes to the east. The counterpart of the Peru Current in the eastern South Atlantic is the Benguela Current that flows northward along the southwest coast of Africa.

All the cold oceanic currents paralleling west coasts of continents in both hemispheres are flowing in such directions that the Coriolis Force, plus frictional drag of winds, tend to throw them offshore. Within 100–200 kilometers of the coasts vertical circulations in the water are set up that bring cold water up from below to replace the surface water that is moving away from the coasts out to sea. It is obvious that in many cases the coldness of the surface water in these currents is due more to upwelling from below than to horizontal movements from higher latitudes, because cold pools of water are often found adjacent to land promontories, with warmer water on either side. For instance, during summer the water off the coast near San Francisco is often the coldest along the entire North American west coast. Obviously, the local factor of upwelling in that region is more significant than the southward horizontal movement of water from Alaska.

The four primary cold currents of the world—Peru, Benguela, California, and Canaries—are positioned under the equatorward eastern portions of the subtropical high pressure cells and further stabilize the atmosphere in those locations by cooling from below. Combined with the heating of atmospheric subsidence from above, they produce an extremely intense temperature inversion at or a few hundred meters above the surface. This limits upward movements of air to the shallow marine layer below the inversion base and induces a paradoxical climatic situation along adjacent coasts—little if any measurable precipitation, but consistently cool, moist surface air with frequent fog and overcast, low stratus clouds.

These four cold currents and the two large warm currents in the Northern Hemisphere are the major climatically significant ocean currents, but many other currents of more localized nature have pronounced climatic effects along their immediate shores. Examples are the cold Labrador Current flowing southward along the coasts of Labrador and Newfoundland in the North Atlantic, the Kamchatka and Kuril currents flowing southwestward along those coasts in the western North Pacific, the Somali Current along the eastern horn of Africa, the West Australian Current flowing northward along the western coast of Australia during much of the summer, and the cold Falkland Current flowing northward along the Patagonian coast through much of the year.

Upper Atmospheric Flow

The atmospheric circulation generally intensifies and simplifies upward through the troposphere. This is primarily due to the so-called "thermal wind effect," which is caused by different rates of decrease of air pressure with height because of different temperatures, and hence thicknesses, of air columns. For instance, the pressure decreases less rapidly with height over the warm equator than over the cold pole. Even if the pressures were the same at the equator and pole at sea level, the pressure at any given height over the equator would therefore be greater than the pressure at the same height over the pole. The higher into the troposphere, the greater the pressure difference (Figure 6.12). Thus, the wind increases from zero at the surface, with equal pressures at equator and pole, to increasingly greater speeds aloft with increasing pressure differences.

The difference in the winds at any two levels is known as the *thermal wind,* for it is dependent on the thickness of air columns between two pressure levels, which is a function of temperature. The thermal wind is a hypothetical flow along isotherms (lines of equal temperature) with low temperature to the left in the Northern Hemisphere and to the right in the Southern Hemisphere. With increasing height in the troposphere, the thermal wind component eventually becomes greater than any surface influence, so that by the time the middle troposphere (approximately the 500-mb. pressure level, or about 5000–6000 meters above sea level) is reached, the complex pattern of pressure cells and resultant winds at the surface has been replaced by one simple pressure gradient from equator to pole, with one large circumpolar whirl circulating from west to east, counterclockwise

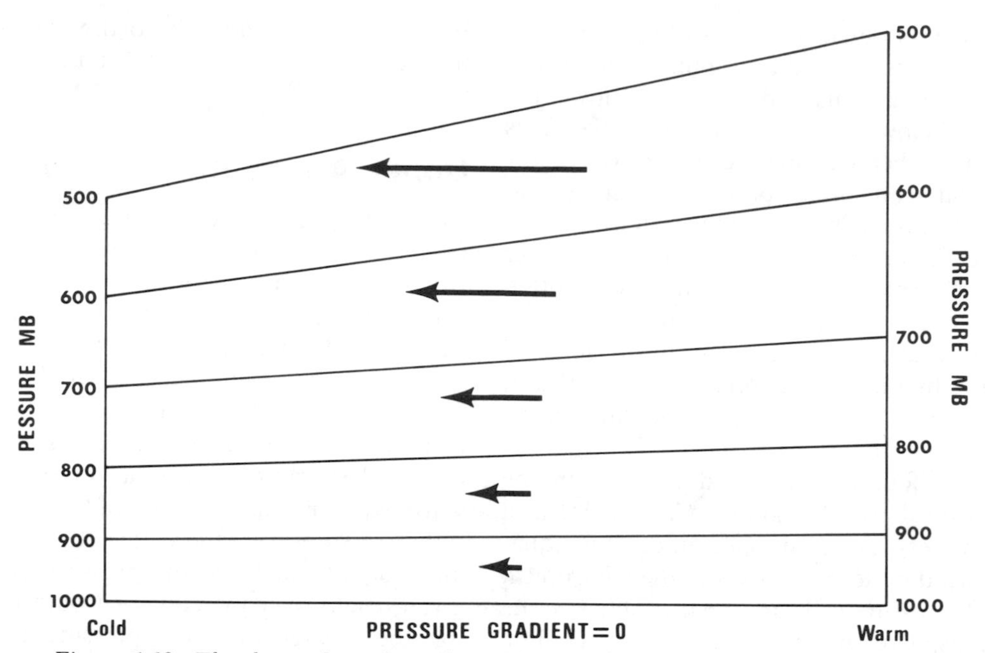

Figure 6.12. The thermal wind. At the surface the pressure is the same everywhere and the wind is calm. As the pressure decreases more rapidly with altitude in the cold air than in the warm air, the pressure gradient increases and the wind becomes stronger. Arrows represent relative magnitudes of the pressure gradient at increasing heights. The wind would blow into the paper (low pressure to the left) in the Northern Hemisphere and out of the paper (low pressure to the right) in the Southern Hemisphere.

in the Northern Hemisphere and clockwise in the Southern Hemisphere (Figures 6.13 to 6.16).

About the only surface features still evident at the 500-millibar level are the subtropical highs, which are displaced farther equatorward than they are at the surface. These high pressure cells, being composed of warm air, decrease in pressure only gradually with height, and therefore extend high into the troposphere. Since the pressure decreases with height most slowly on their warmer equatorward sides, the highs tilt upward toward the warm equatorward air. A cold high, on the other hand, such as that which forms over Asia during winter, is very shallow, since the pressure within it decreases rapidly with height. In fact, the high over Asia is no longer evident even at 850 millibars, only 1300–1500 meters above sea level (Figure 6.17). Low pressures are just the opposite. Cold lows deepen aloft, since the pressure in the center of the low decreases with height faster than it does around the periphery. Warm lows, on the other hand, fill and disappear rapidly aloft, as the pressure decreases with height less rapidly in their centers than around their peripheries. Cold, intense lows tilt upward toward colder temperatures, generally to the northwest in North America or the northeast in Eurasia.

Wind speeds also increase with height through the troposphere, because air density decreases rapidly with height. With the same pressure gradient, the less the mass of air, the faster it will accelerate. Also, drags due to surface friction and turbulence are reduced with height, with resultant increased speeds.

The trend toward faster and more simplified air flow aloft generally continues

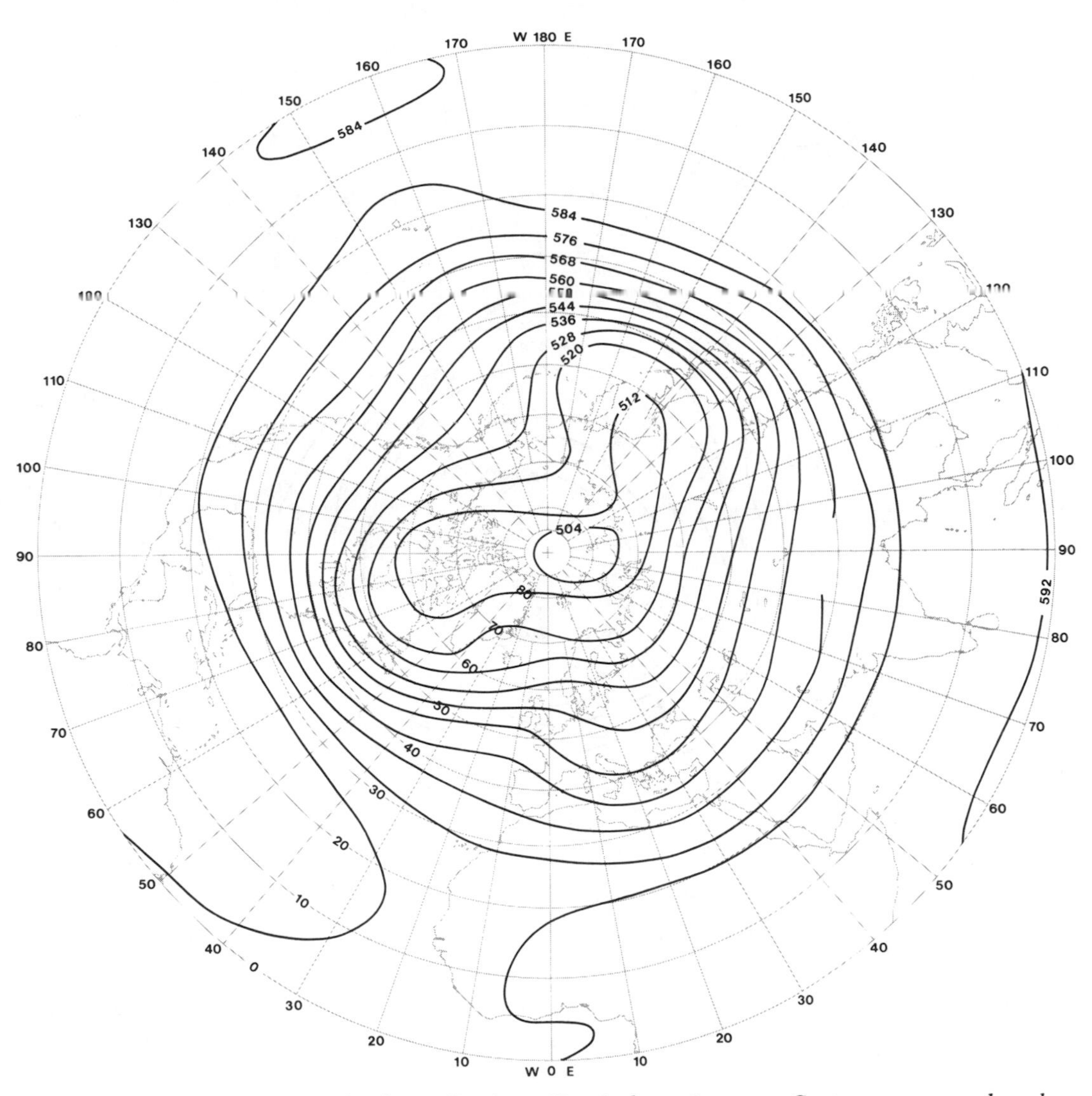

Figure 6.13. Mean 500 mb. chart, Northern Hemisphere, January. Contours are numbered in dekameters (tens of meters). (From NAVAIR 50-IC-52.)

upward through the troposphere to the tropopause, which is usually found between 200 millibars and 100 millibars (11,750–16,750 meters above sea level) during July over the Northern Hemisphere. Shortly above the tropopause, the circulation becomes confused as it adjusts to changed latitudinal temperature patterns. Since the tropopause is reached at lower levels in higher latitudes, the temperature increase with ascent in the stratosphere begins at lower levels at higher latitudes than it does at lower latitudes, and eventually a temperature reversal takes place, with warmer temperatures over high latitudes and colder temperatures over low latitudes. The thermal wind then changes the mid-stratospheric circulation to east-to-west (Figure 6.18).

Such a reversal with height does not take place during winter, when the polar area is constantly in the dark and little ultra-

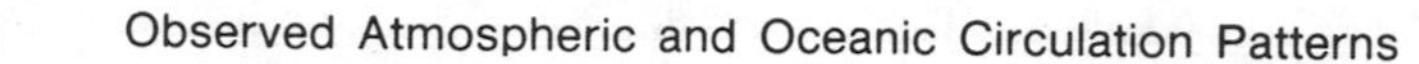

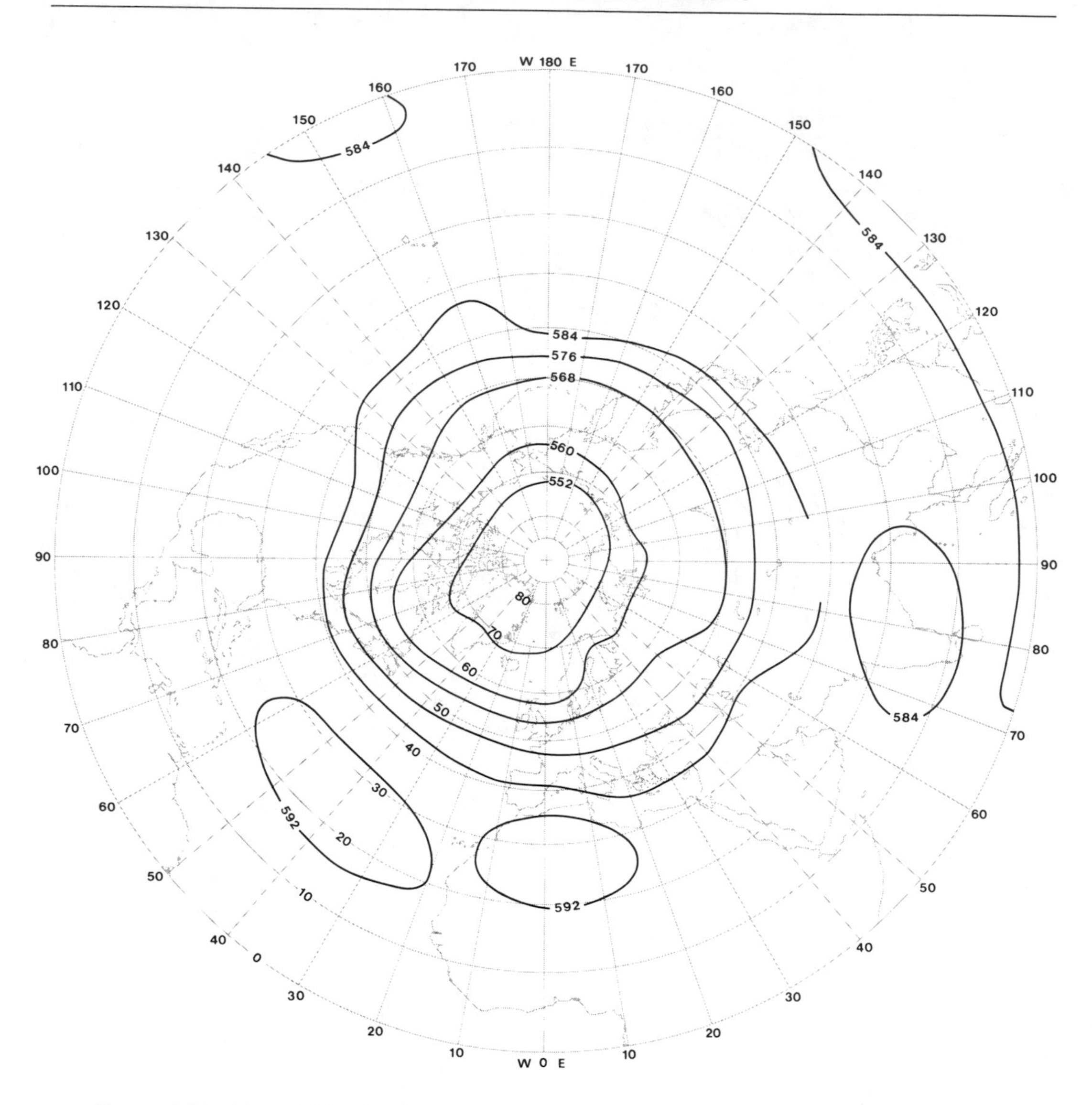

Figure 6.14. Mean 500 mb. chart, Northern Hemisphere, July. Contours in dekameters. (From NAVAIR 50-IC-52.)

violet light is absorbed at the level of the stratopause. The stratosphere at this time of year remains colder at every height over the pole than over the equator. Thus, during winter westerly winds extend upward to the outer limit of the atmosphere over the entire hemisphere, except for the equatorial region itself.

Figure 6.18 may give the impression that the westerlies, which expand latitudinally aloft to at least the tropopause level in all seasons, dominate the circulation of the atmosphere. It must be remembered, however, that most of the mass of the atmosphere is at lower altitudes, and at the surface of the earth easterly winds occupy more area than westerly winds do. Also remember that the earth is much larger around at the equator than at the poles and that half the earth's surface lies between

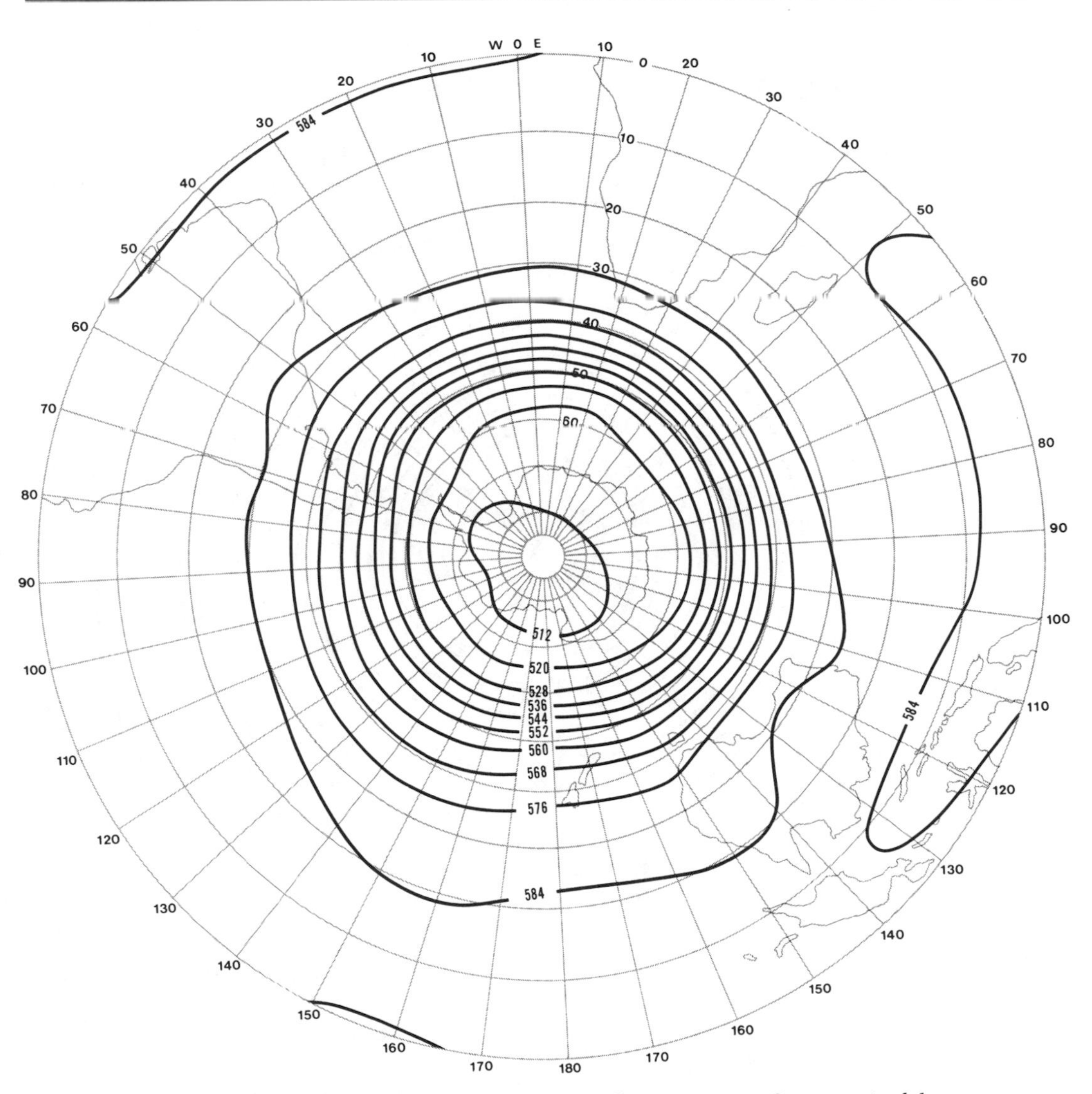

Figure 6.15. Mean 500 mb. chart, Southern Hemisphere, January. Contours in dekameters. (From NAVAIR 50-IC-55.)

30°N and 30°S latitude. At all times the easterly trades span this latitudinal zone, and many times overlap it to 35° latitude or more. And although the polar easterlies are very shallow, they do generally prevail at the surface from the poles equatorward to about 65° or 70° latitude. Therefore, a considerably greater portion of the earth's surface is affected by easterly winds than by westerly. The east-west movement of this denser air near the surface counterbalances the greater volume of westerlies in the middle and upper troposphere and above.

Much still remains to be learned and explained about the circulation of the high atmosphere. For instance, in the equatorial stratosphere there appears to be a 2.2-year cycle of alternating layers of easterly and westerly components, which descend through the stratosphere at the rate of about one kilometer per month. This 26-month

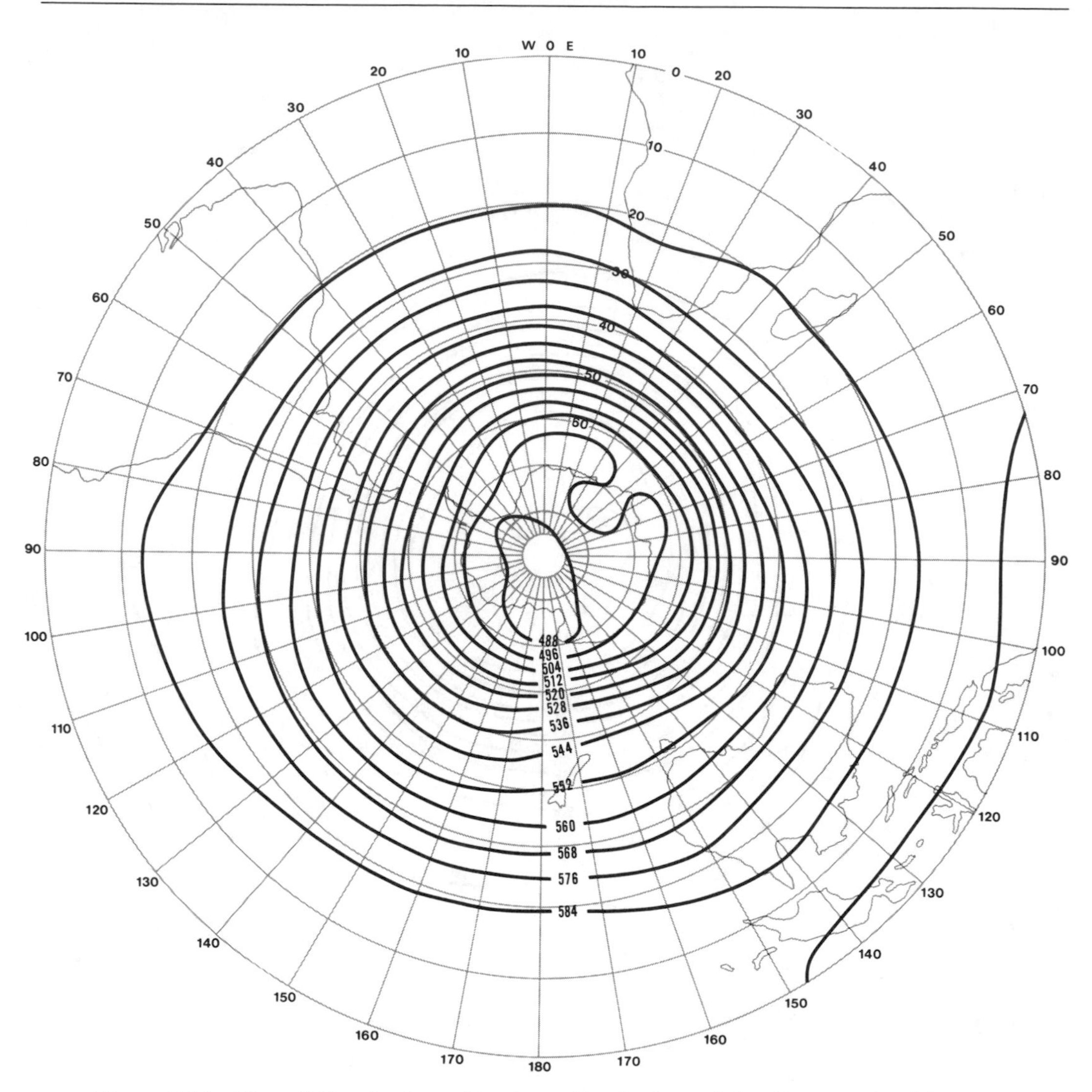

Figure 6.16. Mean 500 mb. chart, Southern Hemisphere, July. Contours in dekameters. (From NAVAIR 50-IC-55.)

cycle, of course, does not correspond to any astronomical period such as the year, and therefore piques much curiosity.

Figure 6.19 attempts to demonstrate the vertical flow of the atmosphere at different latitudes. It illustrates that between the equator and 30° latitude the dominant feature in the troposphere is a Hadley cell with rising air currents over the equator and descending air currents in subtropical zones. These vertical movements initiate equatorward flow at the surface and poleward flow in the upper troposphere. Since the troposphere is warm over the equator, it extends to very great heights and is considerably higher in the tropics than farther poleward. In fact, the tropopause typically is broken at two latitudes, one near the descending poleward arm of the Hadley cell and the other in the subpolar zones around 60° latitude.

The mid-latitude circulation is more complicated, being characterized by traveling waves and vortices that ultimately

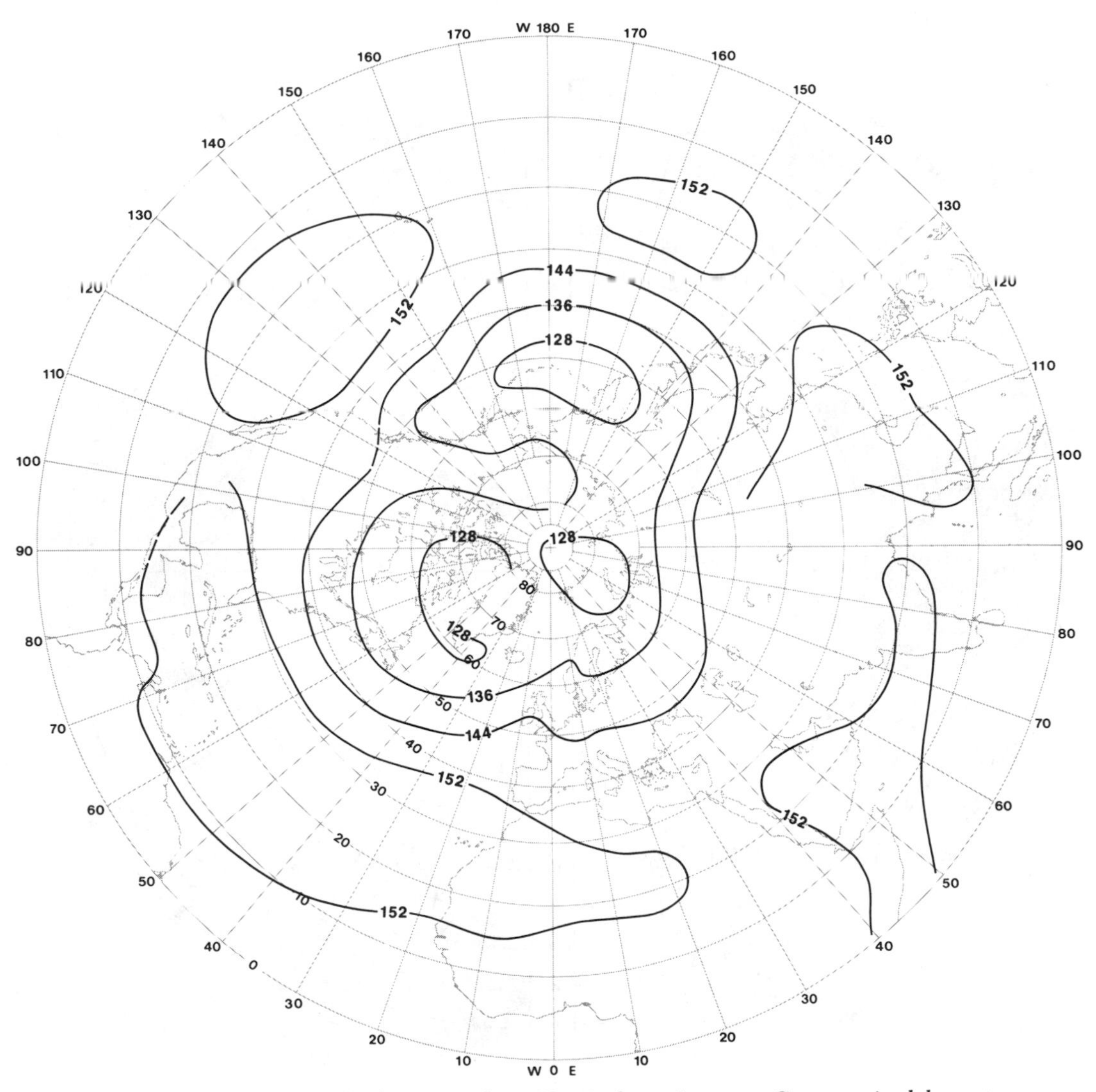

Figure 6.17. Mean 850 mb. chart, Northern Hemisphere, January. Contours in dekameters. (From NAVAIR 50-1C-52.)

transport heat and westerly momentum from the poleward edges of the Hadley cells to drive the westerly Ferrel cells of the middle latitudes. In the vertical, there is no complete circulation; westerly winds generally prevail at all altitudes throughout the troposphere. But averaged over time, there is a gradual movement upward of warm moist air from lower latitudes and a subsiding of polar air from higher latitudes along a baroclinic zone that slopes equatorward from the top of the troposphere to the earth's surface.* (This so-called *polar front* will be discussed in detail in Chapters 9 and 10.) North of 60° latitude there is little vertical movement in the

*An unstable zone in which air pressure surfaces do not coincide with air density surfaces.

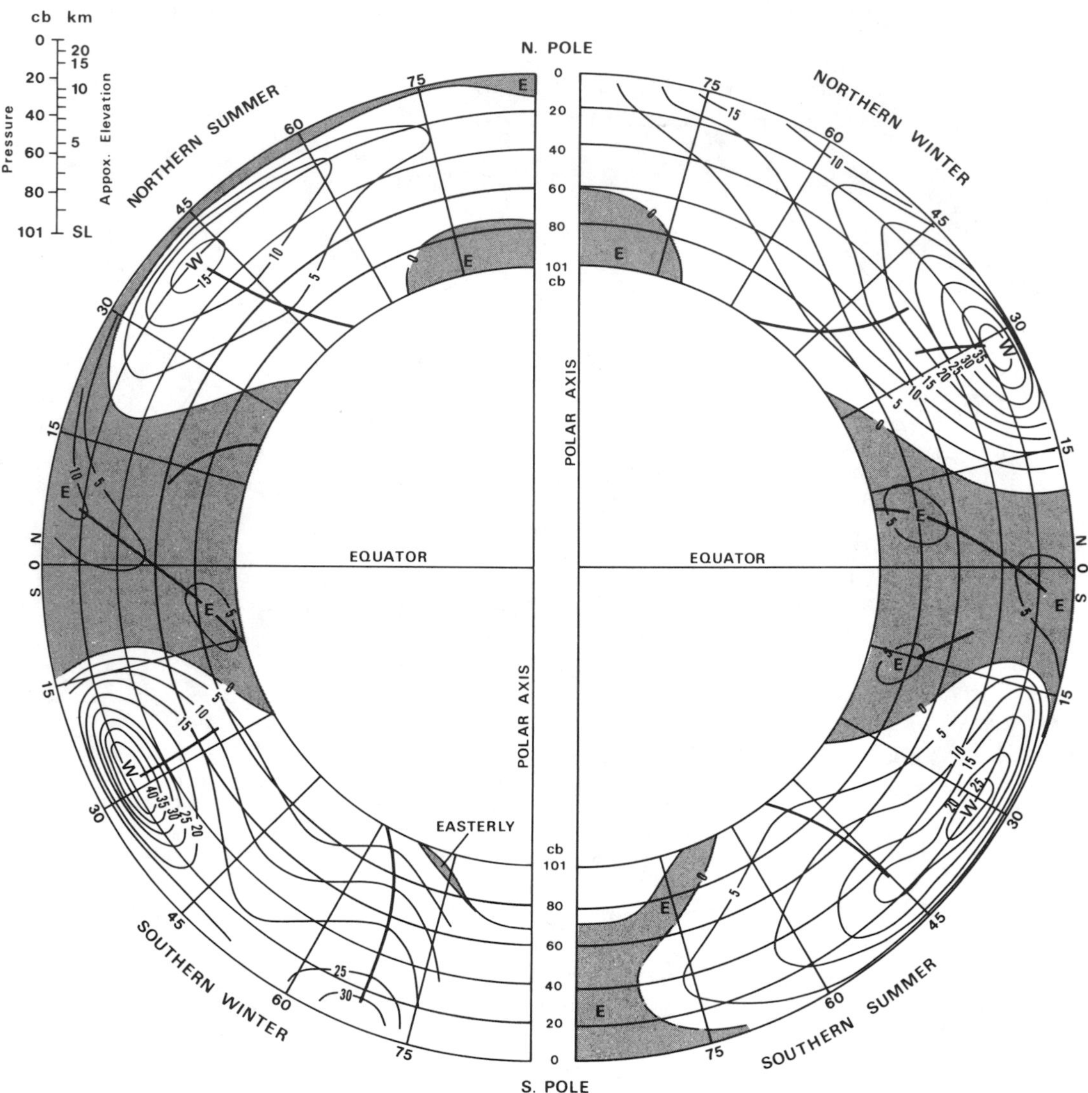

Figure 6.18. Mean zonal (west-east component) winds around the earth, by latitude and altitude. Isotachs (lines of equal air speed) are labeled in meters per second. (After Y. Mintz, "The Observed Zonal Circulation of the Atmosphere," Bulletin, *American Meteorological Society [1954]:209.*

troposphere. The air there is dominated by horizontal mixing that generally is not as vigorous as in the middle latitudes.

Jet Streams

At the subtropical and subpolar breaks in the tropopause, large horizontal temperature gradients set up high-speed flows of air that have become known as the jet streams. According to the World Meteorological Organization, any speed exceeding 30 meters per second (67 mph or 108 kilometers per hour) may be called a jet stream. Speeds vary along the axis of a jet stream, particularly along the polar front jet. Strongest jet streams have been recorded during winter over southern Japan, where speeds up to 140 meters per second (500 kilometers per hour, or 310 mph—270 knots)

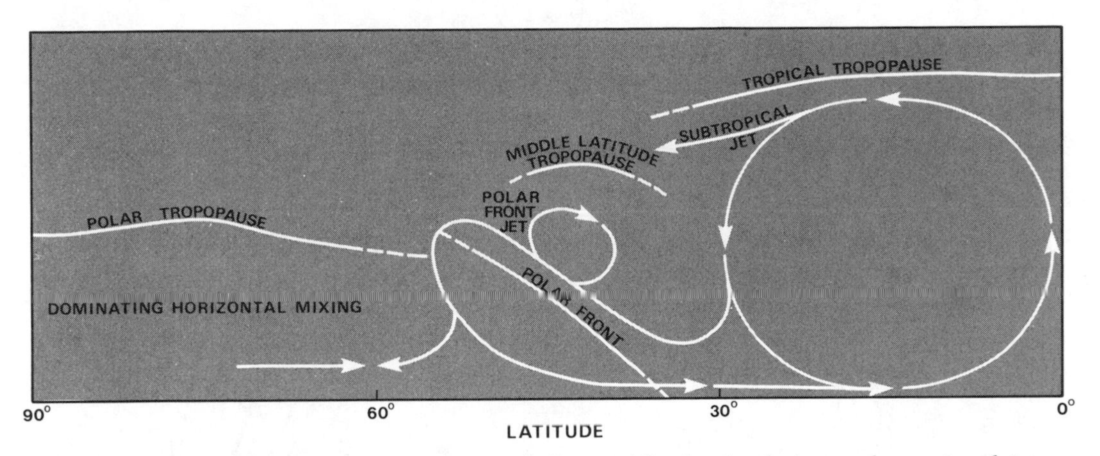

Figure 6.19. Schematic representation of the meridionic circulation and associated jet cores. (After E. Palmén, "The Role of Atmospheric Disturbances in the General Circulation, "Quarterly Journal of the Royal Meteorological Society, *Vol. 77 [1951], pp. 337–54.)*

have been observed. Secondary maxima generally occur off the east coast of North America and in the central part of North Africa (Figure 6.20). Along the edges of jet streams, wind sheers produce sharp turbulence as small vortices form between the fast moving air streams and the more slow-moving air around them. This accounts for the rough air encountered by aircraft upon entering or leaving jet streams.

Two jet streams, the subtropical jet and the polar front jet, are typical during winter in the Northern Hemisphere. The subtropical jet is located, on the average, at about 28°N latitude at about the 200-millibar level, between the tropical tropopause (at approximately 100 mb) and the mid-latitude tropopause (at approximately 250 mb). The polar front jet lies near the 300-millibar level between the mid-latitude tropopause and the polar tropopause. The polar front jet meanders and shifts its position much more than the subtropical jet, so that, while it is usually the stronger of the two, it tends to become obscured when averaged with all jet stream occurrences over the period of a month or more. Thus, most representations of the average position of the wintertime jet in the Northern Hemisphere approximate the position of the subtropical jet centered over the subtropical high pressure zone at the surface (Figure 6.20). But on daily maps during winter, most of the time the two jets are discernible, and the highest wind speeds are usually encountered at higher latitudes, averaging about 50°N (Figure 6.21). During summer in the Northern Hemisphere, the subtropical jet tends to disappear, and the average position of jet streams in the Northern Hemisphere shifts poleward (Figure 6.22).

Although observations are scarce in the Southern Hemisphere, the jet streams there appear to be similar to those in the north, with perhaps less fluctuation in position and intensity than in the north because of smaller land masses in the Southern Hemisphere. The strongest westerly winds are found south of Africa and over the Indian Ocean. During winter a separate belt of subtropical jets is found near 30°S, which is most pronounced over eastern Australia. This subtropical jet apparently disappears during the Southern Hemisphere summer. All jet streams tend to be weaker in summer than in winter, as is the entire circulation of the troposphere, since the latitudinal temperature gradient is reduced between equator and pole during summer.

Standing Waves

The circumpolar west-to-east circulation of the middle and upper troposphere is not

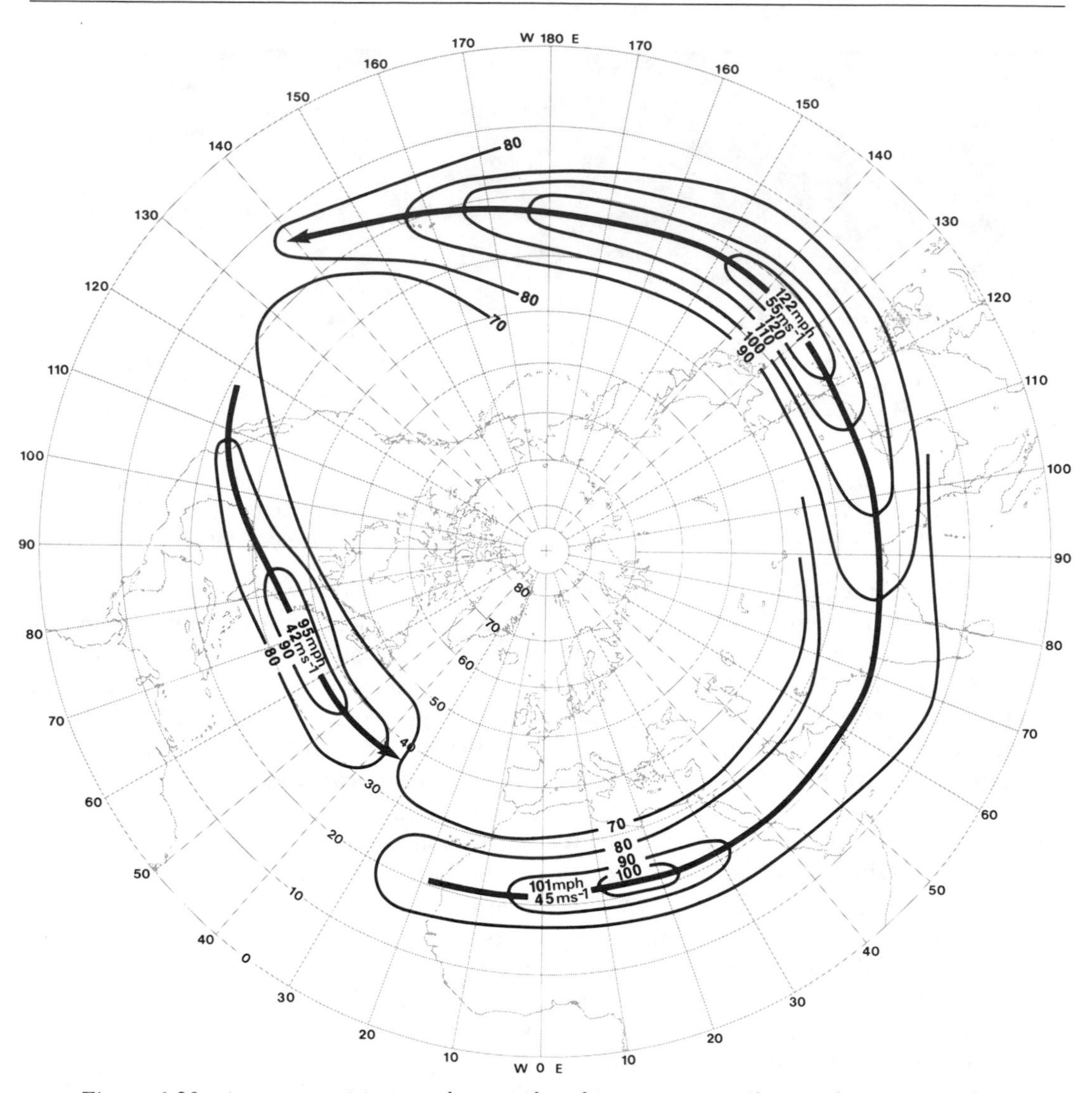

Figure 6.20. Average positions and strengths of jet streams in the Northern Hemisphere, January. (After Namias and Clapp.)

exactly circular, but meanders through equatorward-extending troughs and pole-ward-extending ridges, some of which are quasi-stationary in space over protracted periods of time. These are the so-called *standing waves* of the upper troposphere or the "Rossby waves," named after Carl Rossby who, in 1938, first described them adequately and dealt with them quantitatively. These waves are most pronounced in winter. They generally undergo sequences of events that increase their meridional magnitudes until they become so elongated that they become cut off. Then isolated pools of cold air remain for a while in lower latitudes, and warm pools remain in higher latitudes, while the west-east flow reestablishes itself with only minor undulations to begin the sequence all over again (Figure 6.23).

As can be seen on Figure 6.13, the January average flow shows two very pronounced troughs, one over the eastern Canadian Archipelago and another across east-

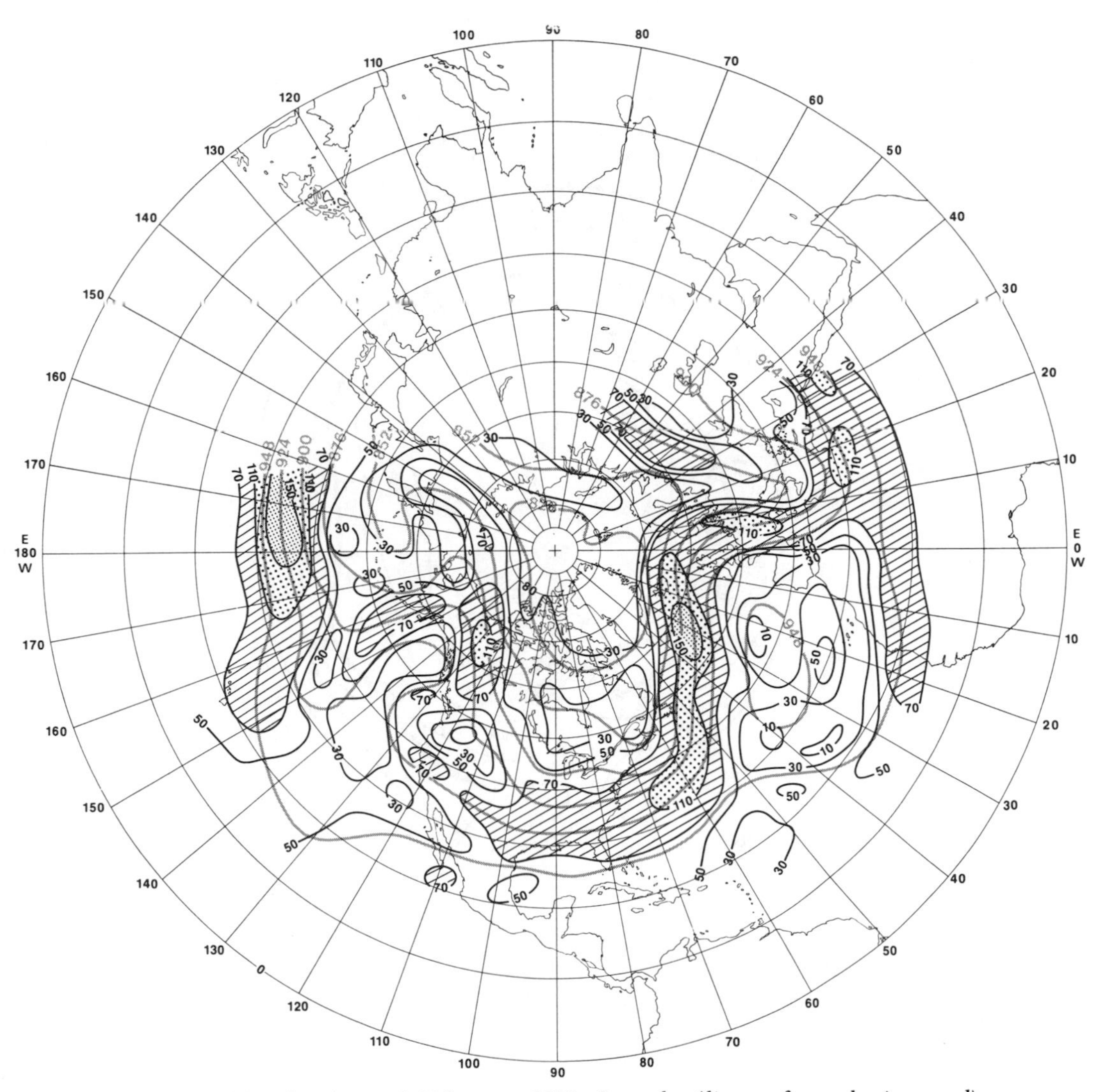

Figure 6.21. 300 mb. chart, 6 February 1981. Isotachs (lines of equal air speed) are black lines numbered in knots (1.15 statute mph). Regions with air speeds greater than 70 knots are shaded. Gray lines are contours in dekameters. Segments of both the polar and tropical jets are discernible particularly over the central Pacific and northwestern Canada, the northern and central Atlantic, and northern Europe and north Africa.

ern Siberia to the Sea of Okhotsk, and a lesser trough through eastern Europe. Relatively pronounced ridges appear over the eastern North Pacific, extending northward into the Bering Strait, and over the eastern North Atlantic, culminating over central Greenland, while a very weak, broad ridge extends over much of the central part of Asia, particularly northward in Siberia. In the upper reaches of the troposphere the jet streams follow these sinuous routes of the standing waves around the hemisphere.

As with the jet streams, monthly averages tend to obscure some features that are apparent on daily maps. While mean monthly 500-millibar maps show an average

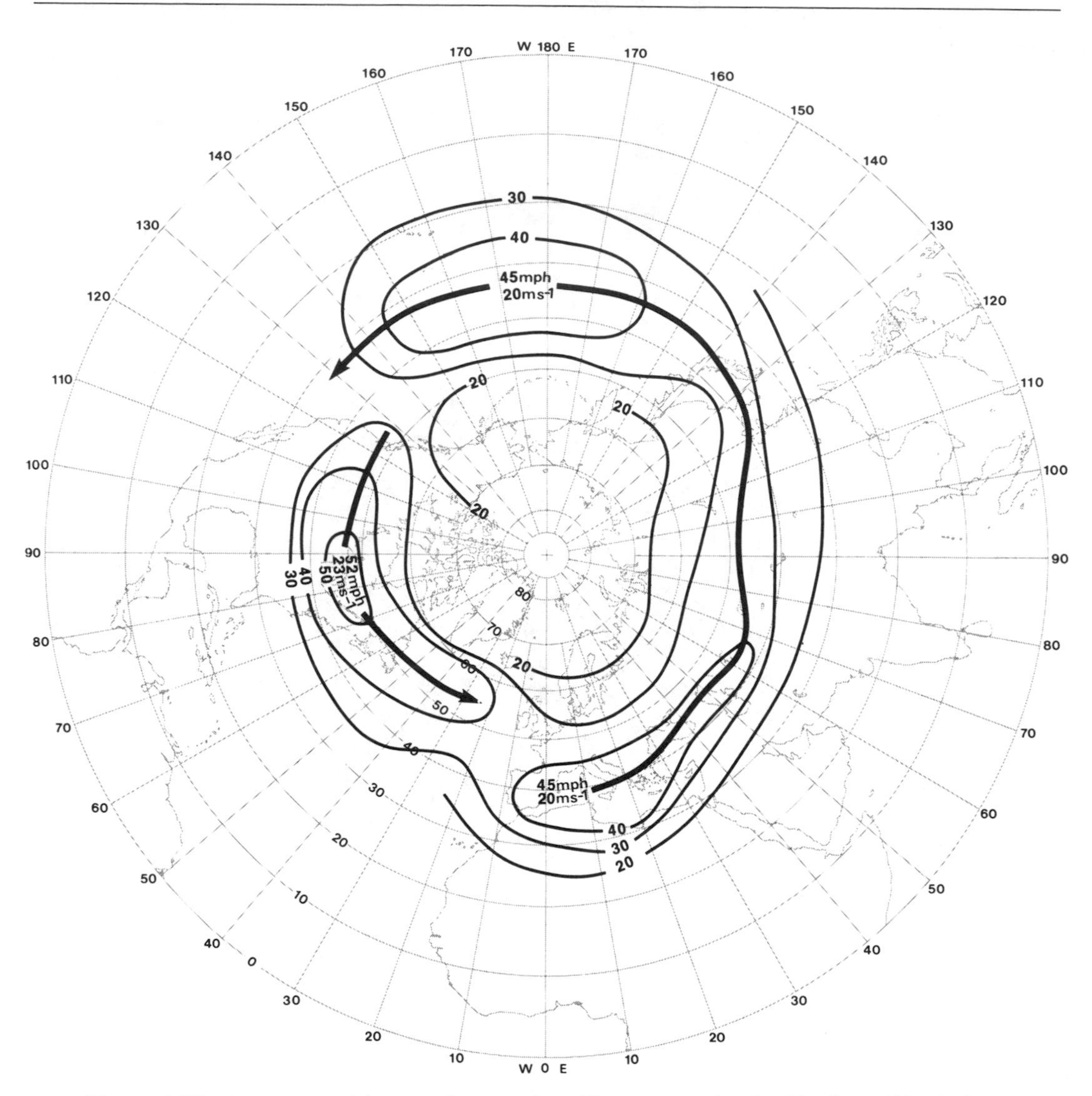

Figure 6.22. Average positions and strengths of jet streams in the Northern Hemisphere, July. (After Namias and Clapp.)

of three standing waves around the Northern Hemisphere, most daily maps show five, some less, some more. The 500-millibar map for 12 April 1983 illustrates a rather extreme case of eight fairly distinguishable troughs around 50°N latitude, plus some short waves in between (Figure 6.24).

The origin of these standing waves is undoubtedly tied to topography and heat sources and sinks, such as the thermal differences between sea and land. Perhaps the greatest influence on the Northern Hemisphere flow is the Rocky Mountains in western North America, which lie essentially perpendicular to the zonal flow and probably cause the deep average monthly trough over eastern North America. The weak average trough over eastern Europe may be simply a resonance wave, a downstream ripple of the anchor wave over eastern North America. But the tem-

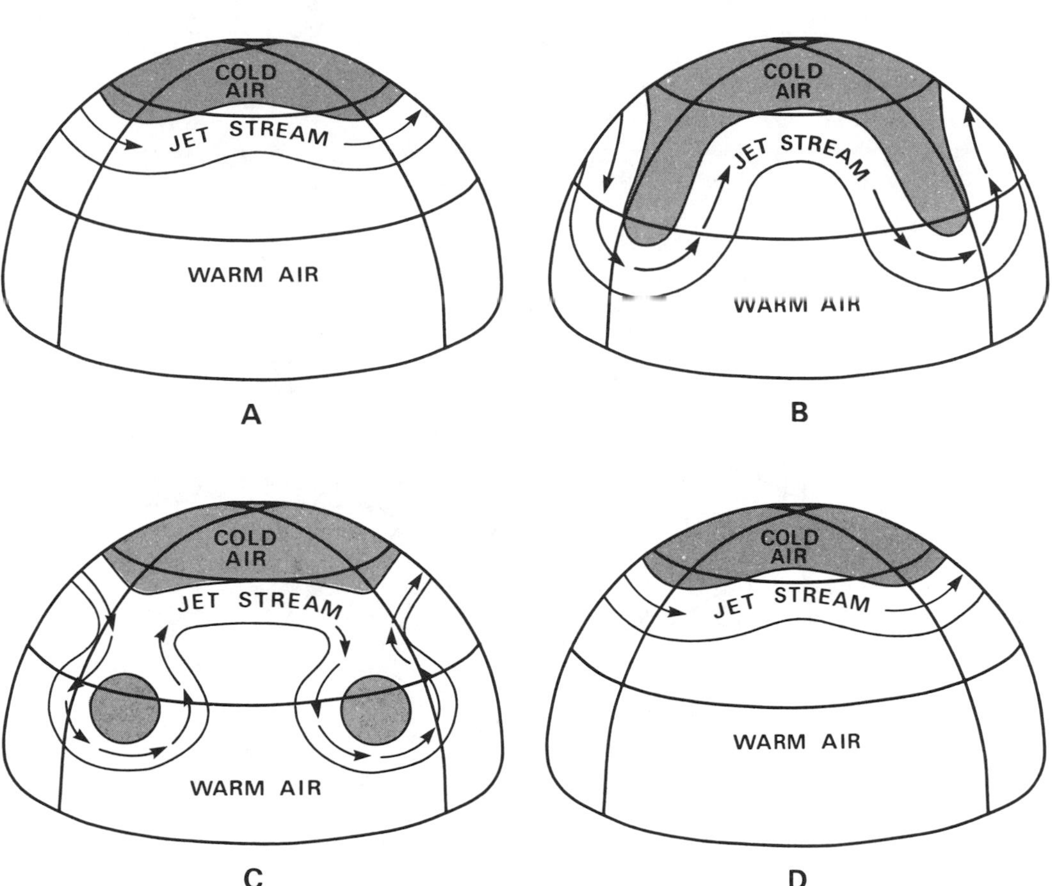

Figure 6.23. Schematic representation of a wave sequence (index cycle) in the upper troposphere.

perature differences between sea surfaces and land surfaces in coastal regions may be just as important as topographic barriers. For instance, whenever sea surface temperatures are high in the eastern Pacific, a pronounced ridge forms in the upper troposphere over the eastern Pacific and western North America, which has profound effects on the weather downstream across most of North America.

In the Southern Hemisphere, probably the greatest influence is the Andes mountain chain in western South America. Nonetheless, since there is much less land in the higher middle latitudes in the Southern Hemisphere, the air flow there seems to be much more zonal (west-east). This apparent simplicity may be partially due to lack of data in the Southern Hemisphere. Plots of balloon paths indicate the presence of five waves in the Southern Hemisphere, similar to the daily situation in the Northern Hemisphere (Figures 6.25).

The standing waves of the middle and upper troposphere remain quasi-stationary, while shorter waves and vortices below move along their contours. It may take weeks or even months for the long waves to shift their positions significantly, and thus they become conservative factors in the atmospheric flow and allow for some extended forecasts that could not be predicted from the rapidly fluctuating flow patterns of the surface air. Although the

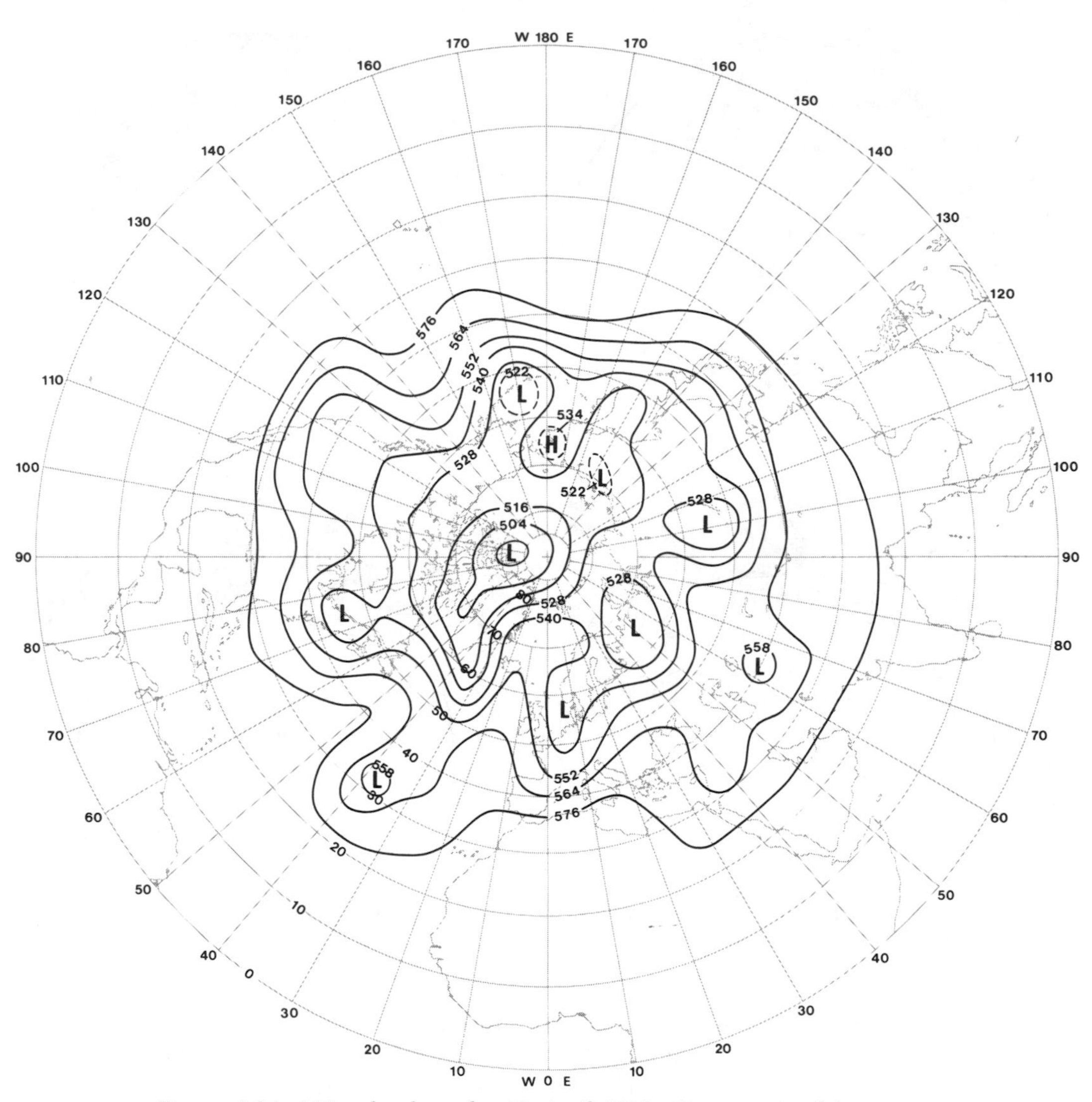

Figure 6.24. 500 mb. chart for 12 April 1983. Contours in dekameters.

standing waves of the upper troposphere tend to remain in semiconstant forms over prolonged periods of time, they sometimes undergo rapid transitions into different formations, which then bring about abrupt shifts in general weather conditions over broad sections of the earth's surface. A region that might have experienced a prolonged rainy spell of several weeks' duration may next find itself in a prolonged spell of droughty conditions. Or a similar contrast of temperatures might take place. The standing waves can slowly shift their positions either downstream, with the wind, or upstream, against the wind. The direction of motion largely relates to the length of the wave—the longer waves tending to regress upstream, while the shorter waves move downstream. Usually, shorter waves move faster than long waves.

Vorticity

The meandering of the upper air streams is a natural phenomenon that relates to the spin of air particles across the face of the

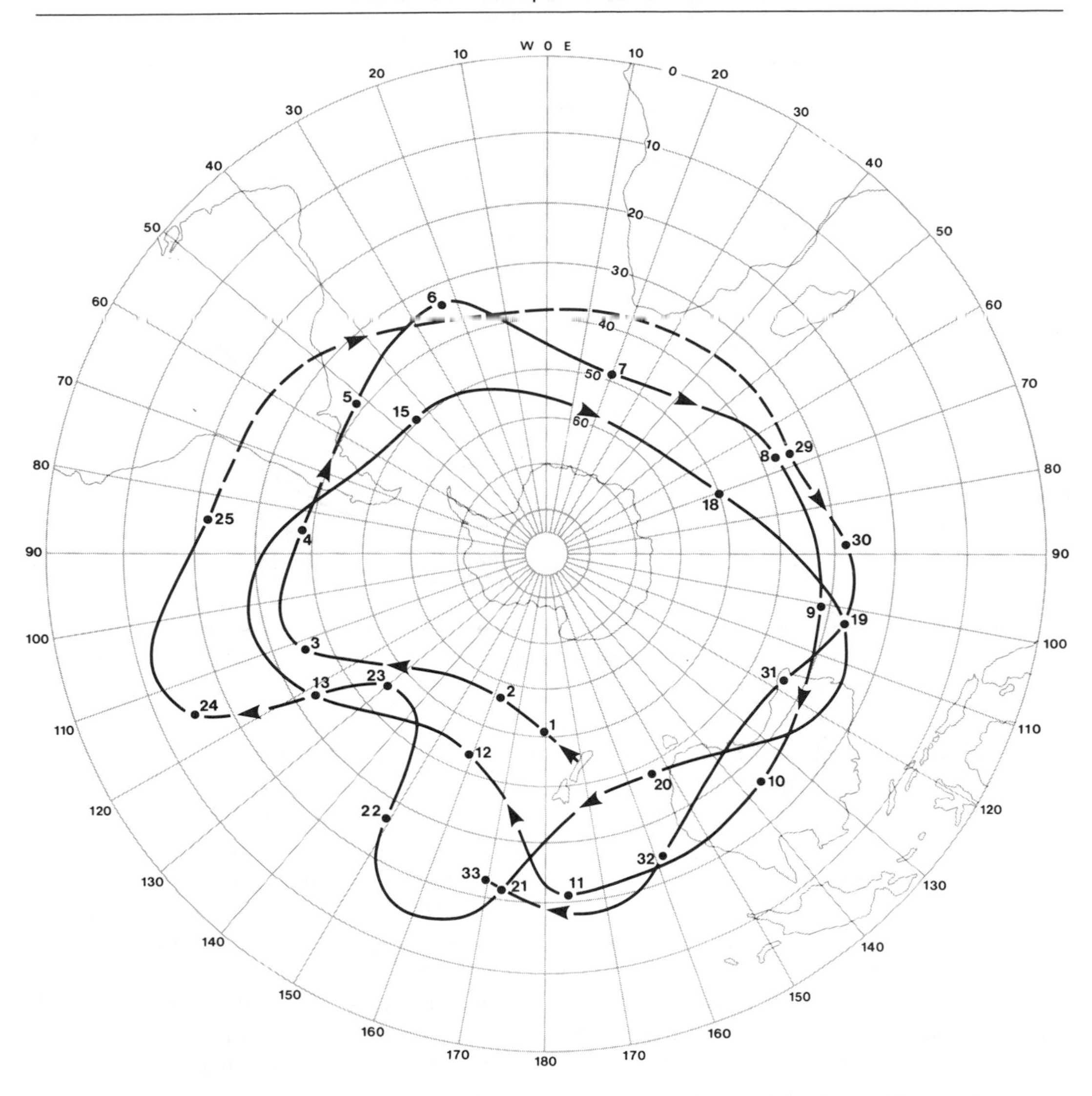

Figure 6.25. Path and daily positions of a constant-level balloon launched from Christchurch, New Zealand, 30 March 1966. The balloon drifted for 33 days at an altitude of about 12 kilometers, during which it circled the Southern Hemisphere three times. (From National Center for Atmospheric Research.)

earth. The amount of spin per unit area is known as *vorticity*. This has two components, one relating to the Coriolis Force, which increases with latitude, and one relating to the curvature of the flow itself, known as *relative vorticity*. *Positive relative vorticity* is defined as increasing cyclonic flow (counterclockwise in the Northern Hemisphere and clockwise in the Southern Hemisphere), and *negative vorticity* is defined as decreasing cyclonic flow or increasing anticyclonic flow (clockwise in the Northern Hemisphere and counterclockwise in the Southern Hemisphere). If an air stream is moving with some meridional component, say from southwest to northeast in the Northern Hemisphere, it will be moving into higher latitudes, which will increase the Coriolis Force, which increases that component of vorticity. For the ab-

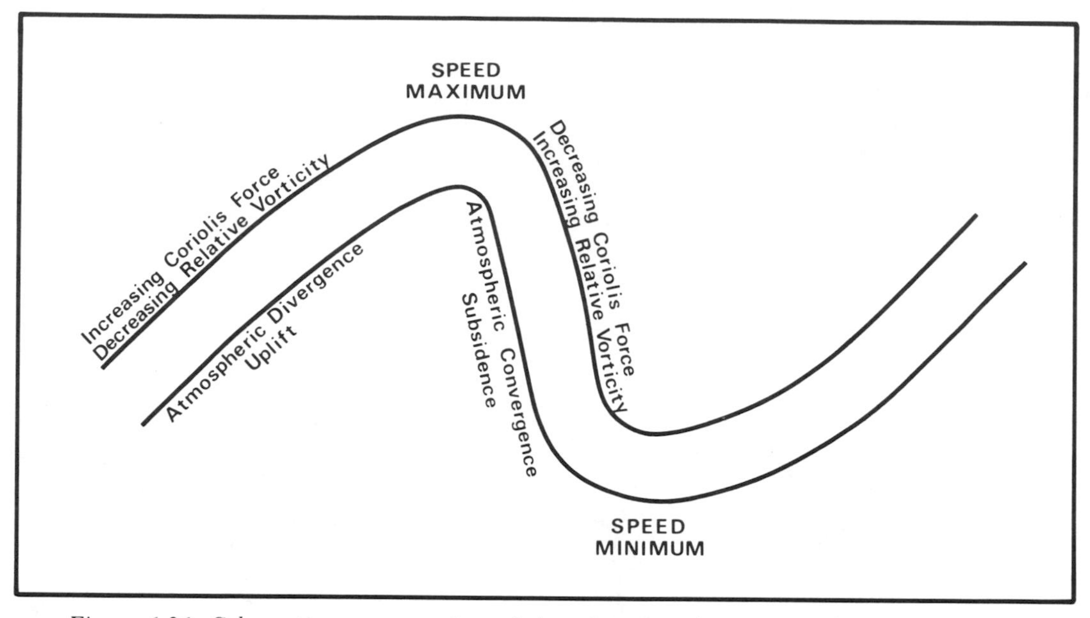

Figure 6.26. Schematic representation of changing Coriolis Force and relative vorticity in flow along a standing wave.

solute vorticity (the total of the two components) to remain constant, the relative vorticity due to curvature must decrease. Thus, the northeasterly moving flow must become more anticyclonic, or curve eastward and ultimately southeastward (Figure 6.26). As the flow starts to move southward again, the vorticity component due to the Coriolis Force begins to decrease, and therefore the relative vorticity due to curvature must increase once more, thus turning the flow in a cyclonic (counterclockwise) direction, which ultimately sends the air back to the northeast again. Thus, the air moves downstream in a wavy pattern with the relative vorticity constantly adjusting to the changing latitude, and the meandering flow tends to perpetuate itself.

Divergence and Convergence

The standing waves of the upper troposphere are important not only for steering, and thus predicting, the future motion of storms and other more localized features in the lower troposphere, but also because they aid in predicting the development of such features; i.e., whether they are going to strengthen or weaken. Since for a given pressure gradient the wind moves faster in anticyclonic flow (around a high pressure ridge) than in cyclonic flow (around a low pressure trough), the air, in flowing along a meandering wave pattern, will be constantly changing its wind speed. This will produce speed convergences and divergences in the flow that will induce vertical motion in the air.

For instance, in Figure 6.26, the air moving from southwest to northeast and undergoing increasing anticyclonic curvature will tend to speed up, while the air moving from northwest to southeast and coming into influences of increased cyclonic curvature will tend to slow down. Therefore, the air will experience a divergence or pulling apart in the horizontal plane in the southwest-to-northeast portion of the wave, and a piling up or convergence in the northwest-to-southeast portion of the wave.

The air deficit in the region of divergence

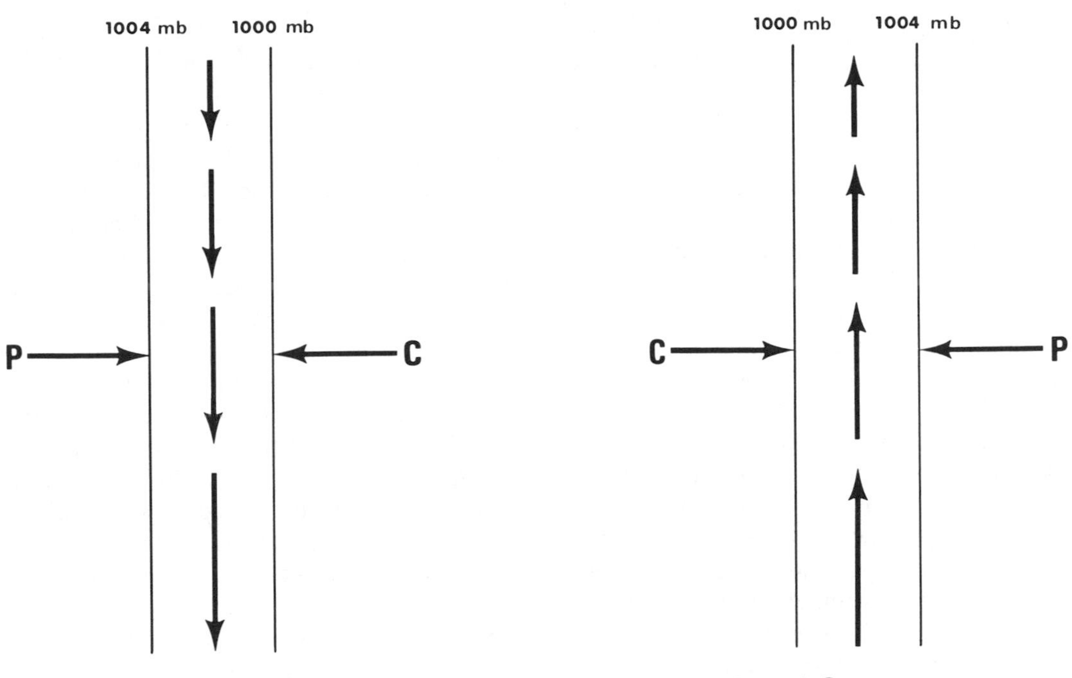

Figure 6.27. Speed divergence and convergence induced by latitudinal changes in Coriolis Force with north-south flow in a field of constant pressure gradient, Northern Hemisphere.

will have to be compensated for by vertical motion. And since this is in the middle troposphere, air can move into the layer from both above and below. Since most of the storm development and cloud formation is below this level, the important movement is the vertical motion rising from below. This will tend to strengthen storm development, which will induce cloud formation and possible precipitation. (These processes will be discussed more fully after the topic of moisture has been treated.) The portion of the wave that is experiencing convergence will have an excess of air that must move vertically downward below and upward above. Again, it is the downward motion below that becomes important to the consideration of weather conditions, and this subsiding air would weaken any storm moving along this poriton of the track and tend to produce clearing skies and fair weather.

Thus, other things being equal, the weather conditions in opposite portions of standing waves tend to be opposite in character: increasing storminess, cloudiness, and precipitation in southwest-northeast-oriented eastern branches of troughs or western branches of ridges, and clearer, drier weather in northwest-southeast-oriented western branches of troughs or eastern branches of ridges in the Northern Hemisphere. Since the standing waves do shift slowly with time, over distances of several hundred or even several thousand kilometers, an area might experience a prolonged period of clouds and rain as the wave occupies one position, and then experience a prolonged period of drought as the wave shifts to another position.

Since essentially opposite types of weather can be expected simultaneously in places about one-half wavelength apart, the Soviets, with their large west-east territory, made use of this knowledge during the

1950s to develop the so-called "Virgin Lands" east of the Urals, which are essentially one-half wavelength away from the older farming regions in the Ukraine. These two areas, which are both droughty, seldom experience extreme drought simultaneously. When it is exceptionally dry in the Ukraine, it is usually exceptionally moist east of the Urals in western Siberia and northern Kazakhstan, and vice versa. By planting both areas with similar crops, the Soviets can assure themselves of a fairly stable year-to-year overall production.

Since cyclones tend to intensify on poleward-moving limbs of waves, and since low pressures tilt toward colder temperatures, the traveling low pressure vortices of the middle latitudes eventually end up moving poleward and ultimately merge in the general region of the subpolar lows to cause constant regeneration of those systems. Conversely, anticyclones intensify on equatorward-moving limbs of waves and tilt toward warmer temperatures, so high pressure cells ultimately move equatorward to merge with the subtropical high pressure belts and constantly regenerate those systems.

Horizontal convergence or divergence can be produced either by directional movements that come together or spread apart (see Figures 6.8 and 6.9) or by speed variations within flows, either of which can be achieved in either straight-line flow or curvilinear flow. Often the speed variations within an air stream are brought about by changing Coriolis Force with changing latitude. For instance, assume straight north-south-oriented isobars in the Northern Hemisphere, as shown in Figure 6.27. In the left-hand diagram the pressure gradient force (P) is directed toward the east and is constant all along the isobars, since they remain equally spaced. Assuming frictionless flow, the wind will flow from north to south with the Coriolis Force (C) directed westward equal in magnitude to the pressure gradient force. As the air moves into lower and lower latitudes, the Coriolis Force tends to decrease because of the decreasing latitude, and this allows the wind to be pulled partially toward the pressure gradient. If the air motion has a component directed down the pressure gradient, it will accelerate. This will increase the speed of the wind, which will increase the Coriolis Force, which will pull the wind back to the original direction, and again the Coriolis Force will balance the pressure gradient force. The net result is an increase in wind speed but no change in direction. Hence, in an equatorward-flowing air stream, horizontal divergence will occur that will induce vertical motion toward that air stream. If the air is at the ground and this is a surface wind, the only direction the air can come from is above, and therefore the horizontal divergence induces subsidence which causes compression, heating, and fair weather. Conversely, air moving from equator to pole along straight, equally spaced isobars will constantly slow down, and air from behind will tend to pile up on air in front, which produces horizontal convergence, which at the surface would produce uplift (right-hand diagram).

It is interesting to note that with the same pressure gradient, the wind will have a higher velocity at low latitudes than at high latitudes. At low latitudes, because of reduced Coriolis Force, the wind must reach higher speed in order for the Coriolis Force to be large enough to balance the pressure gradient force. In reality, however, pressure gradients usually are not constant over extended latitudinal distances, but diminish toward lower latitudes, so that in most cases the winds are quite weak in low latitudes and there are few isobars on pressure maps in equatorial regions. Since the Coriolis Force reduces to zero at the equator, the concept of geostrophic wind breaks down in equatorial areas.

In the middle troposphere, as stated a

few paragraphs back, the same convergence and divergence would produce subsidence and uplift respective of the air below the air stream. At the top of the troposphere, if segments of jet streams are present, strong divergence is most frequently observed in the left-front quadrant of a jet core, followed by the right-rear quadrant. Thus, most uplift, and hence cloud formation and precipitation, will occur to the right of the axis where air enters a jet core and to the left of the axis where air leaves a jet core.

7 | ATMOSPHERIC MOISTURE

Moisture is the third basic ingredient, along with heat and motion, that determines the weather over the earth. It not only supplies the water for cloud formation and precipitation, but also, through changes of state, is intimately involved with great quantities of energy in the form of latent heat. All the moisture in the atmosphere is derived from the earth's surface through the process of evaporation. A great deal of heat is required to change the state of water from liquid to gas. This is known as *latent* (hidden) heat because it does not effect a change in temperature. This heat is not lost forever, but is released again once the water changes back from gas to liquid. The latent heat released may then be utilized to effect a temperature change in the air, or for some other purpose. Also, latent heat is involved in the change of state from solid to liquid and vice versa.

Latent Heat

To envision the quantity of heat involved in the change of state of water, recall that the heat unit, the calorie, is defined as the heat that is required to raise the temperature of 1 gram of water 1°C. To raise the temperature of 1 gram of liquid water from the freezing point, 0°C, to the boiling point, 100°C, would require 100 calories of heat. But to change that 1 gram of water at the boiling point to water vapor at the boiling point would require 540 calories of heat. Also, if water at the freezing point were initially in the frozen state, it would take 80 calories per gram to melt the ice into water at 0°C. Therefore, it can be seen that it requires almost as much heat to melt ice as it does to raise the temperature of water from freezing to boiling, and it requires 5.4 times as much heat to evaporate water at the boiling point as it does to raise its temperature from freezing to boiling (Figure 7.1).

Thus, evaporation and condensation involve truly great quantities of heat. Under natural conditions evaporation usually takes place much below the boiling point of water. This requires more heat. At 0°C, the latent heat of evaporation is 596 calories per gram. Sometimes there is a direct change from solid to gas, as in the case of evaporation of snow. This is known as *sublimation.* The latent heat of sublimation is the sum of the latent heat of fusion (or melting) and the latent heat of evaporation (or condensation), or a total of 676 calories per gram if the sublimation takes place at 0°C. If ice at 0°C is melted and then the liquid water is heated to the boiling point and evaporated, the total amount of heat necessary would amount to $80 + 100 + 540 = 720$ calories.

Water Vapor

The water vapor in the atmosphere can be expressed in a variety of ways. All are significant, since some serve some purposes and others serve other purposes. *Absolute humidity* is defined as mass of water vapor per volume of air. This is not a conservative number, since as the air moves about it expands and contracts. Thus, the absolute humidity can change drastically without any addition or subtraction of moisture.

A more conservative measure is *specific humidity,* which is defined as the mass of water vapor per mass of air. A closely related value is the *mixing ratio,* which is the mass of the water vapor compared to the mass of the dry air. Since air is always much more abundant than water vapor, there is little difference between *specific humidity* (which is the mass of water vapor divided by the total mass of the air including the water vapor) and the mixing ratio (the ratio of the water vapor to the dry air). These values are quite useful in technical work, such as weather forecasting, as they are measures of such things as latent heat energy available in the air due to its water vapor content, availability of precipitable water, atmospheric stability characteristics, and so forth.

To the layman, however, the fact that the air at a given time might contain 10 grams of water vapor per kilogram of dry air probably holds little meaning. More meaningful might be the degree of saturation of the air, which noticeably affects the physiological workings of the human body and all other living organisms, as well as cloud formation, precipitation processes, and so forth. This expression, known as *relative humidity,* is the ratio of the actual humidity of the air compared to the saturation humidity of the air, both expressed in the same units. Thus, if the water vapor content of the air has been measured to be 7 grams of water vapor per kilogram of dry air, and the saturation mixing ratio of the air is known to be 10 grams per kilogram, the relative humidity is 70 percent.

The saturation mixing ratio of the air depends on temperature; it increases at an increasing rate with increasing temperature

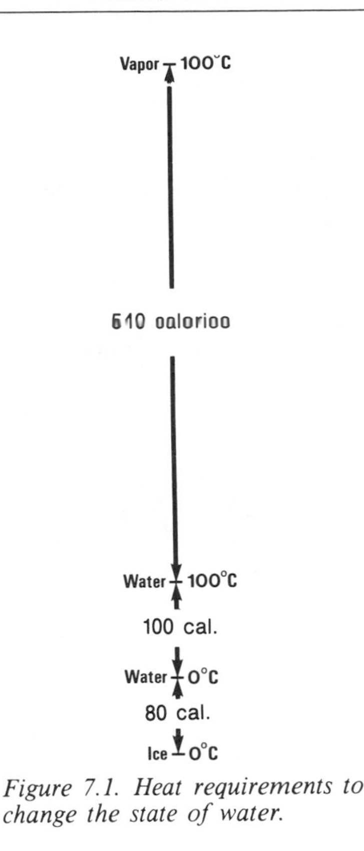

Figure 7.1. Heat requirements to change the state of water.

(Figure 7.2). Thus, a specific change in temperature will effect a greater change in saturation mixing ratio at higher temperatures than at lower temperatures. This fact will become very important when stability characteristics of the air are considered in the next chapter.

Relative humidity can be changed either by adding moisture to or subtracting moisture from the air, or by changing the air temperature. Temperature changes normally take place more quickly than moisture changes, and so short-term relative humidity changes are usually related primarily to temperature changes. For instance, relative humidity is usually rather high early in the morning when the temperature is near its minimum point, but drops as the air heats up during the morning. Thus, in Figure 7.2, if the air on a given morning is saturated at 10°C and warms up during the morning to 20°C without any addition or subtraction of moisture, its relative humidity would have dropped to little more than 50 percent. Conversely, the relative humidity tends to increase during the night as the temperature decreases.

If the air contains much water vapor and the night is clear and still so that much heat is radiated outward from the earth's surface, the temperature drop may be enough to reduce the capacity of the air to the point where it can barely hold all the moisture that it has in it. The air then becomes saturated, and if cooling continues, some of the moisture must be condensed into liquid droplets and may settle on surface features in the form of dew. Thus, the temperature at which the air becomes saturated is referred to as the *dew point,* which is another measure of the amount of moisture in the air. Although the dew point is expressed as a temperature, it is a measure of the actual amount of moisture in the air. This is the expression commonly used in weather reports. With the temperature it gives a rough indication of the degree of saturation of the air and the temperature at which the air will become saturated if cooling takes place.

For certain purposes it becomes most useful to express the water vapor content of the air in terms of the pressure it exerts. Since the air is simply a mixture of gases, its total pressure can be divided into its component parts: that due to nitrogen, that due to oxygen, and so forth, and that due to water vapor. It is the *vapor pressure* that determines the movement of moisture across an interface, such as a water surface in contact with air above it. If the vapor pressure of the water surface is greater than the saturation vapor pressure of the air above it, moisture will move upward from the water surface into the air, whether the air is saturated or not. This phenomenon is often observed over relatively warm water bodies, such as lakes and rivers, early in winter before they freeze. If cold air moves across the warmer water surface, moisture will be injected upward into the cold air and will probably supersaturate the cold air, which will cause recondensation that produces what is known as steam fog.

The evaporation rate from a given surface depends upon the vapor pressure difference between the evaporating surface and the medium into which the moisture is moving. Vapor pressure difference also accounts for moisture movement from small water droplets to large water droplets and from a water surface to an ice surface at the same temperature, since the vapor pressure at the surface of a water droplet is inversely proportional to the radius of the droplet, and it is greater over a water surface than over an ice surface at the same temperature. These facts will be taken up again in Chapter 13 in a discussion of cloud formation and precipitation. At the same time it will be stressed that the air not only contains water vapor but also water in liquid and solid states in droplets that are small enough to be suspended in the air by the motion of the air. These liquid and solid droplets are visible in clouds and fog; water vapor cannot be seen. Thus, most of the time not only an exchange of moisture is taking place at the atmosphere-earth interface but also changes of state of moisture from solid to liquid to gas, and vice versa, are constantly taking place within the atmosphere. Water is unique among the atmospheric gases in that it occurs in all three states within the normal range of temperature within the atmosphere.

Since all moisture is derived from the earth's surface through the process of evaporation, the moisture content of the air is much greater in the lower troposphere than at higher altitudes. Little is found above the tropopause, since vertical mixing is generally limited above that level. Within the troposphere may be found distinct layers, produced by movements of air, some being relatively dry and others containing much moisture.

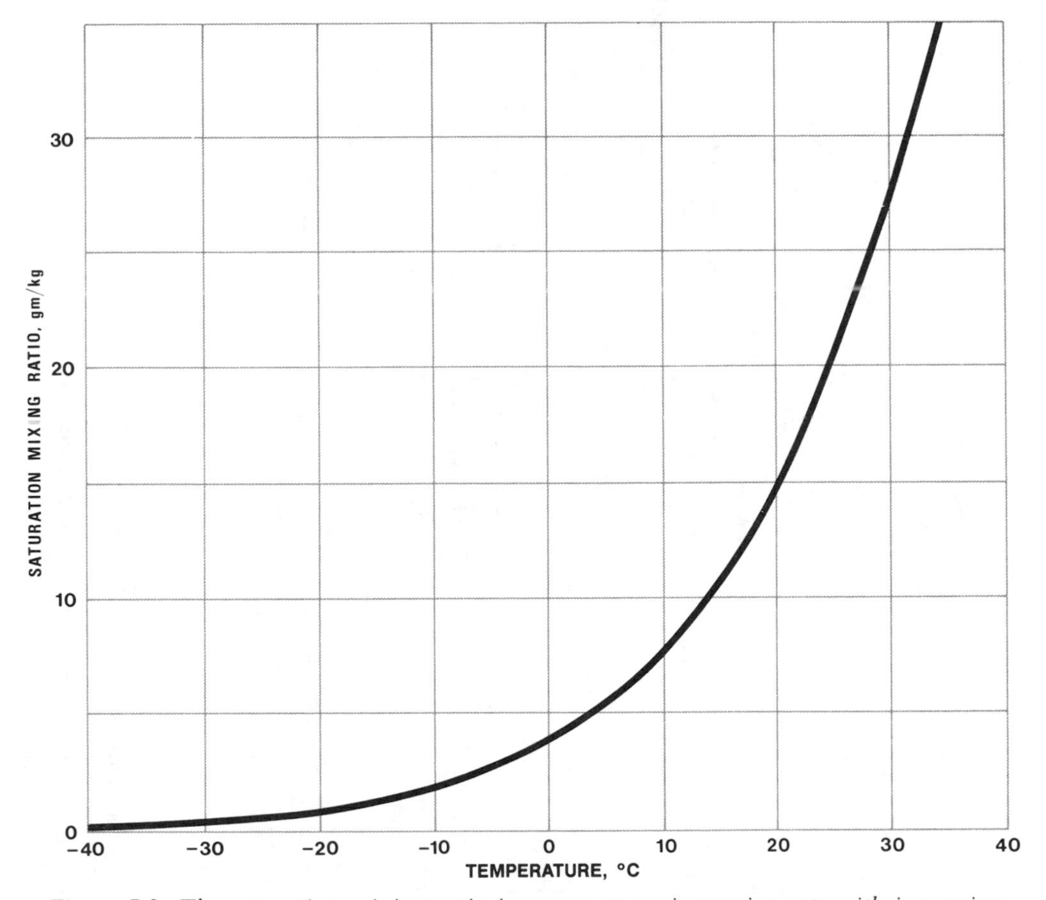

Figure 7.2. The saturation mixing ratio increases at an increasing rate with increasing temperature.

Geographical Distribution of Evaporation and Atmospheric Humidity

The amount of evaporation that takes place over a surface depends upon the availability of moisture at the surface and the vapor pressure gradient between the surface and the air above it. Since plenty of moisture is always available at the surface of the oceans, in general there is more evaporation over the oceans than over the continents at similar latitudes (see Figure 4.8). In a few land areas, particularly those covered by lush tropical rainforests, evaporation plus the transpiration of moisture through plants can exceed evaporation over adjacent oceans, since the tree leaves present more total evaporating surface to the air than the ocean surface does. These are the exceptions rather than the rule, however. Such forested areas exist in equatorial regions with great amounts of cloudiness and high atmospheric humidity, which tend to reduce vapor pressure gradients, and hence evaporation. Among the ocean areas, the subtropical oceans overlain by the subtropical atmospheric high pressure cells experience the greatest evaporation, since the air there is generally clear and dry. Intense heating from the sun coupled with continual replenishment of fresh, dry air through sub-

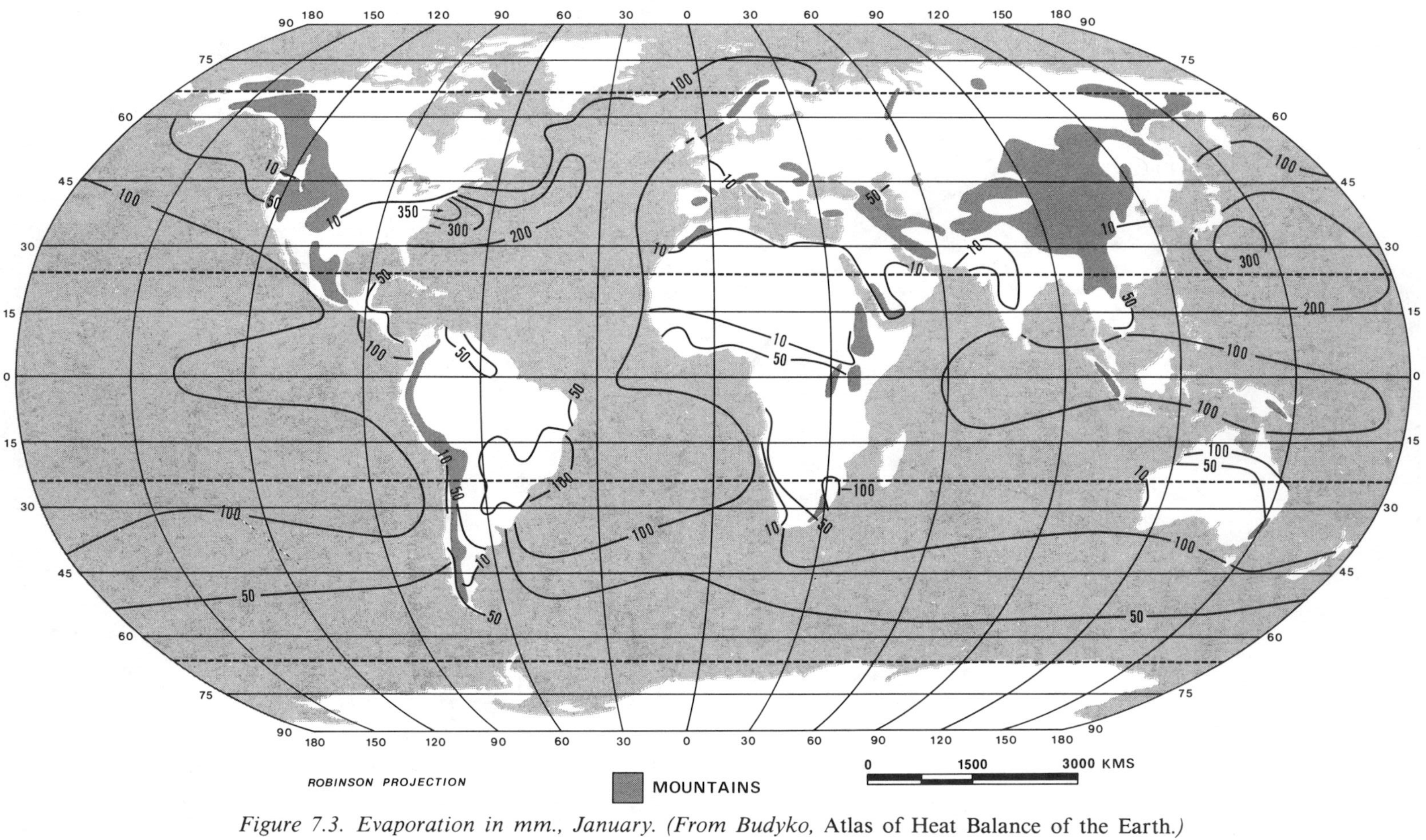

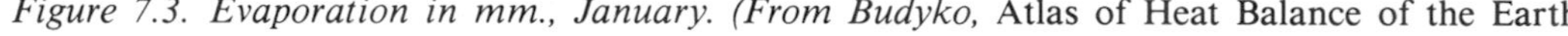

Figure 7.3. Evaporation in mm., January. (From Budyko, Atlas of Heat Balance of the Earth.*)*

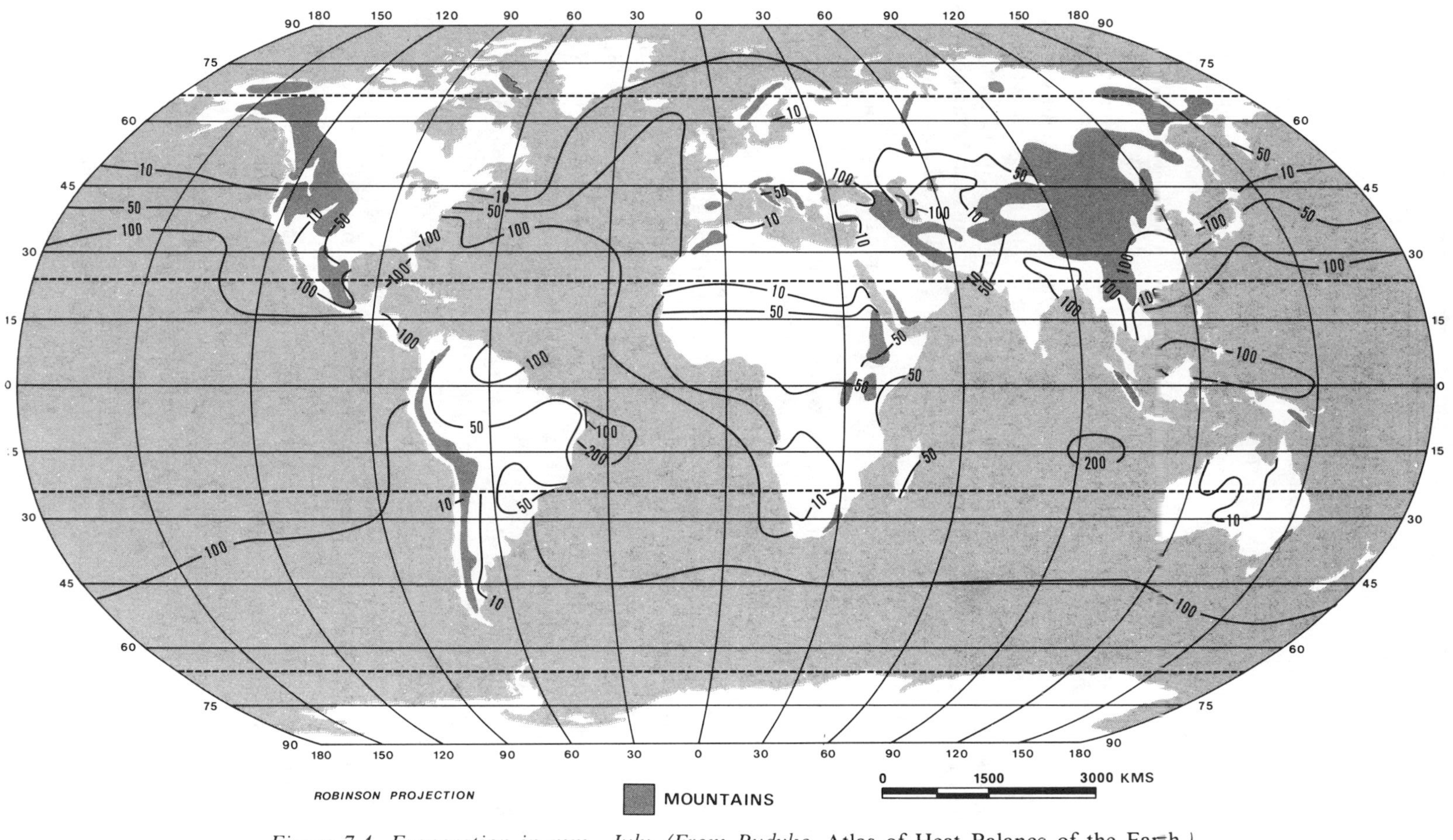

Figure 7.4. Evaporation in mm., July. (From Budyko, Atlas of Heat Balance of the Earth.*)*

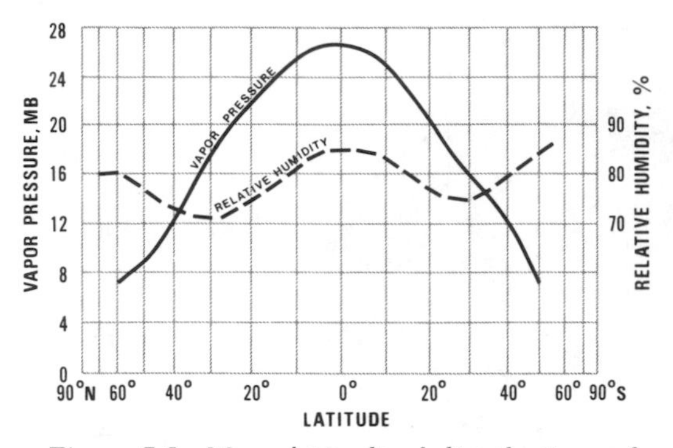

Figure 7.5. Mean latitudinal distributions of atmospheric vapor pressure and relative humidity. (From Howard J. Critchfield, GENERAL CLIMATOLOGY, *4th ed., © 1983, p. 45. Reprinted by permission of Prentice-Hall, Inc., Englewood Cliffs, N.J.)*

sidence from above produce low vapor pressure in the surface air relative to the underlying sea surface.

Nevertheless, the largest amount of evaporation on earth occurs at a somewhat higher latitude, off the east coast of North America between North Carolina and New Jersey. A similar situation occurs off the southeast coast of Japan, but it is not quite as pronounced. The high annual values in these areas are accounted for entirely by winter conditions. As can be seen in Figure 7.3, these areas are very conspicuous in January, but in July they do not appear at all (Figure 7.4). During winter, the vapor pressure difference between the water surface and the air above is maintained high, because the warm ocean currents flowing north-northeastward along these coasts are in stark contrast to fresh, cold, polar air that much of the time blows off the continents across the water currents. Since the Gulf Stream has a much closer contact with the mainland coast of North America than the Japan Current does with the mainland of Asia, the contrast is greater in the North American area, and hence the evaporation is greater. During summer when the air has about the same temperature as the water, or is even warmer, the evaporation is not outstanding in these areas.

Since the water vapor, once in the air, moves with the wind, its flux can be deduced crudely from annual wind direction and speed patterns relative to the major evaporation areas. Unlike the heat flux of the atmosphere, which moves simply from low latitudes to high latitudes with maximum values in the middle latitudes, the moisture flux has a more complex pattern, because the greatest amounts of water vapor are gained not at the equator but in the subtropics on either side. From these areas moisture moves both equatorward with the trade winds and poleward with the westerlies, reaching maximum flux around 10° and 40° latitude in each hemisphere, similar to the flux of latent heat shown in Figure 4.11.

The major part of the moisture flux in the atmosphere converges on the equatorial zone, so that on the average the specific humidity is highest at the equator and drops off steadily toward the poles, where the air is too cold to hold much moisture. The latitudinal distribution of relative humidity is quite different, however, since it depends not only on moisture content, but also on moisture-holding capacity. The lesser amount of specific humidity at high latitudes is still sufficient to bring the air near saturation in cool temperatures that reduce the capacity there to hold moisture. Therefore, average relative humidity is high both in the equatorial region and in the polar regions (Figure 7.5). It is lowest precisely in the subtropical regions of greatest evaporation, since moisture is moved outward in both directions from these zones by the winds, and temperatures are generally high.

Figure 7.6 shows the world distribution of average relative humidity in July. Here it can be seen that the subtropical high pressure zones stand out as belts of low relative humidity, while the equatorial zone between them, as well as parts of the middle and higher latitudes, mainly over the oceans, show higher relative humidities. The monsoon region of southeast Asia shows par-

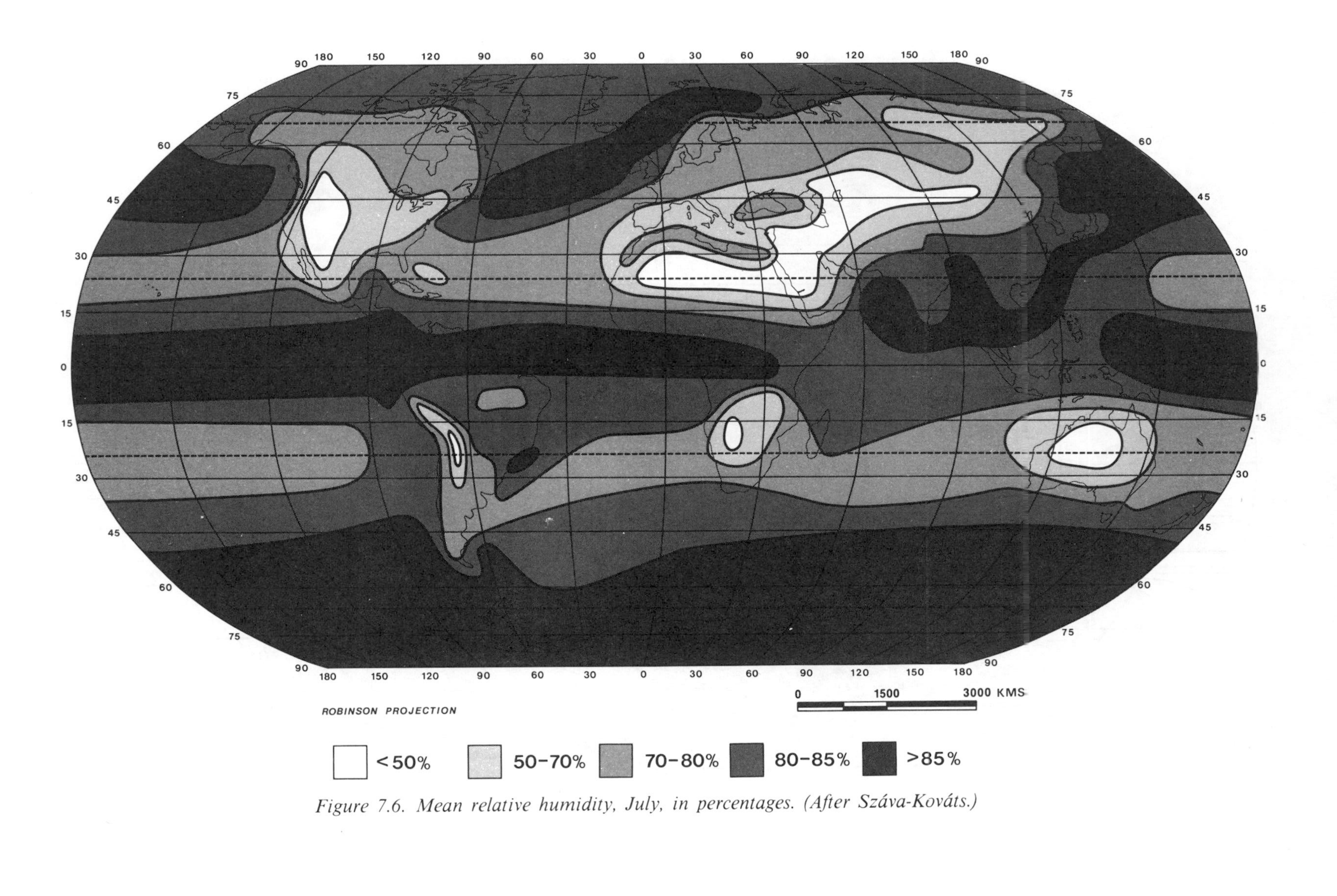

Figure 7.6. Mean relative humidity, July, in percentages. (After Száva-Kováts.)

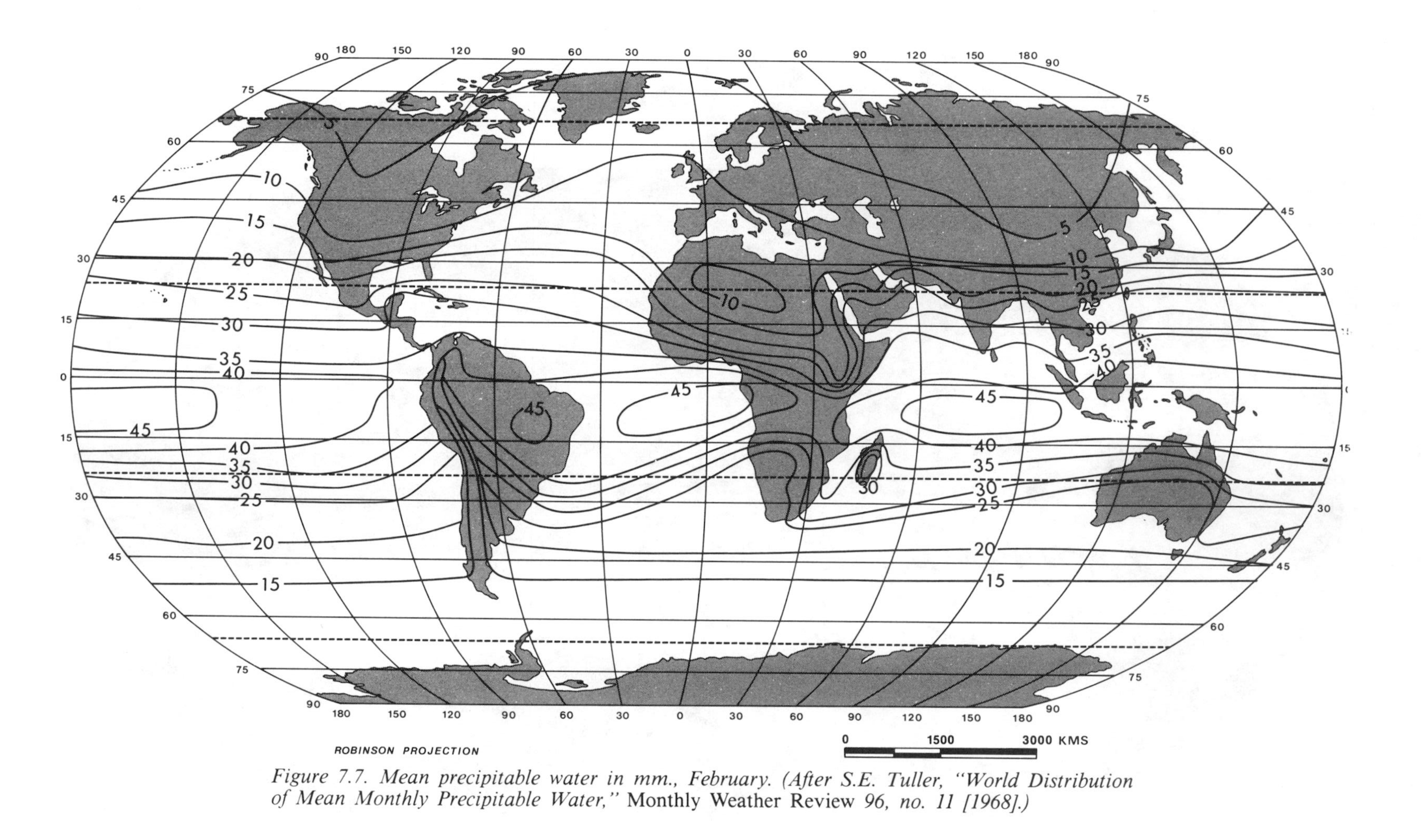

Figure 7.7. Mean precipitable water in mm., February. (After S.E. Tuller, "World Distribution of Mean Monthly Precipitable Water," Monthly Weather Review *96, no. 11 [1968].)*

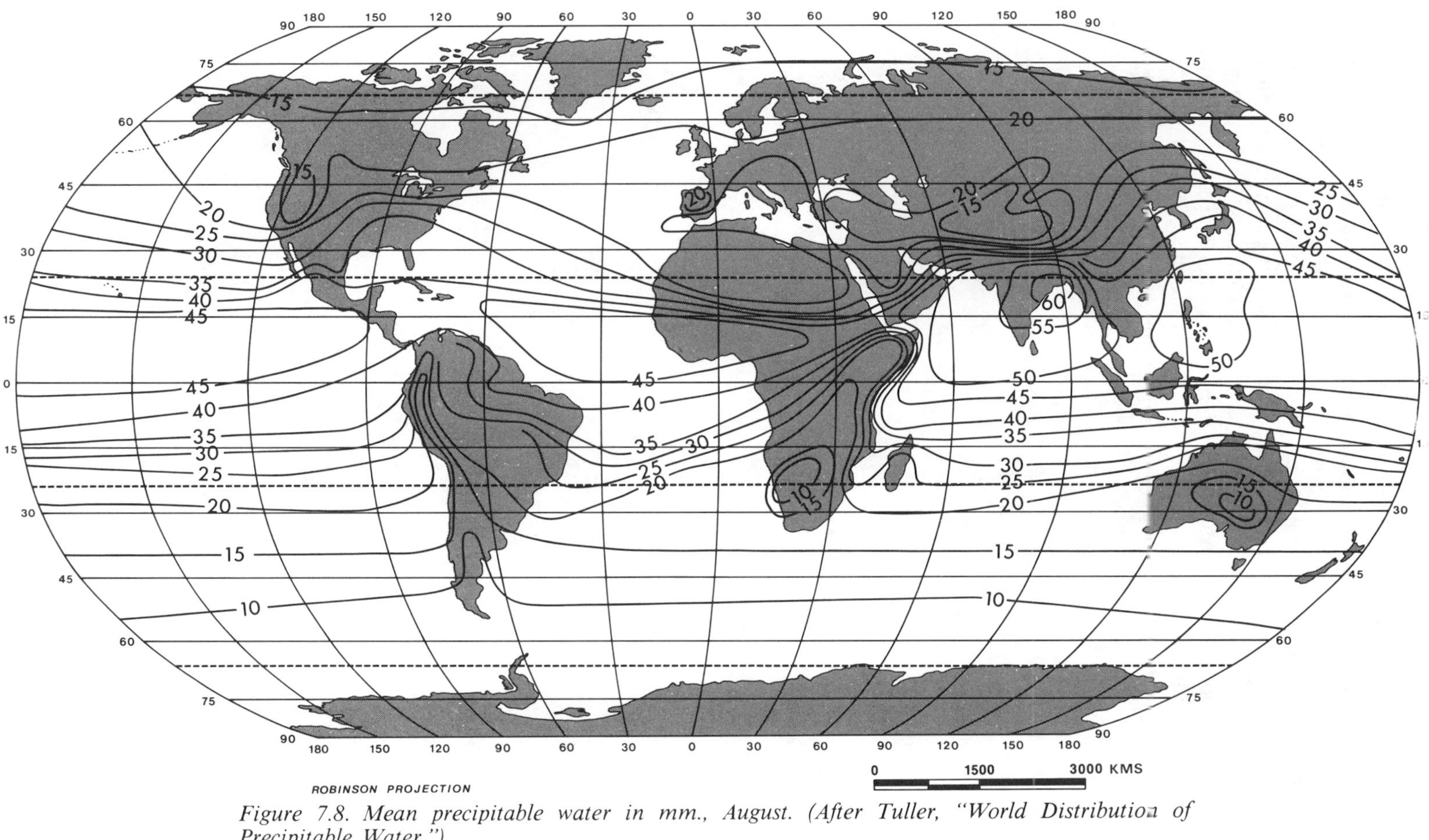

Figure 7.8. Mean precipitable water in mm., August. (After Tuller, "World Distribution of Precipitable Water.")

ticularly high relative humidities at this time of year. In general, the interiors of middle-latitude landmasses, particularly in western North America and central Asia, are occupied by air with low relative humidities comparable to those in the subtropical deserts of the Sahara and Australia.

Maps of precipitable water (Figures 7.7 and 7.8) indicate the actual amount of moisture in the air in terms of amounts of precipitation that would fall if all the moisture in the air were precipitated. Here it can be seen that the intertropical convergence zone, which shifts latitudinally with seasons, generally has greatest amounts of atmospheric moisture, and amounts decrease poleward, with lowest amounts in the winter hemisphere. Areas with consistently converging air streams crossing warm ocean surfaces show particularly high values, such as northeast India during August.

The consequences of these various distributions of atmospheric humidity will become evident in subsequent chapters when atmospheric stability, storm development, clouds, and precipitation are discussed.

8 ADIABATIC PROCESSES AND ATMOSPHERIC STABILITY

It has already been stated that temperature normally decreases with height at a fairly consistent rate throughout the troposphere. On the average this amounts to about 6.5°C per kilometer (3.5°F per thousand feet). One might question how such a situation can occur, with warm air below and cold air above. Since warm air is lighter than cold air it would seem logical that the warm air would tend to rise and the cold air to sink, thereby bringing about a stable situation that would resist motion. The answer lies in the vertical decrease in atmospheric pressure with height and the increase in potential energy of position above the earth's surface.

As the warm air rises, it comes under constantly decreasing atmospheric pressure, which allows it to expand. The work expended by the rising air to expand its volume and increase its potential energy above the earth's surface requires energy, which must be derived from some source. This comes from the internal energy (molecular activity) of the air itself, which reduces the molecular motion of the rising air, and hence, by definition, its temperature. This change of temperature without addition or subtraction of heat energy from external sources is known as an *adiabatic* process.

It can be demonstrated mathematically that the temperature of vertically moving air changes at the constant rate of approximately 10°C per kilometer (5.5°F per thousand feet). This is assuming the air is not saturated and no latent heat is being released by the process of condensation. This is known as the *dry adiabatic lapse rate.* Since this rate is greater than the average, or environmental, lapse rate mentioned before, it can be seen that under normal conditions the temperature of a rising parcel of air will decrease more rapidly than the temperature of the air around it. So the rising parcel everywhere is colder than its environment, hence more dense, and tends to settle back down to the level from which it came. Thus, the average lapse rate is a stable condition in spite of the fact that the temperature decreases with altitude.

Temperature changes in the free atmosphere over short periods of time are essentially adiabatic. Since air is a poor conductor of heat, each air parcel tends to retain its own thermal identity. Therefore any parcel of air that is moving vertically can be expected to change its temperature at the adiabatic rate, decreasing 10°C per kilometer as it rises, or increasing 10°C per kilometer as it descends.

Over the long run this is not so true, and some heat energy is exchanged from one parcel of air to another. Thus the average condition is not adiabatic. Normally, temperature decreases upward through the atmosphere at a rate of about

6.5°C per kilometer, and the atmosphere will probably not be experiencing much vertical motion. This is the lapse rate that is used to change recorded temperatures to equivalent sea level temperatures when constructing temperature charts for the world. The differences in elevation of land surfaces are reconciled by adding 6.5°C for every kilometer (3.5°F for every 1000 feet) above sea level. Thus, if a station at 2 kilometers elevation records a temperature of 10°C, the temperature would be recorded on a sea-level equivalent map as

$$10^\circ + (6.5 \times 2) = 23^\circ\text{C}.$$

However, if an air parcel with an initial temperature of 10°C should descend 2 kilometers to sea level, its resultant temperature would be

$$10^\circ + (10 \times 2) = 30^\circ\text{C}.$$

Thus, atmospheric subsidence greatly warms the air and typically creates a temperature inversion (increase of temperature with height) at the bottom of the subsiding layer.

Near the surface of the earth, temperature changes in the atmosphere are primarily nonadiabatic, or diabatic, because of the intensification of other transfer processes, such as radiation exchanges, conduction, and turbulence. Superadiabatic lapse rates are common in air immediately above the earth's surface when the insolation is intense and the surface is dry, since sensible heat is added to the surface air more rapidly than it can be mixed upward. Such conditions do not usually extend very far upward into the atmosphere, although over very hot desert surfaces they commonly extend hundreds or even several thousands of meters. Such conditions typically produce extreme turbulence in the lower troposphere.

Rising air will eventually cool to its dew point, after which condensation begins. The level at which this takes place is known as the *condensation level* in the atmosphere and is usually marked by the bases of cloud formations. Above this level latent heat of condensation will be released that will slow down the cooling caused by the lift and expansion of the air. Then the air cools at the *wet* (pseudo) *adiabatic lapse rate.* This is not a constant rate, since the amount of moisture condensed, and hence latent heat released, depends on the temperature of the air at which the process takes place. In the lower troposphere there is usually enough water vapor in the air to release a significant amount of latent heat when condensation is caused by rising air. But higher in the troposphere, the amount of water vapor in the air is so small that the latent heat release becomes insignificant even though the air is still saturated. Thus, in very warm air near sea level, the wet adiabatic lapse rate may be as little as 4°C per kilometer (2.2°F per 1000 feet), but in the middle troposphere the wet adiabatic lapse rate commonly approaches the dry adiabatic lapse rate. At a temperature of −40°C (−40°F), the wet adiabatic lapse rate is more than 9°C per kilometer (5°F per 1000 feet).

Atmospheric Stability

The stability characteristics of the atmosphere are very important, since they determine what kinds of vertical motions are likely to take place, and hence whether or not severe storms might develop. Absolutely stable air resists all vertical motion. Although it might be forced to rise by some external process, such as the wind blowing over a mountain range, it would never rise under its own volition, initiated only by energy contained within the air itself. Unstable air, on the other hand, would not resist motion, and once set in motion would continue that motion until changed by some outside force. Usually the state of the atmosphere is somewhere between these extremes. It might be potentially unstable, provided some outside force carried it through some process to a stage that crossed the threshold of instability within the atmosphere.

The stability state of the atmosphere is determined by the availability of energy sources that are functions of the temperature distribution with height and the water vapor content of the air, which represents an energy source in the form of latent heat. These two parameters (temperature and humidity) are observed through a vertical cross-section of much of the troposphere by a device known as a radiosonde—a small box filled with instruments to measure pressure, temperature, and humidity and a small radio transmitter to radio the information back to the ground as the box is carried aloft by a large hydrogen-filled balloon. The radio signals are received at the surface and translated into temperature and mixing ratio data for corresponding pressure levels. The upper winds may be measured at the same time by receiving the signals with directional antennas. The balloon normally rises to 15–20 kilometers, at which height it has expanded under reduced atmospheric pressure to its elastic limit, whereupon it bursts and falls back to earth. The radiosonde is protected from destruction by a small parachute that opens on descent.

The data thus gathered are used to construct (a) vertical profiles of temperature and humidity through the troposphere above given points on the earth's surface and (b) constant pressure charts that show heights, winds, and temperatures over large portions of the earth at standard pressure levels, such as the maps shown in Figures 6.13 to 6.17. They are also used to construct charts that illustrate the thickness of atmospheric layers between given pressure levels, which can indicate something about thermal winds and, over time, reveal areas of divergence and convergence, vertical motions, and changes in stability characteristics.

The vertical cross-sections of temperature and humidity above given points reveal the stability characteristics of the atmosphere. The degree of stability can best be illustrated by running through an example of the procedure used every day by weather forecasters utilizing some form of thermodynamic diagram. The diagram used here is the so-called "pseudo-adiabatic chart," which is composed of two principal coordinates, atmospheric pressure, represented by horizontal lines, and temperature, represented by vertical lines. Upon these two coordinates are superimposed three or four other sets of lines that are mathematically related to pressure and temperature (Figure 8.1).

The slanting light gray lines are known as the dry adiabatic lines. These slope at an angle that equals the dry adiabatic lapse rate, 10°C per kilometer (5.5°F per thousand feet). The slightly sloping darker-gray lines are the saturation mixing ratio lines. They show how many grams of moisture per kilogram of dry air are needed to saturate the air at any pressure and temperature shown on the graph. The dashed gray lines are the wet (pseudo) adiabatic lines representing the rate of change of temperature with height for saturated air that is moving upward and condensing moisture as it cools. The temperature then decreases less rapidly than the dry adiabatic rate, because the latent heat of condensation is being released into the air. At lower elevations (higher temperatures) the wet adiabatic lapse rate is much less than the dry adiabatic lapse rate (on the order of 4°C per kilometer) because of the great amount of water vapor being condensed. But at higher elevations (cooler temperatures) where saturation mixing ratios are low, the wet adiabatic lapse rate approaches the dry adiabatic lapse rate, as the latent heat of condensation becomes negligible. Thus, the wet adiabatic lines are not straight lines, but curve upward to the left on the graph. The sixth set of lines,

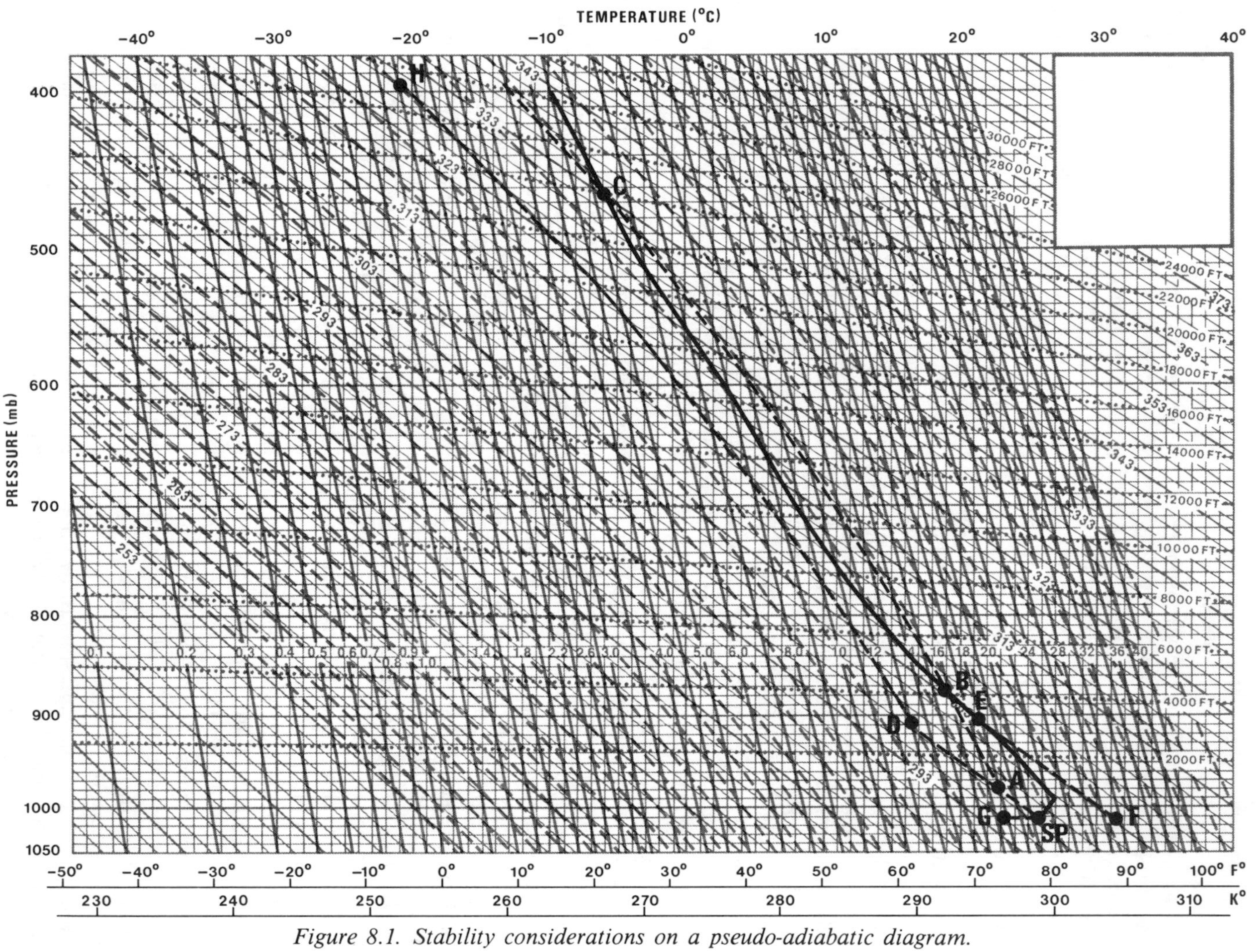

Figure 8.1. Stability considerations on a pseudo-adiabatic diagram.

the dotted gray lines that slope upward slightly toward the left, give a crude indication of the height above sea level according to standard pressure and temperature relationships.

The solid black line on the graph represents the actual lapse rate curve, at a given time over a given point, that has resulted from observations of pressures and temperatures at different levels by a radiosonde run. It can be seen that the surface temperature was 26°C (approximately 78°F) and there was a slight temperature inversion at the surface that extended upward for only a few hundred meters. Above the inversion for a short distance, the air had almost a dry adiabatic lapse rate, but then the lapse rate decreased and remained fairly consistent to the 400-millibar level, where the temperature fell to −10°C.

The moisture content was also recorded at each point, but for this first approximation of stability characteristics, only the mixing ratio at the surface point is needed. Let us assume initially that the actual mixing ratio measured at the surface point was 18 grams of moisture per kilogram of dry air. From this diagram relative humidities and dew points can easily be computed, because the equations that relate them to pressures and temperatures have already been represented graphically. It can be seen that the saturation mixing ratio for the surface point is approximately 21 grams per kilogram. Since it is known that the actual mixing ratio was 18, one simply divides 18 by 21 to get the relative humidity, which is approximately 86 percent. To get the dew point, one moves the surface point to the left at the constant pressure of 1010 millibars until the 18-gram saturation mixing ratio line is crossed, and then reads directly below, point G on the adiabatic diagram. It can be seen that the dew point is about 23.4°C, or almost 74°F. Since all three temperature scales are given on the graph, it is easy to convert from one temperature to another.

To determine the stability characteristics of the air under the process of mechanical lifting, raise the surface point through the rest of the air and observe how its temperature changes. Initially, as long as the air is not saturated, the temperature change parallels the dry adiabatic lines. But at point A on the graph, the temperature path of the rising surface parcel crosses the 18-gram line and the rising air becomes saturated. Above this point the temperature decreases at the wet adiabatic rate, as shown by the dashed line ABC. At point B the temperature path of the rising parcel crosses the originally plotted lapse rate curve. This means that the temperature of the rising parcel becomes warmer than the surrounding air, and therefore no more outside force is necessary to keep the surface parcel moving upward: above point B, which is known as the level of free convection, the rising air is buoyed up by the fact that it is warmer, hence lighter, than its environment at every level. Hence, the air no longer resists upward motion and becomes unstable. The motivating energy is the latent heat of condensation that is constantly being released as the air rises. The air, thus, is conditionally unstable. On the condition that some mechanical force (mountain range, front) is provided to lift the air to point B, the air becomes unstable, and a thunderstorm will likely develop.

The temperature path of the rising parcel recrosses the plotted lapse rate curve at point C, and once again the rising parcel becomes colder than its surroundings and resists upward motion. Assuming no outside force is available at this height, upward motion will cease at point C, and this will mark the top of cumulus-type clouds. The temperature path of the rising parcel recrosses the plotted lapse rate curve because the wet adiabatic lines are curved upward to the left. If the capacity of the air to hold moisture decreased at a constant rate with decreasing temperature, such would not happen, and theoretically there would be no limit to the upward motion of the surface air, hence no upper limit to the buildup of clouds. But since the capacity of the air to hold moisture decreases at a decreasing rate with decreasing temperature, the amount of moisture, and hence the latent heat release, eventually becomes negligible with cold temperatures, and an upper limit to instability is thus set.

To indicate the significance of moisture content to stability, let us assume that the surface air has a mixing ratio of only 13 grams per kilogram. As the surface air parcel rises the temperature will again initially change at the dry adiabatic rate. But this time the temperature path of the rising parcel will parallel the dry adiabatic lines up to point D, where the 13-gram line crosses. The condensation level, and hence the base of the cloud formation, has now lifted from A to D. Above D the air will condense moisture and the temperature will change at the wet adiabatic rate, line DH. The temperature path of the rising parcel never crosses the original plotted lapse rate curve. Hence, the air resists motion all the way up and never becomes unstable; the air is now absolutely stable. It could still be forced to rise by some outside mechanism, such as a mountain range or a front, when clouds would form above the condensation level, and perhaps even precipitation would take place; but the air will always remain stable, and the cloud type now will be stratus rather than cumulus.

The previous two examples assume that a mechanical lift initiates the upward motion of the surface air. Nevertheless, strong heating at the surface may cause convective activity to extend from the surface upward to the level of free convection, above which the air continues on an unstable course. To test for that, mark the point where the mixing ratio of the surface air crosses the plotted lapse rate curve. In our initial example, the 18-gram line crosses the plotted curve at point E. This is the condensation level for the surface air if lifted by convective currents only. It also now becomes the level of free convection, since above that point the air will follow a temperature path to the right of the plotted lapse rate curve. To see how high the surface temperature would have to get for the air to be lifted to point E, drop a line parallel to the dry adiabatic line to the surface pressure at point F. An adiabatic lapse rate has now been hypothetically created from F to E along which any heated air can move without resistance, because it will always be at the same temperature as its surroundings at each level in its rising path. At point E condensation begins, the latent heat of condensation is added to the air, and the temperature path of the rising parcel then parallels the wet adiabatic curves, which puts it to the right of all the other curves on the graph (not shown).

Thus, this air could be made unstable if the surface temperature were raised to about 32°C (89°F). With lift due to surface heating, the condensation level is higher than it was with mechanical lift, because the air is first heated and then cooled adiabatically as convective currents set in. But the condensation level now is also the level of free convection, which is lower in the atmosphere than it was with mechanical lift. Therefore, instability begins at a lower level and extends to a higher level, since the temperature path of the rising parcel is now farther to the right on the diagram. This is the kind of instability that may occur on a hot summer day over southeastern United States in warm, moist air coming from the Gulf of Mexico. Typically, this kind of air produces many scattered thunderstorms throughout humid summer afternoons.

Initially stable air may be made unstable by lifting a thick layer that has a large humidity lapse rate. The wet air in the lower part of the layer may reach its condensation level before the drier upper part of the layer does, which means that with continued lift the lower air cools at the wet adiabatic rate, while the upper air is still cooling at the dry adiabatic rate. In essence, the lower air is being warmed relative to the upper air by the release of the latent heat of condensation. Such a condition is known as *convective instability* and can be responsible for severe thunderstorm turbulence and tornado formation in areas such as the Great Plains of North America where, during summer, dry air from the southwestern deserts frequently overruns moist air from the Gulf of Mexico.

Lifting of a thick layer of air will decrease stability even if the moisture content does not vary throughout the layer. Since the atmosphere becomes less dense with height, the top of a column of air must rise more than the bottom to experience the same pressure decrease. Since adiabatic temperature changes are directly related to changes in height, not pressure, the top of the column will cool more than the bottom. Thus the lapse rate will be steepened. The opposite is true with descending (subsiding) air. Active subsidence typically produces a temperature inversion, hence extreme stability, at the base of the subsiding column.

9 | AIR MASSES AND FRONTS

Air Masses

Since the air in the lower troposphere outside the equatorial doldrums circulates primarily around cells quasi-parallel to isobars, at least above the friction layer, such cells become persistent through time, and the air within them becomes thoroughly mixed and homogeneous in the horizontal plane and distinct from adjacent circulations with which there is little air exchange. This is particularly true in high pressure cells, where friction at the surface causes the air to diverge gradually away from the center of the cell, thus inducing subsidence in the vertical dimension and a spreading out of similar air over a large area at the surface. On the other hand, low pressure cells at the surface are areas of convergence of different kinds of air from various points on the periphery of the circulation system, and therefore usually do not lead to homogeneity of air over a wide area.

Some cells of circulation are persistent in their configuration and location. Some are even permanent, such as the subtropical high pressure cells over the oceans of both hemispheres. Others are more transitory, retaining their characteristics only for a number of days, or weeks at most; and during that time they may move long distances and come into contact with varying surfaces. Such shifting of positions brings about exchanges of heat and moisture between the surface and the air, which ultimately modify the air significantly.

Air within a circulation cell that shows a relatively high degree of homogeneity over a broad horizontal area is known as an air mass. It is distinguishable by its temperature, moisture, and stability characteristics, which, in combination with its temperature relative to the earth's surface, produce predictable associations of weather. Of course, if a circulation system occupies an area hundreds, or even thousands, of kilometers in radius, the air will not be completely homogeneous from one side to another. Temperatures will be higher on the equatorial edge of the air mass than on the poleward edge, and moisture and stability characteristics may vary also. Nonetheless, there is much less variation within an air mass than between it and adjacent air masses.

Over very broad circulation systems, such as the subtropical highs over the oceans, very different types of air may develop at opposite ends. For example, the air on the eastern end of the North Atlantic subtropical high has circulated for thousands of kilometers across the middle and northern portions of the North Atlantic over relatively cool water and has then turned southward around the eastern end of the high, where it has undergone subsidence that has produced a strong temperature inversion at or near the surface. Such conditions produce extreme atmospheric stability, rainlessness, cool surface temperatures, near atmospheric saturation, and much fog and low stratus clouds. As the air turns westward it circulates across thousands of kilometers of very warm water, which injects heat and moisture into the air and lifts the inversion layer on its westward course. As it turns northward around the western end of the high, the air is very warm with a high mixing ratio and unstable or conditionally unstable characteristics that induce much vertical turbulence, cumulus cloud buildup, and thundershowers. The surface air is warm and muggy, with high temperatures combined with high relative humidities. Since the subtropical highs tilt eastward and equatorward, the strongest subsidence is on the eastern-equatorward periphery of the cell, and the weakest subsidence or uplift is on the western-poleward side of the cell.

Source Regions

Two conditions are necessary for air masses to form: (a) a large, relatively homogeneous portion of the earth's surface, and (b) air stagnation or long trajectory over that surface sufficient for a significant thickness of the air to take on the characteristics of the surface. Exchanges of heat and moisture must take place between the earth's surface and the air above so equilibrium between the two can be reached. This takes time—at least several days, and perhaps weeks. Portions of the earth's surface that satisfy these two conditions are known as air mass source regions. Examples are the oceanic areas dominated by the subtropical high pressure cells and certain land areas. Particularly in winter, when they do not receive much sunlight, large portions of the land in high latitudes are covered with snow that presents a fairly homogeneous albedo and radiation emission. Certain desert areas may act as source regions during summer months.

Air may tend to stagnate over certain areas of the earth's surface because of the character of the general circulation of the atmosphere, which forms nodes of circulation in some latitudes and in some land-sea and topographic configurations. Stagnation may take place in other areas because of interior locations between mountain ranges, and so forth. Examples are the intermontane region in southwestern United States and northern Mexico which develops very hot dry air during summer, and the similar Central Asian area east of the Caspian Sea.

During winter the Mackenzie River valley in northwestern Canada receives streams of air from various directions off the Arctic and Pacific oceans, after which the air stagnates and cools over the snow-covered land surface. This process ultimately results in a cold, stable air mass with a low mixing ratio that may extend two or three kilometers into the air and produce high surface pressures. When the air pressure has built up enough, the air mass typically bulges southward, and a cell buds off from the main air mass to move out of the source region southeastward into central United States. Such a discrete cell may separate from the main body of an air mass and move en masse thousands of kilometers downstream, where it slowly modifies over continuously changing surfaces and eventually loses its identity.

Sometimes broad streams of air flood areas downstream with a given type of air without forming discrete circulation cells apart from the main bodies of parent air masses in their source regions. Such are the warm, moist, unstable air masses that move poleward along the western edges of subtropical highs and the relatively cool, stable air masses that move equatorward along their eastern edges.

Air masses are designated according to their source regions as being maritime or continental in origin, and equatorial, tropical, polar, or arctic (or antarctic). Thus, an air mass may be designated as mT (maritime tropical), cT (continental tropical), mP (maritime polar), cP (continental polar), and so forth. In the middle latitudes where the contrasts between adjacent air masses are greatest, the primary air masses are mT, cP, and mP, with cT significant in dry land areas. Such designations have become rather standard, although some of them are sort of misnomers. For instance, the polar air masses do not originate at the poles, but in the subpolar and higher middle latitudes, and are typically found throughout the middle latitudes, while the polar areas are dominated by arctic or antarctic air. Soviet climatologists refer to temperate air masses in the middle latitudes rather than to polar air masses, but this is an ambiguous designation also, since the word "temperate" has been used to encompass a multitude of conditions.

Weather Associations

Air masses are typically associated with certain types of weather that correspond to their water content, stability characteristics, and temperatures relative to the surfaces over which they are traveling. For instance, cP air during midwinter in its source region will have taken on the cold temperatures of the underlying snow-covered surface, which will have led to low mixing ratio content, the formation of a temperature inversion based at the surface, and extreme atmospheric stability. In spite of the fact that the surface temperature inversion with its resultant stability essentially rules out convectional mixing, the cooling effects from the surface continue to extend upward into the air, because of the peculiar reflective and radiative properties of snow. Fresh snow reflects at least 90 percent of incident short-wave radiation, which is already minimal at this time of year at high latitudes. And it radiates almost perfectly (as a black body) at longer wavelengths in the infrared, where all the terrestrial radiation is because of cold temperatures. Thus, the snow assures that little heat is received and much heat is lost.

There is net radiation upward until very low surface temperatures are reached (commonly −50°–−70°C), despite the fact that the air one or two kilometers above the surface is commonly 20° to 30°C warmer than the snow surface. The snow radiates heat upward, and the warm inversion layer of air radiates heat downward into the colder air below. Because of its colder temperature, the snow radiates heat less intently at any wavelength than does the warm air above. But the snow radiates heat across a full spectrum of wavelengths according to its temperature, while the atmosphere, being made up of a mixture of gases, radiates heat only at certain wavelengths that correspond to the radiation and absorption characteristics of the gases in the air. Therefore, the spectrum of radiation coming from the upper air is far from complete, and the sum total of radiation upward from the snow surface is greater until the temperature of the snow surface reaches very low values. After that, an equilibrium is reached, and surface air temperatures remain essentially constant until some advection of different kinds of air upsets the situation, when the process is repeated to reestablish an equilibrium temperature.

Such equilibrium temperatures are generally reached initially in December in places such as northwestern Canada and Siberia, and from then on the temperature varies little until it begins to warm up again with increased insolation as the spring equinox is approached. Thus, the inhabitants of Siberia talk about their "coreless" winters, which do not have temperatures arranged symmetrically about a minimum point in January or February but, instead, consistent temperatures with some nonperiodic variations from December, or even late November, through early March.

When cP air moves out of its source region in northwestern Canada southeastward into the United States it moves across a surface with (usually) constantly increasing temperatures that are everywhere warmer than the surface temperature of the air. Thus, it is classified as a "k" (cold) air mass. The term "cold" is relative, meaning simply that the air mass is colder than the surface over which it is traveling. The air mass will be modified as it moves over the warmer surface, receiving heat, and probably moisture, from the underlying surface as it goes. The heating at the surface will cause convective currents to set up in the lower air, which cause a vigorous mixing of moisture and pollutants upward, making good surface visibility.

If the air is much colder than the surface, the convection may be vigorous enough to produce small cumulus clouds with well defined bases at a common level—the condensation level of the surface air—and tops that are limited by the high degree of stability of the upper air, which has not yet been modified from its cP origin. Generally such cumulus are no more than a few hundred meters thick, with much blue sky in between individual clouds. They have become known as "cumulus of fair weather," although this may seem a contradiction in terms, since if there are clouds the weather is not entirely fair. The connotation is that these cumulus are not going to result in much because the air a few hundred meters above the ground is very stable. They are produced entirely by daytime heating at the surface and will dissipate soon after sundown. Such is the weather usually associated with k air masses—crisp, clear surface visibility, and cumulus of fair weather.

Occasionally, if the temperature differential between a k air mass and the surface over which it is traveling is very great, and if enough moisture is available, the cumulus buildup may reach showery proportions, and the weather for the day will be marked by sharp bursts of showers that last only a few minutes at a time, interspersed by clear, sunny weather in between. One short shower after another may continue through much of the day. This is particularly common in the U.S. Midwest during spring, when the cP air coming south from Canada is still cold and the ground surface over which it is moving has already warmed up greatly. If the temperature is cold enough there may be snow showers, but this is more common in the Midwest during the onset of winter, when the first cold blasts of Canadian air move over a still relatively warm surface. Graupel (snow pellets) are common occurrences with such weather. The showery weather, whether it is rain or snow, is known as instability showers, since it is caused by low-level instability due to surface heating. Such weather is commonly associated with the leading edge of a continental polar outbreak in midwestern United States after the passage of a cold front associated with a cyclonic storm.

Figure 9.1 shows a typical lapse rate curve for a cPk air mass moving southward out of Canada into midwestern United States. Initially the entire air mass is quite stable, as illustrated by line D-C and above. but as the air moves southward over warmer land the surface air heats to a temperature of 10°C, as shown at point A, which establishes a dry adiabatic lapse rate upward to about 850 millibars (1.6 kilometers above the surface). The surface air can now move freely throughout this lowest layer, with moving air parcels remaining at the same temperature as their surroundings at each level. At point B the surface air, with a mixing ratio of 6 grams per kilogram, becomes saturated, and above that point it cools at the wet adiabatic rate. This causes an unstable layer between points B and C, above which the rising air once again becomes colder than its surroundings. Thus, cumulus of fair weather are formed with their bases around 1500 feet (point B) and their tops around 7500 feet (point C). They cannot grow any taller because the upper air is still quite stable, not having been modified by processes going on at the surface. Often a small letter "s" is added to the symbol cPk to designate that the air is still stable aloft (cPks).

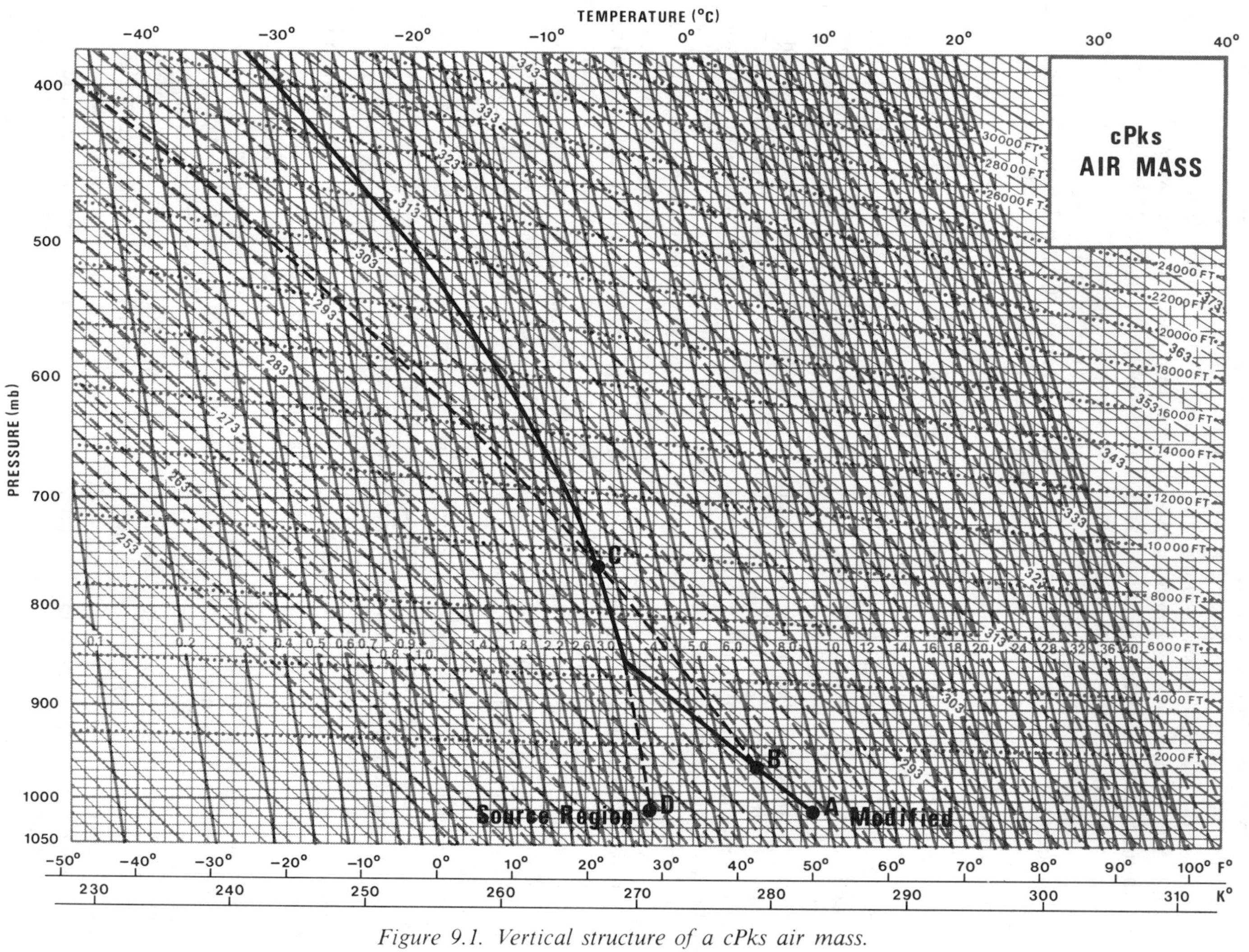

Figure 9.1. Vertical structure of a cPks air mass.

Maritime tropical air moving up the Mississippi Valley from the Gulf of Mexico is usually a "w" (warm) air mass, particularly during cooler seasons. Modification takes place by heat loss from the warmer air to the cooler surface, thereby stabilizing and stratifying the surface air. This often leads to poor surface visibility, haze, and even fog, or large amounts of dew, depending upon the motion of the air. The mT air is commonly warm, humid, and perhaps conditionally unstable, but the cooling at the surface stabilizes the surface air and rules out convective activity.

During summer, maritime tropical air moving northward up the Mississippi Valley from the Gulf of Mexico may actually be a "k" air mass, even though its temperatures may be more than 30°C (86°F). The air over the Gulf at that time of year typically has surface temperatures of 86°F but the land in the Mississippi Valley, perhaps even as far north as Wisconsin and Minnesota, may be between 90° and 100°F, or even above 100°. Thus, as the air moves northward from the Gulf it is heated at the surface; this process may easily set off convectional thunderstorms since the air commonly is conditionally unstable and has such high relative humidity that it does not have to rise very high to reach the condensation level, after which the latent heat released is usually enough to buoy the air on upward.

Figure 9.2 illustrates a typical maritime tropical air mass moving northward from the Gulf of Mexico up the Mississippi Valley. Initially, over the Gulf, its surface temperature is 30°C (86°F), as shown at point A. In late spring or early summer as it moves northward over the land it encounters cooler temperatures, and the surface air temperature cools to 26°C, illustrated by point B, which is below the dew point, assuming a surface mixing ratio of 22 grams per kilogram. This produces a surface temperature inversion, surface stability, and poor visibility, perhaps fog.

Later in summer, the same air mass might prove to be an mTk air mass as it encounters warmer temperatures over the land as it moves northward. The surface temperature might warm to 37°C (98°F), which would then produce a dry adiabatic lapse rate from point C up to point D, the condensation level where the 22 grams of moisture in the air saturates the air. Point D also becomes the level of free convection, since the rising air (dashed line) above that point everywhere is warmer than its surroundings. Since the initial lapse rate curve slopes at an angle intermediate between the wet adiabatic lines and the dry adiabatic lines, the air mass is conditionally unstable, and surface heating during a hot summer afternoon may be enough to trigger thunderstorms. If the surface temperature reaches 37°C or above, this will happen.

In the first instance, the cooler land stabilized the lower 2000 feet of the air mass, but the air still remained conditionally unstable aloft. This conditional instability was released in the second instance when the warmer land heated the surface air. Thus, the same air mass can be either a k air mass or a w air mass, depending upon its relation to the temperature at the surface, and very different types of weather will result. In either case, the small letter "u" might be added behind the mTk or the mTw, to indicate unstable conditions aloft.

Continental tropical air, such as that which forms in southwestern United States during summer and in extensive subtropical deserts such as the Sahara, has exceedingly high surface temperatures and low relative humidities. The great heating at the surface causes strong convectional activity up to perhaps two or three kilometers above the ground, but above that extensions of subtropical highs generally predominate and form temperature inversions that suppress convection. Since the surface air has such low relative humidity, the convective activity usually is not enough to raise the surface air to its condensation level, and therefore the air remains cloudless even

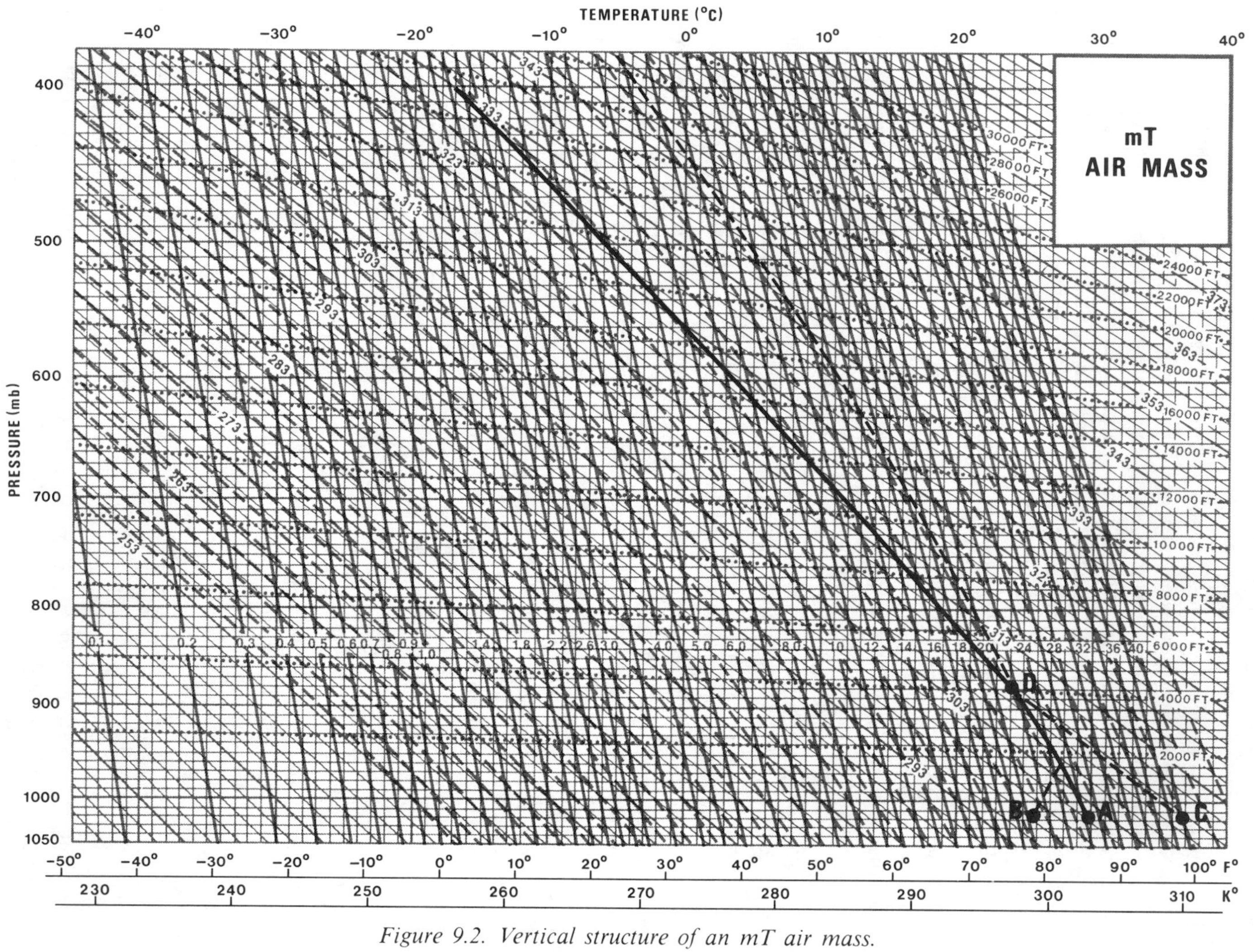

Figure 9.2. Vertical structure of an mT air mass.

though the first two or three kilometers of air are extremely turbulent. cT air is usually a summer phenomenon only.

Maritime polar air moving around the poleward edges of the subtropical high pressure cells is typically cool, moist, and stable as it moves equatorward along the western coasts of North America, Europe, and the Southern Hemisphere continents in the middle latitudes. If it penetrates far enough inland, as it does over Europe, it may modify into a fairly unstable air mass, with thundershower activity during summer, or become colder, drier, and more stable, similar to cP air during winter. mP air is usually a k air mass in summer and a w air mass in winter. Such air does not penetrate far inland on the North or South American continents, since in both cases high mountains rim the Pacific.

Maritime polar air dominates the west coasts of these continents much of the year and brings much stratus clouds and fog to the coastal areas, but thunder is seldom heard on the west coast of North America. This may change as the air rises up the windward slopes of mountains, such as the Cascades and the Sierra Nevada in Washington, Oregon, and California, where the air may eventually reach its level of free convection and thereby spawn thunderstorms, the lightning from which frequently sets off forest fires. But by the time the mP air has crossed such mountain ranges and descended their leeward slopes, it is so modified that it can no longer be identified as maritime polar air. The greatest stability of maritime polar air is found in the equatorward extremities of the eastern ends of the subtropical highs adjacent to such coasts as southern and Baja California, Peru, and northwestern and southwestern Africa.

Fronts

Although certain types of weather are associated with air masses, as described above, most weather (other than clear weather) is associated with boundary zones between air masses. Since most identifiable air masses are associated with high pressure cells, the typical weather within air masses, particularly in their source regions before they begin to be modified by changing temperatures at the surface, is clear, calm weather associated with subsiding air. Although scattered showers may be associated with k air masses or fog with w air masses, these are generally mild weather phenomena compared to the well-organized and intense cloud systems, turbulence, and heavy, prolonged precipitation that may be associated with boundary surfaces between air masses where warmer air is forced to rise over colder air. The colder air mass acts as a wedge over which the warmer air rises, expands, and cools adiabatically, which reduces its capacity to hold moisture to the point where condensation must take place, clouds form, and precipitation falls. Such activity generally produces broad areas of continuous overcast skies and precipitation, and if the warm air is conditionally unstable it may also produce extreme turbulence manifested by thunderstorm activity. Because most hazardous weather is found along these boundary zones rather than within the air masses themselves, weather forecasters generally concentrate most of their attention on the boundary zones, particularly in the middle latitudes where air mass contrasts are greatest and, hence, fronts are most intense.

These boundary zones have become known as fronts because their advances and retreats between air masses reminded the Scandinavian meteorologists of the day-to-day shifts in the battle fronts in western Europe during World War I, at the time when the air mass and frontal theory of the atmosphere was being developed. On a surface weather map fronts are shown simply as lines, but it must be remembered that they are zones, usually several tens of kilometers wide, and they are curved surfaces that extend upward into the atmosphere. The line on the surface weather map represents only the intersection of the frontal surface with the ground.

Distribution

There are three main frontal systems on earth. Their approximate mean positions for January and July are shown on Figures 9.3 and 9.4, along with the air mass types they separate. They shift seasonally, along with the wind and pressure belts, as a result of the seasonal shift of insolation across the face of the earth. And within seasons they shift drastically, so that any daily map might look quite different from these average maps.

1. The first of the three frontal zones is the *intertropical front*, or better termed the *intertropical convergence zone* (ITCZ), since it is not a very distinct front in terms of air mass contrasts across it. It is a broad zone of convergence between the northeast trade winds in the Northern Hemisphere and the southeast trade winds in the Southern Hemisphere. Both of these wind systems carry maritime tropical unstable air masses, so there is little contrast in temperature, humidity, and stability conditions across the front. But because both air masses are warm, moist, and unstable, they have greater capabilities for producing clouds and precipitation, and copious amounts of precipitation fall along this front more than any other. The ITCZ is generally in an area of little surface observational record, with much of the area over oceans, so not much is known about its structure. It is often a diffuse area of convergence with little discernible frontal surface. Probably much of the time the converging surface air is rising approximately vertically without any perceptible frontal slope. In these tropical regions of little surface weather information, the front is often located on daily weather maps simply by the associated cloud bands and precipitation, observed from satellites which, of course, is the reverse of the intent for locating fronts on maps—to understand and predict the distributions of clouds and precipitation.

As can be seen in Figures 9.3 and 9.4, the convergence zone is generally continuous over the oceans and fairly constant in its location, particularly in the Pacific and Atlantic oceans. In the Indian Ocean its latitudinal position fluctuates widely by season, because of the unusual development of the monsoons in this part of the world. During summer the intertropical convergence zone is pushed north of the Tropic of Cancer in northern India, Indonesia, and southern China by the strong southwesterly monsoon winds that blow into thermal lows over the Asian continent at that time of year. The convergence zone is broken across equatorial South America by the high Andes. In both South America and south Africa during the Southern Hemisphere summer (January), the convergence zone tends to become oriented more north-south than east-west in the interiors of these land masses. Over the Pacific and Atlantic oceans, the ITCZ tends to lie north of the geographical equator at all times because of the generally stronger atmospheric circulation in the Southern Hemisphere.

2. The second set of fronts is the *polar front* in each hemisphere. This again is a misnomer, since these fronts are in the middle latitudes rather than in the polar regions. They are by far the most distinct fronts in terms of air mass contrasts. They generally separate cP air from mT air, particularly in the continental areas of the middle latitudes of the Northern Hemisphere. In oceanic areas they sometimes separate one variant of mP air from another variant of mP air. This may also occur over land areas where maritime polar air has free access, such as in western Europe, where the polar front typically separates fresh mP air on the west from a modified form of mP air on the east. The polar front may also separate some variant of mP air

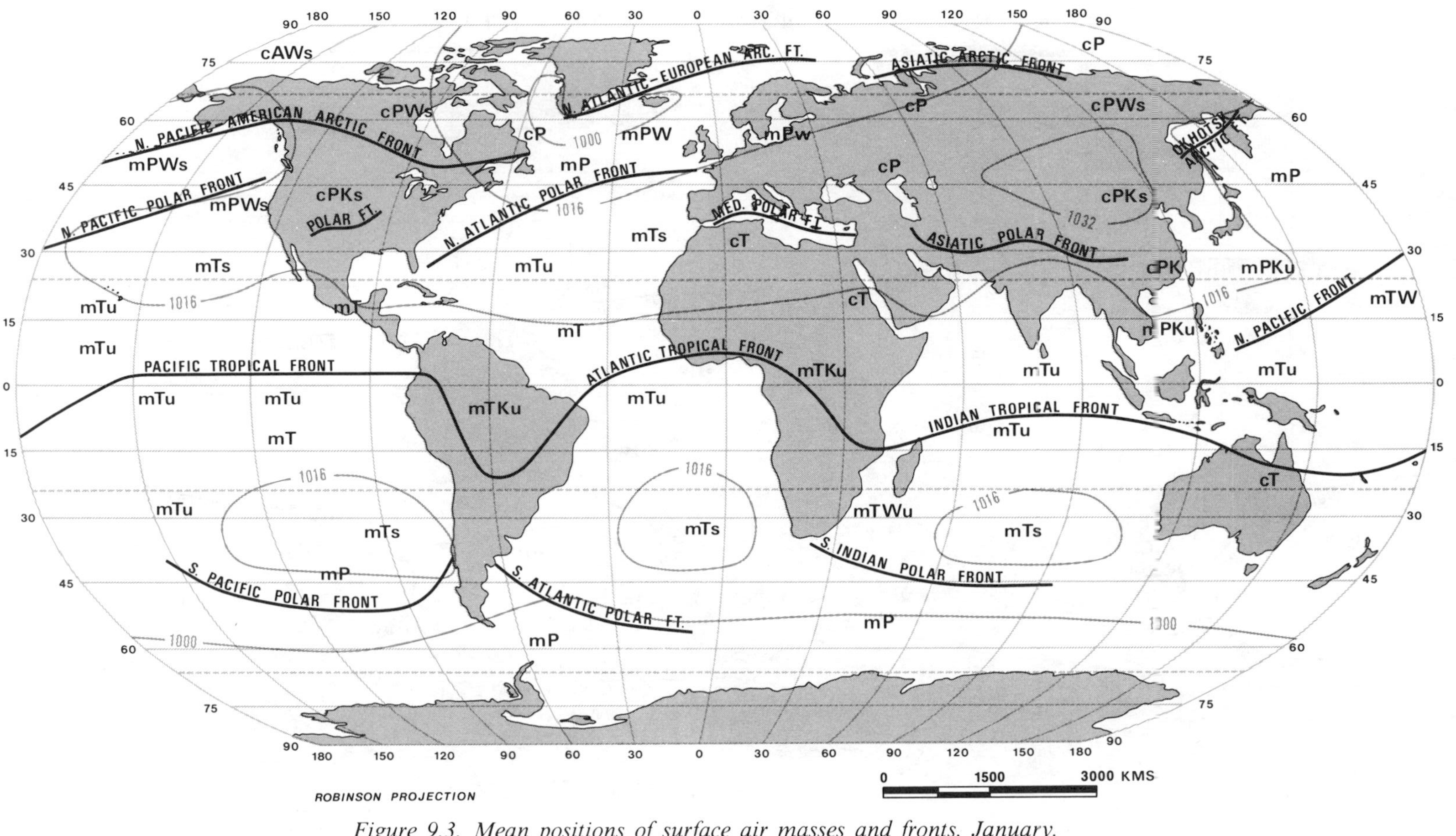

Figure 9.3. Mean positions of surface air masses and fronts, January.

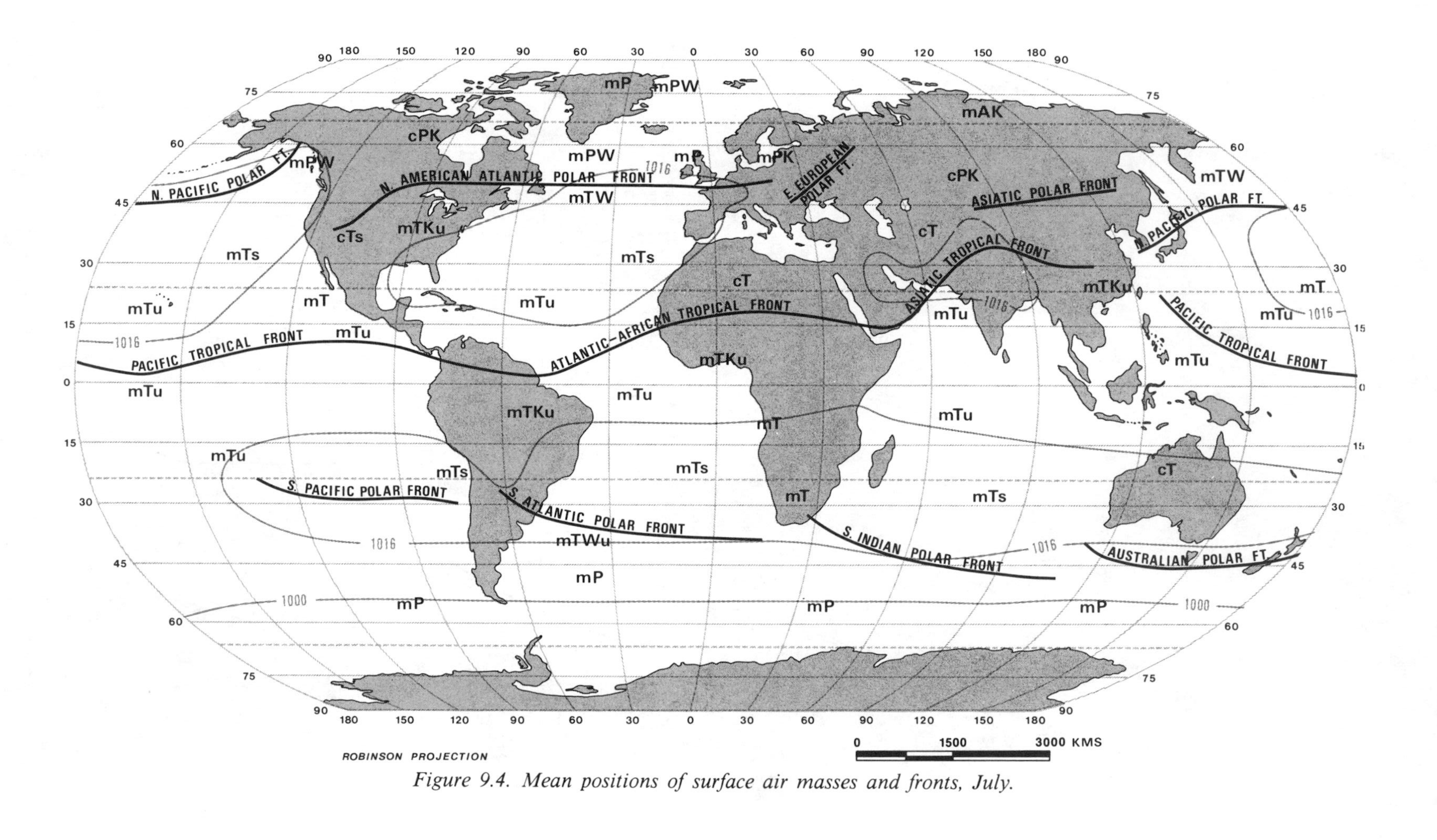

Figure 9.4. Mean positions of surface air masses and fronts, July.

from some variant of mT air. This usually occurs when the polar front moves into lower latitudes, almost always during winter when, of course, the mT air is not very classical tropical air.

The polar front is very discontinuous and distorted in its orientation. Its position fluctuates latitudinally much more than the intertropical convergence zone; at any season it may be found anywhere in North America from southern Canada southward across the entire United States into the Gulf of Mexico and Central America. The polar front's presence is much less felt in the interior of Asia. In the Southern Hemisphere it is more continuous around the earth, since there is little land in the middle latitudes. But the air mass contrasts across it are usually less than in the Northern Hemisphere, since all air masses associated with it are usually some form of maritime polar. Nevertheless, since all the air masses are of a maritime origin, the polar front in the Southern Hemisphere spawns much storminess, cloudiness, and precipitation.

3. The third set of fronts includes the *Arctic and Antarctic fronts* in the two hemispheres. They usually separate some variant of continental polar or arctic air from some variant of maritime polar or arctic air. They are very discontinuous and difficult to locate. Like the intertropical convergence zone, they frequently lie in uninhabited areas where little information on surface air pressures and winds is available. But unlike the ITCZ, the Arctic front does not produce much precipitation, since the air masses associated with it always have low temperatures and therefore little capacity for holding moisture. Segments of the Arctic front may spawn cyclonic storms that may produce rather frequent precipitation, usually in the form of snow, but never very much in amount. Segments of the Arctic and Antarctic fronts usually form somewhere near the coasts of northern North America, Eurasia, and the Antarctic. In those positions they lie in regions of contrasting surface temperatures that in summer are considerably warmer over the land than over the adjacent seas, and in winter are colder over the land than over the adjacent seas. Therefore these coastal areas are quite windy and may experience numerous blizzards.

Of all the fronts, the polar front has attracted by far the most attention, since it lies in the middle latitudes and strongly affects many of the more-developed countries of the world. The typical characteristics of fronts described in the following section relate specifically to the polar front, particularly in North America.

Characteristics

Fronts are labeled "cold" or "warm" according to the direction they are moving. If the cold air is advancing at the surface at the expense of the warm air, then the front is a cold front; if the warm air is advancing at the expense of the cold air, it is a warm front. If the front is stalled or, more likely, oscillating back and forth with little net change in position, it is called a stationary front. In any case, the warm air is being forced to rise over the cold air. The warm air is cooled adiabatically by the lift, and the resultant condensation forms, cloud formations, turbulence, and precipitation types and amounts relate primarily to the characteristics of the warm air mass. They also relate somewhat to the vertical structure of the front and the atmospheric circulation associated with it.

Although the type of front is determined by the direction of motion, each type of front tends to take on a certain form and associated weather phenomena as a result of the characteristics of the motion taking place. A cold front tends to have an initial steep slope that tapers rapidly aloft and usually reaches heights of no more than three kilometers, while a warm front tends to have a more-constant slope and eventually extends to considerably greater heights. The difference in the initial slopes of the two fronts may be due largely to frictional effects between the air and the ground. Since the cold air under a cold front is pushing forward, it is retarded by friction with the ground, while the cold air under a warm front is being dragged back, and friction with the ground tends to elongate its surface extent.

Slopes of fronts are usually no more than 1 or 2 percent, if that, which would be impossible to show in a small diagram drawn accurately to scale. Thus, in the next chapter, figures showing vertical cross sections of fronts will greatly exaggerate the vertical dimension with respect to the horizontal. Although the air rising up a frontal surface may move 100 kilometers or more horizontally while moving only one kilometer vertically, vertical motion is the significant one, since it produces adiabatic cooling of the warm air and resultant condensation and associated phenomena.

10 FLOW PERTURBATIONS IN THE MIDDLE LATITUDES

Waves

A frontal surface is a zone of relatively rapid change in atmospheric density. Although it is not an absolute discontinuity, such as the interface between the atmosphere and a standing water body, it nevertheless acts similarly in producing wave motion. Just as the wind whips up waves on the ocean surface, so also do differential air movements along frontal surfaces within the atmosphere. But since most of the air motion is horizontal, the waves created within the atmosphere along frontal surfaces tend to have much greater horizontal dimensions than vertical ones.

Figure 10.1 illustrates the life cycle of wave formation along a frontal surface in the Northern Hemisphere. Initially, a straight front separates cold air to the north from warm air to the south. In addition to the temperature contrasts and perhaps differences in other characteristics, such as humidity, the front separates two opposing air flows that tend to set up waves along the frontal surface. Around the apex of the wave, a counterclockwise flow develops that serves as a sort of third cog to take up the friction between the opposing air flows, as shown in portion C of Figure 10.1. The right half of the front now moves as a warm front, with the warm air advancing at the expense of the cold air, while the left half of the front moves as a cold front with the cold air advancing. As the counterclockwise circulation intensifies, the wave increases in magnitude and shortens in length while the cold front catches up with the warm front (D).

In part E an occlusion process begins, with one front riding up the slope of the other front, which squeezes the warm air completely off the ground in the apex of the wave. The occlusion produces maximum uplift of the warm air and thereby creates the most intense stage of the storm; but the occluded stage is also the beginning of the end of the wave formation, since the air mass contrast across the front at the ground tends to be destroyed by the occlusion process. Cold air is now on both sides of the front at the surface.

Either side can be the colder, depending upon the situation. Often the western side is colder because that is the region of most active intrusion of fresh air from the cold air mass, the area where the cold air has had the least amount of time to be modified by contact with the ground. In certain instances, however, the air on the eastern side of the front may be colder because it is typically farther poleward in association with waves that are moving from southwest to northeast. If the air to the west under the cold front is colder than the air to the east under the warm front, the warm front rides up the slope of the cold front and the occlusion is a cold-front-type occlusion,

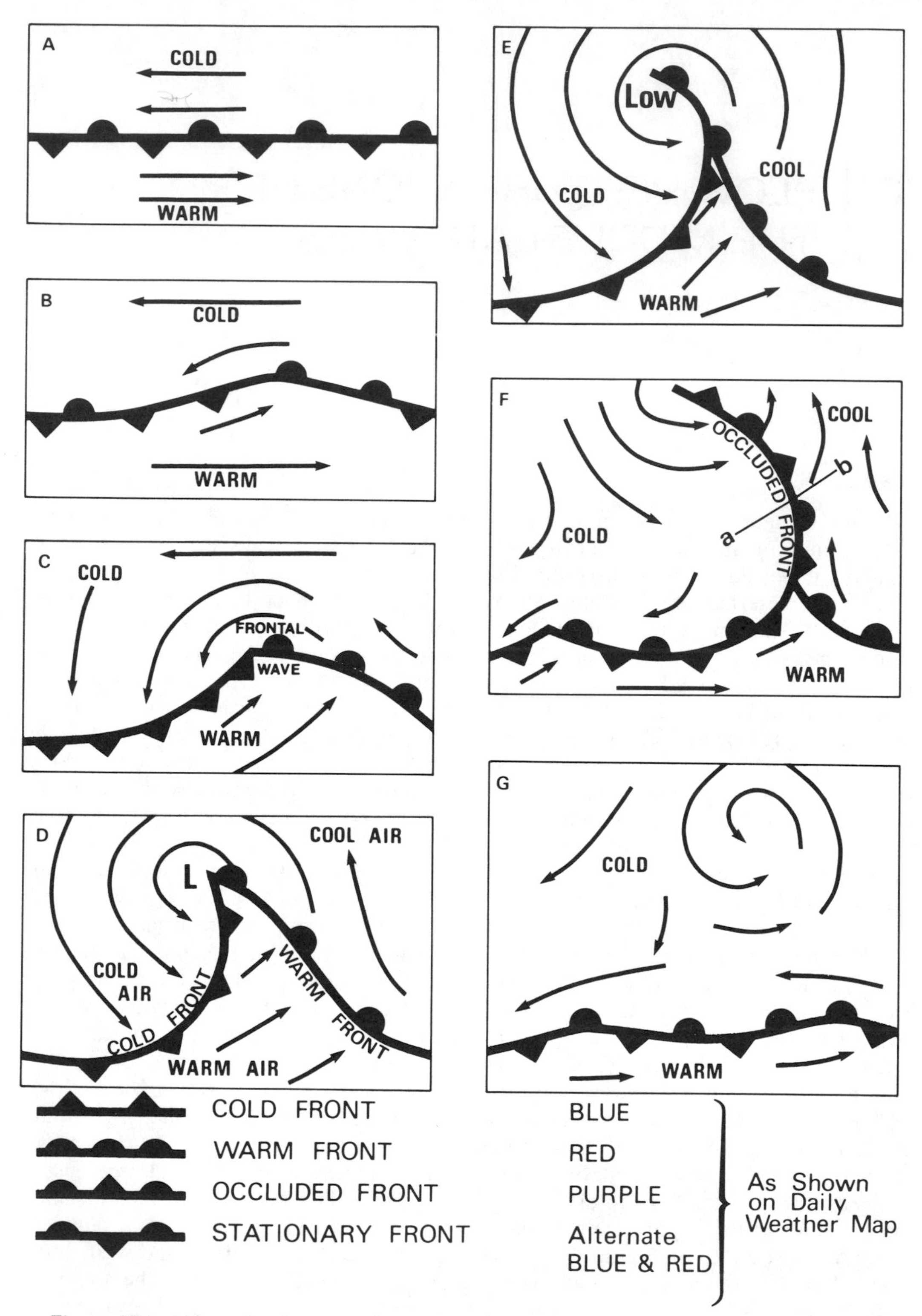

Figure 10.1. Life cycle of a wave formation on a middle-latitude front in the Northern Hemisphere.

since the cold front stays on the ground. Figure 10.2 shows this in a vertical cross section along the line ab in part F of Figure 10.1. But if the air on the eastern side of the occlusion under the warm front is colder than the air on the western side of the occlusion under the cold front, the warm front stays on the ground and the occlusion is a warm front type occlusion (see Figure 10.2).

As the warm air is lifted higher and higher by the occluding process, the occlusion begins to die (*frontolysis*) at the extreme tip and dissipates gradually toward the original wave apex (part G, Figure 10.1). For a time the counterclockwise circulation continues, and thin clouds remain aloft in the contact zone between the lower cold air and the warm air above it. But eventually all this dissipates and only the original front remains. By this time a secondary wave may appear along the trailing stationary front upstream from the previous wave. This often forms as the old occluded low pressure center begins to dissipate (part F). This wave will likely grow and occlude to repeat the cycle of the previous wave. In so doing it may move up the front to join the parent low pressure center. Frequently it traverses approximately the same territory, perhaps at a little lower latitude than the original wave.

Cyclones

When a low pressure center around the apex of a wave becomes encircled by closed isobars, it is known as a cyclone. This is primarily a phenomenon of the middle and higher latitudes, where air mass contrasts are great enough to produce distinct fronts and the Coriolis Force is strong enough to cause a rotary motion. In some instances the term "cyclone" has been used ambiguously. In interior North America pioneers built "cyclone cellars" (underground shelters from tornadoes) and confused the term "cyclone" with the term "tornado," which refers to a violent, small whirl of wind. In the Bay of Bengal area east of India, the term "cyclone" has been used to denote a hurricane-type storm. Within the context of this book, unless otherwise specified, "cyclone" will refer to middle- and higher-latitude low pressure centers with frontal systems in them.

A typical mature, partially occluded Northern Hemisphere cyclonic storm is shown in Figure 10.3. The type of weather varies between different sectors of the storm, and the type of weather that can be expected as a cyclonic storm approaches depends upon the part of the cyclone that is going to pass through one's location. The most sustained lift of warm air occurs up the warm front, since the flow of warm air in the warm sector of the storm between the fronts is usually directly up the slope of the warm front. This produces a large area of continuous overcast skies and probably precipitation, usually extending 200–300 kilometers ahead of the surface position of the warm front. Thus, the northeastern quadrant of a cyclonic storm in the Northern Hemisphere is usually the cloudiest and the one that will produce the most prolonged and widespread precipitation.

Surface winds in this portion of the storm are from the east or southeast, a well-known sign of an approaching storm in the middle latitudes of the Northern Hemisphere and one that has often befuddled laymen into thinking that the worst weather in these regions comes from the east. These easterly winds are a part of the storm itself, however, and do not denote the direction of movement of the entire storm. One must keep in mind that there are two motions to be considered, a circulatory motion around the low pressure center, and a translational motion from some westerly direction to some easterly direction, typically from southwest to northeast in the eastern portion of the United States. Thus, in Figure 10.3, although the first harbingers of the storm are

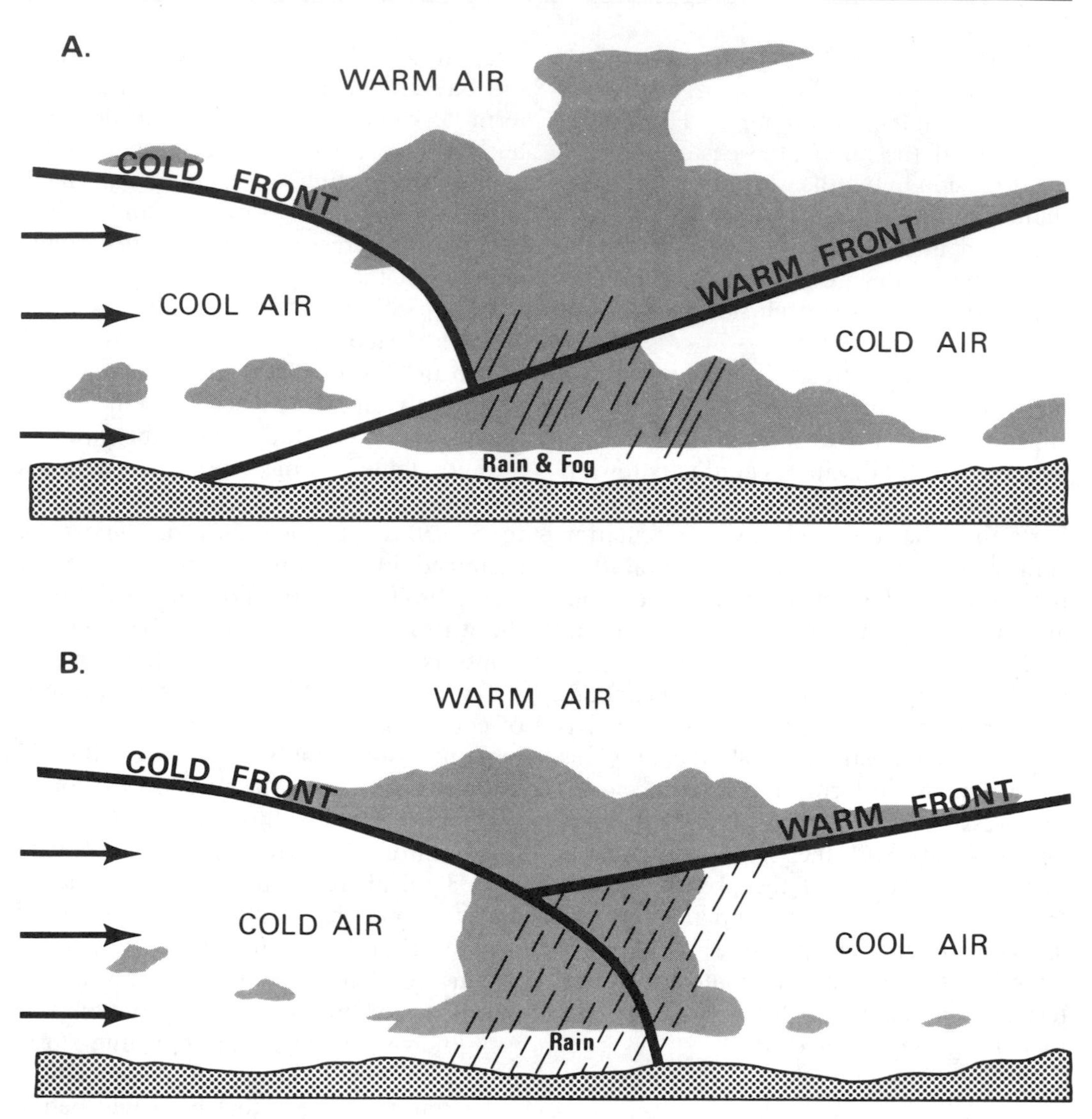

Figure 10.2. Vertical cross-section of occlusion along line ab in part F, Figure 10.1. (A) *Warm-front-type occlusion;* (B) *cold-front-type occlusion.*

easterly surface winds, the storm is moving in from the southwest.

The southeastern portion of a cyclone is the warm sector. In eastern United States the warm air mass is some variant of maritime tropical air with relatively high temperatures, high humidities, and conditional instability. Weather in this sector is typified by scattered clouds, commonly of a cumulus nature, and associated scattered showers, perhaps thunderstorms. But most severe thunderstorms are generally located along the cold front. Since this front has an abrupt lower slope, the warm air undergoes its most active rise here, which frequently triggers instability that produces severe thunderstorms, often in a continuous line all along the front. But since the front tapers off rather rapidly in the upper air, these storms do not extend far behind the front, typically no more than 30–40 kilometers at most, beyond which the weather

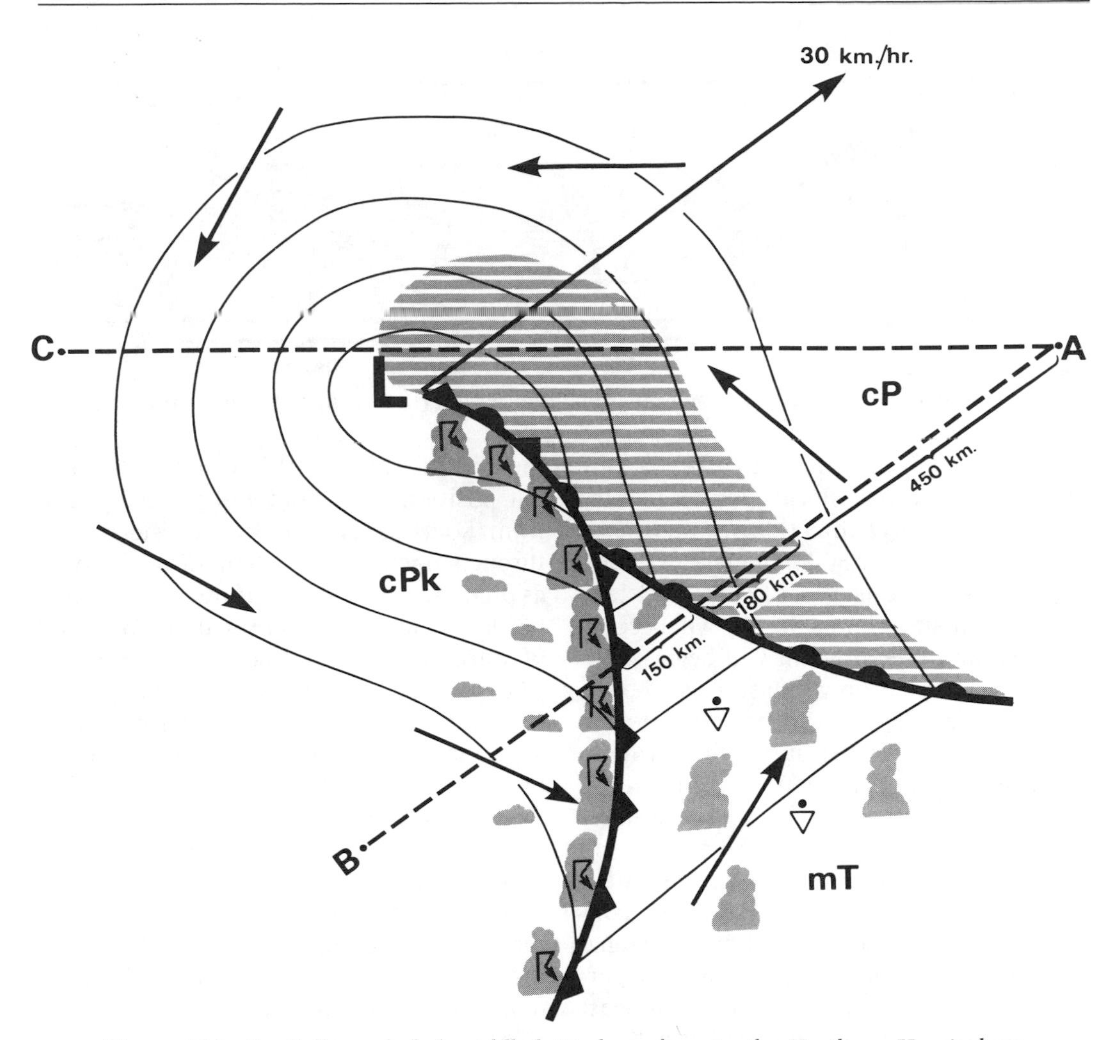

Figure 10.3. Partially occluded middle-latitude cyclone in the Northern Hemisphere.

rapidly clears in cold, fresh air. Since there is usually a k air mass behind the cold front, surface turbulence may produce small, flattish cumulus of fair weather for some distance behind the front—primarily daytime phenomena that dissipate at nightfall.

Since the warm air generally is moving away from the cold front and is only being underridden because the cold front is moving faster, no sustained flow of warm air continues at high levels above the cold front. In fact, on occasion the upper air, typically from some westerly direction, may be moving faster than the cold front itself, so that the air moves down the slope of the cold front rather than up it. Then there exists what is known as a *katabatic* front rather than an *anabatic* front, which is the more typical. On a significant number of occasions, a katabatic cold front may sweep through an area very quickly without any associated weather, the only indications of a frontal passage being a brief period of brisk winds and rapidly falling temperatures. In general, the more rapidly moving a front or a cyclonic system, the weaker the vertical development and the less intense the weather associated with it.

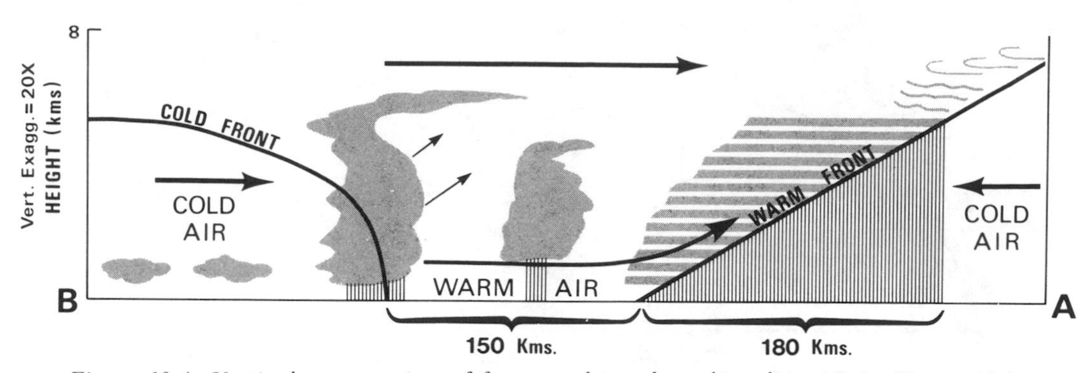

Figure 10.4. Vertical cross-section of fronts and weather along line AB in Figure 10.3.

Figure 10.3 illustrates that if we are situated at A and the storm center is moving in a northeasterly direction, as shown by the heavy arrow, at 30 kilometers per hour, a predictable sequence of events will take place. In about 15 hours the sky will become overcast and precipitation will begin. Continuous precipitation will last for six hours, during which the sky will be overcast with stratus clouds, the surface wind will be southeasterly, and the temperature will be cool. Then the weather will change fairly abruptly. The overcast will break up, and the clouds will change from stratus to cumulus with blue sky in between; the continuous rain will be replaced by occasional showers; and the temperature will rise perceptibly, with a wind shift from southeast to southwest. This type of weather will last for five hours while the warm sector of the storm passes. Next, the wind will become very gusty and shift abruptly from southwest to northwest. Severe thundershowers will probably occur for an hour or so, temperatures will fall precipitously, and they will plunge even more after the thunderstorms have passed as clear, cold weather moves in from the west. This sequence of events is illustrated by the vertical cross-section shown in Figure 10.4, which was taken along the line AB in Figure 10.3.

Six hours after the above forecast, a new weather map becomes available and shows that the storm center has now changed its movement from northeasterly to due east. Now the portion of the cyclone shown by the line AC (Figure 10.3) will pass through our position. The wind will gradually shift from southeasterly to northwesterly, but through north, rather than through south as predicted. The temperature will gradually become colder, but there will be no period of warm weather and no frontal passages. The sky will remain overcast and precipitation will occur for about twice as long a period as predicted. The new sequence of events is depicted by Figure 10.5. Our forecast is a complete bust! And the only change was the storm's direction of forward motion, by about 45°.

This could have been predicted by looking at the upper-air charts and noting that the storm path was approaching the crest of a ridge in the upper troposphere (Figure 10.6). As was explained in Chapter 6, circulation systems in the lower troposphere tend to be steered by flow in the upper troposphere, generally following the standing wave patterns in the circumpolar whirl from west to east. Figure 10.6 shows that successive positions of the center of the surface low conform closely to the western limb of the upper tropospheric ridge. As long as the storm is in this portion of the standing wave, it will move along a fairly straight path toward the northeast. But as it approaches the crest of the ridge it will turn toward the east and, later, toward the southeast. This change in direction will also cause the storm to weaken, as was explained in Chapter 6.

Waves appear to form on the polar and

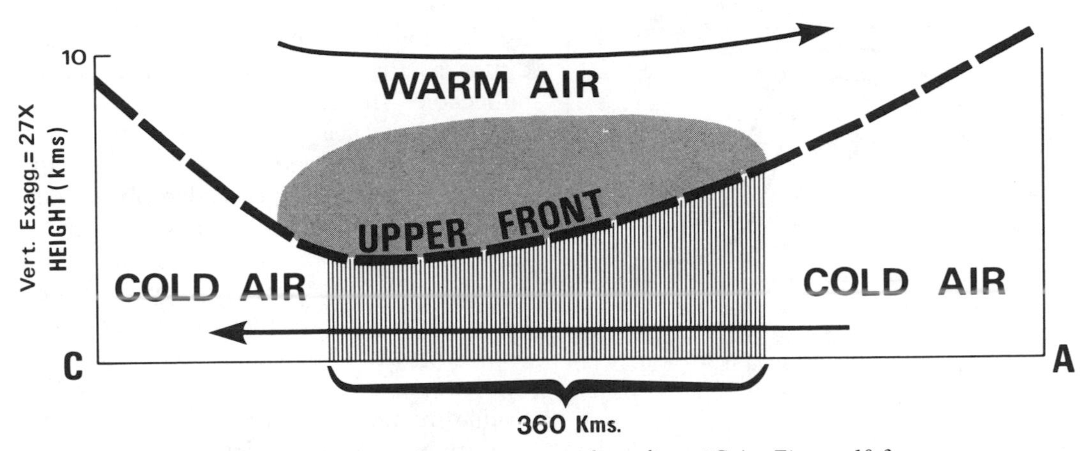

Figure 10.5. Vertical cross-section along line AC in Figure 10.3.

arctic fronts in preferred areas of cyclogenesis (birth of cyclones), and resultant cyclones follow preferred paths across the face of the earth. These are shown for the Northern Hemisphere in a very general way for the two extreme months, January and July, in Figures 10.7 and 10.8.

In North America during winter, preferred areas of cyclogenesis lie just to the east of the Rocky Mountains in Alberta or Colorado or in east Texas and along the Atlantic Coast, especially north of Cape Hatteras in North Carolina. Wherever they originate, their routes eventually converge on the New England–Maritime Province area of northeastern United States and southeastern Canada, as they move into the North Atlantic to merge with the Icelandic Low, which is well developed at that time of year.

Farther east, cyclonic storms move out of the Icelandic Low eastward and southeastward to cross Europe in three preferred paths: (a) along the Arctic coast of Scandinavia and the Soviet Union, (b) up the Baltic to the Arctic coast of the Soviet Union, and (c) through the Mediterranean Sea after which some continue eastward into the Middle East, and some swing northeastward across the Soviet Union toward the gulf of the Ob River in western Siberia, where they converge with storms following the Arctic and Baltic routes. The Ob Gulf plays much the same role as the Gulf of the St. Lawrence in North America, a convergence region of cyclone routes. East of the Ob, the storms usually die out along the Siberian fringes of the Arctic.

Few cyclonic storms penetrate the deep interior of Asia. But they form along Far Eastern segments of the polar front, usually off the Asian mainland south and east of Japan. They swing northeastward into the Aleutian Low and then farther eastward to penetrate the mountainous western portions of North America.

During July tracks are similar but are generally shifted poleward, and frequencies of storms along the routes are reduced. In Europe the Mediterranean route is largely abandoned. The Baltic cyclones penetrate Siberia as a broad, diffuse thermal low replaces the thermal high of winter. In the Far East the weakened polar front shifts westward across the Japanese islands and portions of the mainland, especially along the Mongolian-Soviet border, which becomes a significant area of cyclogenesis during mid- and late summer. Cyclones forming along the Mongolian Front move eastward across the Sea of Okhotsk into the northern Pacific.

It can be seen from these generalized maps that cyclones tend to edge their way poleward and converge in the subpolar low areas. Anticyclones (high pressure cells), on

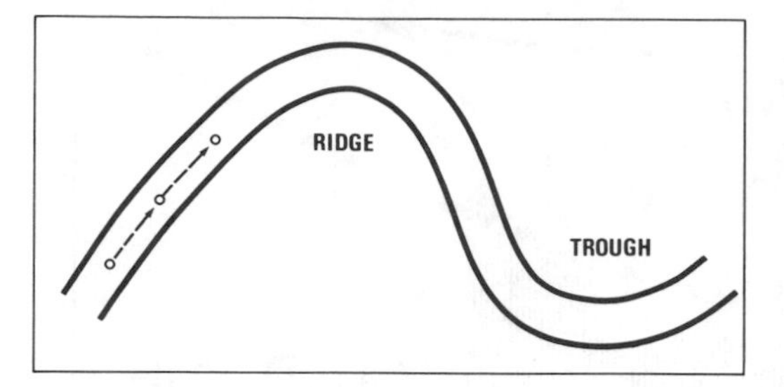

Figure 10.6. Small circles depict successive positions of surface low relative to wave pattern of upper tropospheric flow.

the other hand, in the higher middle latitudes work their way equatorward to join the subtropical high pressure belt (Figures 10.9 and 10.10).

Thunderstorms and Tornadoes

As was demonstrated in Chapter 8, thunderstorms may develop if the air is conditionally or convectionally unstable. For instability to be triggered, the air must be lifted to its level of free convection, either by heating at the surface, cyclonic lift over a front, or orographic lift (over a terrane barrier). Once the level of free convection has been reached, vertical movement is perpetuated by the release of latent heat of condensation, and vertical cells of circulation form that are somewhat self-contained thermodynamic mechanisms. These convective cells produce the commonly observed cauliflower-shaped cumulus clouds that may grow rapidly upward until lightning and thunder occur and a full-blown thunderstorm develops.

Figure 10.11 shows the progressive stages of the development of a thunderstorm. Initially, as the cumulus cloud is growing, the vertical air movement throughout the entire cloud is upward. To satisfy mass continuity, of course, an equal amount of downward motion must occur somewhere, and this takes place around the cloud formation. Thus, cumulus clouds are typically separated from each other by clear air where the downdrafts occur. The downdrafts are usually much more diffuse than the updrafts, so the downward motion is not as noticeable. In its initial stages, each cumulus cell typically is no more than one kilometer in diameter, if that, while the space between cells is considerably greater.

As the cumulus cell grows well above the freezing level in the atmosphere and individual cloud droplets increase in size, the mature stage is reached and precipitation begins. This usually occurs first on the leading edge of the storm, according to its direction of motion. The falling precipitation produces a frictional drag on the air in that portion of the cloud and also cools the air significantly, both by contact and by partial evaporation of some of the falling precipitation which extracts the heat of evaporation from the surrounding air molecules. This results in a strong downdraft in the front edge of the cell. Usually the first noticeable characteristic of a thunderstorm's approach is a sudden shift in wind direction, accompanied by a blast of cold air descending from the roll cloud that is boiling along at the base of the leading edge of the cell, a short distance above the ground. This is known as a "gust front."

About this time the top of the cloud becomes glaciated (composed of ice crystals), and the typical "anvil" forms which is composed of milky-white cirrus clouds made up of ice crystals. This shape, which has been likened to the old-time blacksmith anvil, is formed as the upper tropospheric winds extend the glaciated portion of the cloud tops a long distance downwind underneath the tropopause inversion.

Now the cell has reached its advanced stage, and heavy precipitation throughout the cloud has produced downdrafts everywhere. This is the stage of heaviest precipitation, but it is also the beginning of the end, since no more warm, moist air is being brought in from below to release its latent heat to keep the convection going. In the course of ten or fifteen minutes the bulk of the precipitation falls, and the

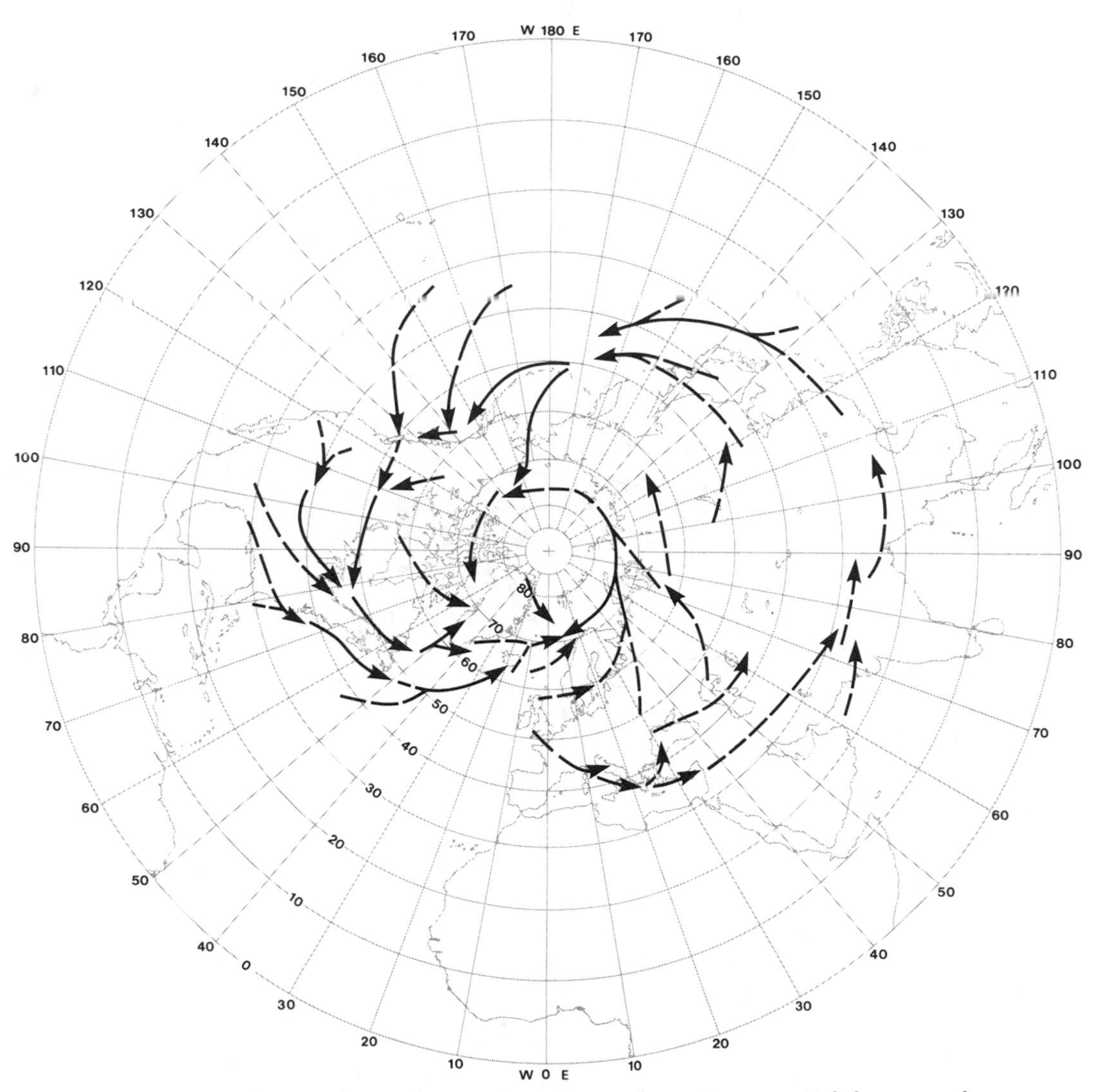

Figure 10.7. Cyclone paths in the Northern Hemisphere, January. Solid arrows denote primary tracks; dashed arrows, secondary ones. (After William H. Klein, Principal Tracks and Mean Frequencies of Cyclones and Anticyclones in the Northern Hemisphere, *U.S. Weather Bureau, Research Paper No. 40 [Washington, 1957].)*

convection cell slowly dissipates as other cells grow around it. The precipitation timespan and the intensity of the cell may be greatly enhanced if the convection cell is tilted so the precipitation at the leading edge does not fall through the updraft in the main body of the cell. The most severe thunderstorms often develop a double-vortex structure, counterclockwise in their equatorward portion and clockwise in their poleward portion.

Thunderstorms are by far the most numerous of the violent, turbulent storms. It has been estimated that more than 40,000 of them take place on earth every day. Since their development depends upon conditionally unstable air and high humidity content, most are found in association with maritime tropical (mT) air masses. As can be seen in Figure 10.12, thunderstorms occur most frequently in the equatorial region, particularly on the continent of

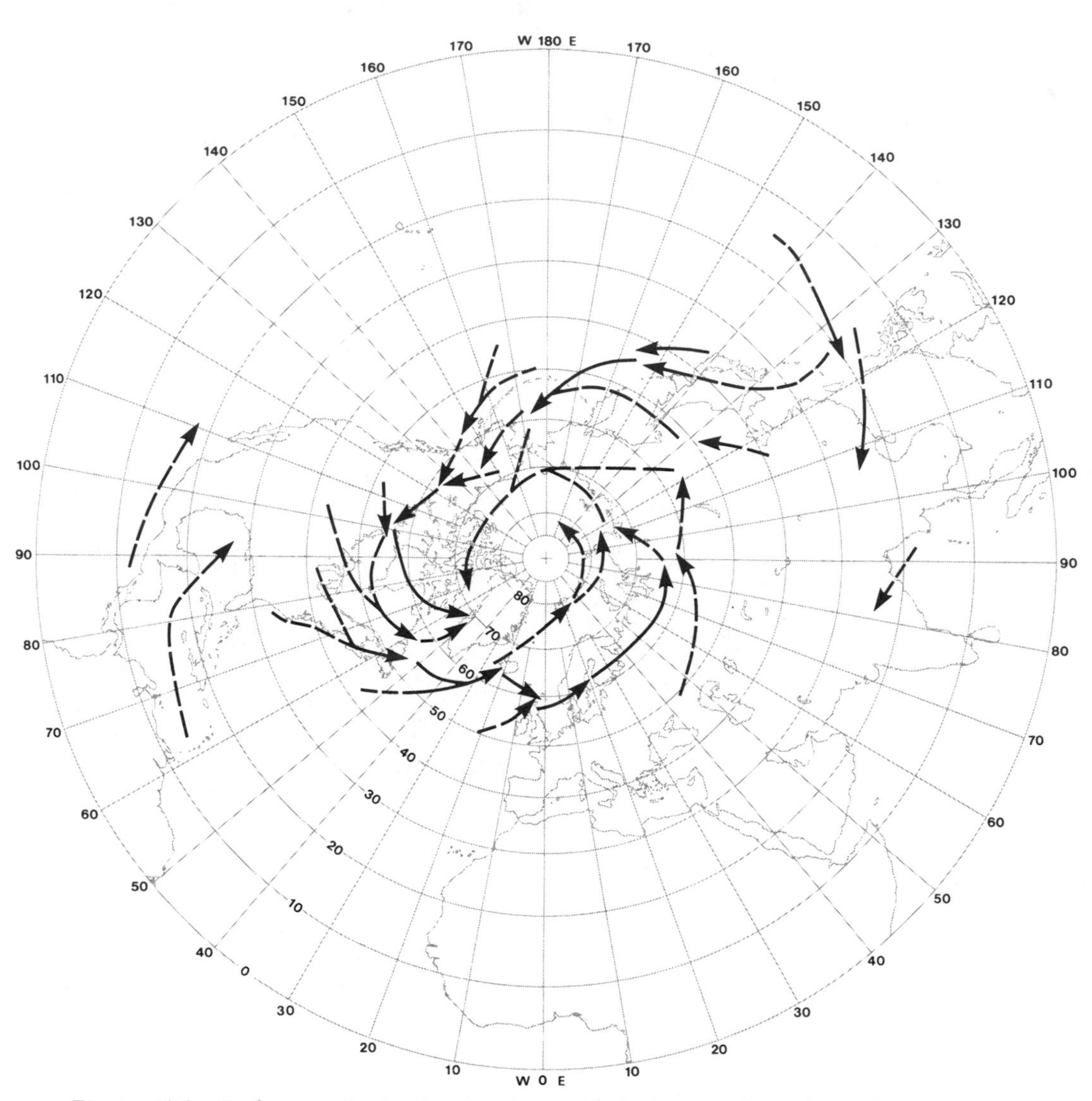

Figure 10.8. Cyclone paths in the Northern Hemisphere, July. (After Klein, Cyclones and Anticyclones in the Northern Hemisphere.*)*

Africa. Within the United States they occur most frequently in the Southeast, which is dominated by warm, moist, conditionally unstable air along the western edge of the North Atlantic High.

Although tropical and subtropical thunderstorms probably produce the greatest amounts of precipitation, they are not necessarily the most turbulent. Extremely turbulent thunderstorms, attested to by large hail and occasional tornadoes, are found in portions of the middle latitudes, particularly in central United States and north-central India. Such turbulence is usually related to a rapid decrease of absolute moisture content with height, which sets the stage for convective instability, triggered either by orographic or cyclonic uplift and strong vertical wind shears.

Individual thunderstorm cells represent

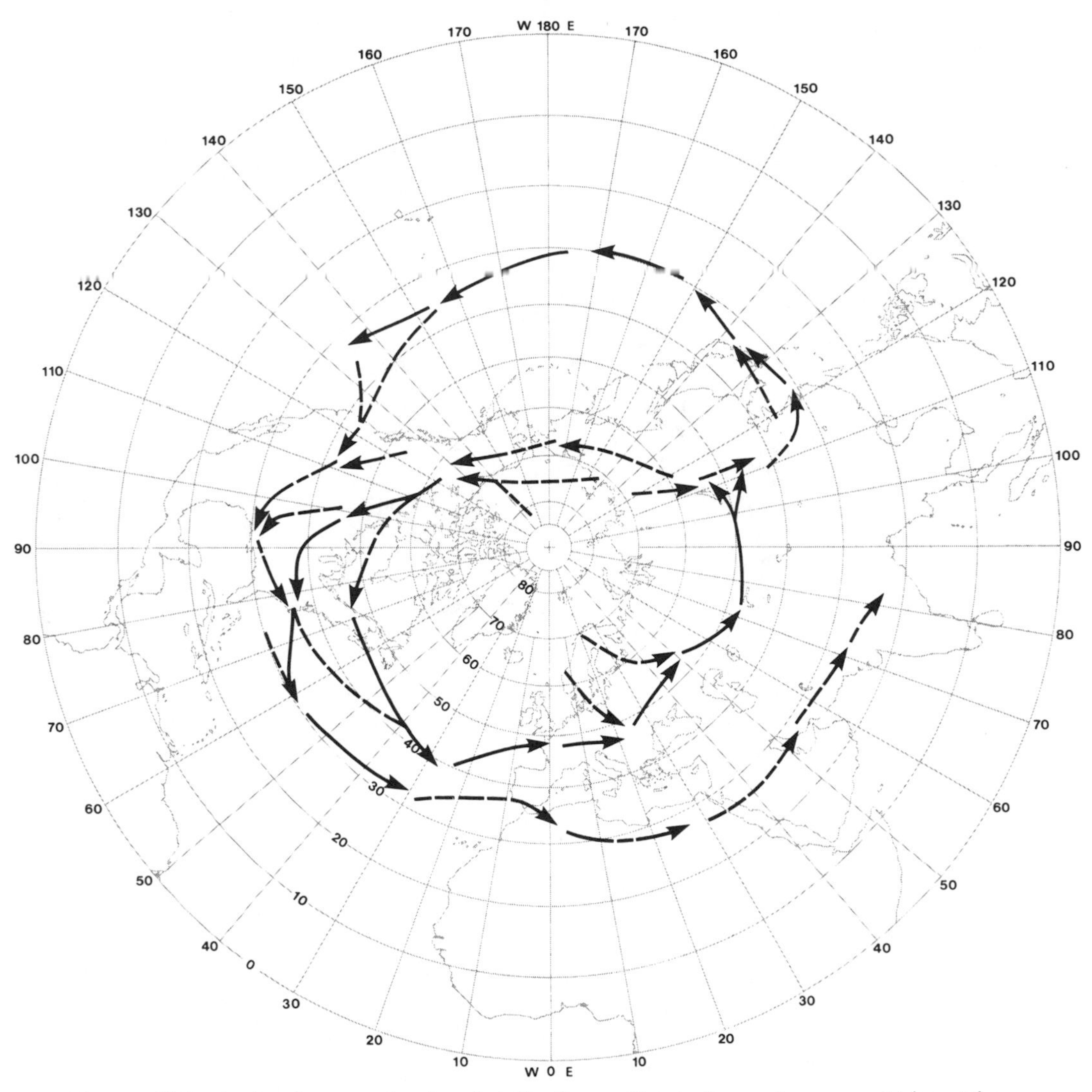

Figure 10.9. Anticyclone paths in the Northern Hemisphere, January. (After Klein, Cyclones and Anticyclones in the Northern Hemisphere.*)*

small circulation systems in the atmosphere that are usually imbedded in broader-scale systems, such as a cyclonic storm, the intertropical convergence zone, or a conditionally unstable k air mass, such as the maritime tropical air moving from the Gulf of Mexico into southeastern United States during summer. Individually, they cover only a few kilometers in diameter at best, but frequently they coalesce into thunderstorm masses consisting of several cells, and occasionally they form lines of almost continuous thunderstorm cells that move in unison as squall lines.

Squall Lines

In central United States squall lines sometimes form in warm sectors of cyclonic storms 50–300 kilometers ahead of cold

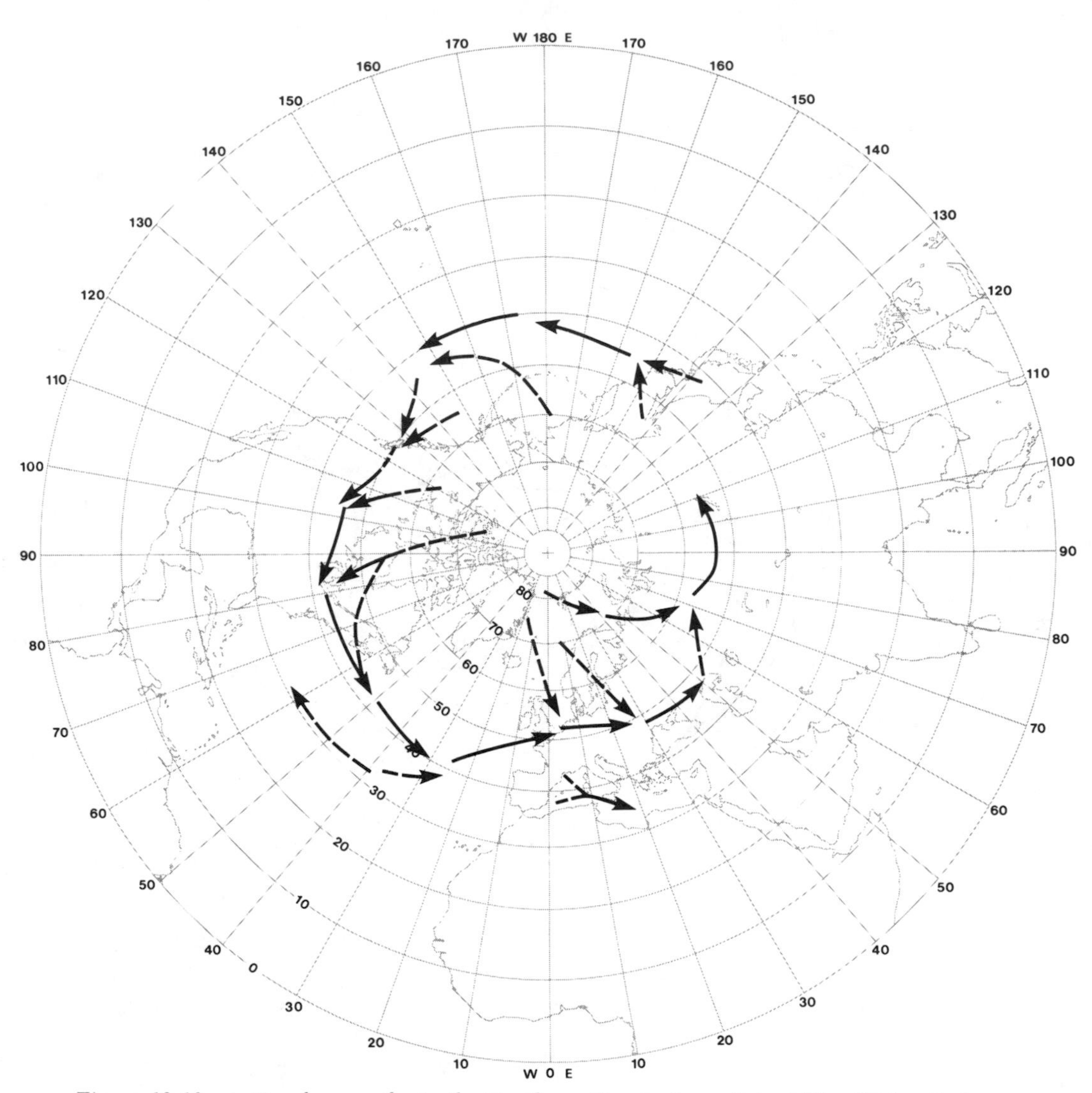

Figure 10.10. Anticyclone paths in the Northern Hemisphere, July. (After Klein, Cyclones and Anticyclones in the Northern Hemisphere.*)*

fronts (Figure 10.13). Downdrafts in the leading edges of thunderstorm cells associated with cold fronts appear to give rise to small high pressure centers that push air (cooled by evaporation or precipitation) ahead of cold fronts in the form of pseudofronts, which underride warm air ahead and trigger instability. Squall lines in this region also appear to be related to low-level jets of moist air blowing from south to north across the Great Plains, particularly from Texas to Nebraska. Similar occurrences are found in east-central Argentina.

Squall lines move with about the same speed and direction as the cold fronts behind them, and their approach is often mistaken as the approach of the cold front, which confuses weather predictions. In some cases squall lines move faster than the cold

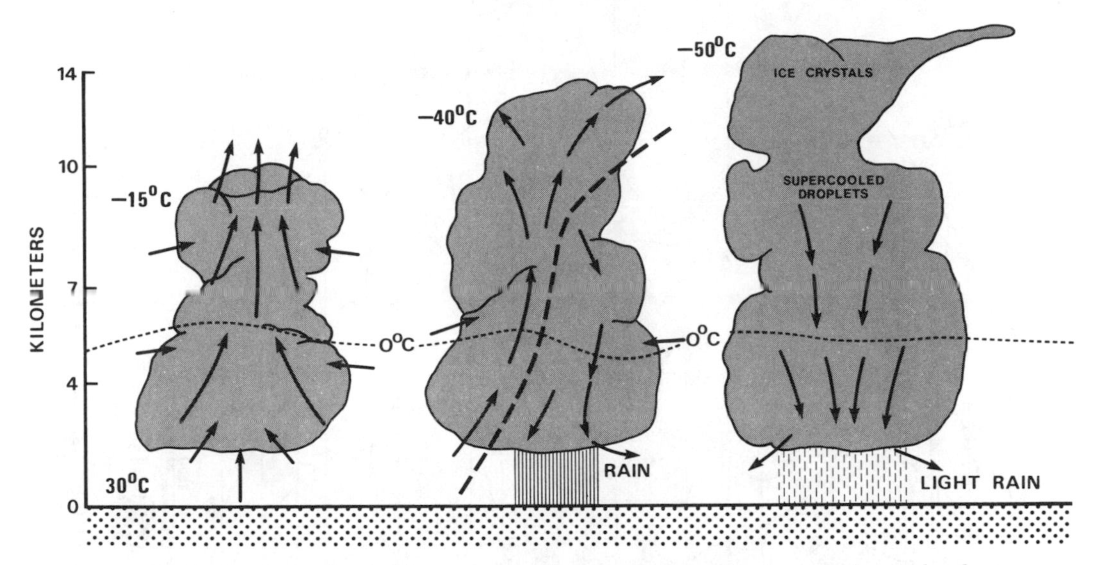

Figure 10.11. Stages in the development of a thunderstorm cell. Left: *cumulus or developing stage.* Center: *mature stage.* Right: *old or dissipating stage.*

fronts behind them and eventually are no longer directly associated with the cold fronts. One of the most remarkable occurrences of squall lines is in west Africa in the Lake Chad–Niger River area, where they become the primary cause of rainfall during summer. Here they occur in lower latitudes unassociated with the polar front and its cyclonic storms. Although they are imbedded at the surface in a southwest monsoonal flow, they move westward with the winds aloft that flow from east to west.

Tornadoes

Squall lines that form near the apexes of warm sectors in cyclonic storms provide some of the most turbulent weather in interior North America and on extreme occasions may spawn tornadoes, the most violent of all weather phenomena. The characteristic funnel cloud builds downward from low-hanging cumulonimbus clouds associated with severe thunderstorm cells and forms a very small vortex, typically less than 400 meters in diameter, which produces extreme wind speeds, perhaps as much as 225 meters per second (500 mph). At the surface the air rotating around the vortex moves inward toward the center and then upward, carrying with it debris as large as automobiles and major portions of houses. Extreme pressure gradients are produced over short distances, perhaps as much as 250-millibars change in a few hundred meters. These add to the destruction, since tightly closed buildings may not allow air inside to escape fast enough to equalize the pressure within the building and the pressure dropping outside with the passage of the storm. Tornadoes often leave paths of destruction in urban areas where entire blocks of houses appear to have literally exploded outward, with their roofs lifted off and walls laid out from the centers of the buildings.

A typical tornado in the United States moves northeastward with a forward speed of perhaps 20 meters per second (45 mph). It frequently skips, with the funnel cloud temporarily receding upward into the base of the cumulonimbus cloud and later dropping down again. The fleeting lifespan and erratic movement make it difficult to trace

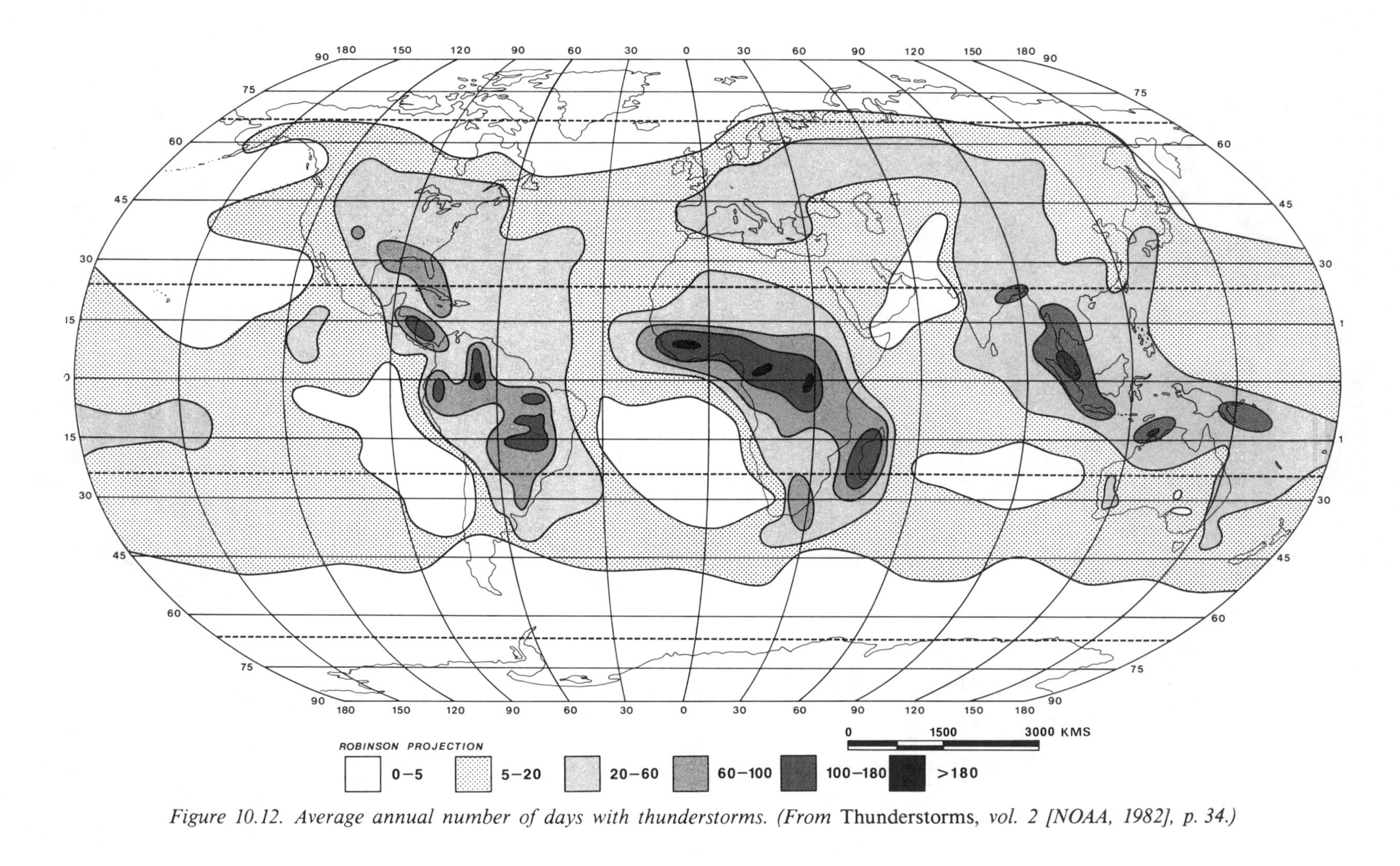

Figure 10.12. Average annual number of days with thunderstorms. (From Thunderstorms, *vol. 2 [NOAA, 1982], p. 34.)*

the path of any single tornado vortex. Although tornadoes are nearly totally destructive where they hit, their routes are narrow and discontinuous, and their appearances do not significantly alter average statistics of winds and other climatic phenomena, even in the regions of their most frequent occurrence. But if they do hit, they are phenomena to be reckoned with. In an attempt to predict their occurrence and warn people in affected areas, the National Weather Service operates a severe-storm warning center in Kansas City, Missouri, with a laboratory near Oklahoma City, near the center of most frequent tornado occurrence.

Most of the tornadoes on earth occur in the United States east of the Rocky Mountains. Although all states have recorded tornadoes, the maximum number occur in the southern and central Great Plains, the Midwest, and the southeastern United States. Outside North America, Australia experiences tornadoes most frequently, particularly in the southeastern part of the continent. Only scattered occurrences of tornadoes have been reported in other parts of the world.

In central and southern United States tornadoes are most likely to develop near stalled sections of the polar front during periods of most intense air mass contrasts, which usually occur during spring. In the states of maximum occurrence, Oklahoma and Kansas, three-fourths of the tornadoes take place during the months of April, May, and June. Most occur during late afternoon and evening during the period of maximum diurnal convective activity. In southeastern United States, particularly along the Gulf Coast, multiple small tornadoes are commonly associated with hurricanes during late summer and autumn.

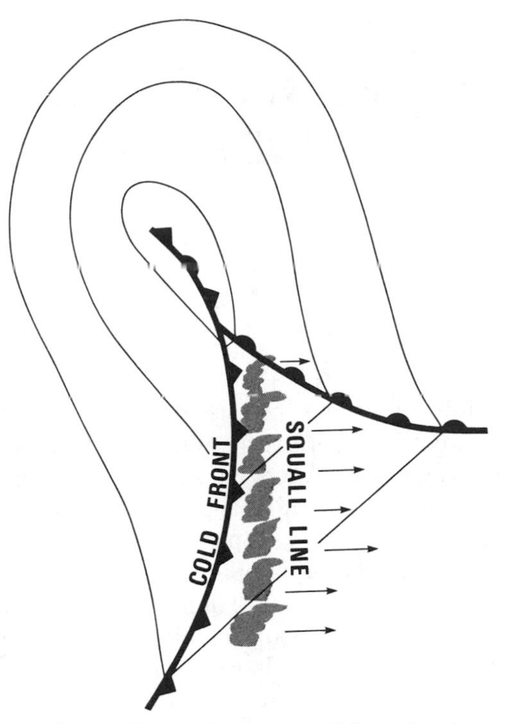

Figure 10.13. Typical squall line ahead of cold front in warm sector of cyclone in central United States.

11 FLOW PERTURBATIONS IN LOWER LATITUDES

In lower latitudes, because of reduced temperature and pressure gradients and reduced Coriolis Force, the geostrophic wind concept breaks down, and the air tends to move more and more perpendicular to the isobars as the equator is approached. As a consequence, perturbations in the air flow that may bring about notable weather take on different forms than they do in middle and higher latitudes.

The Trade Wind Inversion

As mentioned earlier, the air circulating around a subtropical high pressure cell undergoes divergence and subsidence, which is strongest on the equatorward-eastern edge of the cell. Thus, a strong subsidence inversion is formed not far above the surface along western coasts of continents in subtropical latitudes, such as southern and Baja California, northern Chile and Peru, and northwest and southwest Africa. As the air turns westward and moves into progressively lower latitudes along the equatorward margins of these highs, the surface air picks up heat and moisture from the tropical sea, which produces increasing instability in the marine layer underneath the inversion and lifts the inversion layer to greater and greater heights. For instance, during summer on the west coast of North America near San Francisco, the height of the base of the inversion averages about 400 meters above sea level, while at Hawaii 4500 kilometers to the southwest, it averages more than 2000 meters above sea level. Farther west the inversion disappears altogether as the air becomes conditionally unstable along the western edge of the high.

In the equatorward-eastern quadrants of the subtropical highs where subsidence is strongest and the inversion base is lowest above the surface, the marine layer of air under the inversion is usually too thin for convection to produce clouds of significant vertical development. In these places fog and low overcast stratus clouds are the rule. If the inversion is very low above the surface, the marine layer may not be thick enough even to produce stratus clouds, as is often the case in southern California during middle and late summer. But as the inversion lifts westward, convective activity within the marine layer typically produces cumulus clouds that build up to the base of the inversion. These are the well-known "trade wind cumulus" that dominate the eastern halves of subtropical ocean areas. They are very regular formations when viewed from above, such as from the high volcanic peaks on the Big Island of Hawaii, where one can observe them moving in from the northeast, their bases regularly positioned at the condensation level a few hundred meters above the surface of the sea, and their tops distinctly limited by the base of the inversion, perhaps 2000–2500

Figure 11.1. Trade wind cumulus moving with the northeast trades onto the Big Island of Hawaii below the summits of Mauna Loa (distance) *and Mauna Kea* (foreground). *Note the barren surface of the dry summit of Mauna Kea.*

meters above the sea (Figure 11.1). From such cloud formations, intermediate slopes of windward sides of mountains may receive copious precipitation, upward of 7500 millimeters (300 inches) per year or more, while lower and higher slopes remain dry. It has been estimated that the top of Mauna Kea on Hawaii at an elevation of 4206 meters (13,796 feet) receives no more than 375 millimeters (15 inches) of precipitation per year because it is above most of the clouds. Similar conditions exist at Tenerife in the Canary Islands off the northwest coast of Africa.

Easterly Waves

Along the equatorward-western edges of the subtropical highs imbedded in the conditionally unstable trade winds appear open wave formations, 2000 kilometers or more in length, that move westward with the trades at forward speeds averaging about 5 meters per second (11 mph). Therefore, an entire wave takes four or five days to pass a position. Although these are very subtle features in the surface isobaric pattern, which is usually drawn with little available data anyway, they do impart an aperiodic element to an otherwise rather regular diurnal cycle of weather events in the tropics (Figure 11.2). The easterly waves are associated with divergence, subsidence, and clear skies on their leading edges, followed by convergence, uplift, and increased thunderstorm activity after their wave crests pass. Thus, the passing of an entire easterly wave over a region may result in two to three days of less than normal clouds and precipitation, followed by two to three days of increased thunderstorm activity and precipitation. The waves show no tendency to go through a cycle of occlusion, dissipation, and regeneration, as they do in the middle latitudes. Little is known about the origin and fate of these minor perturbations in the trade wind flows.

Hurricanes

Although pressure gradients and winds are usually weak and relatively unorganized in the tropics, occasionally rotary motions are initiated which, under the influence of a weak Coriolis parameter, develop into high wind speeds. Since these disturbances originate in tropical oceanic areas where few surface observations exist, little is known about the initiation of the cellular circulations with pronounced low pressure centers. Satellite photography during the last couple of decades has helped to locate these storms earlier in their lifespans, but still they are not observed until they are fairly well developed. Some of them may originate as easterly wave perturbations or open wave formations on the intertropical front, where it has migrated a considerable distance from the equator. These storms do not occur right on the equator, presumably because there is no Coriolis Force to initiate the rotary motion. Most of them seem to originate 5°–10° latitude on either side of the equator and move westward with the trade winds, gradually working their way poleward until they come under the influence of the westerlies and then curve into middle latitudes. Figure 11.3 shows typical routes of travel of tropical storms that have developed to hurricane strength.

The driving energy seems to be primarily the latent heat of condensation. These storms develop only over ocean areas that have temperatures higher than 27°C (80°F). Air converging toward a low pressure center over a warm water surface carries great amounts of energy in the form of latent heat tied up in the moisture that has been evaporated from a broad ocean surface and concentrated in a small area. Rising motion around the center of the low cools the air adiabatically and releases this energy through the process of condensation. As long as the moisture source, and hence the energy source, is available, the storm con-

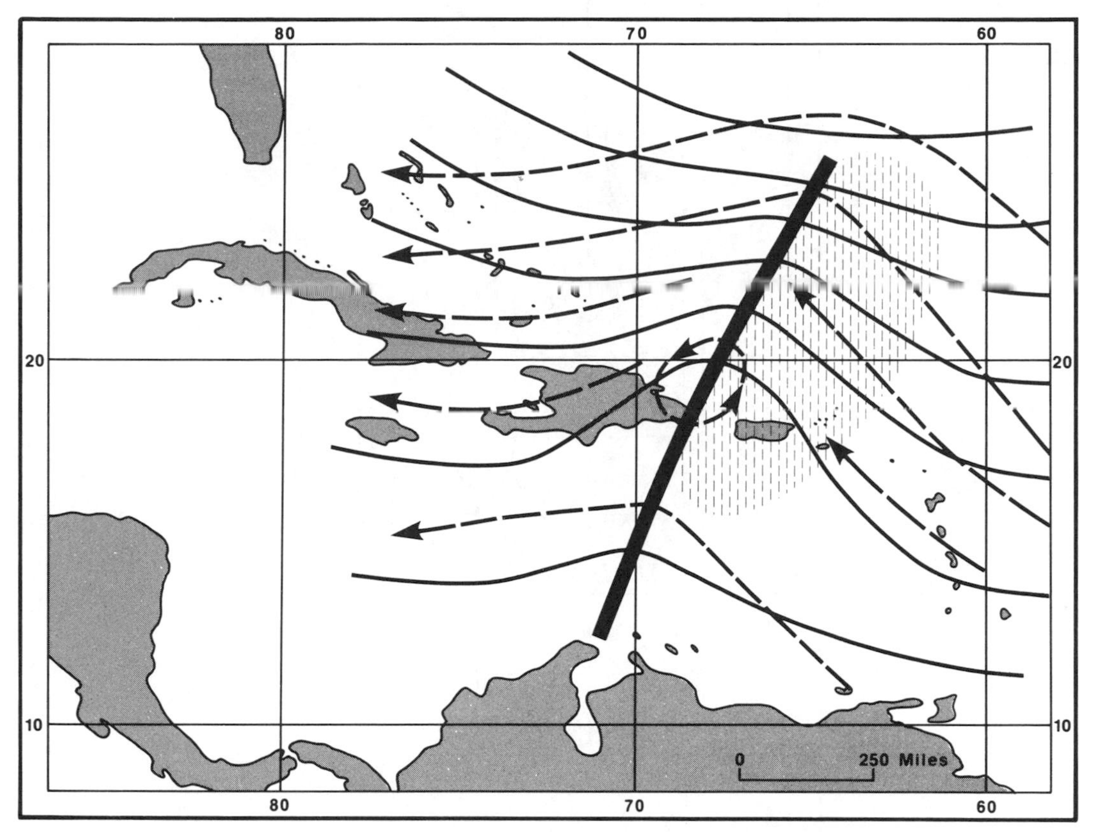

Figure 11.2. Model of an easterly wave in the Caribbean. Solid lines represent isobars at the surface; dashed lines represent streamlines at the 500 mb. level. The heavy line marks the trough line. Stippling indicates area of increased precipitation.

tinues to develop. But when the storm leaves the warm ocean surface, either by moving into higher latitudes or over a continental mainland, it quickly loses the infeed of new energy, and the wind speed dies down almost immediately. The low pressure center with its associated widespread cloudiness and heavy precipitation may move far inland, but the strong winds generally are limited to within 30–40 kilometers of a seacoast. For instance, in the United States a hurricane storm may come onto the mainland from the Gulf of Mexico, move up the Mississippi Valley, curve northeastward up the Ohio Valley into the New York or New England area, and drop several hundred millimeters of rain over huge areas, thereby causing widespread damage from flooding. But local inhabitants are not aware that they are undergoing a hurricane passage, because the wind speeds are no higher than they are with ordinary extratropical cyclonic storms common to the region. In fact, such storms frequently join up with extratropical cyclones coming across the central United States, which causes them to deepen and stagnate, thereby prolonging their influence over the northeastern states. The amount of precipitation released in such a situation can be phenomenal. Hurricane Agnes, in June 1972, dumped about 300 millimeters (one foot) of rain over much of eastern Pennsylvania and adjoining states, causing tremendous

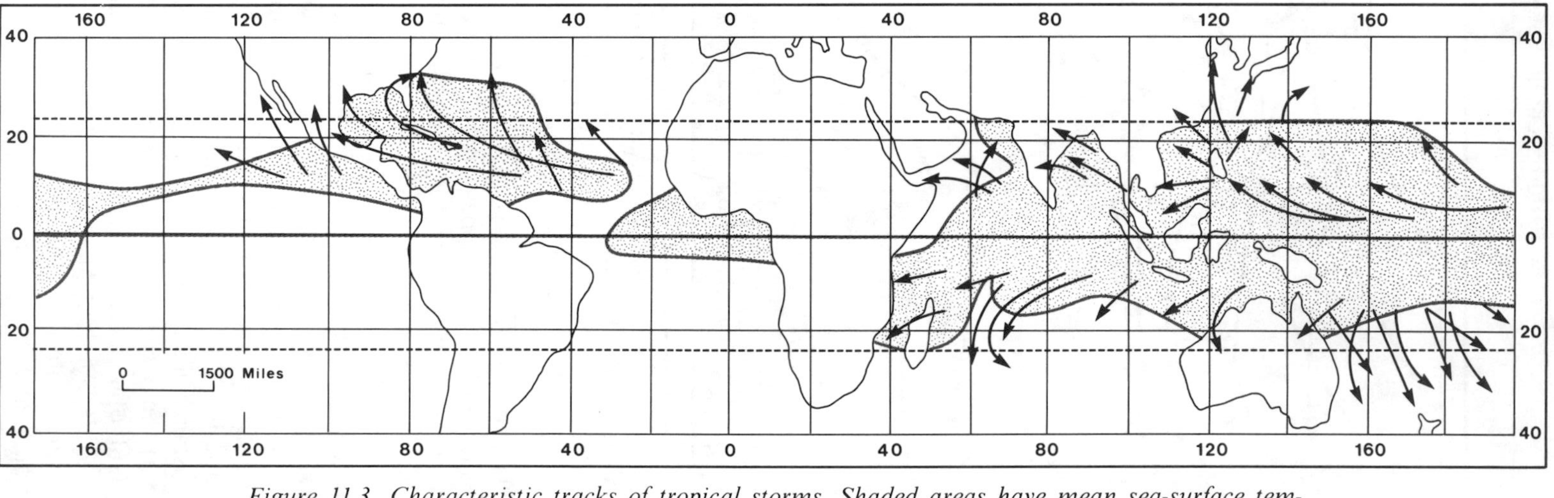

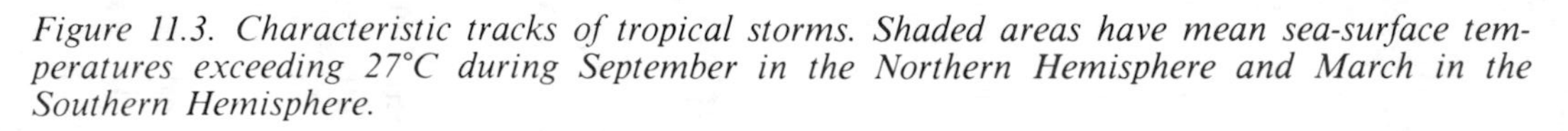
Figure 11.3. Characteristic tracks of tropical storms. Shaded areas have mean sea-surface temperatures exceeding 27°C during September in the Northern Hemisphere and March in the Southern Hemisphere.

amounts of damage from flooding and mudslides.

In coastal areas, in addition to the high winds, one of the most damaging characteristics of a hurricane storm is the surge of wind-driven ocean water as the storm approaches the coast. It is not uncommon for the water level to rise more than 6 meters (20 feet), which, along flat coastal plains, may drive the water far inland. During the night of 17–18 August 1969, Hurricane Camille drove a 7-meter tide ahead of it across the Louisiana-Mississippi coast and literally drowned the small town of Cameron, Louisiana. Such occurrences have been particularly disastrous in the densely populated Ganges delta at the head of the Bay of Bengal, where such storms repeatedly have caused hundreds of thousands of deaths by drowning and untold property damage. The most memorable case, on 12 November 1970, had an official death toll exceeding 200,000 people and, as a result of its millions of homeless refugees, eventually touched off the Pakistani civil war that resulted in the establishment of the new country of Bangladesh.

According to official definitions of the World Meteorological Organization, closed cyclonic circulations in the tropics with wind speeds less than 17 meters per second (38 mph) are tropical depressions. When the wind speed rises to 17–32 meters per second they are called tropical storms. When wind speeds reach more than 32 meters per second (75 mph) they are called something else, depending on the region. In the Caribbean area they are known as hurricanes and in the Pacific, generally, as typhoons, but in different portions of the Pacific they are known by different names. In the Philippines they are known as *baguios,* after a small resort city that in 1911 received 1150 millimeters (46 inches) of rain in 24 hours during the passage of such a storm. In the Indian Ocean area, including the Bay of Bengal and the Arabian Sea on either side of the Indian peninsula, they are known as cyclones, as they commonly are in Australia. Hence, once again the term "cyclone" is used in a different context. They are all the same kind of storm, although usually of different intensities in different areas.

These storms are most frequent and most intense in the western North Pacific, where wind speeds of 70–90 meters per second (150–200 mph) are not uncommon, although the majority of them have wind speeds somewhat less than that. During the height of their season (July–October) it is not uncommon to find two, or even three, such storms in different stages of development following one another across the western part of the North Pacific, 1500–3000 kilometers apart.

The second area of most frequent and most intense occurrence of these tropical storms is the western part of the North Atlantic and adjacent Caribbean Sea and Gulf of Mexico. Here wind speeds very rarely, perhaps only once in 10 or 20 years, reach 90 meters per second (200 mph). Other regions in descending order of occurrence are the northern Indian Ocean, particularly the Bay of Bengal, but to some extent the Arabian Sea; the southern Indian Ocean in its western portion next to Madagascar Island off southeastern Africa; the western South Pacific, particularly to the east of Australia, but also all along the northern coast; and the very eastern North Pacific southwest of Mexico, which has fewer and usually weaker storms. The outstanding exception to such locations is the western south Atlantic next to Brazil, which has never experienced hurricane-type storms (see Figure 11.3). The sea-surface temperature there is usually too cool for the formation of such storms, and the position and shape of the south American continent are such that this coastal area juts well into the central portions of the South Atlantic high pressure cell, so that atmospheric stability is generally greater in this area than in the others mentioned.

In the Northern Hemisphere, highest sea surface temperatures are reached during June–November, and these are the months in which hurricanes develop most frequently, although occasionally they occur in adjacent months of the year. In the north Indian Ocean, which is consistently warm, they have occurred every month of the year. In the Southern Hemisphere, sea-surface temperatures are highest from December through March, and that is the season of most frequent tropical storms.

A fully mature hurricane may range in diameter from as small as 100 kilometers to more than 1500 kilometers. Wind speeds increase from the periphery toward the center, where sea-level pressure may reduce to less than 900 millibars. The lowest sea-level pressures ever recorded are in the centers of such storms. (Pressure may reach even lower levels in the centers of tornadoes, but no measurements or adequate estimates have been possible in those cases.) The official world's lowest sea-level air pressure was recorded in the eye of Typhoon June in the western North Pacific in November 1975, where the central pressure dropped to 25.90 inches (876 mb).

In the immediate center of the storm the wind dies down, and it becomes relatively calm and clear in what is known as the eye of the storm. It is obvious that there is in the center of the hurricane a chimney of descending air that produces subsidence and relatively clear weather, accompanied by only light winds. This may cover a circular area approximately 15–45 kilometers in diameter. A distinctive wall of clouds surrounds this eye, and in this portion of the storm the greatest uplift, highest wind speeds, and maximum precipitation occur.

Around the eye wall the clouds arrange themselves in spiral bands, as shown in the satellite photo in Figure 11.4. Throughout much of the storm there appear to be alternating spiral bands of uplift and subsidence, the bands of uplift being marked by the bands of clouds, and the bands of subsidence by the relatively clear areas in between. General subsidence appears around the periphery of the storm, particularly on its leading edge, which has prompted the old adage "the calm before the storm." Often during an entire day as a hurricane approaches, the weather is unusually calm, clear, and sunny.

Hurricanes frequently extend up to the tropopause. Because of the subsidence and accompanying adiabatic heating in the eye, the center of the storm remains warmer than its surroundings right up to the tropopause. Thus, the hurricane is a warm-core cyclone and must decrease in intensity with height, eventually becoming an anticyclone aloft.

Equatorial Westerlies

Between the trade winds, a belt of surface westerly winds is often found near the equator, especially in the summer hemisphere. The most extensive zone stretches from just west of Africa eastward into the western Pacific. A limited zone intersects the equatorial west coast of south America. Here and in equatorial western Africa, westerly winds prevail throughout the year. But in eastern Africa, southeast Asia, and northern Australia, they shift latitudes widely with the seasons. Their most poleward extent is found during the Northern Hemisphere summer in northwestern India-Pakistan, where they sometimes penetrate to 30°N latitude as the Indian monsoon. Over the Indian Ocean they often extend upward to 5–6 kilometers, but in the rest of the world they tend to be shallower. Over Africa they generally extend only to 2–3 kilometers height. The westerlies have a meridional component directed poleward and tend to be more unstable than the tropical easterlies with equatorward components. The westerlies usually bring heavy rainfall to areas consistently affected by them, particularly along mountainous windward coasts, such

Figure 11.4. Satellite photo of Hurricane Allen in the western Gulf of Mexico, 8 August 1980. Note the spiral cloud bands and the eye.

as the mid-section of the western coast of Colombia in South America, portions of the Guinea-Cameroon coast of Africa, and the West Ghats and Khasi Hills of India.

The Southeast Asian Monsoon

One of the most outstanding features of the air flow in low latitudes is the great seasonal reversal of winds and resultant weather that occurs throughout much of the north Indian Ocean, India, the countries of southeastern Asia, much of China, and parts of adjacent countries, as well as the western fringes of the North Pacific Ocean. To some extent it also affects much of Indonesia, where it extends south of the equator into northern Australia. Discussion here will be confined to its most classic development over the Indian subcontinent and adjacent water bodies—the Arabian Sea to the west and the Bay of Bengal to the east.

Here, particularly, its specific features appear to be due to much more than just the seasonal reversal of land-sea heating differentials. The suddenness of the "burst" of the rainy period during early June in southwestern India, the deflection of the southwest surface winds at the head of the Bay of Bengal to southeasterlies up the

Ganges Valley, and the progression of "monsoon depressions" from the Bay of Bengal northwestward across India are some of the features imbedded in the general summer monsoon environment that cannot be explained by simple land-sea temperature differences.

The highlands to the north appear to play a major role. The massive Tibet Plateau, ranging in average elevation from approximately 4000 meters in its eastern portions to 6000 meters in its western portions, with still-higher surmounting and surrounding mountain ranges, pokes well into the upper troposphere to split the westerly air streams of the east Asian region. During winter two jet streams occupy positions to the south and north of this highland mass. The southerly one down the Ganges Valley of northern India is by far the stronger of the two. But in spring it begins to weaken, and by late May or early June it disappears altogether or flops over the highlands to join the jet on the northern side. This event allows for the "burst" of the summer monsoon rains over much of peninsular India, associated with southwesterly airflow at the surface.

An easterly jet then develops near the 150-millibar level in the upper troposphere which, during summer, is centered at about 15°N latitude and spreads from central India eastward and westward to eventually occupy an extensive area all the way from the western Pacific westward through the northern portions of the Indian Ocean, the southwestern tip of Arabia, and across north Africa to the eastern Atlantic Ocean. This appears to stem, partially at least, from the formation of a thermal high pressure cell in the mid- and upper troposphere over Tibet and surrounding highlands. This upland surface absorbs much heat from the strong insolation coming through the thin, clear air at this altitude, which produces temperatures in the air above the highlands much warmer than they would be at similar altitudes in the free atmosphere without an insolation-absorbing surface. Much heat energy is also injected into the upper troposphere along the southeastern fringes of Tibet by the release of latent heat from copious precipitation occurring along the southern mountain slopes in northeastern India at this time of year. The warm air temperatures create a thermal high above the highlands which, rotating clockwise, brings on the formation of an easterly jet along its southern flank.

The easterly air flow aloft over India deflects the southwesterly surface winds at the head of the Bay of Bengal and causes the surface air to proceed up the Ganges Valley from southeast to northwest. It also guides the "monsoon depressions," weak enclosed lows originating in the Bay of Bengal area, northwestward across India.

What appears to be a segment of the intertropical front becomes positioned far north during the summer in the Ganges-Indus Plain of northern India-Pakistan. But in the Indus Valley it does not bring on precipitation. The extreme dryness of the Thar Desert in this region and the rapid decrease of precipitation northward along the west coast of peninsular India are apparently explained by the abnormal slope of the frontal surface in this region—upward to the south—as the hot desert air from the north overrides the less hot marine air from the south off the Arabian Sea. The frontal surface acts as a lid and limits cumulus buildup in the marine air underneath. Only when the marine layer becomes thick enough, far south of the surface front, do showers commence, usually off the coast well to the south of Pakistan.

During winter the surface air generally flows from northeast to southwest across much of India. The westerly jet becomes reestablished over the Ganges Plain and brings in some weak surface lows from the Middle East, which produce small but critically beneficial amounts of rain to Pakistan and northwestern India upon which the winter wheat of this area depends.

12 | WIND

In this chapter wind will be considered as a climatic element, as it influences objects at the earth's surface. It is usually perceived as a significant climatic element only when it is persistent enough or strong enough to be annoying or damaging, or distinctive enough to be a characteristic part of the climate of an area.

Climatic Representation of Winds

Although surface winds are measured fairly accurately and consistently every hour, as are precipitation and temperature, the wind is more difficult to deal with statistically since it is a vector consisting of two components, direction and speed. The most complete way to depict winds at a place over a period of time is in devices such as frequency tables or wind roses, showing the percentage of all time when the wind blows from each direction of a 16-point compass. These usually represent only direction, although average speed from each direction can be added. In any case, such representations are bulky and difficult to compare among different stations.

Prevailing Wind

One of the easiest climatic expressions of wind is the so-called "prevailing" wind that represents all winds at a point by the direction from which they blow most frequently. But if the winds are variable, the prevailing wind might not be particularly representative, since it might blow less than the majority of the time. Figures 6.2 and 6.3 illustrate prevailing winds of the world by directions of arrows. The widths of the arrows indicate percentages of all time that the winds blow from the prevailing directions and thus show how prevailing the prevailing winds are. It is immediately obvious from these maps that the winds are most consistent in direction in the trade wind belts on either side of the meteorological equator, and probably least consistent over land masses in middle latitudes. These maps indicate nothing about wind speeds, which could be depicted by varying the lengths of arrows. More commonly, direction, constancy, and speed are depicted by varying direction, length, and width of arrows, respectively.

Resultant Wind

Perhaps more representative of the total air movement at a given point is the so-called "resultant" wind. This is the vector addition of all the winds at a given point over a period of time, such as a month, for which the winds are being depicted. Theoretically, this represents the net movement of air during the period. Resultant winds can be determined graphically by connecting heads and tails of arrows whose directions and lengths represent the directions and speeds of individual wind observations. Or resultant winds can be derived statistically by dividing each observed wind direction and speed into north-south and east-west components, averaging all these components, and then recombining them into a resultant wind vector.

As mentioned, the resultant wind theoretically represents the net motion of air across the point over a period of time. Consequently, it might not represent any wind that blows, and in a region of variable winds might significantly underrepresent the windiness of the area. To take an oversimplified case as an example, if the wind in a mountain valley is confined to blowing either up or down the valley and does both with equal frequency and equal speed, the resultant wind is zero. Yet calm conditions may never occur in the valley, and in fact rather strong winds might be experienced much of the time.

Where winds blow equally frequently from opposite directions, a prevailing wind cannot be designated, either. In such cases, as far as influences of the wind on surface objects are concerned, it might be more useful simply to average wind speeds and forget about directions. Wind direction is very significant for weather analysis and forecasting, but perhaps wind speed is more significant as far as effects on external objects are concerned.

Resultant wind arrows have been utilized in so-called *streamline analyses*. Individual arrows computed for resultant winds are connected by smoothly curved lines that theoretically depict net air flow over broad parts of the earth's surface and clearly reveal lines and vortices of convergence and divergence, as shown in Figures 6.8 and 6.9. Such streamlines do not relate directly to either constancy of direction or speed of the winds. In mountainous areas it is unrealistic to draw streamlines that ignore details of the topography.

Characteristic Wind

In some areas, occasional characteristic winds, while too infrequent to alter prevailing or resultant wind determinations, may nevertheless have profound effects on the inhabitants of the regions. Examples are the foehn winds of southern California and many other mountainous areas of the world, which blow only occasionally but can bring on the highest temperatures, lowest humidities, clearest skies, and greatest fire hazards of the year. Other examples are extreme winds associated with occasional storms, such as tornadoes and hurricanes, which have already been discussed in Chapters 10 and 11. These phenomena have such profound effects that, although often too infrequent to alter significantly long-term averages of wind directions or speeds, they must be taken into account in the climate of a region. Thus, in a full climatic description of an area one must include the characteristic winds determined by inspection of day-to-day records rather than long-term averages.

Local Winds

Characteristic of many areas are local winds imposed upon the general circulation by topographic influences. Most common are sea and land breezes, mountain and valley breezes, foehns, bora, and katabatic winds.

Sea-Land Breezes

Sea and land breezes tend to set up in a diurnal regime along the coast of any major water body because of different temperature reactions to heat inputs and losses. During daylight hours the air over land warms more rapidly than over adjacent water and eventually becomes warmer and lighter than the air over the sea. A vertical circulation cell is set up, with sea air moving inland at the surface and land air moving seaward aloft. This so-called "sea breeze" will usually begin blowing inland late in the morning and continue through the afternoon. Such breezes may become particularly strong, 10 meters per second (22 mph) or more, along coasts paralleled by cold ocean currents, such as the coast of southern California or the Peruvian coast of South America. In such areas they may penetrate 30 or more kilometers inland and in some cases considerably more, if the inland topography contributes to them. At night the land cools more rapidly than the sea, and after midnight the air over the land becomes colder than that over the sea, so a surface land breeze is set up. This is usually less well developed than the sea breeze during the day.

Along subtropical coasts paralleled by cool ocean currents, the land-sea breeze regime may be a diurnal occurrence practically every day of the year and become the dominant wind feature in the immediate shore area. On other ocean coasts it may occur only occasionally, when the general circulation over the area is weak. Large lakes, such as the Great Lakes of North America, may also cause diurnal regimes, but they are usually much less pronounced and penetrate no more than a few kilometers inland.

Monsoons

In many coastal areas of the world, land-sea exchanges of air may be set up on a seasonal basis. Such seasonal reversals of winds are known as monsoons. By far the best known are the monsoons of southeastern Asia, but as was mentioned in Chapter 11, the seasonal reversal of air flow there is due to more than just seasonal reversals in the heating of land and sea. Conforming more closely to a seasonal reversal of wind caused by land-sea temperature differences is the Arctic coast of Siberia. During winter, cold winds blow from south to north off the cold land toward the relatively warmer air over the Arctic ice. Even though the Arctic Ocean is frozen over right up to the coast during winter, considerable heat is conducted upward from the unfrozen subsurface water through the approximately three-meter-thick surface ice to maintain surface air temperatures around −35°C, which is usually 10 to 15 degrees warmer than over the Siberian landmass to the south. During summer, winds blow from north to south from the relatively cold fringes of the Arctic seas onto the warmer landmass. Ice floes keep water temperatures only three or four degrees above freezing during summer, while surface air temperatures over the land average 10°–15°C.

Mountain-Valley Breezes

In mountainous topography, diurnal regimes of wind may be set up (much as they are along coasts) by the differential heating between valley floors and mountain slopes. During the day, convection currents from heated valley floors move up mountain slopes in the form of valley breezes. During the night, upper slopes of mountains cool rapidly due to radiational heat loss through thin air, and the cooled air moves downslope under the force of gravity to settle into valley floors as mountain breezes. Mountain breezes at night are generally the best developed of this regime, and they may become quite brisk if funneled through constricted areas such as mouths of canyons.

Where mountainous topography and coastlines are in juxtaposition, the two regimes can combine to pull sea air far inland during the day. This is particularly conspicuous in areas such as southern California, where air that has crossed the Pacific Coast at Santa Monica in the morning may be found in the afternoon at Palm Springs, 130 kilometers (80 miles) inland. Pollution control agencies in southern California, tracing the movement of individual particles injected into the air, have found that this much movement does indeed occur during the day. Of course, by the time it has reached Palm Springs the air has warmed to as much as 40°C (104°F) or more, and therefore has lost the characteristic of a sea breeze.

Foehns and Boras

In mountainous areas, other types of local winds may occur that are related to the general circulation over the region rather than to local heating differences. Such are *foehn* and *bora* winds—downslope winds on lee sides of mountains that are crossed by the general circulation of air over the region. These descending air flows warm adiabatically at the rate of 10°C per kilometer (5.5°F per thousand feet). If the air is not too cold to begin with, it arrives at the base of mountains as hot, exceedingly dry, clear air that can increase surface temperatures as much as 15° or 20°C in a few minutes and drop the relative humidity to 10 percent or less, thus creating extreme fire hazards. Temperatures are warmer than they were at the same elevations on the windward slopes because latent heat of condensation was added as the air dropped precipitation as it surmounted the windward slopes.

If the air is funneled through constrictions such as canyons, these winds can reach velocities of as much as 45 meters per second (100 mph) or more, but in open areas the air flow is generally mild, and the increase in temperature and drop in relative humidity are the most discernible characteristics. Such conditions are known as "foehn" winds, a German term meaning a fall or descending wind down a mountain slope. Foehn winds occur in many mountainous areas of the world, sometimes on a seasonal basis if the general circulation of air has a seasonal character. During winter they can be very beneficial because they can cause snow covers to evaporate (sublime), thus keeping pastures open for grazing. The nomadic herders of Central Asia have a saying that "One day of foehn is worth two weeks of sunshine." On the other hand, foehns may induce rapid thawing of snow and disastrous avalanches in mountainous regions, such as the Alps and Caucasus. In some regions effects are so spectacular that foehns are given special names, such as the "Santa Ana" of southern California and the "Chinook" of the northern Great Plains along the eastern foothills of the Rocky Mountains.

If the descending air originally is much colder at the tops of the mountains than the air normally is at the bases of lee sides, the air may arrive at low elevations colder than the normal temperatures, even though the air has warmed adiabatically during its descent. Such is the case during winter when cold air builds up over the Swiss Plateau of Europe and then descends the southern slopes of the Alps through several structural valleys. It frequently arrives on the Mediterranean and Adriatic coasts of France, Italy, and Yugoslavia as relatively cold winds with speeds as high as 20–30 meters per second (45–70 mph). This is very disturbing to vacationers along the Riviera at this time of year. Such winds have been termed "bora" winds, a term derived from the region around the head of the Adriatic Sea. In some areas such winds are known by local names, such as the "mistral" in southern France.

Katabatic Winds

Another kind of descending wind is the katabatic wind, which is caused simply by the force of gravity, cold air descending under its own weight. This is common in areas such as the fringes of the Antarctic and Greenland ice caps, where cold air from the high ice plateaus constantly slides downslope to the sea. These winds are typically gusty, since after they flow for a short while the cold air on the uplands becomes exhausted. After it rebuilds, more air slides down again. Therefore, in the fringes of these ice caps extreme blizzard conditions may occur where first the wind is relatively calm and then suddenly may blow with a speed of 25 meters per second or more, often changing direction abruptly.

Extreme Wind Speeds

Since heat and easterly momentum must be transferred from low latitudes to high latitudes with greatest flux through the middle latitudes, the atmospheric circulation by necessity is generally most vigorous in the middle latitudes. Thus, average wind speeds are generally higher in the middle latitudes than at either low or high latitudes. Of course, there are exceptions under certain conditions, such as the large temperature contrasts between land and sea along some high-latitudes coasts or on occasions when significant pressure gradients occur in lower latitudes. Remember that the Coriolis Force is very weak in low latitudes, and therefore even a weak pressure gradient can cause excessive winds, such as those found in hurricanes. The highest wind speeds on earth are found in tornadoes, which are largely middle-latitude phenomena, primarily in the United States east of the Rocky Mountains. Here it has been estimated that speeds as high as 225 meters per second (500 mph) may occur, although no one has stood there with a three-cup anemometer and measured them.

The highest measured surface wind speeds on earth have been recorded at an elevation of 1917 meters (6,288 feet) on the top of Mount Washington in New Hampshire, where the wind has reached a speed of 103 meters per second (231 mph). There it has averaged 77 meters per second (173 mph) for one hour's duration and 58 meters per second (129 mph) during an entire day. Over the course of the year the wind there averages 16 meters per second (35 mph). This is somewhat less than an annual average of 19 meters per second (43 mph) at Cape Denison along the fringe of Antarctica. During July 1913, winds there averaged 25 meters per second (55 mph) over the entire month. Such winds and low temperatures produce extreme wind chills. The Soviets have estimated wind chill equivalent temperatures as low as −150°C along the Arctic coast of eastern Siberia during winter, when strong land breezes are the rule.

Surface features, primarily vegetation or the lack thereof, greatly influence average wind conditions. Dry lands with meager vegetation and high surface temperatures are conducive to constant winds which, although they may not be extreme in speed, are often so persistent as to become exhausting. Similar conditions prevail over tundra vegetation in the treeless fringes of the Arctic and mountains above the tree line. In many of these areas the convective activity is strong during daylight hours, and there is often a diurnal regime to wind speeds. They are strongest during the afternoon when convective activity has reached its greatest vertical extent and mixes higher-speed winds from above downward into the friction layer next to the earth. The speeds often die down rapidly as sunset approaches and convection ceases, and may be quite calm through much of the night, when the surface air becomes stratified.

13 FORMS AND DISTRIBUTIONS OF CONDENSATION AND PRECIPITATION

Condensation

The process of condensation that changes water from vapor to liquid renders the water visible. Water in gaseous form (vapor) is not visible; therefore, clouds or fog indicate that water in liquid droplet form is suspended in the air. Condensation can be brought about either by cooling the air below its dew point, as was discussed in Chapter 7, or by injecting moisture into it to cause supersaturation. In the majority of cases, condensation is brought about by cooling the air temperature.

Condensation is a fairly automatic process once the air becomes saturated with water vapor, that is, once the relative humidity reaches 100 percent. Although in theory supersaturation may result to several hundred percent relative humidity, in very calm, clean air free of condensation nuclei, in nature this really does not occur. The air contains so many hygroscopic (water-attracting) particles (such as sea salt, smoke, and dust) that condensation is usually guaranteed at about 100 percent relative humidity. In fact, in very dirty air condensation may begin when the relative humidity is no more than 90 percent.

Air can be cooled either by diabatic processes at its base or by adiabatic processes through lifting, as was discussed in Chapter 8. Through diabatic processes heat is lost by the air to its external environment, such as the surface of the earth and objects on it. If the air's temperature is cooled below the dew point, some visible form of condensation will occur. Such cooling is usually accomplished by either of two methods: (a) *radiational* heat loss from the earth's surface at night or (b) *advection* (horizontal movement) of air over a colder surface.

Dew and Frost

Dew and frost almost always are the result of radiational heat loss from the earth's surface at night. For maximum heat loss to occur to assure the decrease of surface air temperature to below the dew point, skies should be clear and the surface air relatively calm. If clouds exist, they will probably absorb most of the terrestrial radiation and reradiate it back to earth, so the surface air temperature will not reach the dew point during the night. And if there is too much mixing of the surface air, the cooling of air in contact with the colder earth's surface will be mixed upward through too thick a layer of air for any portion of it to be cooled below its dew point. On the other hand, the radiational heat loss from the earth's surface on calm, clear nights will be great, and the cooling of the air in contact with the earth's surface will be limited to a thin layer. Under such conditions a surface air temperature inversion may be established, and dew or frost may form on surface objects.

Whether dew or frost forms depends upon the relationship between the dew point temperature and the freezing temperature. If the dew point is below freezing, the vapor in the air will change directly to tiny ice crystals without going through the liquid state. Since the ice crystals have much air trapped within them, the frost takes on a very whitish, fluffy appearance. This is known as hoar (white) frost. Of course, if the dew point is a little above freezing, the cooling during the night may reduce the air temperature, first, to below the dew point, which causes dew to form, and then further to below the freezing point, which freezes the water that has collected on plants and other objects. This is known as black frost and is generally more damaging to plants than hoar frost.

Since air cooled by radiational heat loss tends to settle into the lowest parts of the topography, dew and frost are found most frequently in low portions of the landscape. Since net radiational losses generally occur only at night, dew and frost are usually quickly evaporated after the sun comes up the next morning.

Fog

If many of the condensed water droplets remain suspended in the air, visibility will be impaired by fog. This will require enough mixing of the surface air to keep the water droplets suspended. But if mixing is too great, the surface cooling will be spread through too thick a layer for the surface air to be cooled below the dew point. Therefore, fog formation, particularly that due to cooling by radiational heat loss at night, places very stringent requirements on atmospheric motion. Because of such strictly limited criteria for fog formation, it is difficult to forecast.

Fog is categorized according to the method by which it was produced. Most fog is due to the cooling of the surface temperature below the dew point, either through radiational heat loss or through the advection of moist air over a cooler surface. Radiation fog is usually thin and spotty, commonly occurring only in the lowest topographic depressions. It typically dissipates shortly after sunrise after insolation once again raises the temperature of the surface air above the dew point. Advection fog, on the other hand, is usually widespread over large continuous areas, thick, and persistent. Since it does not depend upon radiational heating and cooling, it does not show a diurnal regime, but persists as long as the surface air is advected over a cooler surface. This is very prevalent in some coastal areas, particularly those paralleled by cool ocean currents, such as many of the subtropical western coasts of the world. In these areas fog occurs most frequently in summer, when the temperatures of the oceans are coolest relative to those of adjacent land.

Advection may also be a factor in the formation of two other types of fog: upslope fog and warm front fog. When surface winds blow persistently up mountain slopes or up a gradual incline, as they do when they move northwestward from the Gulf Coast of the United States into the western Great Plains of eastern Colorado and Wyoming, adiabatic cooling can reduce the temperature below the dew point and clouds can form that are still on the ground, hence fog. Thus, windward slopes of mountain ranges often record high frequencies of fog.

Thick, persistent fog is often associated with leading edges of warm fronts in cyclonic storms. Warm, moist air rising over colder air produces a vapor pressure gradient directed downward from the moister air above. Since the actual vapor pressure of the warm air is higher than the saturation vapor pressure of the colder air below, moisture will move downward even though the lower air may be saturated. This supersaturates the cool surface air and some of the moisture condenses into liquid water droplets to produce fog. Also, some of the moisture in the warm upper air may be mixed downward into the cool air, and some may fall as rain droplets into the cool air below. All these processes combine to produce a very dense, persistent fog that is often mistaken as simple advection fog in such areas as the Great Lakes region of North America, when warm, moist air is being advected northward from the Gulf of Mexico into a cyclonic system. Such fog is usually strictly demarcated along the warm frontal surface, however; it is found in the cool air underneath the warm front, but not in the warm surface air equatorward from the front's surface position. If the front becomes stationary, the fog may hang in a certain belt for many hours or even days, and then suddenly clear within a few minutes as the front pushes on again. Although advection is a necessary part of its formation, in order to bring the moist air over the frontal surface, the fog itself is produced not by the surface cooling of the warm air, but by supersaturation of the cool air underneath the front by the downward injection of water vapor.

Fog may also be formed by the injection of more water vapor into almost-saturated air when cold air is advected across a relatively warmer water surface. In middle latitudes this is a frequent occurrence over lakes and rivers during early winter, before the water surfaces freeze. Because the vapor pressure of the water surface is greater than the saturation vapor pressure of the cold air above, moisture moves upward from the water surface into the air, even though the air might already be saturated. Almost immediately above the water surface, the water vapor will recondense into liquid water droplets and form fog. Since heat is also being added to the lower air, active turbulence will mix the water droplets upward into the air. Under such conditions it appears that steam is "boiling up" from the water's surface, hence the name "steam fog." With extreme water-air temperature differences, this sort of fog may take on the appearance of low cumulus clouds based near the surface and extending a hundred meters or more into the air. Under such conditions major water bodies can be discerned from a distance by lines of menacing low clouds in an otherwise crystal-clear sky. In sparsely populated regions such as much of Canada and Siberia, most of the settlements are found in river valleys and along lake shores, so climatic records there often show unusually high fog frequencies in these areas during autumn and early winter. But these conditions do not pertain in uplands between streams.

Since fog often is fleeting in existence and scattered in distribution, it is difficult to represent accurately on a world scale, particularly over land where more of the fog is due to radiational cooling. Nevertheless, some general patterns can be discerned. Outstanding in fog occurrence are the subpolar seas along the fringes of the Arctic and Antarctica, especially during summer when open water fringes the ice caps and causes large temperature differentials. Fog is also relatively common along certain subtropical and middle-latitude coastal areas paralleled by cold ocean currents. And a great deal of northwestern Europe experiences persistent advection and frontal fogs during winter, as warm moist air from the Gulf Stream and its extension, the North Atlantic Drift, moves eastward from the Atlantic over the continental landmass.

Clouds

Although cooling at the surface can reduce the air below the dewpoint to form dew, frost, or fog, in such cases only a thin layer of surface air is cooled below the dew point and, hence, the total amount of condensation is limited. The only way a thick layer of air can be cooled enough to cause precipitation is by adiabatic lifting and cooling, as was illustrated briefly in Chapter 8. No matter how dry the surface air, if it can be lifted high enough, its temperature will eventually be reduced below the dew point and the condensation level in the atmosphere will be reached. Once this has happened, the liquid water droplets will become visible as clouds. Lifting may be accomplished either by the wind blowing up a terrain barrier, such as a mountain slope (*orographic* lifting), by advection up a frontal slope (*cyclonic* lifting), or by turbulent air currents set up by heating at the earth's surface (*convective* lifting).

Forms. The form the clouds take is dictated primarily by the stability characteristics of the rising air. If the air is stable, clouds will form if the air is forced to rise up a mountain or frontal slope above its condensation level; these clouds will be straiform, featureless layered clouds that extend upward only as high as the air is forced to rise by some external force. If the air becomes unstable upon lifting, large vertical air currents will develop cumuliform clouds with large vertical extent and billowing shapes.

Stratus and cumulus are the two basic forms of clouds that produce precipitation. A third form, cirrus, are thin, wispy, high clouds composed mostly of ice crystals, but these do not cause precipitation. Since cirrus often mark the leading edges of cyclonic systems, they may be the harbingers of approaching storms, but long before precipitation commences the clouds will have lowered and thickened into stratus or cumulus types.

A fourth term that is applied to clouds is "nimbus." Nimbus does not refer to a particular form, but indicates only that the cloud is precipitating. The term is never used alone, but always in combination with one of the three terms indicating form. Hence, a cumulonimbus cloud is a cumulus cloud that has grown large enough to cause precipitation to occur. By this time, the cloud has usually grown upward beyond the freezing level in the atmosphere, and the top of the cloud has become glaciated and takes on an anvil form, as was described in Chapter 10. Nimbostratus clouds are featureless dark gray clouds that usually produce continuous precipitation over widespread areas, often associated with warm or occluded fronts. Weather services generally recognize about 30 cloud types, but they are all combinations of stratus, cumulus, and cirrus forms and their combinations with nimbus and height designations.

Clouds are the visible portion of the sky and lend much of the character to day-to-day weather. An understanding of their forms relative to the processes that produce them allows for a considerable understanding of the synoptic situation over a general area and perhaps some fairly accurate short-range forecasting, even without access to weather reports from surrounding areas. Clouds are very important climatically, since they greatly affect radiational exchanges between space and the earth's surface. Their tops are highly reflective to sunlight, having albedos comparable to that of fresh snow. Their water droplets are effective absorbers of heat energy, particularly the long wave radiation from the earth's surface. And they are very effective radiators of long-wave radiation. Most of the heat radiated from the earth-atmosphere system to space is radiated from cloud tops. And the amount of cloudiness seems to affect people psychologically: changeability seems to be more stimulating than either prolonged, dreary overcast or searing desert sun.

Distribution. On the average, the subpolar low pressure areas have the greatest amounts of cloud cover and the subtropical high pressure cells, the least. Clouds in these zones vary some in amount and shift latitudinally with seasons (Figures 13.1 to 13.4). They also vary somewhat from day to day according to synoptic situations, but not nearly as much as in other portions of the earth where circulation systems are less

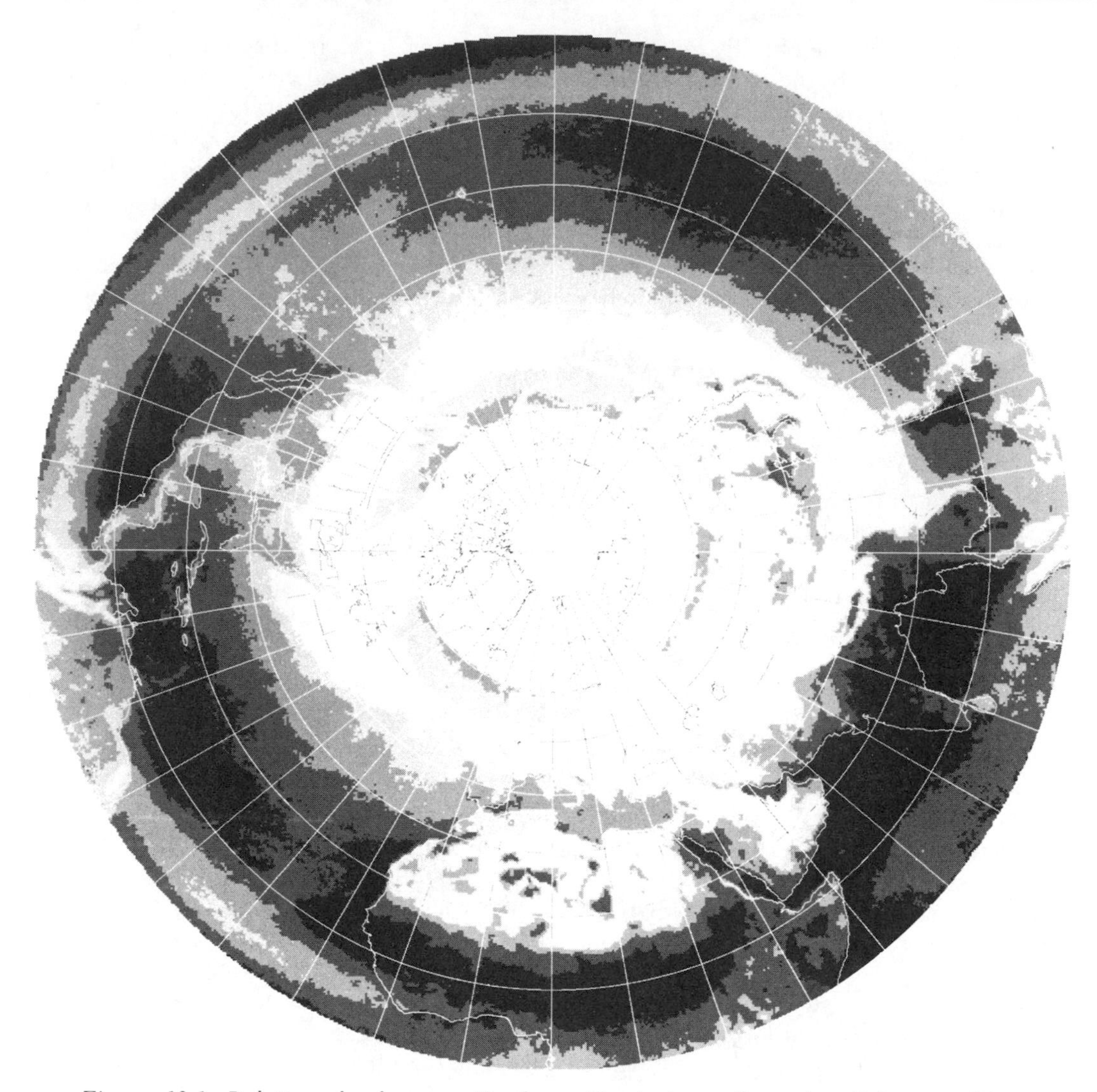

Figure 13.1. Relative cloud cover, Northern Hemisphere, December–February. (From Global Atlas of Relative Cloud Cover, *1967–70 [U.S. Department of Commerce, Washington, 1971].)*

consistent. Other gross features of cloud cover and their seasonal variations can be discerned even on these small maps. Such are the monsoonal effects on cloud cover over India and southeast Asia. Widespread cloudiness results from the southwest monsoon during summer. There is much less cloudiness during winter, although there is some intense cloud formation along the east coast of Vietnam, and to a lesser extent along southeastern India, derived from the northeast monsoon at this time of year. Another conspicuous seasonal reversal takes place over the Mediterranean Sea, which area appears quite clear during summer but is relatively cloudy during winter, although it is still less cloudy than Europe to the north or Africa to the south. In the Southern Hemisphere, conspicuous stratus cloud decks form off the coasts of Peru and northern Chile and off southwest Africa during winter (June–August), but are not

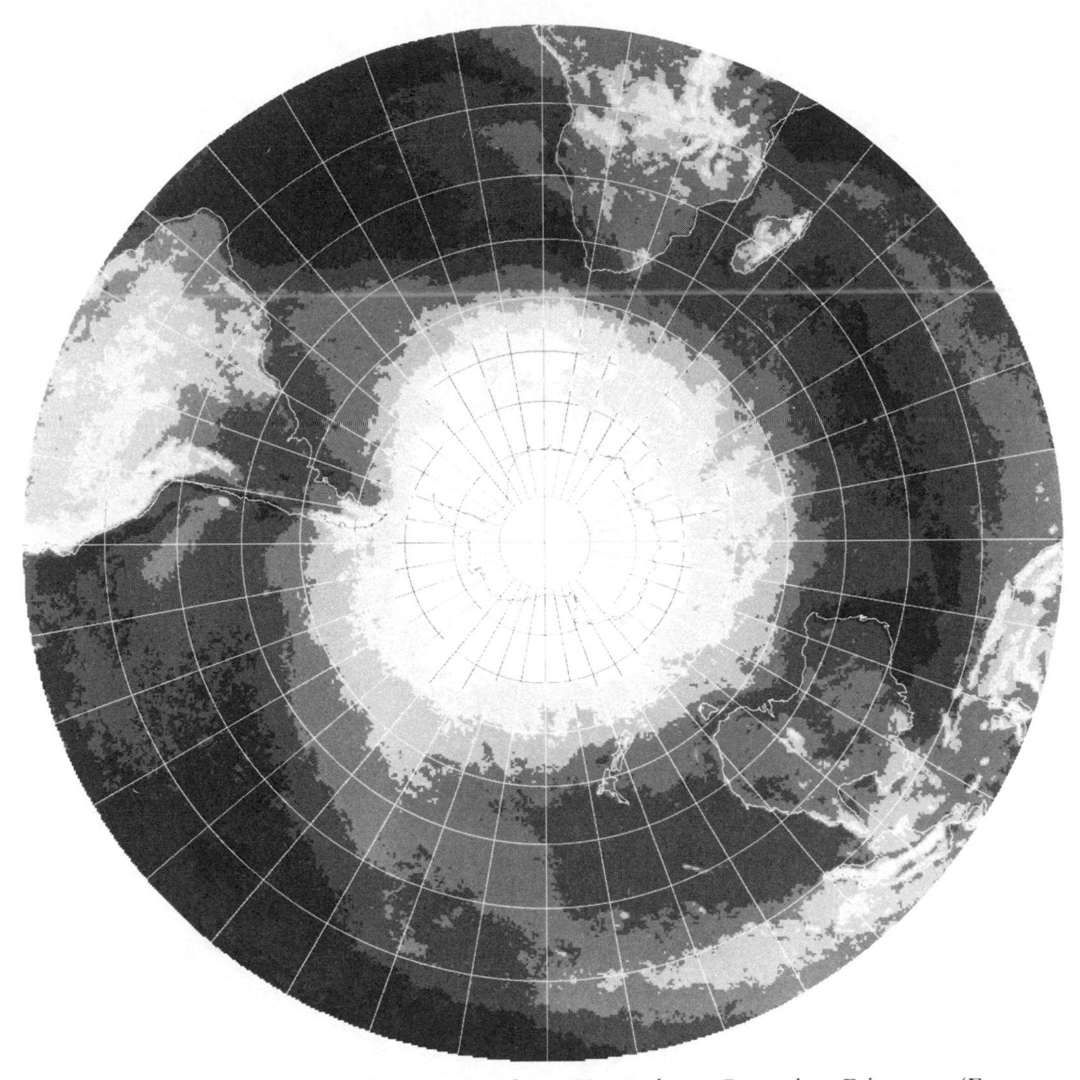

Figure 13.2. Relative cloud cover, Southern Hemisphere, December–February. (From Global Atlas of Relative Cloud Cover.*)*

as conspicuous during summer (December–February), when cumulus clouds tend to build up over the adjacent land.

It must be borne in mind that these polar projections of cloud cover are composites of data derived from satellite sensors over a period of four years, and represent only the amount of sky covered by clouds, not the cloud type or thickness. In general, the clouds in subpolar areas are stratiform that do not extend to great heights. They usually result in overcast conditions and persist endlessly over time with very few breaks. This dreary type of weather is so prevalent in northern European Russia that the Russians have a special term for it—*pasmurno*—which means dull, dreary weather. Likewise, the cloud decks that form under subsidence inversions over the cool ocean currents off western South America, southwest Africa, and other similar coastal areas are stratiform, with thicknesses generally of no more than a few hundred meters. It is difficult to compare the climatic

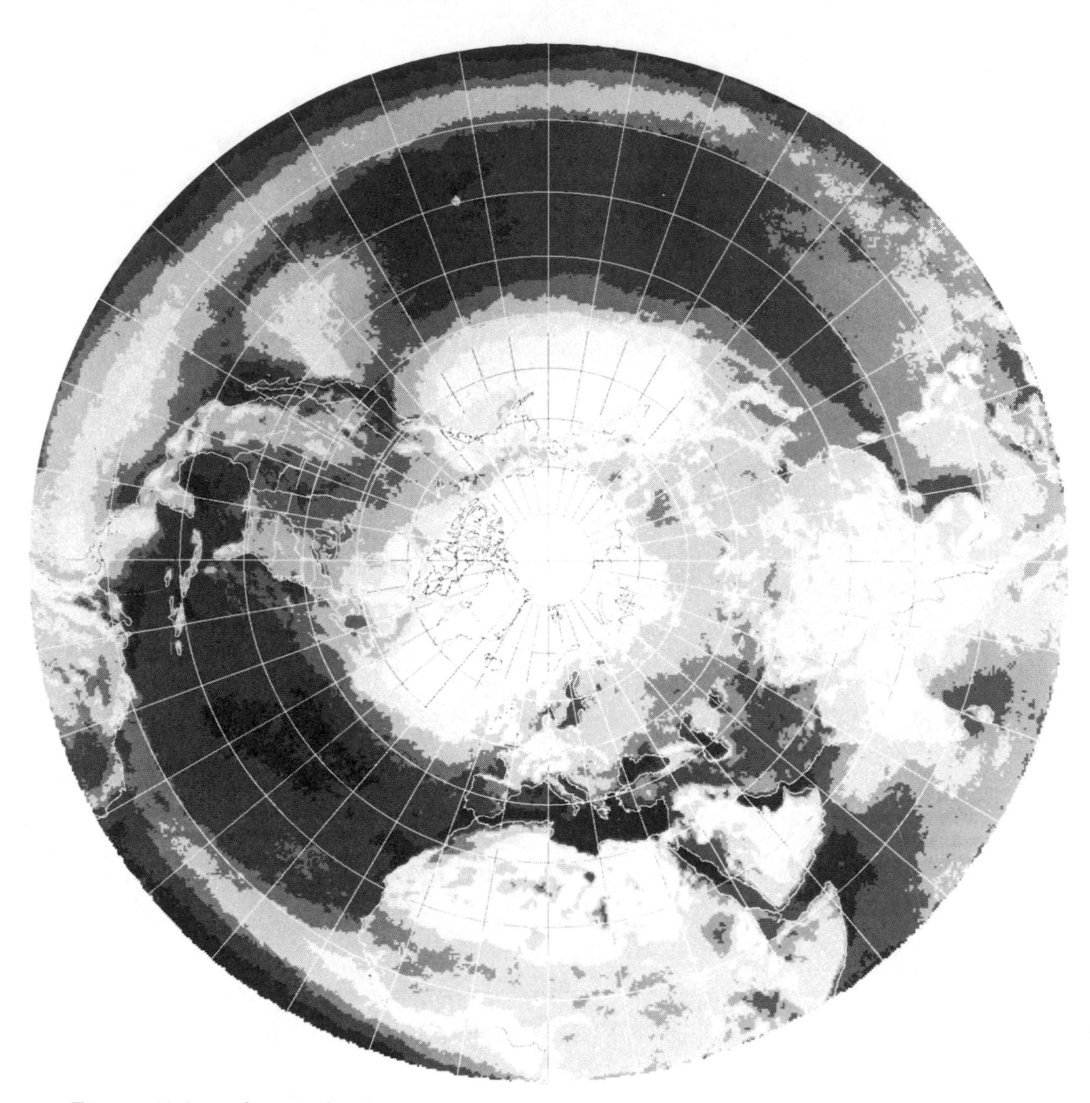

Figure 13.3. Relative cloud cover, Northern Hemisphere, June–August. (From Global Atlas of Relative Cloud Cover.*)*

effects of such clouds with the great vertical development of cumulus extending upward 10,000 meters or more in places in the tropics, subtropics, and lower middle latitudes.

One must also remember that these composites of satellite-sensed light include light reflected from all surfaces below the satellites. By far the most reflective of these surfaces are cloud tops, snow, and ice. Thus, in higher middle latitudes, particularly in the Northern Hemisphere, the fact that the subpolar areas have greater cloudiness during summer than during winter is somewhat masked by light reflected during winter from sea ice and snow. Also during winter, no sunlight falls on the polar areas so no satellite sensing of clouds was possible when these mosaics were compiled. Therefore, large whited-in circles blot out the pattern in these regions. Infrared sensing of clouds is now done routinely, day and night. Desert

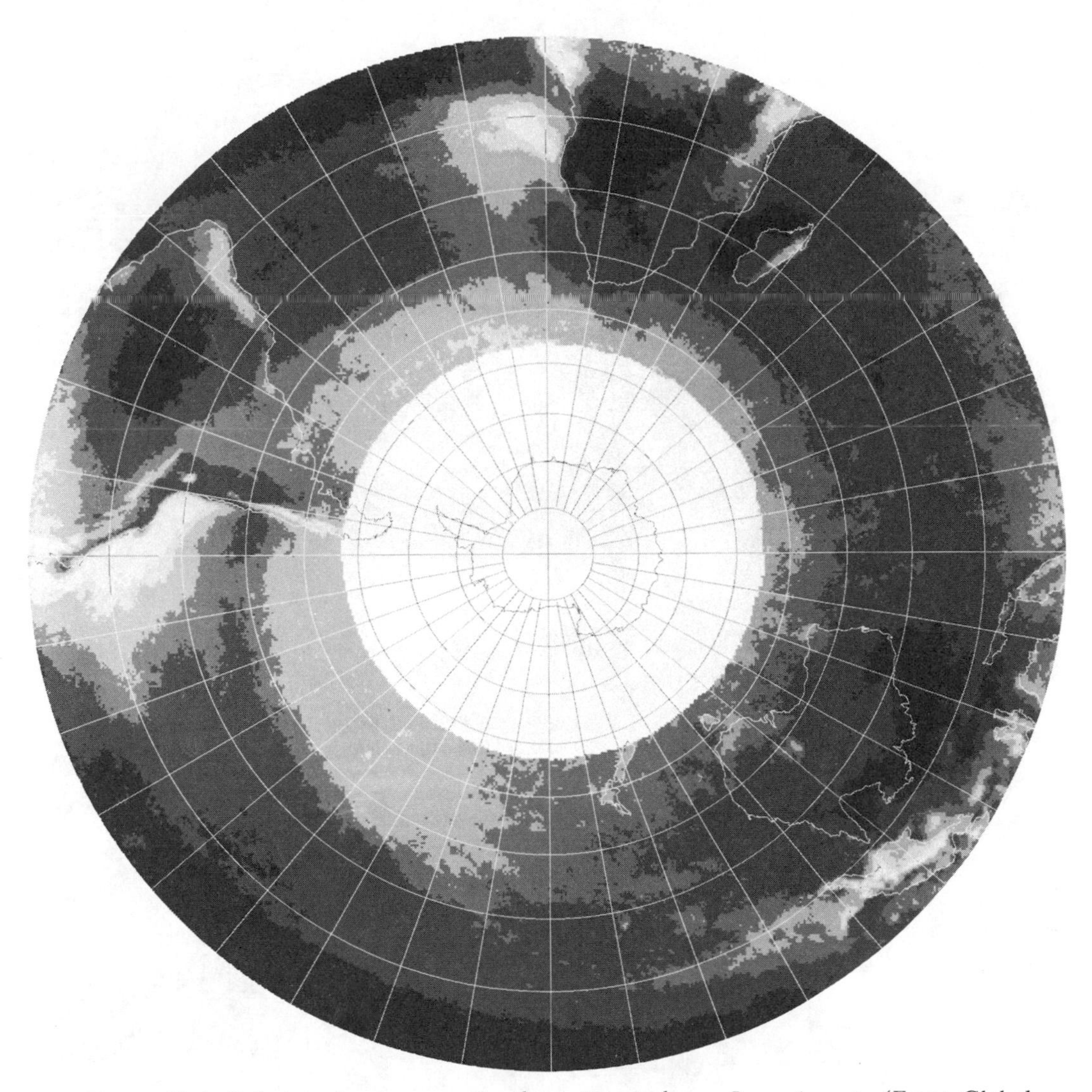

Figure 13.4. Relative cloud cover, Southern Hemisphere, June–August. (From Global Atlas of Relative Cloud Cover.*)*

sand also has a fairly high reflectivity and might account for some of the seemingly high amounts of cloud cover in places such as the Sahara and Arabia.

The lower latitudes are depicted in Figures 13.5 and 13.6. The most-conspicuous pattern in the tropics is the extended band of clouds approximately 5- to 10-degrees latitude wide, extending across the Pacific and Atlantic oceans. This, of course, represents the intertropical convergence zone, and its position north of the equator throughout the year corroborates the discussion in Chapter 6 regarding the northerly position of the meteorological equator relative to the geographical equator, due to the stronger general circulation of the atmosphere in the Southern Hemisphere. Also well depicted on these composites are the extensive developments of stratus cloud decks over cold ocean areas off the subtropical western coasts of continents with

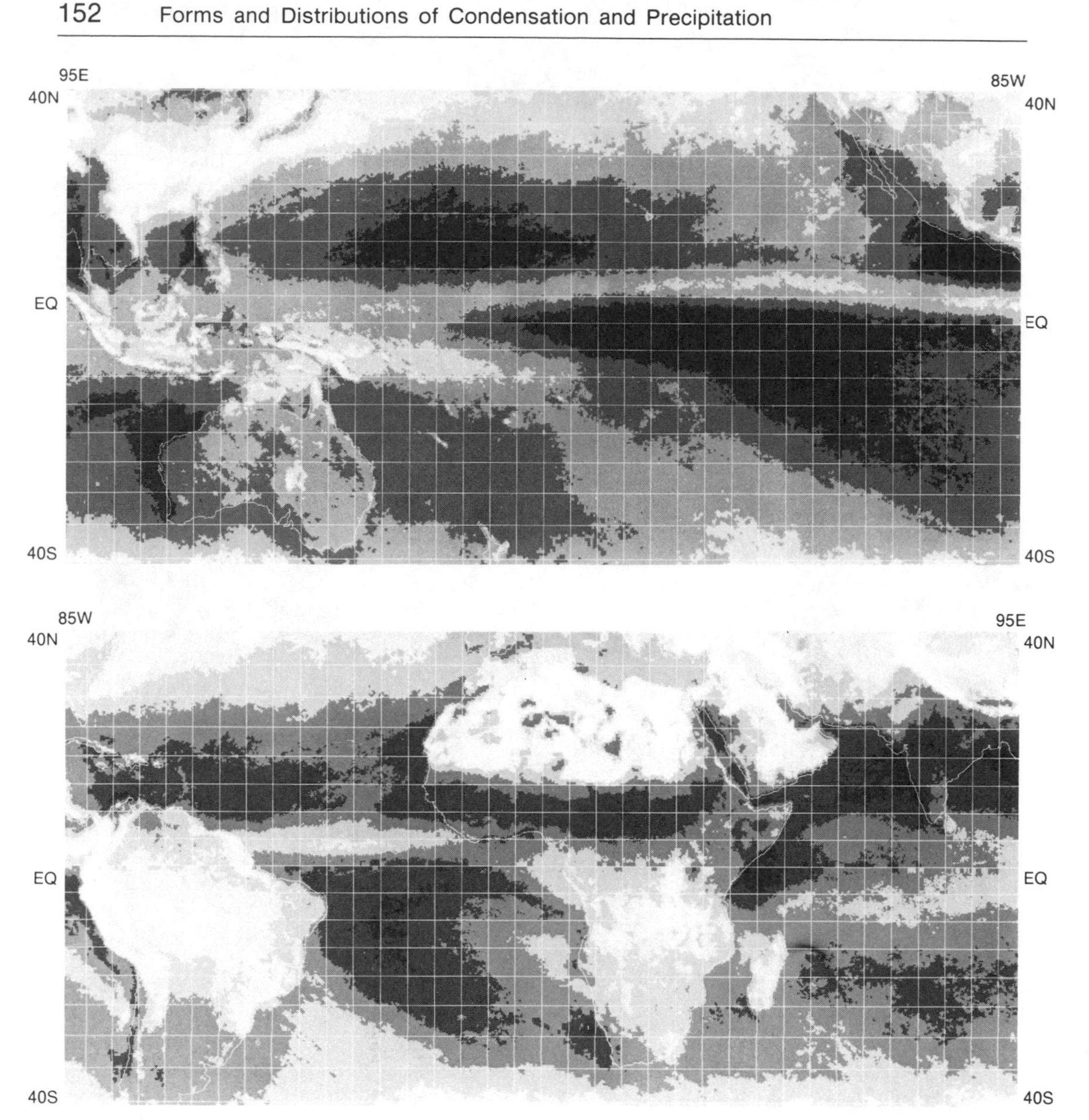

Figure 13.5. Relative cloud cover, December–February. (From Global Atlas of Relative Cloud Cover.*)*

their greater development during winter (June–August) off South America and southwest Africa, and during summer (June–August) off Baja California. Seasonal change in cloudiness over India and southeast Asia is also striking.

Precipitation

Precipitation is not as automatic a process as condensation. One can confidently expect that condensation will take place as 100 percent relative humidity is approached, but in no way can one be sure when precipitation will fall from a cloud. The initiation of this process seems to be related primarily to drop size. Most cloud droplets are quite small and, on falling, evaporate long before they reach the surface of the earth. Precipitation depends upon many of these cloud droplets coalescing into drops that become large enough to fall to earth.

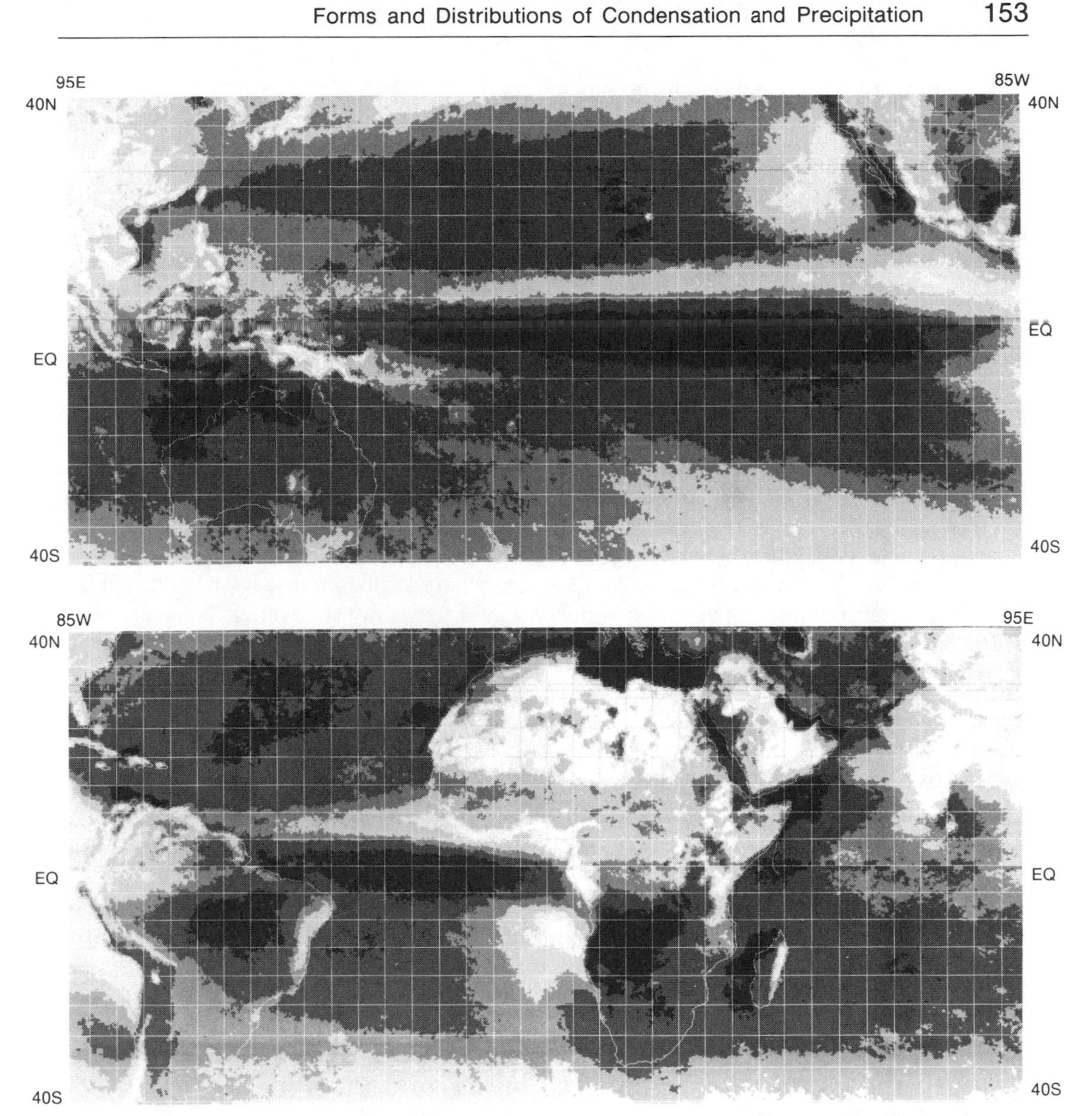

Figure 13.6. Relative cloud cover, June–August. (From Global Atlas of Relative Cloud Cover.*)*

Several physical laws favor this happening. (a) The cloud droplets are in constant motion and undergo constant evaporation and recondensation within the cloud, providing ample opportunity for collisions to take place and for moisture to be transferred from one droplet to another. (b) Aiding this process is the fact that the surface vapor pressure of a drop of water is inversely related to the radius of the drop. Therefore, vapor pressure is higher around the surface of a small droplet than around the surface of a larger drop. This creates a vapor pressure gradient from small drops to large drops which enables larger drops to grow at the expense of smaller ones. (c) Greatly speeding up the process is the fact that above the freezing level in the atmosphere, supercooled water droplets and ice crystals coexist within clouds, and at the same temperature the surface vapor pressure is greater over water than over ice.

Thus, moisture tends to move from water droplets to ice crystals, and ice crystals act as nuclei around which large water drops form.

Some meteorologists contend that no precipitation can occur without ice crystals in the higher levels of the clouds; others point out that considerable amounts of precipitation fall in the tropics when clouds have not built up to the freezing level. It does appear that the availability of appropriate hygroscopic nuclei in sufficient abundance is a major criterion for precipitation to occur. This principle is the one exploited in so-called "cloud seeding" projects that attempt to trigger showers prematurely by injecting nuclei consisting of crystals of dry ice (frozen carbon dioxide) or silver iodide into clouds that contain supercooled water droplets at the right stage to coalesce around such nuclei. Nonetheless, the effectiveness of these methods is still under question.

Forms of Precipitation

Various forms of precipitation can occur, depending upon the relations between (a) the dew point and the freezing point in rising air that is causing moisture to condense, (b) the temperature of the air through which the precipitation must fall on its way to the surface, and (c) upon the temperature of the surface itself. If the dew point of the rising air is above freezing, water droplets will form as the air rises above the condensation level. If nothing happens to them as they fall to earth, they will fall as rain. They may evaporate on the way down and never reach the earth, or they may fall through a colder air mass underneath a front, whereupon they freeze into ice crystals that arrive at the earth as sleet. If the temperature of the cool lower air is around the freezing point, falling rain drops may not freeze until they reach the earth, whereupon they freeze to all surfaces to form glaze ice.

If the dew point of the rising air is below freezing, water vapor will turn directly into ice crystals above the condensation level. These grow into snowflakes in the cloud, and if nothing happens to them as they fall to earth they fall as snow. Many times they melt on the way down and fall as rain. Sometimes falling supercooled water droplets freeze to produce small, round, whitish pellets called "graupel" (sometimes called "soft hail"), which bounce when they hit the ground.

Snow does not fall everywhere on earth (see Figure 13.7). In general, it is a rare occurrence equatorward of about 30° latitude in the Northern Hemisphere and at an even higher latitude in the Southern Hemisphere, except in high mountains where it may occur at any latitude. And anywhere it does fall it soon melts, unless winter temperatures average below freezing. Many countries do not record amounts of snowfall; indeed, the Russians argue that what is important is not how much snow falls, but how much accumulates. This is not entirely true, of course, since occasional heavy, wet snows wreak havoc in areas such as the southeastern United States where, on the average during winter, there is little or no snow cover.

It is difficult to depict the distribution of snow, since snowfall data are inadequate and snow depth on the ground varies drastically over short distances. Yet a snow cover is such a significant control over various other aspects of climate that attempts have been made to map it accurately, and some climatic classification schemes have set major climatic boundaries according to whether or not there is a persistent snow cover during winter. Once a durable snow cover has been established, the entire heat exchange of the surface is altered drastically. (The effects of these altered relationships on temperature magnitudes and seasonal distributions will be taken up in the next chapter.)

An unusual form of precipitation is hail, which consists of spherical or irregular lumps of ice that generally fall during the hottest days of the year. This seemingly

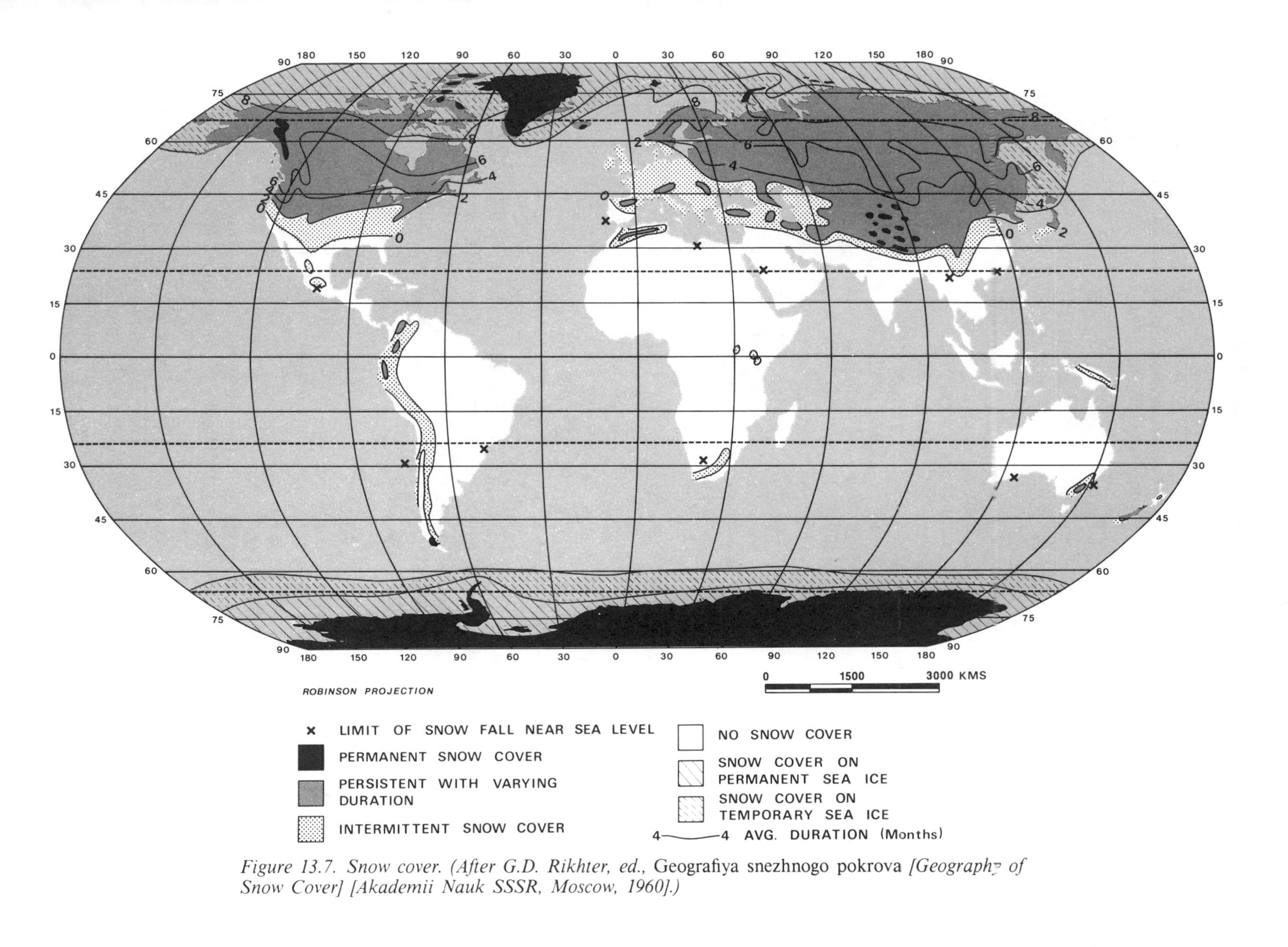

Figure 13.7. Snow cover. (After G.D. Rikhter, ed., Geografiya snezhnogo pokrova *[Geography of Snow Cover] [Akademii Nauk SSSR, Moscow, 1960].)*

paradoxical timing is explained by the necessity for strong updrafts, which are generally present only during very warm periods when the air has a high degree of instability. Hail is formed by rain drops being carried upward beyond the freezing level. Passing through layers of air with differing cloud droplet concentration, they accumulate alternating shells of clear and rime ice. Most hailstones are less than a centimeter (.4 inches) in diameter, but hailstones as large as 14.2 centimeters (5.6 inches) in diameter have been recorded in Kansas. Hail is always associated with severe thunderstorms, and its distribution over the earth relates more to the intensity of thunderstorms than to the frequency of their occurrence (compare Figures 13.8 and 10.12). Places such as northern India with large vertical moisture gradients, hence great convective instability, are especially prone.

Amount and Distribution of Precipitation

The amount of precipitation that falls on the earth's surface during the course of a year varies greatly from one part of the earth to another, from less than 50 millimeters (2 inches) along the subtropical west coasts of South America and southwest Africa and parts of the Sahara and Central Asia, to more than 5000 mm. (200 inches) in some mountainous coastal areas, such as northeast India and Burma, the Cameroon and Guinea coasts of Africa, the western Colombian and southern Chilean coasts of South America, and spots in the Hawaiian islands (Figure 13.9).

In a few places where the atmospheric moisture content is very high and the atmospheric flow consistently converges and rises up mountain slopes, the precipitation may be copious. Such spots are Mt. Waialeale on the island of Kauai, the northwesternmost of the large islands of the Hawaiian chain, which averages 11,675 mm. (460 inches) of rainfall per year; Cherrapunji, at an elevation of 1313 meters on the southern slopes of the Khasi Hills in northeastern India, which averages 11,419 mm. (450 inches); and Debundscha, near the base of Cameroon Mountain just north of the equator at the bend of the continent of western Africa, which averages 10,279 mm. (405 inches). Similar amounts may occur in spots in western Colombia, but comparable records are lacking there.

Areas with the least rainfall are, apparently, Arica in northern Chile, which over a period of 59 years averaged 0.75 mm. (0.03 inches), and Wadi Halfa, Sudan, which over a period of 39 years averaged 3 mm. (0.12 inches). Much of the coast of Peru and northern Chile averages less than 25 mm. (1 inch), as do portions of the southwest coast of Africa. In such places years may pass with no measurable precipitation, and then a rare shower will account for the long-term average.

Since precipitation falls in different forms, mainly rain and snow, with different proportions in different parts of the world, to construct a world distribution map (such as Figure 13.9) to show correct relative amounts of precipitation, all forms must be reduced to water equivalents. The standard practice is to convert snow to water at a ratio of 10 to 1. The wetness of snow, of course, varies considerably, and now many places weigh or melt snow samples to obtain more realistic water equivalents. In places such as the upper slopes of the southern Andes and the coastal mountains of the Alaskan Panhandle and British Columbia, much of the precipitation falls as snow. Thus, the 3000–5000 mm. of precipitation shown on the map in these areas may amount to more than 3000 centimeters (100 feet) of snow in addition to some rain. In such places glaciers accumulate in valleys between mountains, even though temperatures are not extremely cold. But so much snow falls that all cannot melt during summer, and it accumulates over the years and consolidates into glacial ice.

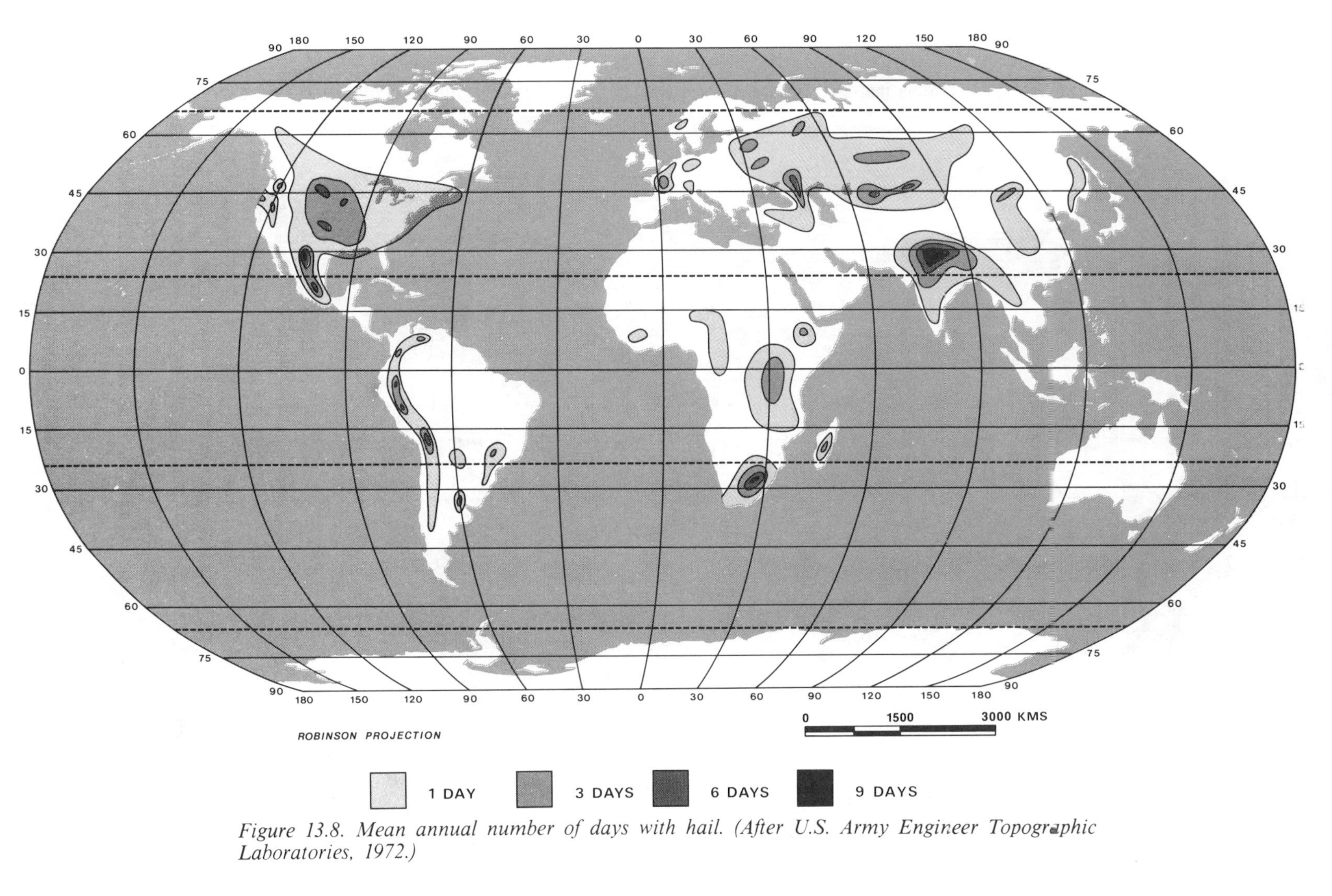

Figure 13.8. Mean annual number of days with hail. (After U.S. Army Engineer Topographic Laboratories, 1972.)

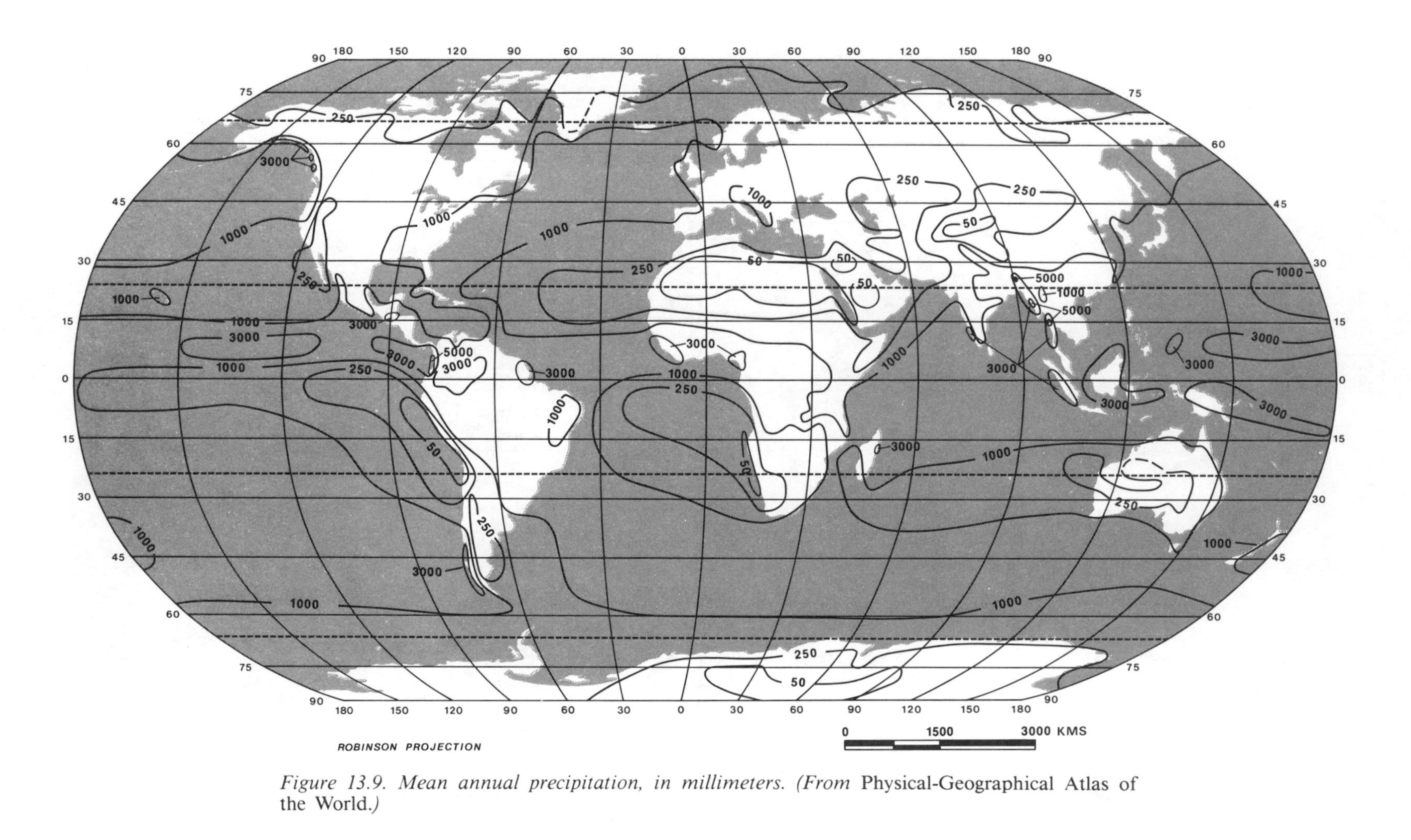

Figure 13.9. Mean annual precipitation, in millimeters. (From Physical-Geographical Atlas of the World.*)*

It is apparent from Figure 13.9 that precipitation totals depend on two major controls: (a) amounts of moisture in the air and (b) lifting mechanisms to cause the water to precipitate out of the air. Except for a few irregularities induced by rugged topography, such as the southern Andes and northern Rockies, by far the greatest amounts of precipitation fall in near equatorial areas where atmospheric moisture is great and the air is generally converging along portions of the intertropical convergence zone, at least part of the year. In places where orographic uplift is added, precipitation may become copious.

Areas receiving less than 500 mm. (20 inches) of precipitation can be categorized into three groups, although areally they merge in some places. First, the subtropical high pressure belts, essentially straddling the tropic circles, account for many of the extensive deserts on earth, such as the Sahara and Kalahari in Africa, the Australian desert, and the coastal desert of western South America. Particularly in the coastal areas of these deserts, the oceans provide enough moisture for precipitation to occur, but the lifting mechanism is lacking. Strong atmospheric subsidence in the subtropical high pressure cells produces adiabatic warming and, in most cases, extreme temperature inversions that stymie precipitation processes. Second, interiors of large continents, such as much of Central Asia north of the high mountains in the Soviet Union, China, and Mongolia, and the intermontane region of western United States, have meager precipitation because of the lack of a moisture source. Ordinarily, the atmosphere in these regions is relatively dry, either because they are so far from the sea or because marine air is blocked by mountain ranges. Third, the high latitudes have only meager precipitation, the polar areas generally have no more precipitation than some of the subtropical deserts. Because the air is always cold in these areas, its capacity to hold moisture is very low, so there is little capacity for precipitation. Even though precipitation can occur fairly frequently in some of these areas, the amounts are not great. Since it is always cold or cool, the ground is either frozen or very wet during a thaw, so such areas are not classified as dry despite their meager precipitation.

Much of the plains areas of the middle latitudes receive 500–1500 mm. (20–60 inches) of precipitation, which is generally favorable for crops adapted to these areas, but seasonal and annual fluctuation in amounts may cause problems.

Seasonality. Latitudinal zones that lie within the influence of the same circulation systems throughout the year tend to have about the same precipitation in all seasons. In contrast, intermediate zones lying between two very different precipitation controls may have great seasonal contrasts, because the belts of high and low pressure, frontal zones, and so forth, shift latitudinally with seasons. Chapter 4 pointed out that the sun's direct rays shift from 23.5°N latitude on 21 June to 23.5°S latitude on 21 December, and the zone of maximum heating shifts even more widely, to about 30°–35° latitude on either side of the equator during the course of the year. Although inertias within the atmosphere keep the wind and pressure belts from shifting nearly that far, and local effects do induce irregularities, the circulation systems that largely control precipitation shift significantly from one season to another in most parts of the world.

These shifts result in the various belts shown in Figure 13.10. For instance, it can be seen that as the intertropical convergence zone shifts about 10° latitude during the year, the zone 5° or so on either side of the equator is within its influence year-round and hence receives rain throughout the year. In actuality, in much of the Atlantic and Pacific oceans, this zone remains north of the geographical equator all year, since the meteorological equator in those

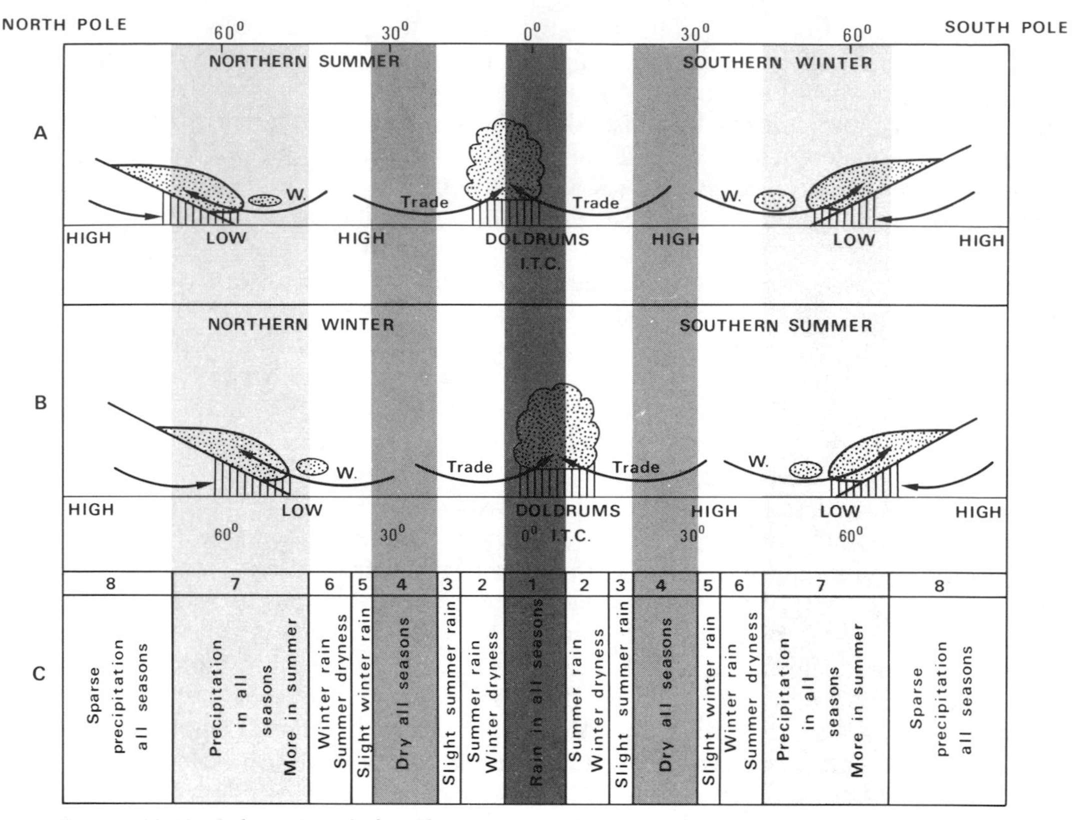

Figure 13.10. Schematic shifts of pressure and wind belts and resultant precipitation during the course of a year. (From Introduction to Meteorology *by Sverre Petterssen. Copyright © 1958 by McGraw-Hill, Inc. Used with permission of the McGraw-Hill Book Company.)*

regions lies significantly to the north of the geographical equator (see Figure 13.9). On either side of this so-called "doldrums" belt of continuous precipitation lie belts that receive rain primarily during summer, when the poleward influence of the intertropical convergence zone shifts into their vicinities. During winter the subtropical high pressure belts shift equatorward over these zones and bring drought. These are the next significant precipitation controls poleward from the intertropical convergence zone, and their influences are felt throughout the year farther poleward, around 20°–30° latitude on either side of the equator, where the climate is dry year-round.

The next significant precipitation controls poleward are the segments of the polar front with its cyclonic storms in the prevailing westerlies in the middle latitudes. These circulation systems affect belts from about 45° to about 65° latitude on either side of the equator and bring precipitation to this latitudinal zone year-round. Just to the equatorward of these belts lie zones about 10°–15° latitude wide (approximately latitudes 27°–32° to 38°–45°) that receive most of their rain during winter when the prevailing westerlies shift equatorward, but during summer they are akin to deserts as the subtropical high pressure belts shift poleward over them. Poleward of about 65° latitude there is usually some precipitation throughout the year, but in all seasons it is meager because of the restricted water-holding capacity of the colder air.

In the middle latitudes a fairly pronounced summer maximum of precipitation is the rule in interiors of continents, since at that time of year convective activity is much greater than during winter. Within these continental interiors, the summer maximum tends to lag later into summer the farther poleward the area, since in the higher middle latitudes (55°–65°) the lateness of snowmelt and ground thaw delays convective activity until later in summer. This adds a double indemnity to agriculture in higher middle latitudes: in these areas severe limitations are set by the shortness of the growing season, which is compounded by the fact that during the growing season the precipitation regime is about the opposite of what it ideally should be. In spring and early summer, the period of rapid growth of crops, when moisture is most needed it may be lacking, and later in summer and fall the harvest season is generally the wettest time of year. Since the harvest season is very short at these latitudes, this dampness may be critical, and the entire harvest may be halted by an early snowfall. Such is the case throughout much of the USSR and Canada.

In coastal areas in middle latitudes, precipitation tends to reach maximum levels in fall and early winter when sea-surface temperatures are generally at their highest, and therefore convective activity is greatest during outbreaks of relatively cold mP air masses in the rears of cyclonic storms.

Variability. In addition to seasonal cycles, precipitation amounts in many parts of the world show great degrees of aperiodic variability, with the greatest absolute differences from year to year occurring in areas of great amounts of rainfall. For instance, Cherrapunji, India, which averaged 11,421 mm. (450 inches) per year over a 74-year period, in the one year from 1 August 1860 to 31 July 1861 received 26,447 mm. (1,042 inches), more than 15,000 mm. more than normal. Obviously such large absolute deviations could not occur in regions of low rainfall that never get that much in the first place. However, on a percentage basis, generally the more arid regions of the world show the highest year-to-year variability (Figure 13.11). Thus, not only do these regions suffer from deficits of precipitation, but the little they do get cannot be relied upon. Such regions are located on the fringes of rain-bringing circulations. Some years the circulation systems penetrate fairly deeply into these dry lands and bring considerable amounts of precipitation, but in other years the circulation systems do not penetrate the regions at all and the areas remain essentially rainless.

Developers planning uses of dry lands have over the years painfully become aware of these inevitable fluctuations, and they have also come to realize that long-term averages do not represent precipitation medians. For example, if a wheat farmer can produce a crop with only 400 mm. of rainfall per year but to survive must produce a good crop every other year, he should be wary of farming in a region with a long-term yearly average of exactly 400 mm. of rain, because there will not be a 50-50 split in years above and below the 400 mm. average. There will be more dry years than wet years. Precipitation is not an infinite continuum in both directions; it cannot be less than zero. Therefore, variations on the dry side of the average cannot be as great in magnitude as variations on the wet side, and a few very wet years can balance a greater number of relatively dry years. Thus, more than half the years are going to fall below normal, and less than half above normal. If the farmer's margin of survival depends on 50 percent of the years being above average, he is going to go bankrupt.

Short-term variations of days', weeks', or months' duration occur in east-west dimensions within latitudinal belts, particularly in the middle latitudes where the standing waves of the middle troposphere so strongly influence formation and movement of surface circulation systems and

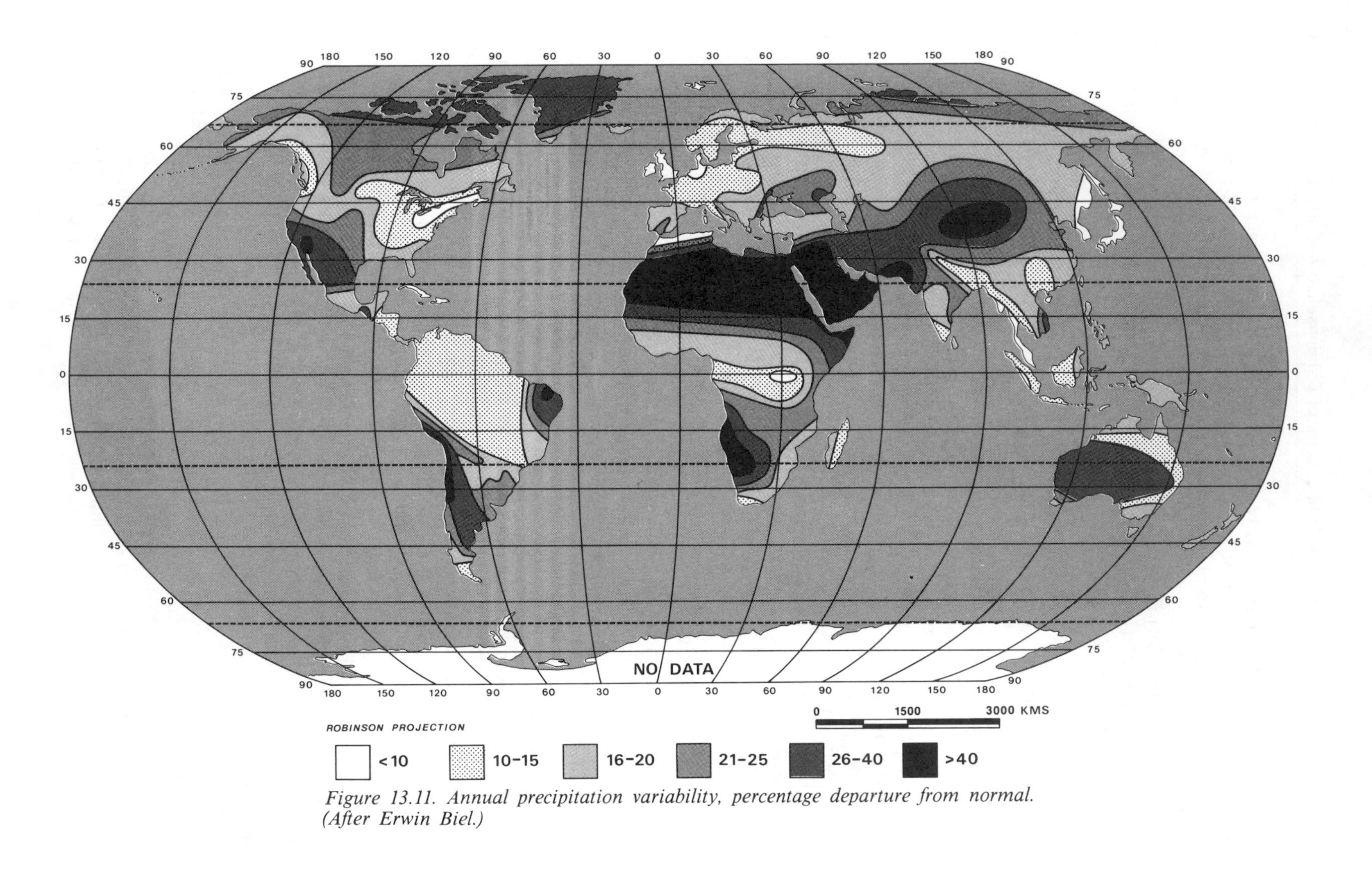

Figure 13.11. Annual precipitation variability, percentage departure from normal. (After Erwin Biel.)

resultant precipitation amounts. It was pointed out in Chapter 6 that generally opposite types of weather will be experienced in opposite limbs of troughs and ridges of these wave patterns. Generally increased precipitation will be experienced on the eastward limbs of troughs (western limbs of ridges), and reduced precipitation will be experienced on western limbs of troughs (eastern limbs of ridges). Since these standing waves do migrate slowly over time, the same point on the earth's surface may find itself first under one limb for several weeks and then under another limb for another several weeks. Thus, a prolonged spell of rainy weather will be followed by a similarly long spell of dry weather, or vice versa. In some places such shifts show a considerable seasonality that brings on unusual occurrences in the weather at about the same time each year. Such phenomena are known as *singularities.* But in most cases such east-west shifts are spurious and are simply integral parts of the climate of much of the earth. Thus, even an area that would appear to be quite humid, judging from annual amounts of precipitation, could have periods of drought imbedded within the year. If such occurrences could be predicted for growing seasons several months ahead of time, it would be a great boon to agriculture, but under present technology this is impossible.

In many mid-latitude interiors of continents where so much of the precipitation depends on summer thunderstorms, great variations in precipitation amounts can occur over short distances because of the intensity of rainfall over small areas. Thus, it is difficult to generalize about drought over large areas during any given growing season.

Intensity

The manner in which precipitation falls is also very important. If it falls in infrequent cloudbursts it may do more harm than good, causing floods and heavy erosion and generally running off the surface rather than soaking in and moistening the soil. On the other hand, if it falls in very prolonged drizzly weather, it is very effective in moistening the soil and providing high humidity conditions for lush plant growth. Much of northwestern Europe receives less than 750 mm. (30 inches) of precipitation per year, but in many areas this is spread over more than 150 days, so that practically every other day gets a little rain. Under the cool, cloudy, humid conditions that prevail in this area the rainfall is quite adequate. On the other hand, in many subtropical regions precipitation falls mainly as short, heavy showers with thunderstorms, interspersed with prolonged periods of very hot, sunny weather when the earth dries up and vegetation suffers. Because of a general lack of vegetative cover in dry regions, such sharp showers run off rapidly and cause erosion and flash floods. Thus, the dry lands of the world are thrice damned: they don't get enough precipitation in the first place, it cannot be relied upon from one year to the next, and when it does come in the form of hard showers it may cause more damage than good. Agriculture in the desert areas is carried on with irrigation where water supply can be completely controlled. An unexpected heavy shower at the wrong time of year may create havoc, particularly during the harvest of perishable fruits and vegetables.

14 | TEMPERATURE

Temperature is undoubtedly the most pervasive environmental factor to which all living things respond. If one lives in an area characterized by changeable weather, probably one of the first thoughts to enter one's mind in the morning is, "what is the temperature outside?" This will determine the mode of dress and will continue to bear on activities throughout the day. Since this very important element is influenced by everything that goes on in the earth-atmosphere system, its discussion has been delayed until all the process principles have been developed.

Summary of Temperature Controls

Chapters 3–11 have illustrated that many things affect the temperature in a given place at any given time. These factors fall into the categories of either radiant heat exchange, air movement, or the change of state of water.

Heat Exchange

On a global scale the radiant heat exchange between the earth-atmosphere system and space is paramount. The magnitude of this exchange and its seasonal variations depend primarily on latitude, which, as has been shown in Chapter 4, affects both the intensity of the exchange (sun's angle above the horizon) and its duration (length of daylight period). Large-scale deviations from this latitudinal pattern of radiational exchange are induced by the state of the atmosphere and the nature of the underlying surface.

Clouds and the moisture content of the air in vapor form are the most important aspects of the state of the atmosphere that affect heat exchanges. Clouds are highly reflective to short-wave radiation from the sun and highly absorbent of long-wave radiation from the earth's surface, thereby greatly increasing the greenhouse effect. A consistently cloudy, moist atmosphere will moderate both extremes of temperature, reducing maximum temperatures during the day and preventing excessively low minimum temperatures during the night. Other constituents of the air, such as carbon dioxide and various pollutants, may have similar effects on heat exchanges, but on much smaller scales than do the various forms of moisture.

Observed temperatures are greatly affected by varying heat exchanges over different types of earth surfaces. On the broadest scale is the difference between land and water. Water surfaces change temperature much more slowly than land surfaces do, for several reasons:

1. Sunlight penetrates water to much greater depths than land, perhaps as much as 75 meters in clear water, whereas there is essentially no penetration into soil.
2. Water is free to mix, which carries the heat to even greater depths.
3. Water surfaces generally have lower

albedos than land, except perhaps at low sun angles.

4. There is usually more evaporation from an open water surface than from a land surface, which uses up more of the incident sunlight in the form of latent heat, which is then unavailable to effect a temperature change at that point on the earth's surface.

5. The specific heat of water is significantly greater than that of land materials. Specific heat is a measure of the amount of heat it takes to excite the molecules of a substance to a given energy level (temperature). Water molecules are about the most sluggish of all molecules, so a lot of heat is needed to alter the molecular motion of water. Conversely, much heat must be lost from a body of water before a significant reduction in motion takes place.

For the above reasons, the water bodies of the world, which cover three-fourths of the earth's surface, act as great heat reservoirs. They temper the extremes of temperature by absorbing great quantities of heat with minimum temperature increases during periods of positive heat balance, and by giving up great quantities of heat to the overlying atmosphere without much reduction in water temperature during periods of negative heat balance.

Similar temperature differences result, at reduced magnitudes, over varied types of land surfaces. Of course, there is little penetration of sunlight into the earth's land surface anywhere and no mixing of heat downward; but the conduction of heat downward, albedo, and evaporation rates of water vary greatly depending on whether that surface is wet soil, dry desert sand, green vegetation, snow cover, or whatever (see Figure 3.6).

During winter, large areas of the earth's surface in the middle and high latitudes are rendered essentially homogeneous to heat exchanges by their snow cover. Once a significant (30 centimeters or more) snow cover has been established, there is little connection between the surface air temperature and the soil temperature beneath the snow. The snow is a very effective insulator since it entraps much air, which benefits wintering plants by protecting them from extreme cold and from rapid fluctuations in temperature. The snow surface, on the other hand, becomes very cold because the snow reflects so much of the meager short-wave insolation incident upon it during winter, and it radiates heat away from the surface almost perfectly, as a black body, in infrared wavelengths. Since little heat is derived from the soil underneath the snow, the snow surface and the air above it become very cold. Typically, a strong surface temperature inversion develops that, in places such as Siberia and Canada, may extend upward two kilometers or more into the air. The heat exchanges between the snow surface and the inversion layer and the resulting surface temperatures during winter have already been discussed in Chapter 9.

During spring, as insolation intensity increases rapidly, surface air temperatures do not rise much above the freezing point until the snow cover has melted and the great amount of meltwater has evaporated, since the excess heat from insolation is being used in the latent heat of melting and evaporation. Thus, a snow cover delays the coming of spring. In higher latitudes a significant surface air temperature rise may be delayed almost until the summer solstice, at which time it may jump up almost instantaneously after the melting and evaporation have been accomplished, since the season of maximum heat input is then at hand.

Thus, in high latitudes, for practical purposes spring becomes nonexistent and the annual temperature curve becomes skewed, since autumn does not experience heat use for the change of state of water. In fact, as the ground and water bodies begin to freeze, autumn may be prolonged somewhat by the release of latent heat to the atmosphere through the fusion process. Thus, because of the unique heat exchange characteristics associated with a snow cover, the annual regime of temperature in high latitudes is characterized by a long, cold, coreless winter; a practically nonexistent spring; a peaked, short summer; and an autumn of reasonable length that experiences a consistently rapid decline in temperature from relatively high summer values to extremely low, early winter values.

Atmospheric Motion

Atmospheric motion, both horizontal (advection) and vertical (convection), often becomes the dominant control of temperature. Particularly in middle latitudes, air masses of widely varying characteristics in juxtaposition to one another across segments of the polar front sweep across the earth and account for nearly all the day-to-day variations in surface air temperatures. The broad-scale control by such things as latitude and land-sea differences set the background for general temperature changes that can be expected seasonally, but the daily variations superimposed upon this background are almost completely a function of the advection of air masses. These not only cause the horizontal translation of sensible temperatures from one place to another, but also greatly affect the atmospheric conditions, particularly cloudiness and humidity, which have such profound effects on insolation receipts and radiational exchanges between the earth and the atmosphere.

Cloud formations, of course, are related to vertical air motion, which also affects temperature by adiabatic heating and cooling. If the locale is a mountainous one, surface air temperature may depend more on vertical air motion than anything else. The heating effects of foehn winds have already been described in Chapters 8 and 12. They may dominate the seasonality of temperature if the general circulation over the mountainous region varies seasonally. In southern California, for instance, Los Angeles experiences its highest temperatures in September and October when a northeasterly general circulation produces foehns down the southern and western slopes of the mountains surrounding the basin on the north and east.

Altitude alone, of course, is an extremely important control on air temperature, even if the air is not in motion. As was seen in Chapters 2 and 8, in the troposphere a temperature decrease of about 6.5°C per kilometer (3.5°F per thousand feet) elevation is normal. This is usually a far greater temperature change than is experienced with any horizontal distance. Thus, if one is attempting to depict temperature distributions that are a result of latitudinal, land-sea, and other broad controls one must eliminate the altitudinal control, or it would dominate all the others and obscure the global distributions one is trying to illuminate. Therefore, many isothermal (lines of equal temperature) maps are "reduced" to sea level by adding 6.5°C per kilometer (or 3.5°F per thousand feet) to the observed temperature at the given altitude.

Change of State of Water

The latent heat involved in the changes of state of water has already been elaborated in Chapter 7 and mentioned several times in the preceding paragraphs in conjunction with other temperature controls. The influence of the presence of water in some form on observed temperatures cannot be overemphasized. Since approximately three-fourths of the earth's surface consists of standing water, and water is also present in practically all types of land surfaces and their vegetative covers, the relationships between inputs and outputs of heat and resultant temperatures are always being modified by moisture considerations. And the moisture content of the air also greatly affects the stability conditions that will, to a great extent, determine the amount of vertical motions of the air. Thus, the moisture situation of an area greatly affects observed temperatures.

Temperature Perception

The water content of the air also affects temperatures perceived by the human body, since the temperature-regulating mechanisms of the body depend to a large extent on the evaporation of perspiration from the surface of the skin. High atmospheric moisture accentuates both extremes of perceived temperature. On a hot, humid day the body is unable to lose heat as rapidly as it needs because the high humidity in the air retards the evaporation of perspiration. Therefore, one feels overheated. Conversely, on a chilly, damp day small droplets of liquid water settle out of the air onto the skin and require the body to lose more heat than it wishes to evaporate the moisture and keep the skin relatively dry. Therefore, the weather feels more chilly than the temperature really is.

Wind also causes temperatures to be perceived as chillier than they actually are because body heat is being lost more rapidly than normal. Although wind chill is difficult to measure, and no two investigators have agreed completely, a standard wind chill table has been adopted by the National Weather Service for use in the United States. This lists equivalent temperatures (with calm conditions) according to actual temperatures and wind speeds. For instance, the equivalent wind chill temperature for a recorded temperature of 0°F would be −22°F if the wind speed were 10 mph, −40°F at 20 mph, −49°F at 30 mph, −54°F at 40 mph, and −56°F at 50 mph. The relationship is a quadratic one, and wind speeds greater than 40 mph have little additional chilling effect.

Another important influence on perceived temperature is sunlight. It has been estimated that direct sunlight may make a sunbather feel as much as 14°C (25°F) warmer than would be the case with the same air temperature but cloudy conditions. Under strong sunlight, the body surface temperature rises well above the air temperature around it. This is true of any absorbing surface, whether animate or inanimate, and is a major factor in microclimatic considerations.

Because of these effects of moisture, wind, and insolation, the human body is a rather poor judge of temperature.

Expressions and Distributions of Temperature

Expressions

Temperature is observed and recorded every hour of every day of every year for thousands of climatic stations all over the world, and it is obvious that over a period of years an astronomical number of individual entries is compiled. Therefore, to comprehend and derive generalizations from the mass of data, it must be statistically reduced in some manner. Often, monthly averages are computed and isothermal maps drawn, particularly for the months having the most extreme temperatures, which in most cases are January and July. In some cases temperatures plotted on these maps are corrected to equivalent sea-level temperatures to show world distributions without the influence of altitude. To illuminate controls other than latitude, such as land-sea differences, maps showing isanomalies of temperature may be drawn to illustrate deviations of temperatures from latitudinal norms.

Mean annual ranges of temperature are computed and often plotted on maps. These show differences between mean temperatures for January and mean temperatures for July. This mean temperature range, of course, is much less than an absolute temperature range, which is the difference between the highest and lowest temperatures ever recorded in an area. On a comparative basis, however, the mean temperature range may be more realistic, since absolute maxima and minima depend, among other things, on the length of record. Mean diurnal range of temperature by month is also important. Thus, temperature data are usually represented by some sort of means and deviations. Sometimes it is important to have a finer mesh of temperature distribution with time. Such might be a frequency distribution of temperatures grouped according to size categories.

Other aspects of temperature may be important for certain purposes. For instance, the freezing point of water is a critical temperature for many things. Since many plants are killed when freezing temperatures occur, it is important to know the growing season of an area in terms of its period of consecutive frost-free days. Obviously, a crop will not grow after a killing frost, even if the temperature warms up. Also, in some cases, it might be important to know the sequence of temperature events. Wintering crops may suffer more from frequent temperature fluctuations above and below the freezing point than they would if the temperature were consistently below freezing.

Certain threshold values of temperature are important in other instances. Although most plants will not be damaged when temperatures remain above freezing, they may not grow unless the temperature rises significantly above freezing. These are threshold temperatures for growth. The degree-day concept is a way of keeping track of heat resources during the period when temperatures remain above the threshold of plant growth, which in much of the world is usually considered to be 10°C (50°F). In the United States, a 24-hour period that averages 60°F would be logged as ten growing degree-days, etc.* Similarly, heating and cooling degree-days are logs of heat resources kept for the convenience of heating and cooling houses. In the United States when outside temperature drops below 65°F, it is assumed that furnaces will turn on in houses and some fuel will be consumed. Conversely, during the warm season when the temperature outside rises above 65°F, it is assumed that air conditioning units will kick in and electricity will be consumed. Fuel and electric companies keep such logs to judge needs for energy.

Spatial Distributions

Figures 14.1 and 14.2 show world distributions of mean surface air temperatures for January and July. Probably the most conspicuous pattern is the latitudinal one. Second, land-sea differences show up as wide deviations from the latitudinal pattern, particularly in coastal areas such as over the North Atlantic northwest of Europe and in the North Pacific along the coast of North America. These land-sea contrasts are particularly distinctive during winter. For instance, the mean January temperature is about the same (−8°C) in the Barents Sea north of Scandinavia at a latitude of almost 80°N as it is in western China in the middle of the Asiatic continent at a latitude of approximately 40°N. At this time of year isotherms in northwestern Europe tend to follow coastlines more closely than parallels of latitude.

A third major influence on temperature—elevation—is also evident on these maps. Steep temperature gradients exist where the relief is great, such as in northern India where the land suddenly rises from the low, flat Ganges Plain into the Himalaya

*Different crops have different thresholds of growth. NOAA agencies in the United States use the 50°F threshold for corn, with an upper limit of 86°F (30°C). A day averaging about 86°F would be credited with only 36 growing degree-days.

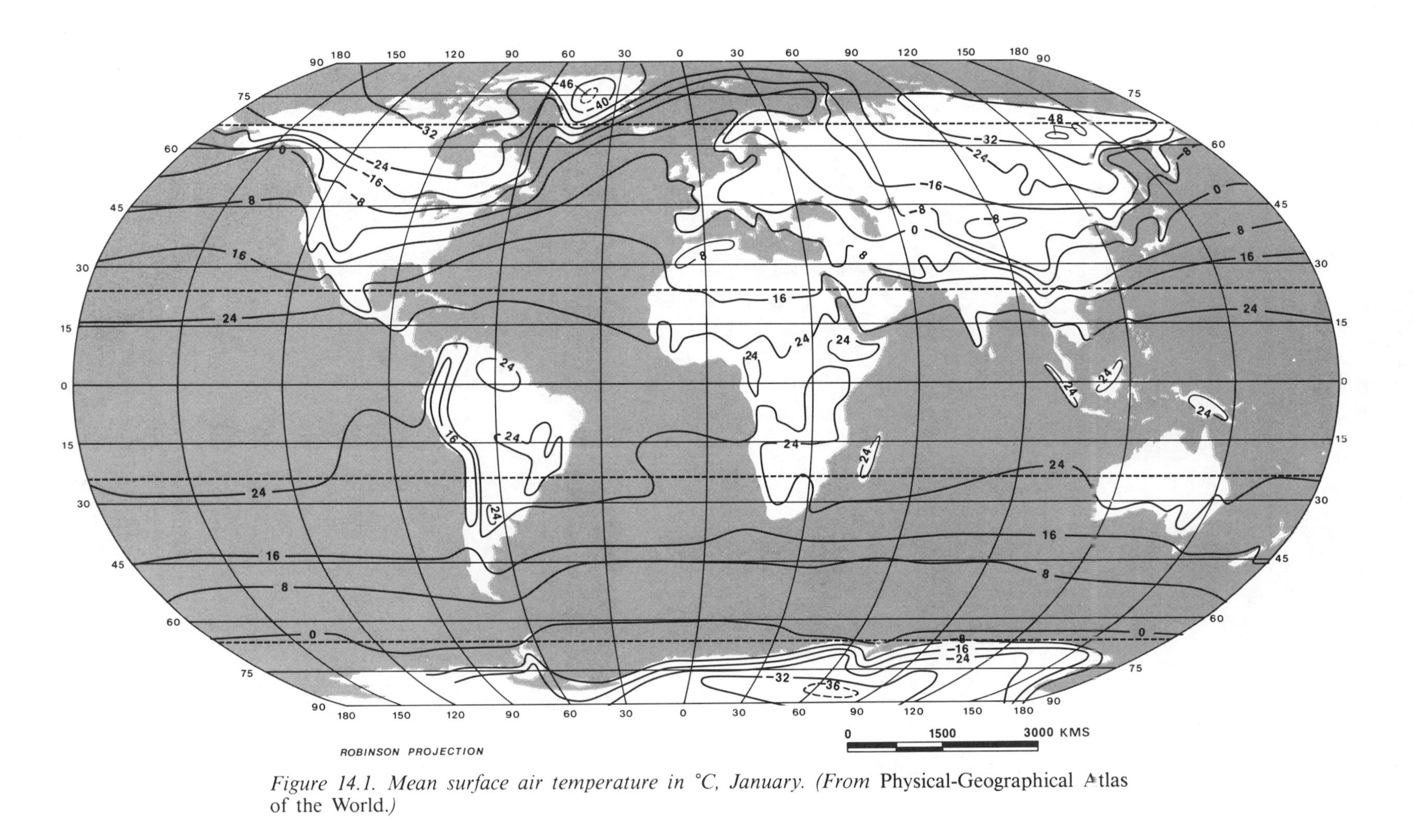

Figure 14.1. Mean surface air temperature in °C, January. (From Physical-Geographical Atlas of the World.*)*

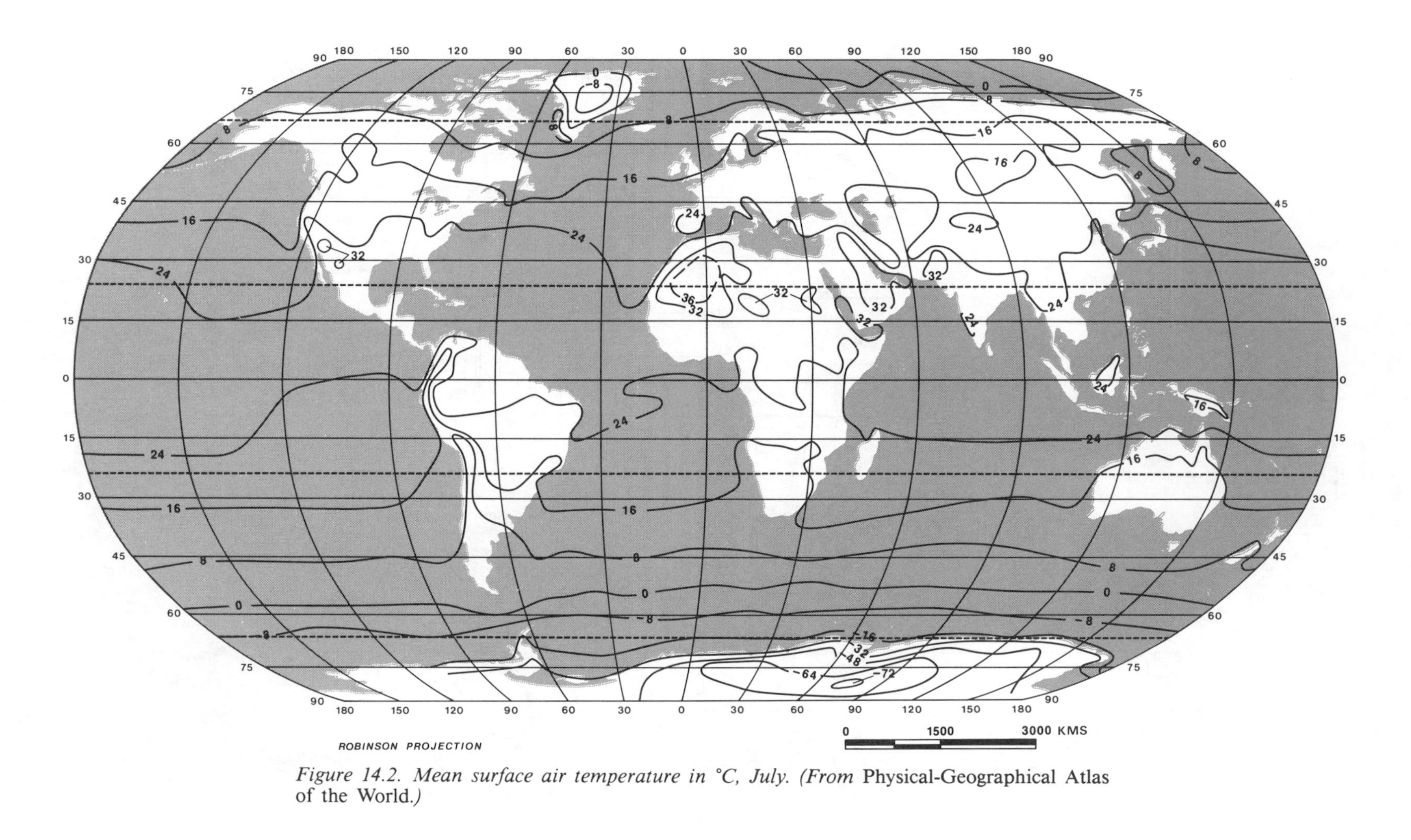

Figure 14.2. Mean surface air temperature in °C, July. (From Physical-Geographical Atlas of the World.*)*

Mountains. The intricate pattern of isotherms throughout much of southern and eastern Asia reflects the influence of topography.

A comparison of the January and July maps reveals significant latitudinal shifts in temperature zones between seasons, and highlights the fact that outside Antarctica the Southern Hemisphere, with its lack of middle-latitude landmasses, generally lacks the severity of winter that is so characteristic of much of the Eurasian and North American landmasses.

Isotherms are generally packed more closely together (temperature gradients are greater) during winter than during summer, particularly in the middle latitudes because, as was explained in Chapter 4, both the angle of sun's ray and length of day decrease poleward during winter, but during summer they decrease in opposite directions. Yet in upper middle latitudes with consistent snow covers during winter, such as much of the European plain of the Soviet Union, mean January isotherms show more north-south trend, related to distance from the sea, than they do latitudinal trend. At these latitudes during winter little insolation is received, and the small amount that is received is largely reflected from the snow's surface. Therefore, resultant temperatures are not much affected by latitude.

Figures 14.3 and 14.4 illustrate the variations of mean sea-level air temperatures from latitudinal norms during January and July. These particularly emphasize land-sea differences, which are accentuated during winter. The highest positive temperature anomaly on earth occurs at this time of year over the sea off the coast of Norway where a northeasterly extension of the Gulf Stream carries warm ocean water far poleward to warm this region. Here the January temperatures of the surface air average 26°C (47°F) above the norm for the Arctic Circle. To a lesser degree, the northeastward drift of the Japan Current in the eastern Pacific produces an abnormally warm area in the Gulf of Alaska. Extremely cold temperatures are found over the land masses in between. The far northeastern portion of the USSR experiences temperature anomalies of as much as −24°C (−43°F) during January. During July anomalies are not as great, since in the Northern Hemisphere the landmasses warm up rapidly and become a little warmer than the seas around them, while in the Southern Hemisphere during its winter season there is little land at high latitudes outside Antarctica. On the July map some of the most conspicuous anomalies are negative deviations from latitudinal norms in the summer hemisphere produced by cool ocean currents such as the California Current off the west coast of North America and the Canaries Current off northwest Africa.

Large landmasses at high latitudes show the greatest annual temperature ranges (Figure 14.5). The same conditions—distance from the sea and air stagnation in enclosed basins—that make for extreme cold in winter are conducive to abnormally high temperatures during long summer days with almost 24 hours of sunlight. By far the greatest temperature ranges on earth are found in the northeastern extremity of the Soviet Union, where the greatest negative temperature anomalies are found in January. Verkhoyansk, in a deep river valley between mountains rising to about 3000 meters on either side, at an elevation of 137 meters records a mean January surface air temperature of −48.9°C, but in July the mean temperature rises to 15.3°C. Thus, Verkhoyansk has a mean annual range of temperature of 64.2°C (115.6°F). Its absolute temperature range amounts to 103°C (185°F), from a minimum of −68°C (−90°F) to a maximum of 35°C (95°F). A little to the south in the mountains at an elevation of 740 meters, Oymyakon has experienced an absolute temperature range of 104°C, from −71 (−96°F) to +33 (91°F). As can be seen in Figure 14.6, these areas can expect minimum temperatures below

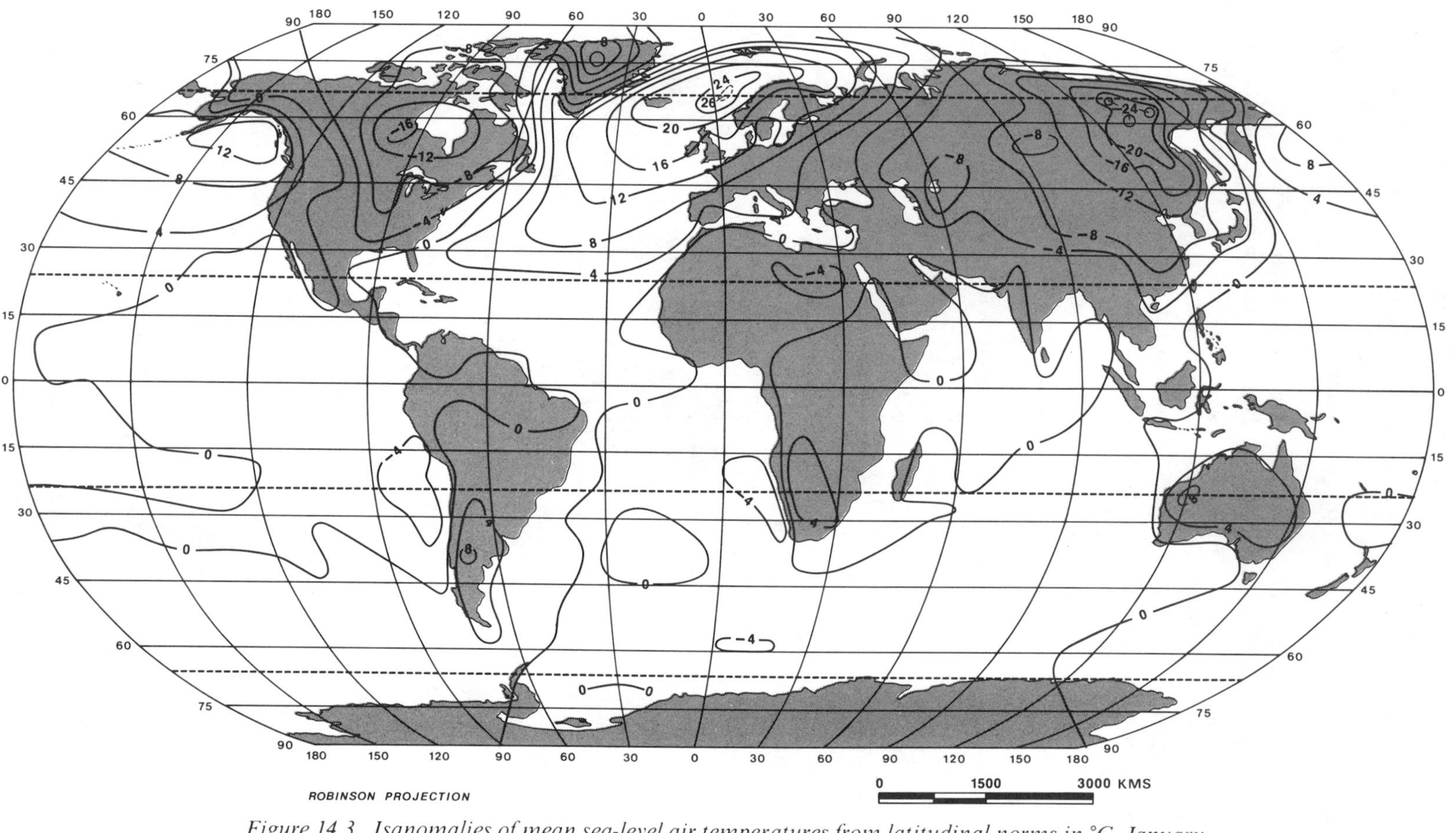

Figure 14.3. Isanomalies of mean sea-level air temperatures from latitudinal norms in °C, January. (From Physical-Geographical Atlas of the World.*)*

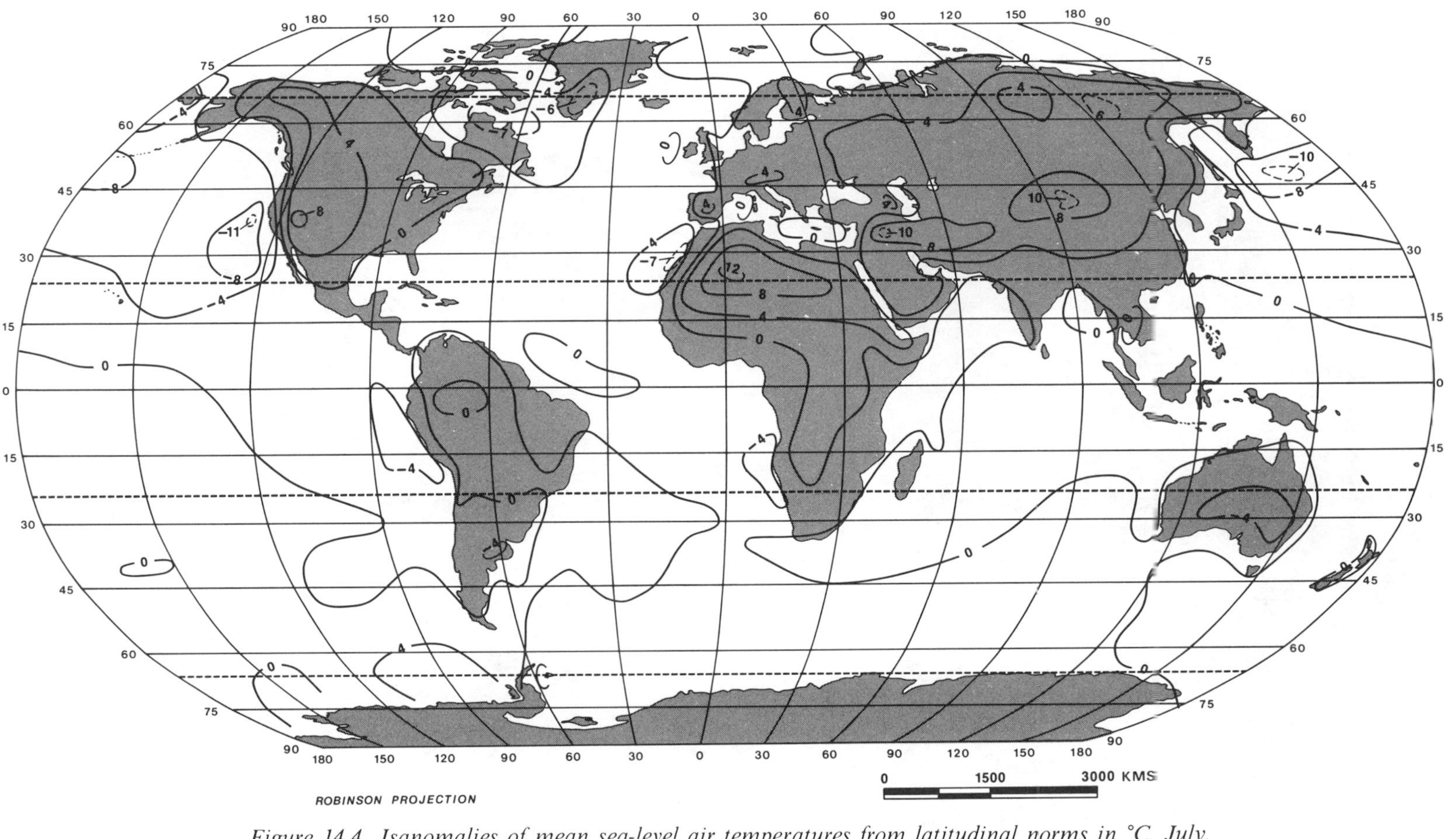

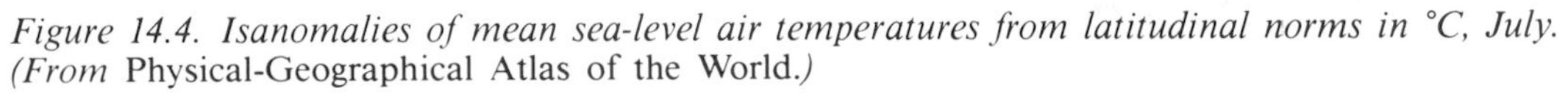

Figure 14.4. Isanomalies of mean sea-level air temperatures from latitudinal norms in °C, July. (From Physical-Geographical Atlas of the World.*)*

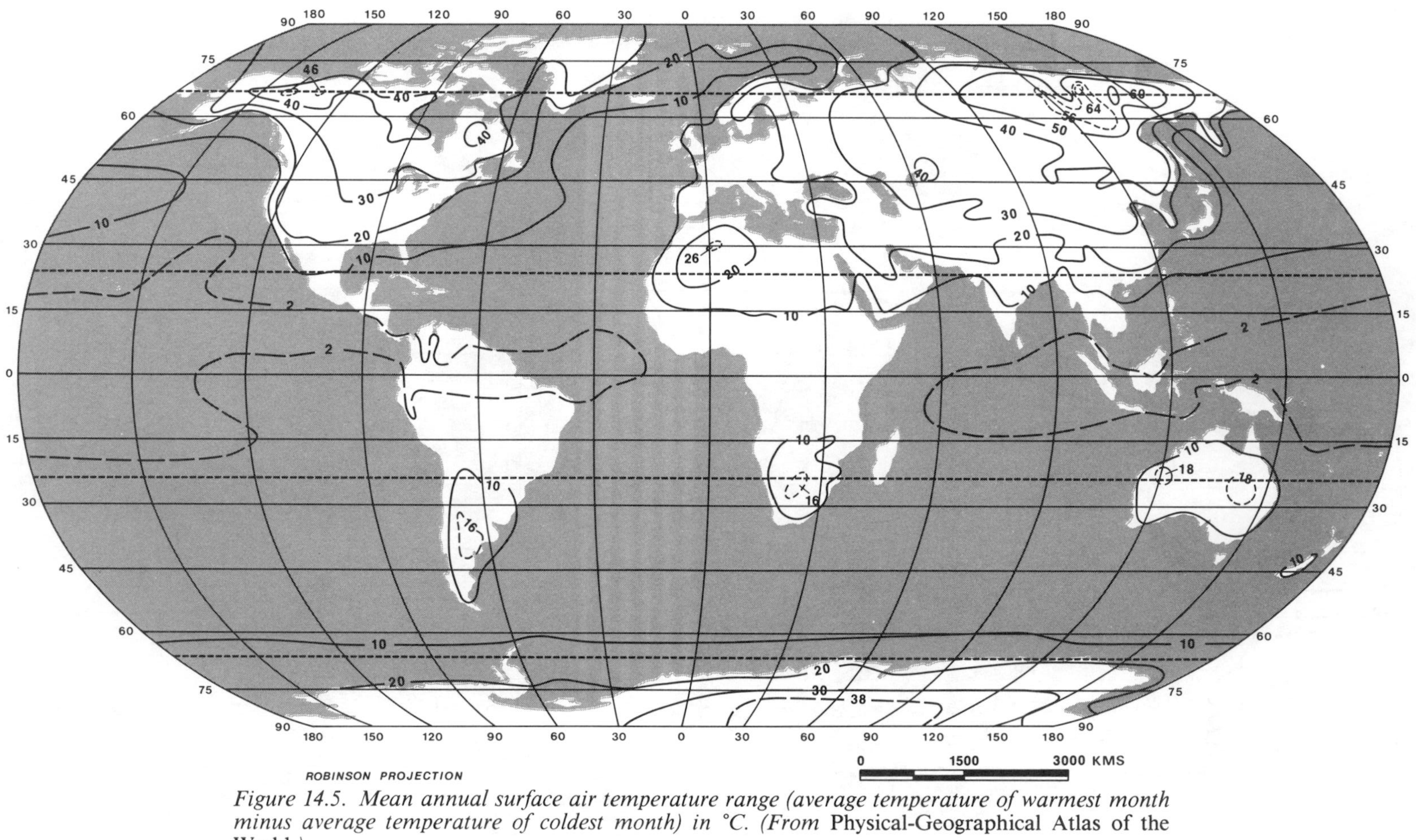

Figure 14.5. Mean annual surface air temperature range (average temperature of warmest month minus average temperature of coldest month) in °C. (From Physical-Geographical Atlas of the World.*)*

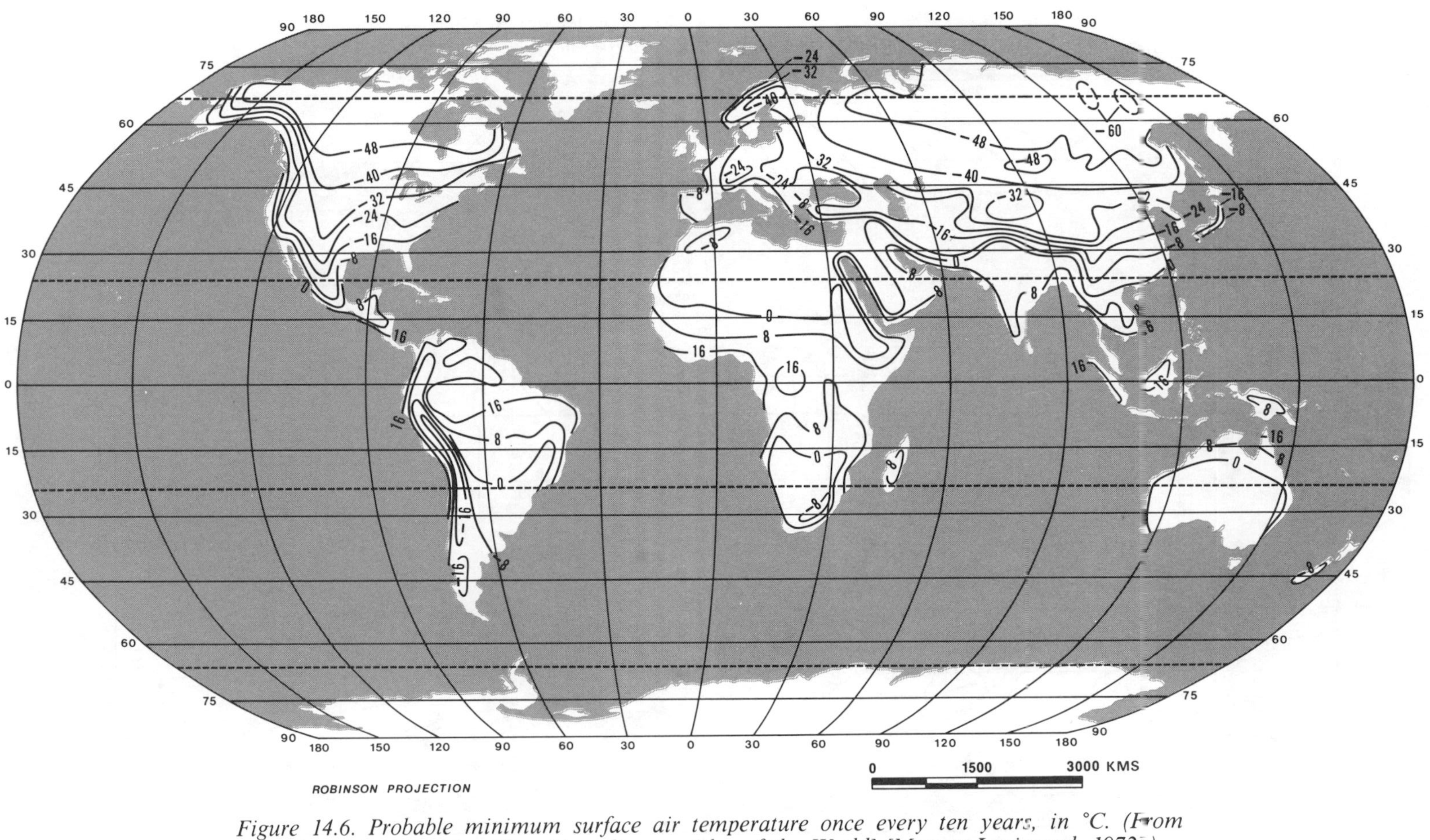

Figure 14.6. Probable minimum surface air temperature once every ten years, in °C. (From Agroklimaticheskiy Atlas Mira *[Agroclimatic Atlas of the World] [Moscow-Leningrad, 1972].)*

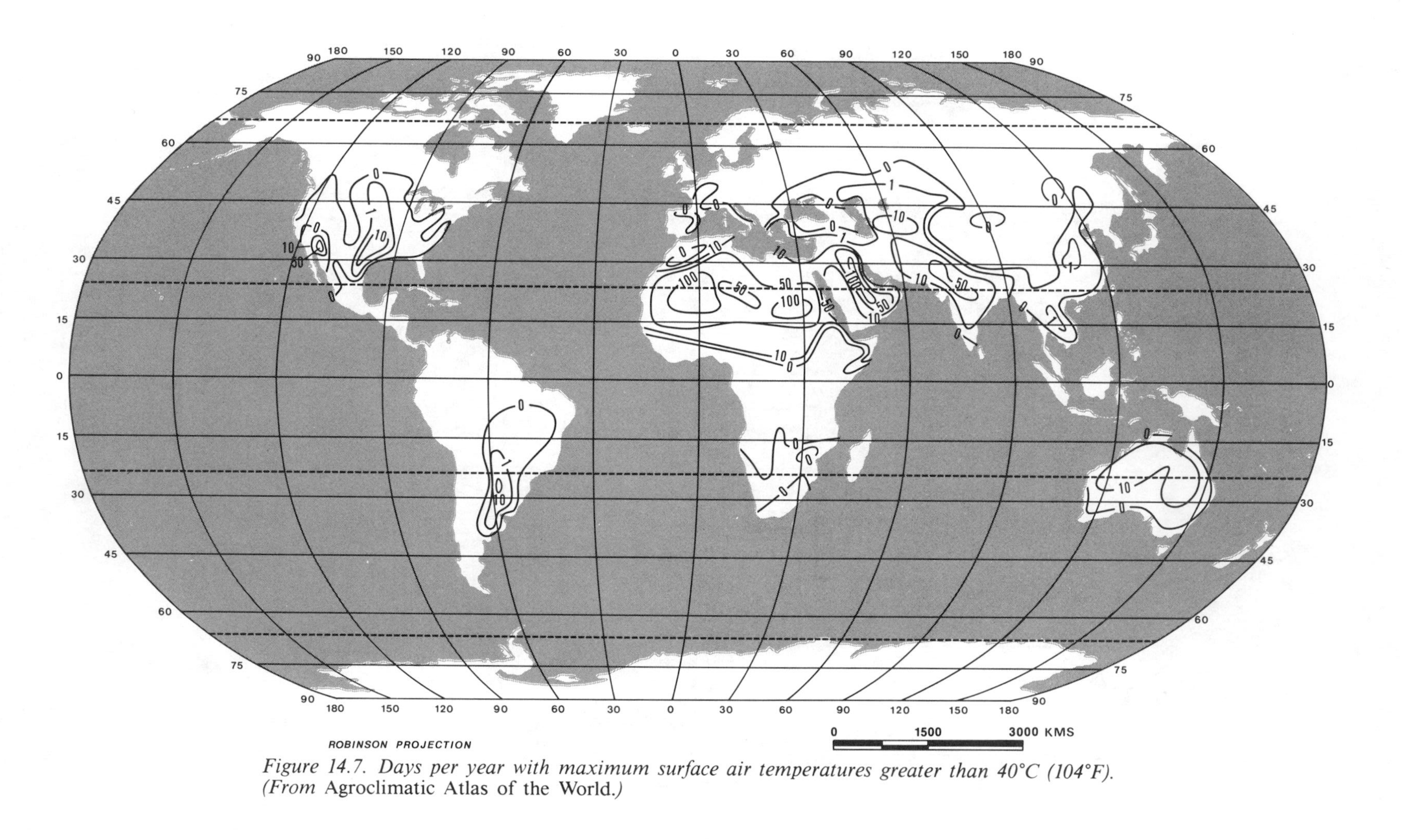

Figure 14.7. Days per year with maximum surface air temperatures greater than 40°C (104°F). (From Agroclimatic Atlas of the World.*)*

−60°C (−76°F) at least once every ten years.

Soviet climatologists have recorded even colder temperatures in Antarctica, where at their research station, Vostok, a temperature as low as −89.2°C (−128.6°F) has been recorded. But this was at an elevation of 3488 meters above sea level. The annual range there is not as great, since summer is not warm on this high ice plateau: the warmest temperature ever recorded there was −21°C (−6°F).

The hottest temperatures on earth are found between 15° and 35° latitude, where maximum amounts of insolation are received at ground surface during the summer solstice under the influence of subtropical high pressure cells in dry desert locations (Figure 14.7). In large portions of the Sahara and Arabia, as well as in a small portion of the Indus Valley in Pakistan, more than 100 days per year experience temperatures greater than 40°C (104°F). The lower Colorado River region in the southern portion of intermontane western United States shows the greatest frequency of days with temperatures greater than 40°C at latitudes higher than 35°. The hottest temperature ever recorded is 58°C (136°F) at Azizzia, Libya, in northern Africa. A temperature nearly as high, 54°C (129°F), has been observed in Death Valley, California.

Temporal Variations

One might logically wonder why the months of extreme temperatures are January and July, rather than December and June during the solstice periods when the earth is receiving either the greatest amount or least amount of insolation from the sun. The answer, of course, lies in the fact that temperature is not the result of insolation alone, but of heat loss as well. As long as incoming radiation is greater than outgoing radiation, the temperature will rise. Although outside the tropics the rate of heat gain is the greatest during the summer solstice, there is still a net gain of heat for another month or so before terrestrial radiation begins to exceed insolation. Thus, the temperature continues to rise, albeit at a slower rate than earlier, until it reaches a peak sometime in mid-July. Conversely, during winter there is usually a net loss of heat until about mid-January.

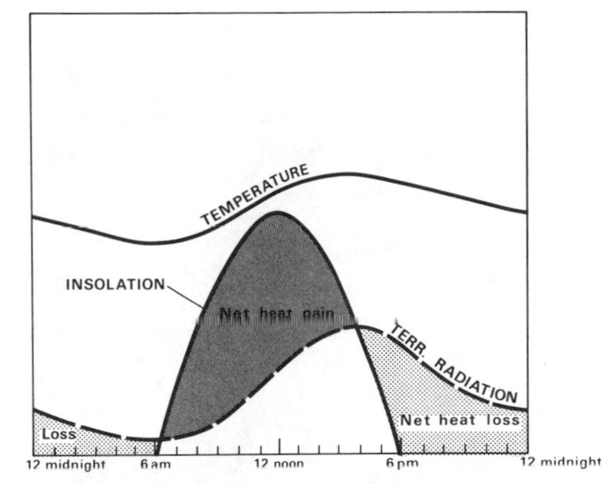

Figure 14.8. Typical relationship between insolation, terrestrial radiation, and surface air temperature during a clear day at an equinox.

The same sort of relationship between heat balance and temperature can be illustrated on a daily basis. Although maximum insolation occurs at noon, usually the highest temperatures during the day are observed sometime during the middle of the afternoon (Figure 14.8). Assuming sunrise at 6 AM and sunset at 6 PM, the insolation rises rapidly after sunrise to a peak at noon and falls rapidly to zero at sunset. The back radiation from the earth's surface depends upon the temperature of the earth and thus starts from a minimum value shortly after sunrise and increases, first slowly, then rapidly, and then slowly again until mid-afternoon, after which it goes into a decline. A net heat gain occurs throughout the morning and early afternoon, so the temperature continues to rise until mid-afternoon, when it reaches a maximum. After that there is a net heat loss that continues through the night, when no insolation is being received. The net loss

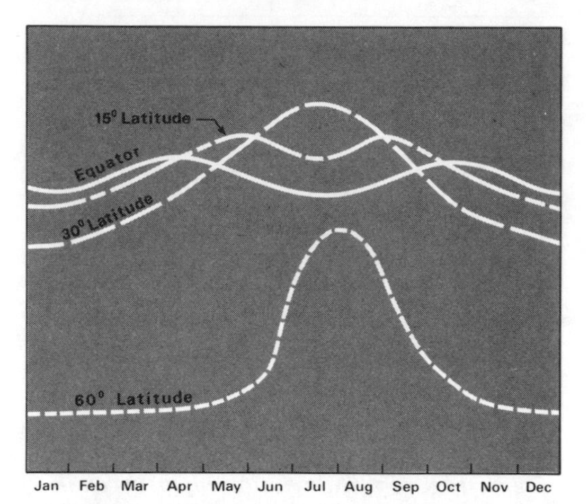

Figure 14.9. Theoretical seasonal temperature regimes and relative magnitudes at various latitudes.

continues shortly beyond sunrise, since little insolation is received from a rising sun low on the horizon. Thus, minimum temperatures generally occur shortly after sunrise.

Figure 14.8, of course, ignores all other influences on temperature such as clouds, advection, and vertical movements of the air. And it is not entirely correct for the heat exchange at the point at which the temperature is being measured, since it assumes an earth-atmosphere interface that is impervious to heat transfer downward. In actuality there is some heat transfer downward, by conduction several centimeters into soil during a day's time, and by light penetration and mixing much deeper in water. Thus, during the heat of the day some heat is stored in the soil, and later in the afternoon, as net radiation becomes negative, heat moves back upward through the soil to the earth-atmosphere interface. This slightly alters the timing of the maximum temperature. Over water it might be quite a bit, particularly on an annual basis, when heat exchange can involve a thickness of water of a hundred meters or more. Typically, along an ocean coast on an annual basis, temperature maxima may be delayed another month beyond what they are in interior land locations, so that August is very often the month of maximum temperature in coastal areas. Over open water surfaces it may be as late as September or even October.

If controls other than heat exchanges at the surface become dominant, the temperature regime may be erratic. For instance, in southern California where foehn winds are most common in fall and early winter, the highest temperatures of the year may be experienced at that time. Near seacoasts the diurnal regime of temperature may be significantly altered by sea breezes that may begin to blow inland in late morning and continue throughout the rest of the day. This typically lops off the effects of radiational exchanges that would normally take place after noon, and leaves a reduced maximum temperature for the day that might thus occur around noon.

Within the tropics, annual regimes of temperature reflect the double passage of the direct rays of the sun during the year unless obscured by local effects. At the equator the temperature is fairly consistent throughout the entire year, since incident angle and day length never change greatly. If solar radiation is the primary control, the temperature should show two maxima—at about or shortly after the equinoxes, when the direct rays of the sun are overhead—and two minima—at about or shortly after the solstices, when the direct rays of the sun are at 23½° latitude on either side of the equator. Poleward the two maxima converge toward each other, since the passage of the direct rays of the sun moving poleward from the equator and returning back to the equator occur closer together in time as the tropic circles are approached. On the tropic circles the two maxima come together and remain as one single maximum and one single minimum to the poles. The amplitude between winter and summer increases poleward, and the summer period shortens as the winter season lengthens (Figure 14.9). Stations within the tropics rarely exhibit such classic re-

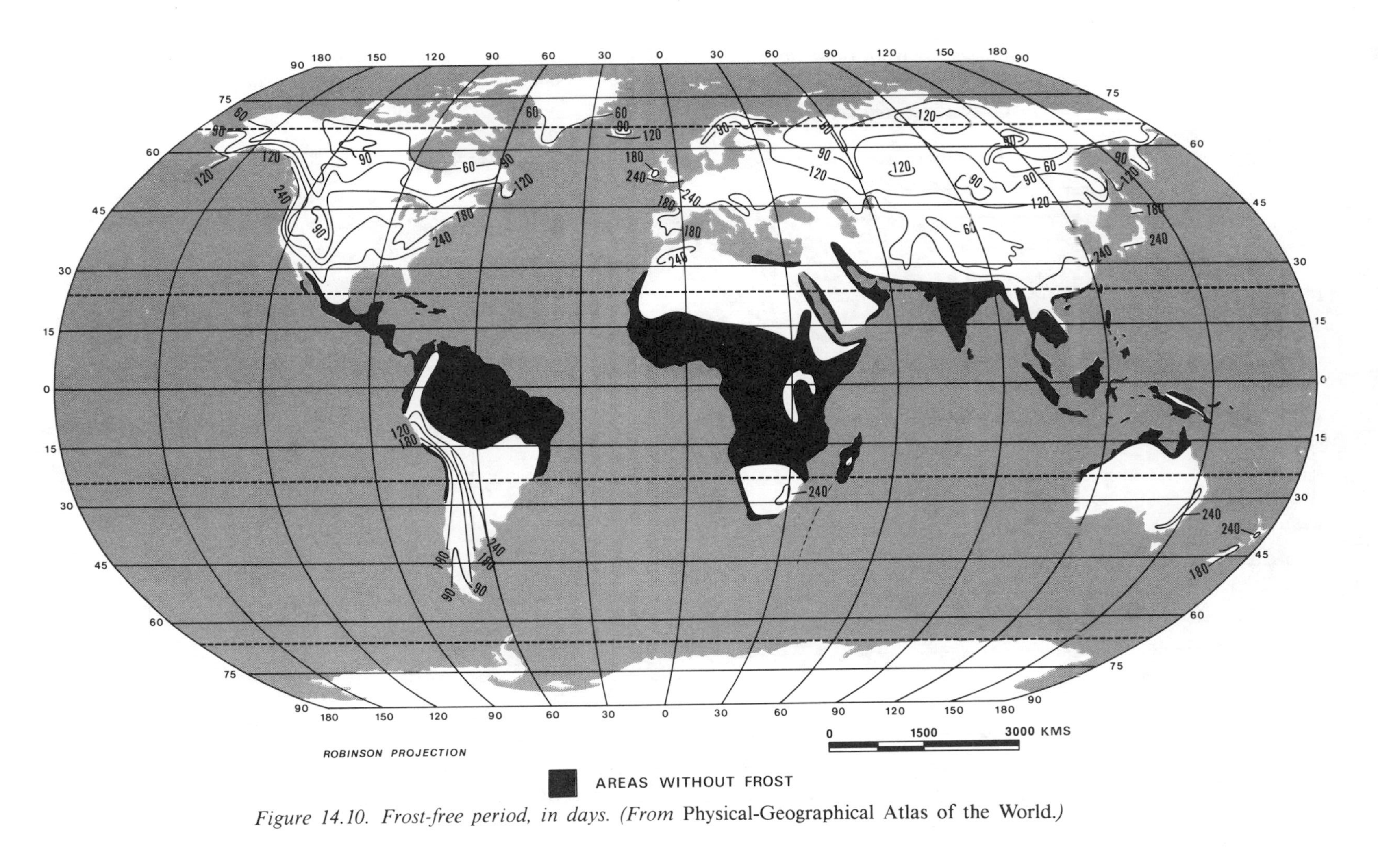

Figure 14.10. Frost-free period, in days. (From Physical-Geographical Atlas of the World.*)*

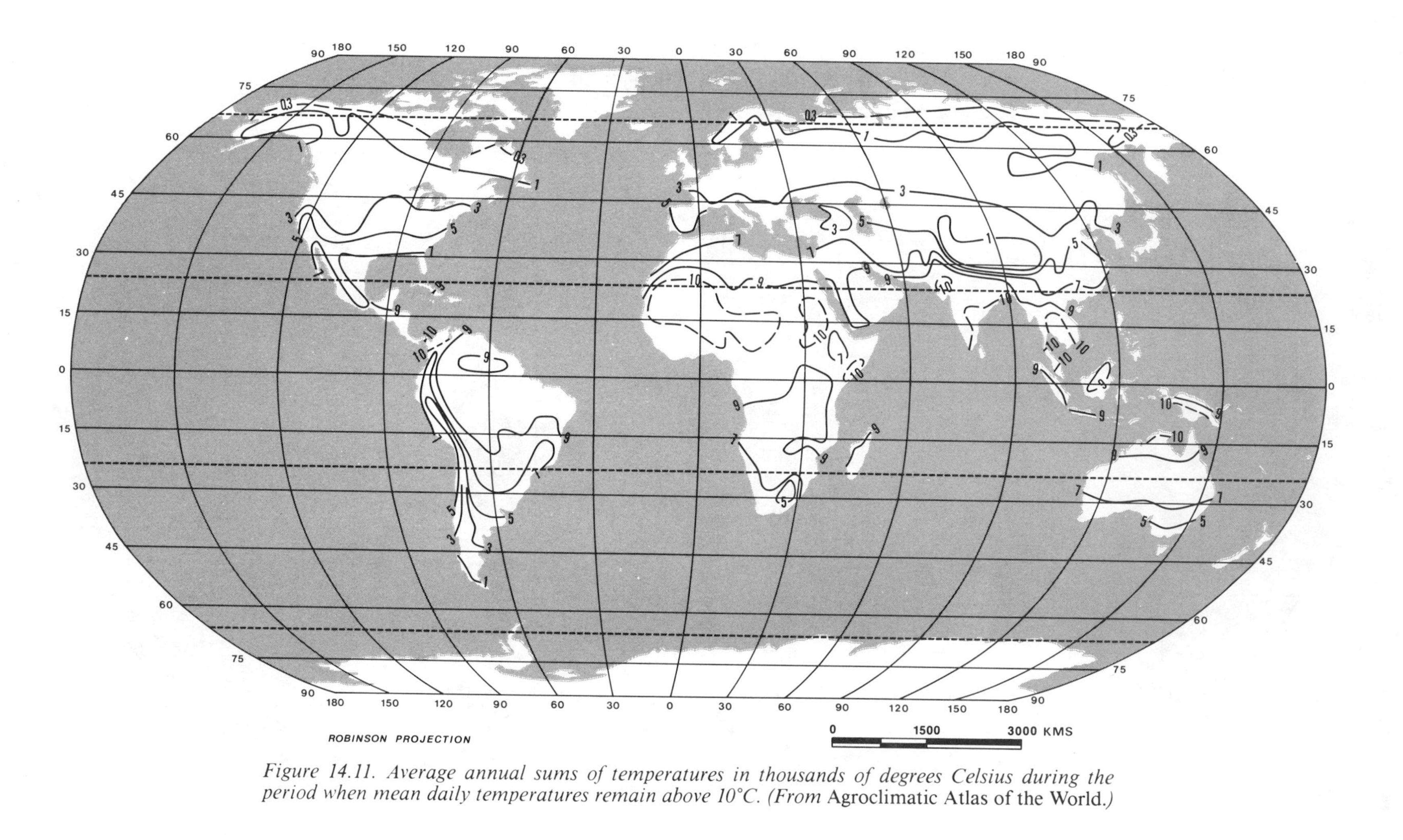

Figure 14.11. Average annual sums of temperatures in thousands of degrees Celsius during the period when mean daily temperatures remain above 10°C. (From Agroclimatic Atlas of the World.*)*

gimes, because all sorts of things beside radiational balances, such as clouds and precipitation, may play major roles. Many tropical stations show somewhat cooler temperatures during rainy seasons and warmer temperatures during dry seasons, regardless of time of year. Since in low latitudes temperature variations are small and irregular, the designations "winter" and "summer" become meaningless. Instead the expressions "high sun" and "low sun" are often used to designate times of year when the noon sun is highest or lowest above the horizon.

Distributions of Certain Threshold Values

As was mentioned earlier, certain values of temperature act as thresholds for certain purposes. The freezing point is probably the most universally important. Figure 14.10 shows the length of frost-free period during the year. Here it can be seen that lowland areas within the tropic circles seldom, if ever, receive frosts. To about 35° latitude on either side of the equator, frosts occur only on occasional years. Beyond about 35° latitude, frosts occur every year, and of course they become more severe poleward. Along the northern fringes of North America and Eurasia, most places have less than a 60-day period when frosts ordinarily do not occur, and some years frosts may occur in mid-summer. Over much of the Arctic and the Greenland and Antarctic ice caps frost is a constant feature. In high mountain regions worldwide, frosts typically occur on most nights of the year. During summer in high mountains the freezing point may be crossed twice each day, warming above freezing during the day and cooling below freezing at night. This so-called periglacial climate of alternate freezing and thawing has profound effects on the landscape.

The length of growing season or frost-free period, of course, is important to agriculture. But also important is the warmth of the frost-free period. To derive some measure for that, the Soviets utilize an index that is derived by adding all temperatures in degrees Celsius during the period when the mean daily temperature remains above 10°C, which they consider to be the threshold value for plant growth. These sums of temperatures are represented in Figure 14.11 and afford comparisons regarding heat resources in different parts of the world. Sums range from less than 300 along the Arctic coasts of Eurasia and North America (and even zero for the ice caps) to more than 10,000 in portions of the tropics. Minimum values of sums of temperatures have been determined for different types of crop growth and are used in the Soviet Union similar to the way the degree-day is used for agriculture in the United States and some other countries.

15 CLIMATIC CLASSIFICATION AND DISTRIBUTION

Most fields of science, and some other disciplines as well, find it necessary to categorize phenomena into classification systems that convert incomprehensible masses of individual pieces of information into groups or relationships. Such sciences as physics and chemistry accomplish this by establishing formulas that are based on laws or principles, each of which hold true for a given body of actions or events. Other sciences, such as the biological and earth sciences, have developed hierarchical systems of categories to group, in some logical fashion, all the phenomena that fall in their purviews of study.

Some of these categorizations have real meaning, such as the biological classification of plants and animals based upon physiological functions and evolutionary delineations, and chemistry's periodic table of the elements, based upon atomic weights and numbers. Such classification schemes provide not only convenient means for conveying information, but also facilitate further research, since once the system is discovered it helps to reveal gaps in information that further research can concentrate upon. Other categorizations are more arbitrary and provide no innate system to be discovered. Individual phenomena are simply grouped into categories to organize information more comprehensively. Since no inherent classification exists in these cases, a number of classifications are equally valid, each being most appropriate for its specified purpose. Such are the classifications in climatology, as well as in other fields of earth and atmospheric sciences.

The classification of climate differs from other fields of earth sciences because climate cannot be observed. It is a mental construct, composed of statistical abstractions of measured elements, on the one hand, and observable occurrences of weather episodes, on the other. To depict fully the climate of an area one must somehow convey both the end results of climate, the statistical means and deviations of measured elements, such as temperature and precipitation, and the reasons behind them—the day-to-day weather events and the atmospheric circulations, local effects, and other things that cause them.

So far discussions of areal distributions in this book have dealt with only one element at a time. Somehow these must be combined to depict the distribution of climate itself. This is not an easy task, since the climate of a place is made up of many incommensurable things, some of which are measured accurately on a regular basis, and some of which are not. This chapter will discuss some of the things that must be considered in a climatic description of an area and some of the options available for this purpose. It will culminate in an outline of a specific classification system, which will be utilized in the climatic descriptions of continents in Chapters 16–21.

Empirical Versus Genetic

One must first decide whether the classification system is to be an empirical scheme, based on observations of such elements as temperature and precipitation, or genetic, based on causes behind the observed facts. Since weather episodes come and go in a few days' time, it is possible to observe, over the passage of years, many complete weather sequences that make up the climate of a region. Therefore, it might seem plausible to classify climate according to the genetic relationships of the weather episodes of which it is composed. On the other hand, a plethora of statistics have been compiled over time from observations of individual elements, such as temperature, precipitation, wind, and so forth. Therefore, good, clean data are available on which to base a rigorous empirical classification scheme. Both types have been attempted, but generally the genetic classifications have been less successful, since they incorporate features that cannot be precisely measured and therefore must be handled subjectively. It is generally agreed that it is better to base a classification scheme on observed fact, rather than on assumed genetic relationships, and then to discuss the causes behind the facts, rather than incorporating them into the classification scheme itself.

Criteria

Next must be decided the criteria upon which the classification will be based. Climate offers a wide range of possibilities. Temperature has numerous facets, which were discussed briefly in Chapter 14. Precipitation has almost as many, which were discussed in Chapter 13. The various condensation forms, particularly clouds, certainly exert considerable influence on people and objects at the surface, as well as on other climatic elements. Atmospheric humidity and wind have important consequences for many things. Such elements as visibility are all-important to certain activities such as transportation. All these climatic elements are measured and recorded regularly and can be manipulated statistically.

Another group of phenomena are not as commensurable. These include variability of day-to-day weather: storminess, individual extreme happenings such as tornadoes and hurricanes, frontal passages, air mass movements, and so forth. Since no two air masses or cyclonic storms are identical, a mere counting of such things is not sufficient. Therefore they are very difficult to deal with statistically.

An adequate description of the climate of an area must use both statistics and words to include facets of both the measureable and immeasurable elements of climate. If this were not the case and all climatic information could be dealt with statistically, there would be no need for a book, since everything could be included in a cleverly devised table. A complete understanding of the climate of an area has to include both a knowledge of statistical means and deviations of measured phenomena and a knowledge of the individual weather episodes that constitute those statistics. In a study of climate, one should arm oneself with statistical compilations in one hand, and daily weather charts in the other. This is indeed how a climatic researcher studies in detail any particular climatic phenomenon or the climate of a small portion of the earth's surface.

But to provide a framework within which the climatic information for the entire earth can be presented, such as this text is trying to do, one cannot possibly handle the complexity that would result from including in a classification scheme all the measured data available, let alone the incommensurable information. Instead, a process of selection must be strictly adhered to. In most cases this has meant that classification schemes have been based only on temperature and precipitation data, while other information was covered in word descriptions about each climatic type.

The best known and most used classification was devised by an ethnic German from Russia, Vladimir Koeppen, during the early part of this century.[1] It is one of the simplest classifications and is based on actual degrees of temperature and millimeters of precipitation, so that final designations can easily be related back to the original data. Some investigators have criticized the Koeppen scheme for its simplicity; they say that it is not detailed enough to be of any use to researchers. This criticism, however, can be applied to any classification scheme that endeavors to provide a framework for the entire earth. No climatic researcher has found any climatic scheme to be of use in his research on small areas or limited topics, other than to provide some perspective for his own piece of work within a world context. Climatic classifications have become tools for teaching more than for research.

Some classifiers have attempted to find more meaningful indices for the thermal and moisture factors of climate than just temperature and precipitation data. The best known is the American C.W. Thornthwaite, who during the 1930s and 1940s devised a classification scheme based on what he called a "precipitation effectiveness index" and a "thermal efficiency index."[2]

It is unfortunate that the derivation of these indices requires a great deal of time, and the final result is so far removed from the original data that it is difficult to relate his climatic types to actual temperatures and precipitation amounts. Also, four maps are required to depict the climate of the earth, instead of only one, as in the Koeppen system. And although the great amount of statistical manipulation creates the illusion of greater accuracy, in the final analysis Thornthwaite's classification scheme is based on no more than temperature and precipitation data. Because of its complexity the system has not caught on in its entirety. But certain aspects of Thornthwaite's work have thrown a great deal of light on some of the thorny problems of classification in general, and also have proved to have considerable practical value to, for example, certain types of agriculture.

The Soviet climatologists A.A. Grigoryev and M.I. Budyko, working primarily during the 1950s and 1960s, devised a scheme that uses a thermal index based upon heat units rather than temperature degrees.[3] Their thermal factor is the radiational heat balance at the earth's surface, and their moisture factor is the so-called "radiation index of aridity"—the ratio between the radiation balance at the earth's surface and the annual precipitation multiplied by the latent heat of evaporation. Although the argument for the use of heat units rather than temperature degrees is probably a valid one, in most cases radiation balance measurements are unfortunately lacking. At best, the radiation balance has to be derived as a residual in equations relating other radiational measurements, which themselves are often lacking. Therefore, Grigoryev and Budyko had to resort to the use of temperatures in order to depict radiation balance. They found a fairly high correlation between the annual radiation balance at the earth's surface and the sum of temperatures during the period when the mean daily temperature remains above 10°C. Therefore, their system utilizes the sum-of-temperatures concept mentioned in Chapter 14. Although this system has attracted some attention in the Soviet Union, and a climatic map of the Soviet Union has been constructed on that basis, it is not well known outside the USSR.

E.E. Federov, another Soviet climatologist working primarily during the 1920s and 1930s, was dissatisfied with the limited number of climatic elements that most classification schemes utilized, and tried to devise a system that included all facets of climate.[4] To achieve this he went back through all the climatic records for all the weather stations in the Soviet Union and characterized each day's weather with a group of symbols that included not only the measurable elements of climate, but also various incommensurables such as the synoptic situation and even the state of the surface of the ground. While his was an admirable effort and he certainly pursued this particular line of thought further than anyone else, he ended with a very complex system (that even he termed "complex climatology") that embodied some subjective decisions regarding incommensurable phenomena. Therefore, it is not completely objective, as are the three systems previously described.

In Federov's system, the climate of a place was depicted as a frequency distribution of weather types; areal distributions had to be shown one weather type at a time. Thus, the depiction of the distribution of climate over an area entailed many maps. Because of its complexity, the system as a whole has not been utilized to any extent; but like the Thornthwaite endeavor, certain aspects of the system, particularly some of the illustrative schematics, have been used in standard Soviet works such as the large *Physical Geographical Atlas of the World,* from which several maps in this text have been derived. But outside the Soviet Union the system is little known.

The four classification schemes mentioned so far are all empirical classifications. There have been attempts at genetic classification schemes, but none has been very successful. These have been based upon such things as air mass frequencies and general circulation systems, all of which are nonmeasurable and noncomparable phenomena. Usage of such criteria by necessity involves much subjective thinking, so different investigators may end up with different results even though they use the same information. Nevertheless, such efforts often have illuminated certain causal relationships that aid in explanations of climatic distributions.

Boundary Conditions

Once the decision has been made regarding the criteria to be used, boundaries need to be decided upon. If the criteria are temperature and precipitation, are any values of these elements more important than any other values? Probably not, although one might argue that the freezing point of water is a more important temperature than any other temperature. Within the field of climate, apparently no values of such elements are more significant than any other values. Therefore, to attach significance to particular values of climatic elements, classifiers have had to go outside the field of climate to some related phenomena, such as wild vegetation or soils, to see if zones of rapid transition in these phenomena match up with any climatic values. If so, perhaps these are significant climatic boundaries, since such natural phenomena are very dependent upon climate for development. For instance, vegetation growing naturally without the interference of humans can be considered an overall climatic indicator that integrates all weather elements over the growing period, the final result of vegetative growth representing the totality of the climate of a region.

Most climatic classifications, at least the empirical ones, have utilized vegetation and other natural phenomena to determine climatic boundaries. Hence, these climatic classifications have become systems that classify the state of the surface of the earth rather than the air immediately above it. This has led to some peculiarities in climatic classifications. One is the labeling as "desert" of coastal strips such as that along the Pacific coast of Peru, which always has near-saturated surface air, much fog and low cloudiness, cool, temperate conditions with limited diurnal temperature ranges, and often a fine drizzle that keeps the streets and sidewalks damp. The only thing "desert" about such a region is its almost complete lack of measurable precipitation and its limited vegetative growth, despite its constantly cool, humid, overcast conditions. Nevertheless, on a global scale, the use of such things as natural vegetation to set climatic boundaries has alleviated much confusion and indecision and has attached some significance to categories that might otherwise have been purely arbitrary.

When Koeppen was beginning to contemplate his climatic classification, botanists were in the final stages of drawing some of the first world maps of natural vegetation, and he was impressed that the earth's vegetation was organized into fairly discrete zones separated by narrow belts of rapid transition. He reasoned that these transition belts must be caused by certain facets of climate that determined the type of plants that could grow. In his classification scheme, published in 1900, he gave names of specific types of vegetation or other related phenomena to all his climatic types. As he refined his system he narrowed his climatic boundaries to a few major ones, determined either by certain temperature values or by temperature-precipitation combinations that seemed to relate to rapid transitions of ecological groupings of plants. For instance, he found that if the temperatures for the coldest month of the year averaged below 18°C (64.4°F) many tropical plants ceased to grow. Thus, he decided that the 18°C isotherm for the coldest month was a significant boundary, the poleward limit of tropical climate. Likewise, he found that for trees to grow at high latitudes, temperatures usually had to average above 10°C (50°F) for at least one month of the year. Therefore, he decided that the 10°C average for the warmest month delineated the poleward margins of tree growth. He found that if the warmest month average was below 0°C (32°F) nothing grew, and ice caps would develop over time. Thus, he decided that the 0°C isotherm for the warmest month was the equatorward limit of ice cap climate.

Koeppen was very impressed with the effects of snow cover, since he had worked during the early part of his career at the Main Geophysical Observatory in St. Petersburg (Leningrad), where the great Russian climatologist A.I. Voyeykov had carried out many years of fundamental research on the effects of snow cover. Koeppen decided that the equatorward edge of a persistent snow cover during winter should be a significant climatic boundary. He found that in Europe this corresponded roughly to the −3°C (26.6°F) isotherm for the coldest month average. (North American climatologists subsequently decided that the equatorward limit of a persistent snow cover in North America corresponded more closely to the 0°C (32°F) isotherm.)

Like other climatologists attempting to formulate classification systems, Koeppen had the greatest difficulty deciding on a humid-dry boundary. This has proved to be the major bugaboo in practically every attempt to classify climate. Certainly no one isohyet (line of equal precipitation) can serve as a boundary, since moisture availability for plant growth depends not only on precipitation inputs, but also on various moisture losses from the soil. Eventually all the precipitation that falls on the earth's surface evaporates and is transferred back into the air again. So attempts have been made to fit moisture losses into some sort of evaporation formula. After much inspection of climatic and vegetation data, Koeppen finally devised three simple equations that related evaporation losses to average annual temperatures and precipitation regimes. He recognized that not only the annual amount of precipitation was important, but also seasonal regime, since if most of the precipitation fell during the warm season more of it would be lost through evaporation and would not serve to moisten the soil. He therefore devised three formulas, one for a fairly even distribution of precipitation through the year, one for precipitation concentrated during the warm season, and one for precipitation concentrated during the cold season.

But it remained to Thornthwaite to shed the greatest light on the problem of moisture loss. He coined the word "evapotranspiration" and attempted to measure it directly by inventing so-called "evaporimeters," large vessels that he embedded in the earth and in which grew indigenous types of vegetation under controlled (measurable) water inputs and losses. He quickly came to the conclusion that it was not the actual evapotranspiration that should be balanced against precipitation but the so-called "potential evapotranspiration"—the amount that would evaporate were there an unlimited amount of water to be evaporated. This quantity became his famous PE (potential evapotranspiration) index. Unfortunately, his moisture-loss considerations, like Koeppen's, take into account only temperature, and not such things as relative humidity and wind speed, which also alter evapotranspiration rates. Therefore, although his estimate of moisture losses might be one of the more accurate ones, it still leaves much to be desired. Subsequent estimates by Soviet climatologists and others have produced a world map of precipitation-evapotranspiration balance (Figure 15.1).

Another of Thornthwaite's innovations is his concept of soil moisture storage. He devised a monthly log of precipitation inputs, soil moisture changes, and evapotranspiration losses, and through this running account of monthly moisture conditions he was able to demonstrate that in many areas plants may actually experience some moisture deficits during the warm season of the year, even though it might be the rainiest season of the year. This component of his work induced Birdseye Foods to support him financially so that he could set up a climatic laboratory at Seabrook, New Jersey, to advise them when to use supplemental irrigation in the growing of critical crops, such as green peas, and other vegetables for the frozen food market. Graphical representations such as that for Raleigh, North Carolina, shown in Figure 15.2, became very useful for these purposes. These graphic representations of moisture inputs, moisture losses, and soil moisture conditions illustrated very clearly the periods of soil moisture deficits occurring right during the summer maximum-rainfall periods. Of course, Figure 15.2 represents the long-term average condition, and the moisture components vary considerably from year to year. By keeping a log of current conditions during each growing season, Thornthwaite could illustrate clearly what the soil moisture conditions were at any particular time and could recommend supplemental irrigation during periods of soil moisture deficits before these deficits had existed long enough to cause curtailment of crop yields, quality, and progression of growth stages.

The Classification Used in This Book

Many of the concepts originated by Koeppen have stood up well over time, and although the simplistic approach of his system often provides only first approximations of reality, this very simplicity and the ease of relating final results to original temperature and precipitation data have resulted in its being the most frequently adopted system for teaching climatology around the world. Like the English language, it might not be the best system on earth, but because more people know it than any other system, it has become the *lingua franca* for the field of climatology, despite its irrational symbolization. Five major types of climate are designated by the five capital letters A, B, C, D, and E. Imbedded in this group are the B climates, which he designated as all the dry climates of the world. The rest are all humid climates

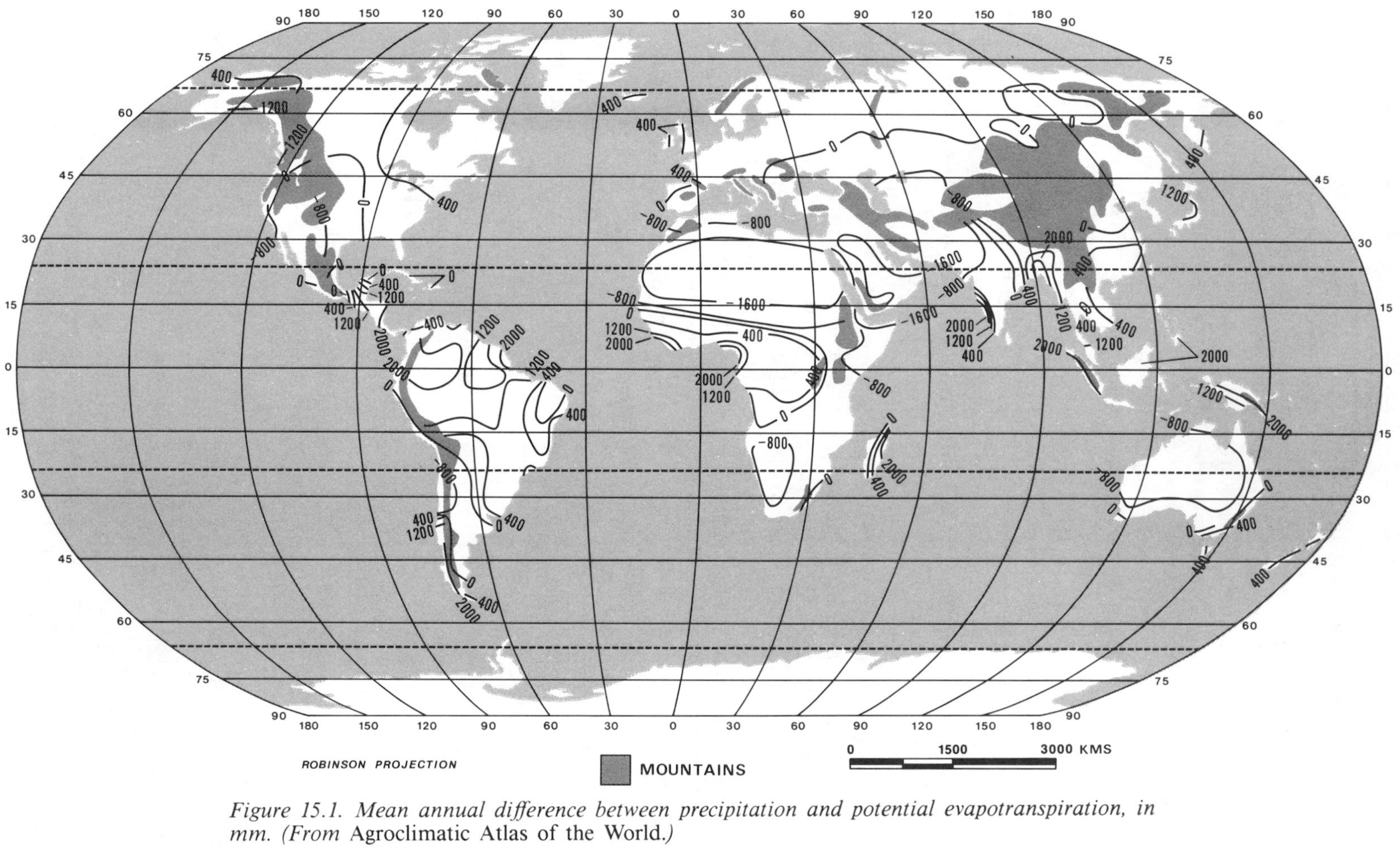

Figure 15.1. Mean annual difference between precipitation and potential evapotranspiration, in mm. (From Agroclimatic Atlas of the World.*)*

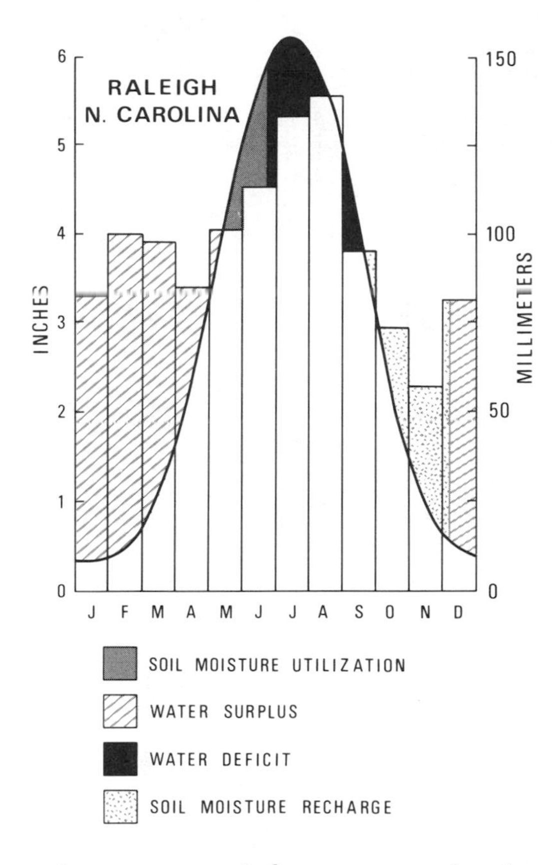

Figure 15.2. Monthly moisture conditions at Raleigh, North Carolina. Bar graphs represent precipitation; the line graph, potential evapotranspiration. Potential evapotranspiration exceeds precipitation from about 10 May through about 20 September, even though summer is the season of maximum precipitation. During the last two-thirds of May and the first three-fourths of June the precipitation deficit is compensated for by plants by utilization of soil moisture (light gray area under line graph). But by 23 June all the available soil moisture is used up, and a soil moisture deficit (drought) obtains through late June, all of July, much of August, and the first two-thirds of September (dark gray area). After that a precipitation surplus is used to recharge the soil moisture (stippled area) until early December. By about 5 December the soil moisture is fully recharged, and a moisture surplus (runoff) obtains through much of December through early May (striped area). (Data from Thornthwaite, Mather, and Carter. [Modified from An Introduction to Climate, *5th ed., by Glenn T. Trewartha and Lyle H. Horn. Copyright © 1980 by McGraw-Hill, Inc. Used with permission of the McGraw-Hill Book Company].)*

and are separated from one another by certain temperature conditions. This unfortunate positioning of the symbol for dry climates in the midst of the consecutive series of letters of the alphabet, the bulk of which represent humid climates delineated on temperature bases, has had the unfortunate result of deluding many people into thinking that Koeppen's primary subdivision of climatic types is based on temperature, when actually the first-order breakdown is on the basis of moisture.[5] Only at the second level of the hierarchy does temperature come into play. It would have been more evident had the humid and dry climates been separated out more obviously and then each subdivided on temperature bases.

Nevertheless, Koeppen's symbolization, probably more than any other facet of his classification, has become locked into everyone's thinking. Anyone who has studied climatology anywhere in the world knows that the A climates are tropical, the B are dry, C are mesothermal or subtropical, and so forth. Thus, this is the nomenclature that has become the *lingua franca* of climatology. Even G.T. Trewartha, who devised a classification system of his own that departs widely from the Koeppen system, has retained essentially the Koeppen nomenclature, with a few well-conceived changes and additions. Trewartha's classification, which first appeared in 1968 in the fourth edition of his *An Introduction to Climate,* will be used here, since he emphasized summer temperatures and lengths of growing seasons more than Koeppen did, and thus probably matched climatic types to such things as vegetative growth and human occupancy better than did the old Koeppen system.[6] Table 15.1 lists the climatic types, their boundary conditions, and associated features. Figure 15.3

Table 15.1 Climatic Classifications

Symbol	Climatic type	Boundary criteria	Rationale	Vegetation
B	**DRY**	R = 2T (s precip. regime) R = 2T + 14 (f precip. regime) R = 2T + 28 (w precip. regime) where R= hypothetical ann. precip. in cm.; T=ann. average temp. in °C; s=summer aridity (wettest winter month more than 3 × precip. of driest summer month); w=winter aridity (wettest summer month more than 5 × precip. of driest winter month); f (*feucht*)=more even distribution of precip. than in s and w.	Potential evapotranspiration exceeds precipitation (essentially the aridity limit of tree growth).	
S	Steppe		Semiarid	Bunchgrass, brush, scattered trees
		Aver. ann. precip.=R/2		
W	Desert (*wüste*)		Arid	Sparse shrubs, cactus
h	Hot (*heiss*)			
		8 mo. aver. temp.≥10°C (50°F)		
k	Cold (*kalt*)			
A	**TROPICAL HUMID**	All months aver. temp. > 18°C (65°F)		
r	Rainy			Rainforest of broadleaf evergreens
		2 mo. aver. precip.= 6 cm. (2.4 in.)		
	Wet-dry			
w	low-sun dry			Savanna grasses
s	high-sun dry			
		Coolest mo. aver. temp. = 18°C (65°F)	Poleward limit of many tropical plants	

Code	Type	Criteria	Limits	Vegetation
C	**SUBTROPICAL**			
f	Humid			Mixed forests
s	Dry summer	Wettest winter mo. 3 × driest summer mo. Driest summer mo. < 3 cm. (1.2 in.)		Chaparral, grass, live oak, eucalyptus
w	Dry winter	Wettest summer mo. 5 × driest winter mo.		Mixed forests
a	Hot			
b	Cool	Warmest mo. temp. = 22°C (72°F)		
		8 mo. aver. temp. ≥ 10°C (50°F)		
D	**TEMPERATE**			
o	Oceanic			Mixed, coniferous, and broadleaf forests
		Cold mo. aver. temp. = 0°C (32°F)	Persistent snow cover (one mo. or more)	
c	Continental			
a	hot			Mixed forests Primarily broadleaf deciduous
		Warmest mo. temp. = 22°C (72°F)		
b	cool			Mixed broadleaf and coniferous
		4 mo. aver. temp. ≥ 10°C (50°F)	Limit of most crop cultivation	
E	**BOREAL**			Coniferous and small-leaved trees (taiga)
		Warmest mo. = 10°C (50°F)	Poleward limit of tree growth	
F	**POLAR**			
t	Tundra			Mosses, lichens, sedges
		Warmest mo. = 0°C (32°F)	Limit of vegetative growth by glacial accumulation	
i	Ice cap			No vegetation, ice covered

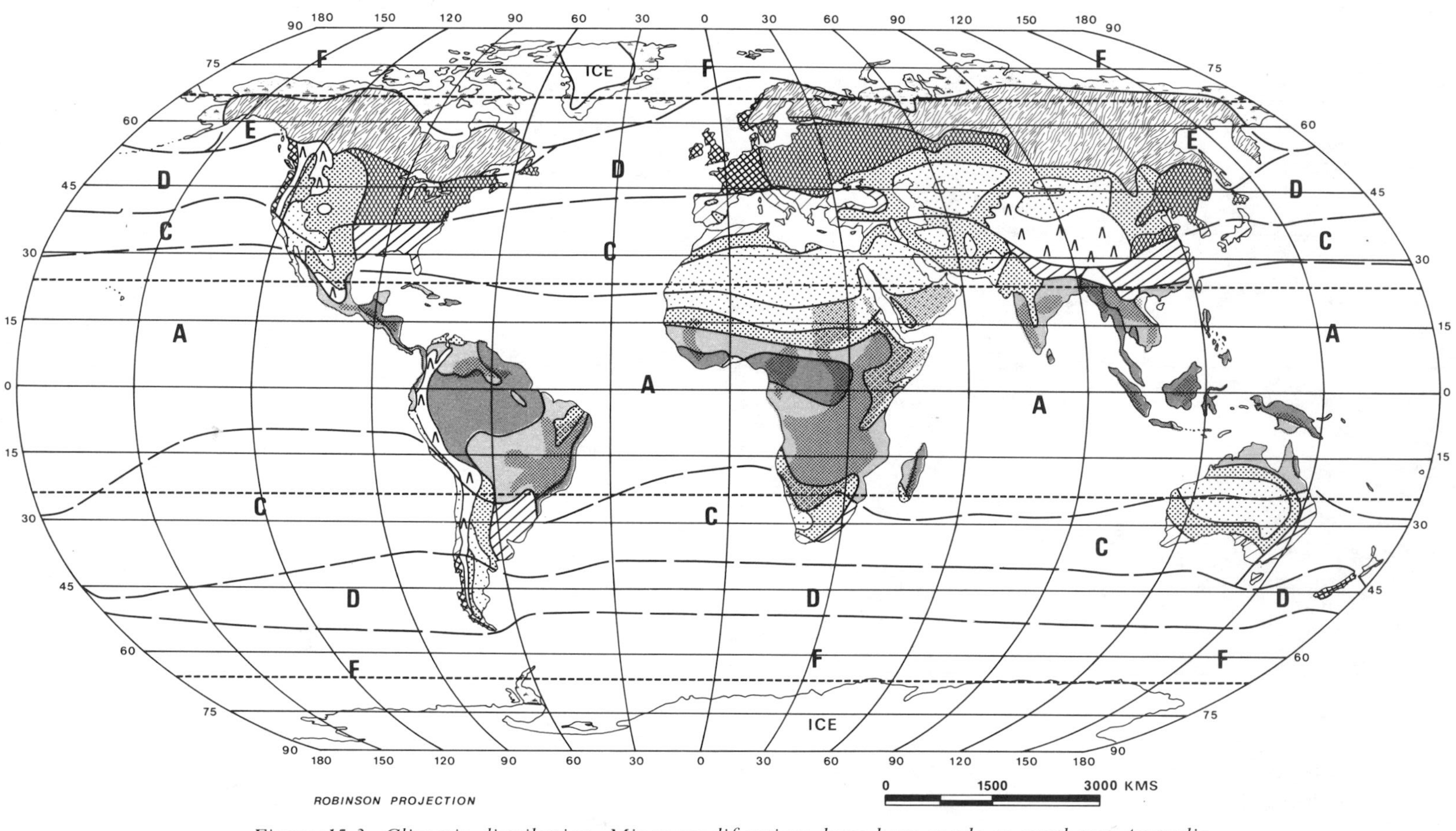

Figure 15.3. Climatic distribution. Minor modifications have been made in southeast Australia-Tasmania and northern India. (Modified from An Introduction to Climate, *5th ed., by Glenn T. Trewartha and Lyle H. Horn. Copyright © 1980 by McGraw-Hill, Inc. Used with permission of the McGraw-Hill Book Company.)*

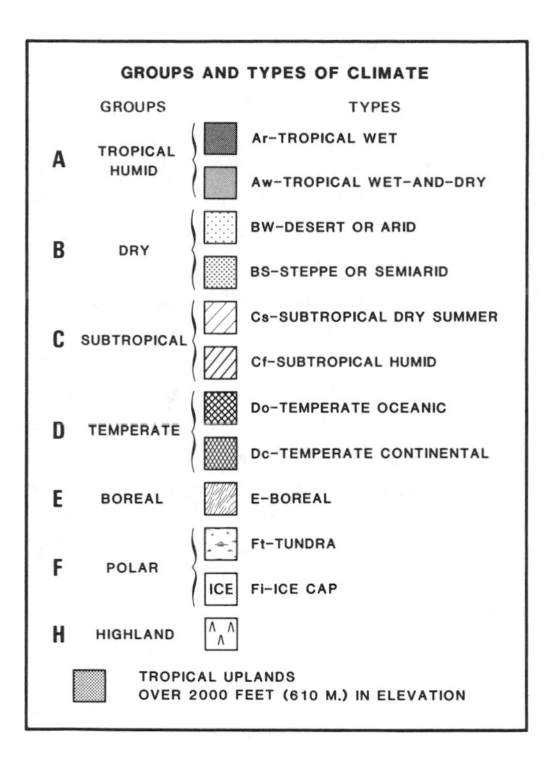

shows their distribution over the earth. Since the B climatic type is distinguished by its dryness, it will be discussed first, and then the humid types will be taken up in turn.

Climatic Types

B: Dry Climates

Climate is considered dry if the annual potential evapotranspiration exceeds the annual precipitation. From inspection of maps of natural vegetation types, temperature, and precipitation, Koeppen derived three approximate formulas, according to seasonality of precipitation, to define the humid-dry boundary, as follows:

$R = 2T$ (precipitation during wettest winter month is more than three times as much as that during the driest summer month),

$R = 2T + 14$ (precipitation evenly distributed through the year), and

$R = 2T + 28$ (precipitation during the wettest summer month is more than five times that during the driest winter month).[7]

In these equations, R = the hypothetical annual precipitation in centimeters that would place the station with the average annual temperature T in °C exactly on the humid-dry boundary. Converted to precipitation in inches (R′) and temperature in °F (T′), the formulas are as follows:

$R' = 0.44T' - 14$ (rainfall concentrated in winter),

$R' = 0.44T' - 8.5$ (rainfall evenly distributed), and

$R' = 0.44T' - 3$ (rainfall concentrated in summer).[8]

Dry climates have been divided according to whether they are semiarid or arid into BS (steppe) and BW (*wüste,* the German word for desert). The boundary is taken as half that of the humid-dry boundary. The dry climates have also been divided into h (from *heiss,* the German word for hot) and k (*kalt* is German for cold), according to whether more or less than eight months have average temperatures above 10°C (50°F).

To illustrate the classification of a borderline station, consider the data for Williston, North Dakota (Number 28 at the end of Chapter 16). Since the wettest summer month, June, has more than five times as much precipitation as the driest winter month, February, the formula for a w-type (winter dry) rainfall regime will be used, $R = 2T + 28$. The annual average temperature is 5°C. Therefore, $2 \times 5 + 28 = 38$ centimeters or 380 millimeters. This is the hypothetical annual precipitation that would put Williston, with an annual average temperature of 5°C, right on the humid-dry boundary. Since the annual precipitation at Williston is only 349 millimeters, the climate is on the dry side of the humid-dry boundary. To determine whether it is steppe or desert, divide the computed R by 2 ($380 \div 2 = 190$). Since the annual precipitation at Williston is above 190, the climate is steppe rather than desert. Since fewer than eight months average above 10°C, the region is a cold steppe. Thus, the climatic classification for Williston, North Dakota, is BSk.

To illustrate the importance of the seasonality of rainfall, consider the case of Los Angeles, California (Number 12, Chapter 16), with an average annual temperature of 18°C, an annual average precipitation of 373 millimeters, and almost all of its rainfall during the winter months. In this case, the summer-dry formula, $R = 2T$, would be used. Thus, $R = 2 \times 18 = 36$ centimeters or 360 millimeters. Since the annual precipitation is slightly more than that, the area is humid, even though May–September has very little rainfall. In this case the climate would classify as Csa, since the coolest month averages below 18°C, more than eight months average above 10°C, the summers are dry, and the warmest summer month averages above 22°C.

The deserts of the world are generally located astride the tropic circles, where the subtropical high pressure cells induce subsidence much of the time, and in the interiors of large continents or behind mountain ranges, where a moisture source is lacking (see Figure 15.3). In North America and Asia dry climates from these two causes merge to produce very large areas of aridity interrupted only here and there by mountains. To some degree this is also true in the southern part of South America. More unusual locations of dry climate occur in limited areas in such places as northern Venezuela, eastern Brazil, and the eastern horn of Africa, which require other explanations that will be taken up in Chapters 16–21. In total, dry climates occupy about 26 percent of the continental areas, of which 14 percent is steppe or semiarid and 12 percent is desert or arid.

Usually, dry climatic areas are characterized by low atmospheric humidities and much sunshine, which lead to great temperature extremes on both diurnal and annual time scales. Nevertheless, a set of desert strips on coasts adjacent to cool ocean currents have very different characteristics, as was mentioned earlier. These will be taken up with their respective continents in Chapters 16, 18, and 19.

Dry areas generally have sparse vegetation covers consisting of *xerophytic* (drought-resistant) species of plants whose forms reflect their physiological necessity to reduce transpiration surfaces. Middle-latitude steppes are typically vegetated by bunch grasses that may grow to heights of 30 centimeters or more, but which usually do not form a continuous sod over the surface because of the lack of sufficient soil moisture. In some places, such as large portions of the intermontane region of the American West, sagebrush and other types of woody plants may replace the grasses. Low-latitude steppes may also be vegetated by sparse grass covers, although the grasses are related more to the tropical grasses of the savanna lands on their equatorward sides, which are less nutritious than the middle-latitude grasses. Also many tropical semiarid regions are vegetated by thorn shrubs and stunted trees.

Desert areas have even more meager vegetation covers. Usually there is much bare soil in between individual plants. In many areas, various species of brush growing a meter or two high are the predominant vegetation. In other areas, various forms of cactus are dominant. Cactus are conspicuously absent throughout most of the deserts of Central Asia, where a unique form of low tree or bush, the saxaul, has become established. In addition to the sparse perennial vegetation, a short-lived ephemeral vegetation of flowering herbaceous plants typically springs up after rainy spells. Few desert areas are completely devoid of vegetation, although this may be true in limited areas where surface materials are constantly shifted by wind, such as in some of the alluvial fan areas of the Sahara or the beach sands of coastal deserts.

A: Tropical Humid Climates

In the tropical humid climates, all months average above 18°C (65°F). They are free of killing frosts, and therefore have year-round growing seasons. Since temperatures are always warm, the tropical climates are not subdivided on that basis, but on precipitation regime. If no more than two months have less than 6 centimeters (2.4 inches) of precipitation, the climate is considered to be rainy year-round (Ar). If more than two months have less than 6 centimeters of precipitation each, the climate is Aw if the dry season comes during winter (low-sun) or As if it comes during summer (high-sun). In the A climates it is not realistic to talk about winter and summer, but rather low-sun, when the noon sun is lowest on the horizon, and high-sun, when it is highest above the horizon. Even low-sun and high-sun seasons may be difficult to identify, since near the equator high-sun occurs twice a year at the equinoxes, and low-sun occurs twice a year at the solstices. Also, there is often little correlation between the temperature regime and the sun regime, since the minor temperature fluctuations that exist may be caused more by the state of the sky, being typically cooler during rainy seasons than during dry seasons. Since throughout most of the tropics rain comes in the form of thundershowers that are largely the products of the intertropical convergence zone, easterly waves, and local convection, in most places maximum rainfall coincides approximately with the high-sun period. Therefore, there are only few and restricted areas of As (summer dry) climate.

The A climates are shown on Figure 15.3 in two categories, the Ar—tropical wet climate—and the Aw—tropical wet-and-dry or savanna climate. Since the earth is much larger around at the equator than farther poleward, A climates occupy a very large part of the earth's surface, approximately 36 percent. However, this region consists more of ocean surface than land surface. A climates occupy about 43 percent of the earth's ocean surface and only 20 percent of its land surface. Therefore, A climates occupy less territory on land than dry climates do.

Ar Climate. The largest areas of Ar climate straddle the equator in the Amazon Basin of South America, the Congo Basin of Africa, and the islands of Indonesia. The intertropical convergence zone between the trade winds generally lies in or near these areas year-round and is associated with heavy rainfall. In low-lying plains annual rainfall typically amounts to 1750–3000 or more millimeters. But in some places, such as the Colombian coast of northwestern South America and the Guinea coast of west Africa, the intertropical convergence zone is consistently located north of the equator and produces exceptionally heavy rainfalls, 5000 millimeters or more per year. The heaviest rainfalls on earth, however, are found farther poleward where consistent air flow, at least part of the year, blows against mountain coasts. Such are the northeast Indian-Bangladesh-Burmese coasts, which receive much rain from consistent southwesterly winds blowing across the Bay of Bengal during summer and early fall, and the northeast coast of some of the Hawaiian Islands which receive the northeast trades year-round. To a lesser extent the combination of rugged topography and consistency of winds also brings heavy rainfall to much of the Caribbean side of Central America and many of the Caribbean islands, eastern Madagascar Island east of South Africa, and many portions of the mountainous Philippines and surrounding island groups.

Vegetation in Ar areas is typically tropical rainforest—dense stands of huge hardwoods of many species reaching heights of 50–60 meters, which consist primarily of broadleaf evergreen trees. On the drier margins of Ar areas many trees drop their leaves during dry spells and allow enough sunshine to get to the ground to produce junglelike tangles of undergrowth and vines.

Aw Climate. The Aw climatic areas generally lie on either side of the Ar areas. The Aw is a hybrid climate, having very different precipitation controls during opposite seasons. During the high-sun season the intertropical convergence zone with its thundershowery precipitation affects the areas much as it does in the Ar areas year-round; during the low-sun season the subtropical high pressure cells shift equatorward over the areas much as they occupy the subtropical desert areas on the poleward side of Aw areas year-round.

Typically, Aw areas receive 1000–1500 millimeters of precipitation during the high-sun season but very little during the low-sun season. Thus, during the wet season the ground is saturated, the rivers are flowing full, and there is a rich vegetative growth. During the dry season the ground bakes as hard as a brick and cracks to great depths, and the vegetation withers and dies. Since most trees must have adequate moisture year-round, tree growth is sparse, and coarse tropical grasses, in many cases higher than a human's head, take over. These grasslands may be interspersed with widely spaced, stunted, drought-resistant trees. In some places thorn forests of low, stunted trees take over.

As Figure 15.3 shows, large portions of Aw climatic areas lie at significant elevations above sea level. In these areas, recorded surface temperatures generally would be too cool to classify as tropical climate. Nonetheless, daily weather episodes and annual rainfall and temperature regimes reflect the tropical latitudinal positions of the areas. Therefore, the climate of these areas does not correspond to any of the other climatic groups, and on Figure 15.3 these areas are shown as having an upland variant of Aw climate.

C: Subtropical Climates

The subtropical climates have at least one month averaging below 18°C (65°F) and eight or more months averaging above 10°C (50°F). The 18°C average for the coolest month limits the growth of many tropical plants, and a monthly average of 10°C is a significant threshold of growth for many domesticated crops. Thus, the subtropical climates have at least an eight-month growing season, and most of the areas have considerably more than that. In a few places, particularly near windward coasts, frosts may never occur and the growing season is year-round. Even in some inland areas, frosts may not occur every year. On the other hand, along the poleward margins of interior areas, temperatures have dropped as low as −25°C. In such places snow may fall frequently, occasionally in quite heavy amounts, but it does not stay on the ground long since midwinter thaws are frequent. On the other hand, windward coastal areas near the equatorward margins of subtropical climates never experience snow.

The C climates are subdivided according to two criteria: (a) the precipitation regime and (b) the summer temperature. C climates are divided into the subcategories f, s, and w, which denote seasonality of precipitation the same as they did in delimiting humid-dry boundaries. The s category of the C climates has the additional criterion that the driest month of summer must have less than 3 centimeters (1.2 inches) of precipitation. In other words, summers must be truly dry, rather than just less than one-third as wet as winter. C climates are also subdivided into a and b, according to whether or not the warmest month temperatures average above or below 22°C

(72°F). Windward coasts, particularly those paralleled by cool ocean currents, generally average below 22°C during summer, while all other C climates in the interiors of continents and on leeward coasts average above 22°C. Thus, warm summers are much more widespread than cool summers.

Theoretically, then, the C climates can be subdivided into six categories—Cfa, Cfb, Csa, Csb, Cwa, and Cwb—but not all differ significantly one from another. In general, the areas having dry winters (w) also have rather cold winters, so the winter precipitation amount is of little consequence to plant growth at that time. Therefore, usually the Cf and Cw climates are grouped together. Also, the Cf and Cw climates are typically located in continental interiors or leeward coastal areas where the summers are almost always warm. In such locations the cool-summer type, if it exists at all, is usually limited to uplands. For these reasons, the four categories Cfa, Cfb, Cwa, and Cwb are grouped together on Figure 15.3 in a single pattern. Where dry winters are pronounced, such as in western China, the Cw symbol is shown, but the area is not differentiated by pattern from the Cf area farther east. Also, Csa and Csb are lumped together, since Csb generally occurs only along cool-water coasts or at elevations above sea level.

In Figure 15.3, then, the C climates are divided only into Cs and Cf climates. These two are positioned very latitudinally, generally on opposite sides of continents.

Cs Climate. The Cs, or subtropical dry-summer, is usually at latitudes between 30° and 45° on western sides of continents that are dominated by the strong atmospheric subsidence of the eastern ends of subtropical highs during the summer, but that during winter may occasionally be influenced by the cyclonic storms of the westerlies when the polar front swings equatorward. Hence, summers are generally dry and sunny, although in cool coastal areas there may be considerable summer fog and low stratus. Winters are still dominated to a large extent by the subtropical highs, but three or four times during winter the polar front may be positioned in the area, during which successive wave formations follow each other along the front to bring almost continuous rain for several days on end. Annual rainfall amounts typically total 375–750 millimeters in lowlands and twice that much or more in adjacent highlands, where much of the winter precipitation falls as snow.

By far the largest region on earth with Cs climate is the Mediterranean borderland, including almost all the Iberian Peninsula, southern France, most of Italy, western Yugoslavia, Greece, southern Bulgaria, the coastal areas of Turkey and Palestine, foothill areas of the Middle East, and the Mediterranean fringe of Africa north of the Atlas Mountains. For this reason, the Cs climate is often called the Mediterranean climate. Nevertheless, the Mediterranean region has the least-typical Cs climate in the world. It extends farther poleward in southern Europe than anywhere else because of the influence of the warm Atlantic, and its summers are not as dry as they are in southern California or central Chile. Small areas of Cs climate also occur in the very southwestern tip of Africa around Capetown and in the southwestern and southeastern parts of Australia.

Since the Cs areas are frequented by long periods of aridity during summer, commonly five or six months with little or no rainfall, natural vegetation must be of the drought-resistant type. This usually means a brushy growth of many species—some evergreen, some deciduous—that collectively is known as *chaparral* in California and *maquis* in the Mediterranean region. Such vegetation has little use other than to reduce erosion and prevent mud slides that occur on the steep slopes that prevail throughout most of the Cs regions of the world. Many areas are also covered by grasses sometimes interspersed by widely spaced trees, such as the live oak in California or the cork oak in the Iberian Peninsula of Europe. Eucalyptus trees are indigenous to the Cs areas of Australia and have been introduced into most of the other Cs areas of the world as ornamental trees to line streets of cities and parks, since they are able to grow to heights of 30 meters or more despite dry summers. In addition, a profusion of flowering ephemeral plants proceed rapidly through a short life cycle after the winter rainy season, much as they do after rainy seasons in many desert areas of the world.

It so happens that the Cs areas have many mountains. Therefore the climate varies locally with elevation and exposure, and heavy snow packs typically form in the mountains during winter and provide irrigation for the plains during summer. Because of the nearly year-round growing seasons and the availability of irrigation water during summer, the Cs climates have become primary areas for the growth of many crops such as citrus, for vineyards, and for a variety of other fruits and vegetables that may be exotic to other parts of the same countries. Also, the mild winters and clear, dry summers have tended to make Cs areas the playgrounds of the world for tourists.

Cf Climate. As can be seen in Figure 15.3, Cf (subtropical humid) climate generally occupies southeastern portions of continents at latitudes around 25°–37°. These areas are influenced strongly by the western ends of subtropical high pressure cells which have much different characteristics than the eastern ends that influence Cs areas. In Cf areas the air circulates poleward after having crossed a broad expanse of low-latitude ocean and therefore is warm, moist, and usually conditionally unstable. Such air during summer brings prolonged sultry conditions, which is perhaps the major characteristic of this type of climate. During summer, Cf areas may be even more oppressive than equatorial areas where temperatures are not quite as high.

On the other hand, the Cf areas do have some relief when cooler, drier continental air from across the polar front occasionally invades the areas. But the continental air is more dominant during winter, and thus brings much colder winters to the Cf areas than the Cs areas experience. This is especially true in the two major Northern Hemisphere areas, southeastern United States and China. The Southern Hemisphere areas, in the pampas of Argentina, Uruguay, southern Brazil, and Paraguay, the southeastern coastal area of Africa, and eastern Australia and the northern island of New Zealand do not have major landmasses positioned poleward that can send continental polar air into these regions. The small region of Cf climate along the eastern Black Sea in the Soviet Transcaucasus and adjacent Turkey, as well as in northern Iran bordering the southern end of the Caspian, also experience fairly mild winters, although temperatures may dip well below freezing.

Typical precipitation amounts in Cf areas range from 1200–1400 millimeters in the wetter eastern parts, to 900 millimeters or less in their drier western parts where they merge into dry climates. Typically, the maximum rainfall comes during summer in the form of thundershowers. The seasonality is very pronounced in China, where the winters tend to be quite dry because of the dominance of the northwesterly winds coming off the continent from the Asiatic High. Interior southeastern United States is an exception to the summer maximum regime. A large area centered on the state of Tennessee has more precipitation during the winter half-year than during the summer half-year.

The Cf areas in southeastern United States and the China-Japan area experience some of the most violent weather on earth. Hurricanes and typhoons frequent the areas during summer and autumn, and in these latitudes they have usually reached their mature stages with winds often stronger than they are farther equatorward in the tropics. Also, in southeastern United States many tornadoes occur at any time of year, but especially during spring.

Since the Cf areas generally grade from quite moist conditions in their eastern portions to subhumid conditions in their western portions, the natural vegetation grades from mixed forest in the east to prairie grasses in the west. Since the pampas of South America are somewhat drier than the other Cf areas of the world, grasses were the predominant original vegetation in the area.

D: Temperate Climates

The D climatic group is characterized by four to seven months with temperatures averaging above 10°C (50°F). "Temperate" is rather a misnomer, since great extremes occur, particularly in the interiors of large continents. The word "temperate" relates only to average conditions, not to extremes. The term is much more appropriate along western coastal areas where marine influences temper extreme conditions. And it is primarily on this basis that the D climates are subdivided into Do (temperate oceanic) and Dc (temperate continental).

The boundary between the Do and Dc climate is the 0°C (32°F) isotherm for the coldest month average. Since the average temperature of freezing coincides closely with the equatorward limits of a snow cover in winter, the Dc areas generally have persistent snow covers for at least a month and much more in northern portions of the regions, whereas Do areas do not have persistent snow covers at lower elevations, although certainly it can snow in these regions. Also, the temperate continental areas are set off from the subtropical humid areas farther equatorward by the same snow cover conditions. This was Koeppen's original criterion to delimit the C and D climates, although he found that in Europe the equatorward edge of the persistent snow cover in winter coincided more with −3°C than with 0°C. In the classification used in this book, however, the D-C boundary has been shifted to summer growth threshold temperatures.

The D climate group can also be subdivided according to summer temperatures and precipitation regimes, the same as C climates are. But within the D climates these subdivisions are not as significant, except perhaps for the summer differentiation into Dca and Dcb, based on the boundary of 22°C for the warmest month average temperature. A Dcs climate essentially does not exist, since in these continental locations at higher middle latitudes there is generally a pronounced summer maximum of precipitation due to greater atmospheric moisture content and added convective activity during that time of year. In fact, in some places, particularly in northern China and adjacent areas of the Far East, summers have so much more precipitation than winter that a Dcw classification would apply. But since these areas have very cold winters, the w category is insignificant and is lumped with the f category with year-round precipitation. Dos does occur along the west coast of North America, which has unusually dry summers for such a high latitude, but the s category is not particularly significant. And most of the Do areas of the world are so marine controlled that they do not have hot summers, so the subcategory a is not applicable to that climatic type. Thus, in Figure 15.3 only two categories, the Do and the Dc, are shown by pattern differentiation. In addition, portions of the Dc areas are divided into a and b, depending upon summer temperatures because these exert considerable influence on types of natural vegetation and agricultural crops.

Do Climate. Do climate is a very marine-controlled climate in higher middle latitudes. Its primary characteristic is its rather low annual temperature range for such a high latitude. Winters are unusually mild, and summers are relatively cool. In general, precipitation is ample and evaporation rates are modest. In mountainous areas precipitation can be copious, upward of 5000 millimeters per year, with much in the form of snow. In spite of relatively mild temperatures, the snow may not all melt during the year, and therefore mountain glaciers have accumulated in regions such as the British Columbian and Alaskan coast in western North America, the southern Andes in South America, and the western coast of the south island of New Zealand and have produced deeply gouged, fiorded coastlines in these regions.

Northwestern Europe, the fourth and largest area of Do climate, has no very high mountains oriented perpendicular to the westerly wind flow, and therefore precipitation is generally more modest. Since the mild Atlantic air has open access across the European landmass, the Do climate grades gradually into the Dc climate in central Europe, whereas in North and South America, the Do climate is limited to the western coastal areas by high mountains across which are found dry climates on the eastern side.

All the Do areas are frequented by high relative humidities, fog, stratus clouds, and cyclonic storms. Much of the precipitation comes in the form of prolonged light-to-moderate intensity, which is very effective in moistening the soil. Therefore, even the lowlands of northwestern Europe look more humid than precipitation amounts would warrant. Consequently, all these areas originally were regions of forest, the types of which varied greatly from one continent to another. In northwestern Europe they were primarily mixed forests, while in the Pacific Northwest of North America and the southern Chilean area of South America they are primarily coniferous evergreens, some of very large size.

Dc Climate. The Dc climate occurs only in the Northern Hemisphere, since in the Southern Hemisphere there is no land at corresponding latitudes to produce continental polar air masses that would result in below-freezing average temperatures during winter. Three large areas cover (a) the northeastern United States and adjacent Canada, (b) central and eastern Europe, extending into southwestern Siberia, and (c) the north China-Manchuria-Korea-northern Japan area in the Far East. Although all three regions have a four-to-seven months growing season, below-freezing average temperatures during winter, and a fairly persistent snow cover during winter, there are very significant differences from place to place, which will be described in Chapters 16 and 17.

E: Boreal Climate

The boreal (northern) climate is defined as having one to three months with temperatures averaging above 10°C (50°F). Thus, it has a short, cool growing season that generally rules out agriculture but does allow the growth of a variety of coniferous trees as well as small-leaved trees, such as birch and aspen. This is the type of vegetation that the Russians refer to as the *taiga.* The taiga is one of the two remaining large virgin forests of the world, the other being the tropical rainforest, which has much different characteristics.

The boreal climate is limited to the Northern Hemisphere, since no place in the Southern Hemisphere has such cold winter temperatures and yet significant summers. The climate stretches across North America from southwestern Alaska clear across central Canada to the Atlantic coast, and across northern Eurasia from Scandinavia to the Pacific. It occupies most of Siberia except for the tundra fringe to the north and the Dc and steppe fringes along the south.

Winter is by far the dominant season in the boreal climate, occupying more than half the year. Summer is short but peaked, the temperatures rise and fall rapidly during the transitional seasons, particularly during spring, which is almost nonexistent. The prolonged winter has led to the formation of permafrost over large sections of the boreal areas. Because of the constantly cool or cold air, moisture-holding capacity is low, and hence precipitation is usually meager.

F: Polar Climates

In the polar climates no month averages above 10°C (50°F) and there is essentially no tree growth. The polar climates are divided into two categories, depending on summer temperature. If the warmest month averages above freezing, a small amount of plant growth can take place, such as mosses, lichens, sedges, and dwarf willow that creep along the ground rather than stand upright. Such vegetation is known by the Finnish word *tundra,* and the climate has become known by the same name. Tundra climate is denoted by Ft. If no months average above freezing, then glacial ice will accumulate over time, since not all the snow can melt during the brief thaws. This, then, leads to ice caps such as those on Greenland and Antarctica, and the climate is known as the ice cap climate, denoted by Fi.

The tundra climate primarily fringes the Arctic on the northern peripheries of North America and Eurasia. Small areas also fringe the ice caps on Antarctica and Greenland. Throughout the tundra and ice cap climates the precipitation is meager because of the low moisture-holding capacity of the cold air. But since the area is frozen much of the year, moisture losses are very low, and when the surface soil does thaw much water stands on the ground because it cannot seep into the frozen subsoil.

In high latitudes sunlight itself becomes a noticeable climatic element because of its extremes of duration—none in winter and continual in summer—and its peculiarities in relation to the horizon and direction with regard to the pole. Extremely long twilight periods are characteristic.

Highland Climate

It is impossible to classify climate in highlands since there is so much local variability, depending not only on altitude but also on exposure to sun, wind, and moisture. In general, radiational effects are more pronounced in highlands where the air is thin and pure, and advection effects are reduced. Annual range in temperature decreases while diurnal range increases. Although average monthly temperatures may classify a mountain area as a certain zonal climate, daily weather types may not be typical for that zone. Therefore, there are no very realistic lowland analogs for highland climates. In highlands, air pressure becomes a significant climatic element because its rapid reduction with elevation strongly influences the physiology of humans and other living organisms.

Notes

1. For a succinct account of Koeppen's classification system see G.T. Trewartha and L.H. Horn, *An Introduction to Climate,* 5th ed. (New York: McGraw-Hill, 1980), pp. 397–403.
2. C.W. Thornthwaite, "An Approach Toward a Rational Classification of Climate," *Geographical Review* 38 (1948):55–94.
3. A.A. Grigoryev and M.I. Budyko, "Classification of the Climates of the U.S.S.R.," *Soviet Geography: Review and Translation* 1 (1960):3–24.
4. P.E. Lydolph, "Federov's Complex Method in Climatology," *Annals of the Association of American Geographers* 49 (1959):120–44.
5. It does fit into the geographical positioning of climatic types, however. The five major categories generally progress poleward, alphabetically, from the equator.
6. Professor Trewartha has graciously allowed the use of his classification system and world climatic map, with minor modifications.
7. This modification of the original Koeppen formula seems to fit vegetation distribution better.
8. Although these three formulas have occasionally been combined into one formula utilizing half-year precipitation amounts rather than monthly amounts, thus apparently simplifying the procedure, such simplified presentations in fact have complicated the computations, since climatic tables generally list monthly rather than half-year averages.

16 | CLIMATE OF NORTH AMERICA

General Controls

The climatic characteristics of North America are primarily determined by its great latitudinal extent, its shape, and its topography (Figure 16.1). The continent spans the entire middle latitudes of the Northern Hemisphere from south of the Tropic of Cancer in central Mexico to well beyond the Arctic Circle in Alaska and northern Canada. Its east-west dimension expands poleward from a narrow isthmus in the south. And its topography consists of a broad *cordilleran* system athwart the westerlies in the western part of the continent, and wide-open plains, east of the Rockies, which engulf most of the rest of the continent except for minor topographic features such as the Appalachian Mountains in eastern United States.

The western mountains have a profound effect on the climate of much of the continent, not only because they block much of the Pacific air at the surface, but also because they probably play a major role in inducing the most pronounced trough anywhere in the world in the standing waves of the upper troposphere, which is usually anchored somewhere over eastern North America and thereby greatly affects the climate of much of the continent. The interior lowlands between the Rockies and the Appalachians allow air to move meridionally freely and rapidly over great latitudinal distances. No other continent experiences such latitudinal exchanges of air, and hence such fresh, contrasting air masses (cP and mT) across the Polar Front. This makes for exceedingly vigorous air flow, severe storms including tornadoes, and great day-to-day changes in weather.

The shape of the west coast, curving westward toward the north around the Gulf of Alaska, induces the formation of an unusually expansive and poleward-centered subtropical high in the eastern Pacific during summer, which greatly affects the precipitation regime along much of the North American west coast. Nowhere else in the world do dry summers extend so far poleward. In the south the Gulf of Mexico extends westward half the width of the North American continent and provides a source of warm, moist, relatively unstable air that accounts for much of the precipitation that falls in North America east of the Rockies and south of Hudson's Bay. The southwest-northwest orientation of the east coast from Florida to Newfoundland maintains close proximity to the largest warm ocean current on earth, the Gulf Stream, flowing northeastward in the Atlantic.

Primary Surface Air Flows and Action Centers

At the surface three primary air streams affect North America: (a) the Pacific air stream, (b) the tropical air stream, and (c)

Figure 16.1. Major relief features of North America. Numbers correspond to data tables at end of chapter.

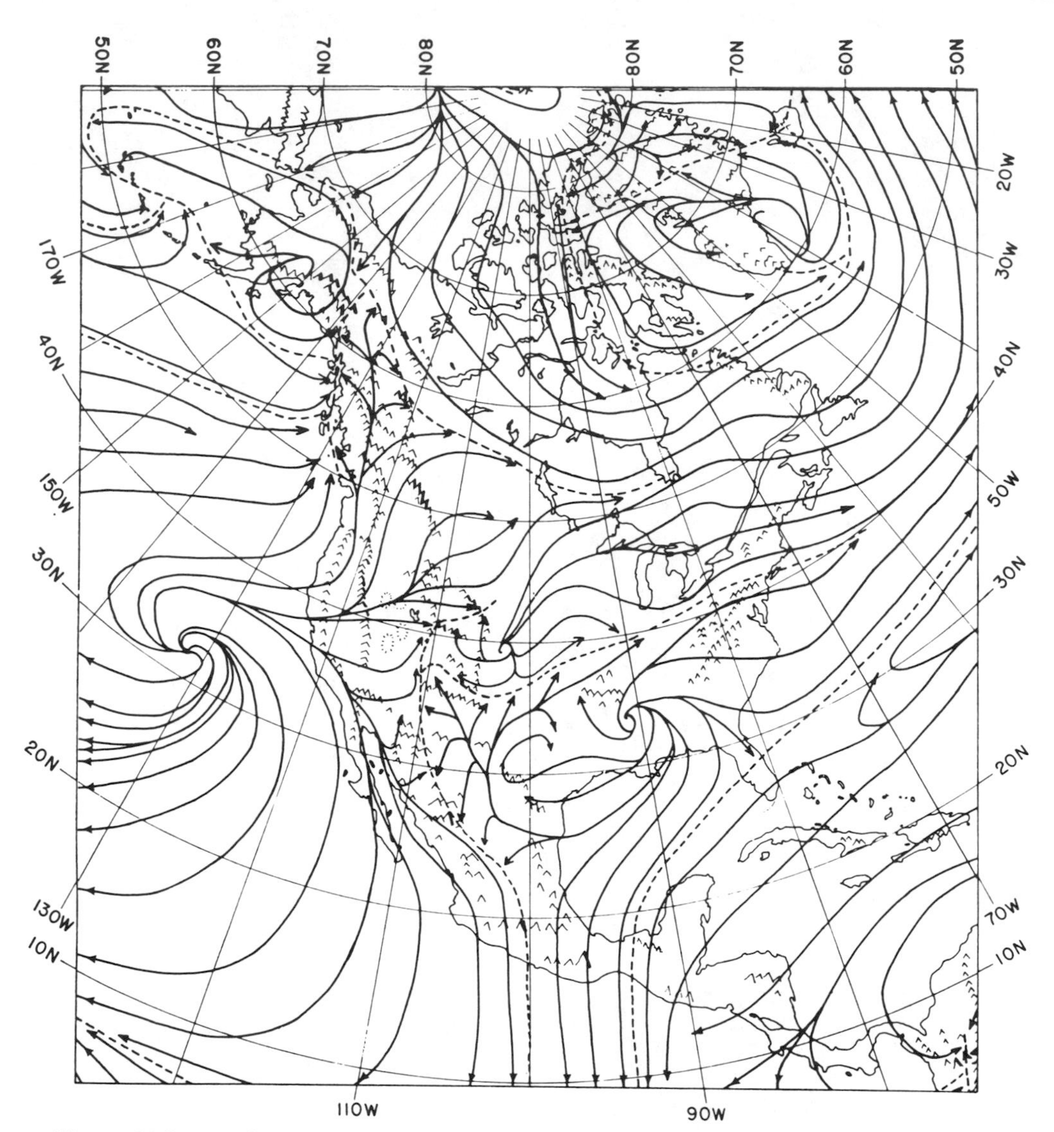

Figure 16.2. Resultant surface streamlines, January. (From R.A. Bryson and F.K. Hare, The Climates of North America, *Vol. 11 of* World Survey of Climatology *[Amsterdam: Elsevier, 1974].)*

the Arctic air stream. Their seasonal flows are illustrated by the streamline maps, Figures 16.2 to 16.5.

Pacific Streams. The Pacific flow really consists of two streams: the northern westerlies, which enter the Pacific out of Asia, primarily in winter, and a more southerly flow from the northeastern periphery of the subtropical high pressure cell in the eastern Pacific. The northern westerlies, originally cold from the Asiatic landmass, cross the relatively warmer but still-cold northern Pacific along the southern periphery of the Aleutian Subpolar Low, where they pick up moisture and arrive at the west coast of North America with high humidity and a near moist adiabatic lapse rate to considerable heights. The more southerly flow undergoes subsidence along the eastern edge of the subtropical anticyclone as it moves southeastward parallel

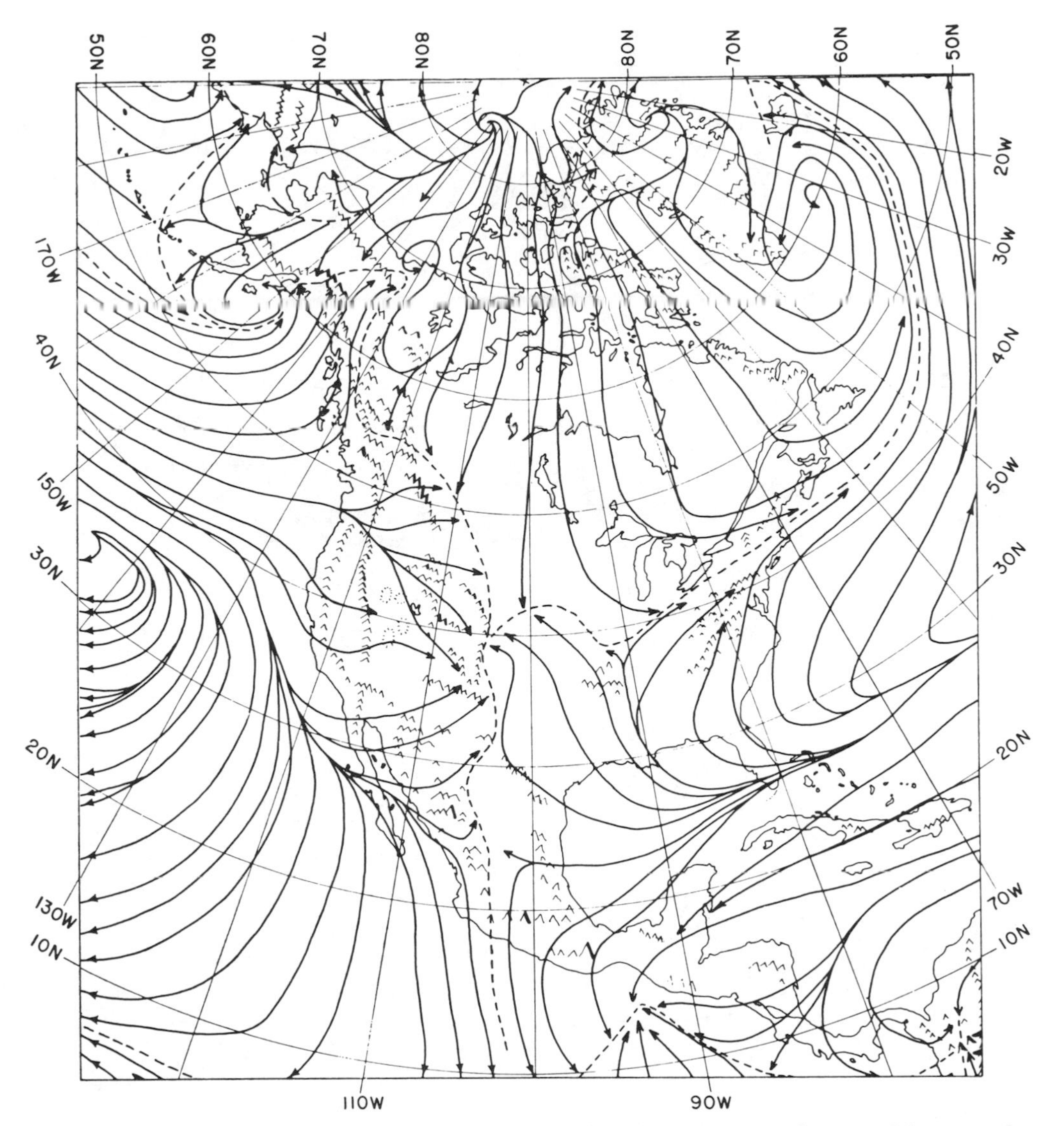

Figure 16.3. Resultant surface streamlines, April. (After Bryson and Hare, Climates of North America, *Elsevier.)*

to the coast of California and Mexico. This flow is present year-round but is best developed during summer, when the anticyclone engulfs much of the eastern Pacific and the Aleutian Low shrinks or even disappears.

Between the northern and southern Pacific air streams usually lies a portion of the Polar Front, particularly during winter when cyclonic storms issuing from the Aleutian Low move along the front across the North American coast. Embedded between two varieties of maritime polar air, the front is not distinct according to air mass differences across it, but since both air masses have high moisture content, storms forming along it produce widespread overcast skies and much rain and snow as they are forced to rise over the mountains of the Alaskan Panhandle, British Columbia, Washington, Oregon, and California.

The Pacific air at the surface is limited

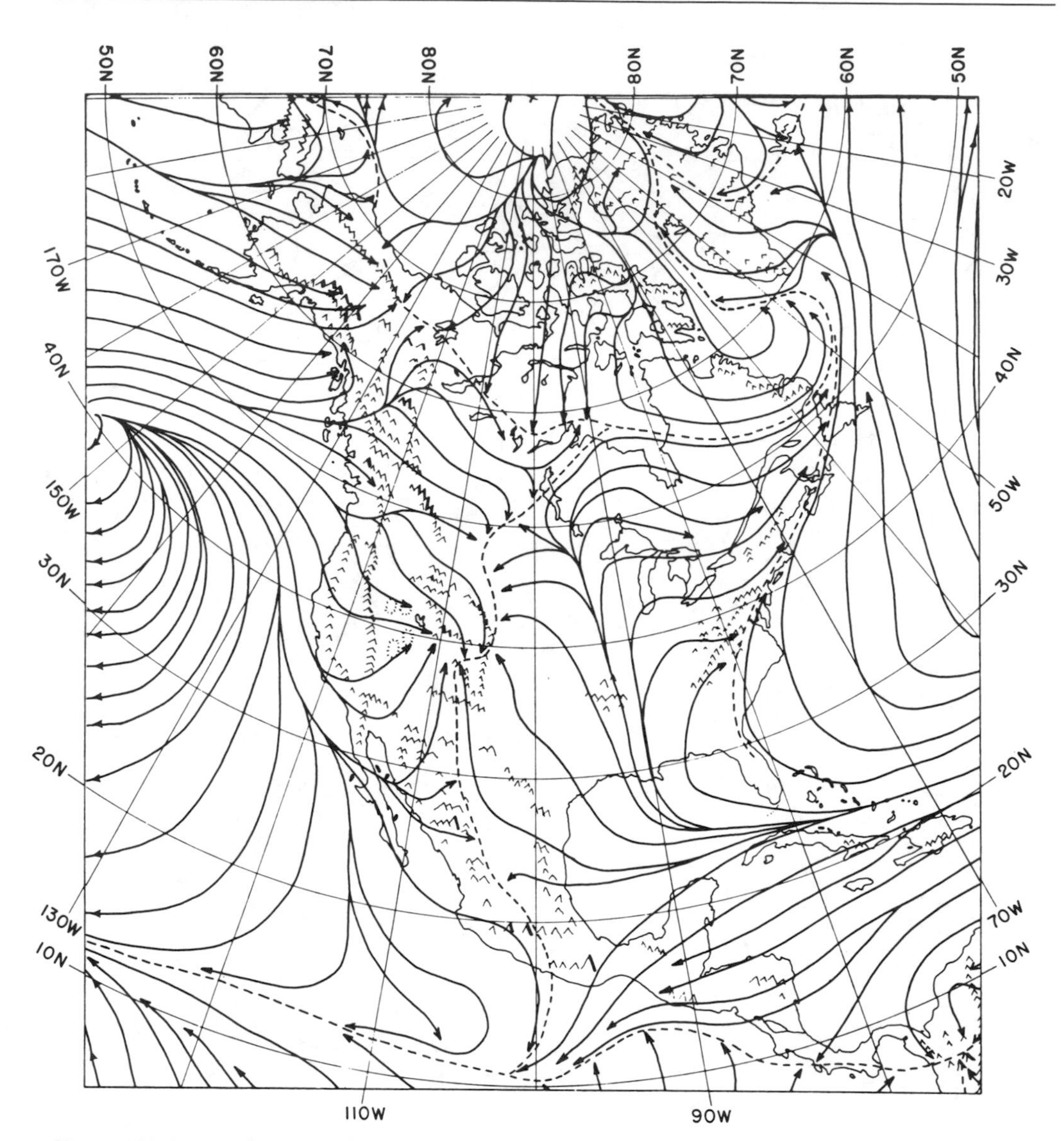

Figure 16.4. Resultant surface streamlines, July. (After Bryson and Hare, Climates of North America, *Elsevier.)*

primarily to the west-coastal area of North America, since mountains generally rise almost immediately east of the coast; but farther aloft the air finds its way through various breaks in the topography to flood much of the intermontane region of western United States between the coast ranges and Cascade–Sierra Nevada on the west and the Rockies on the east. It also typically fills up the narrow plateau and valley surfaces among the Canadian Rockies. It may cross the entire cordilleran system of British Columbia and Alberta, Canada, around latitudes 50° to 60°N, where the westerlies are strongest, particularly during autumn when this air flow generally floods the southern half or more of Canada all the way to the Atlantic (Figure 16.5).

This flow may continue through much of the winter, when it produces the famous

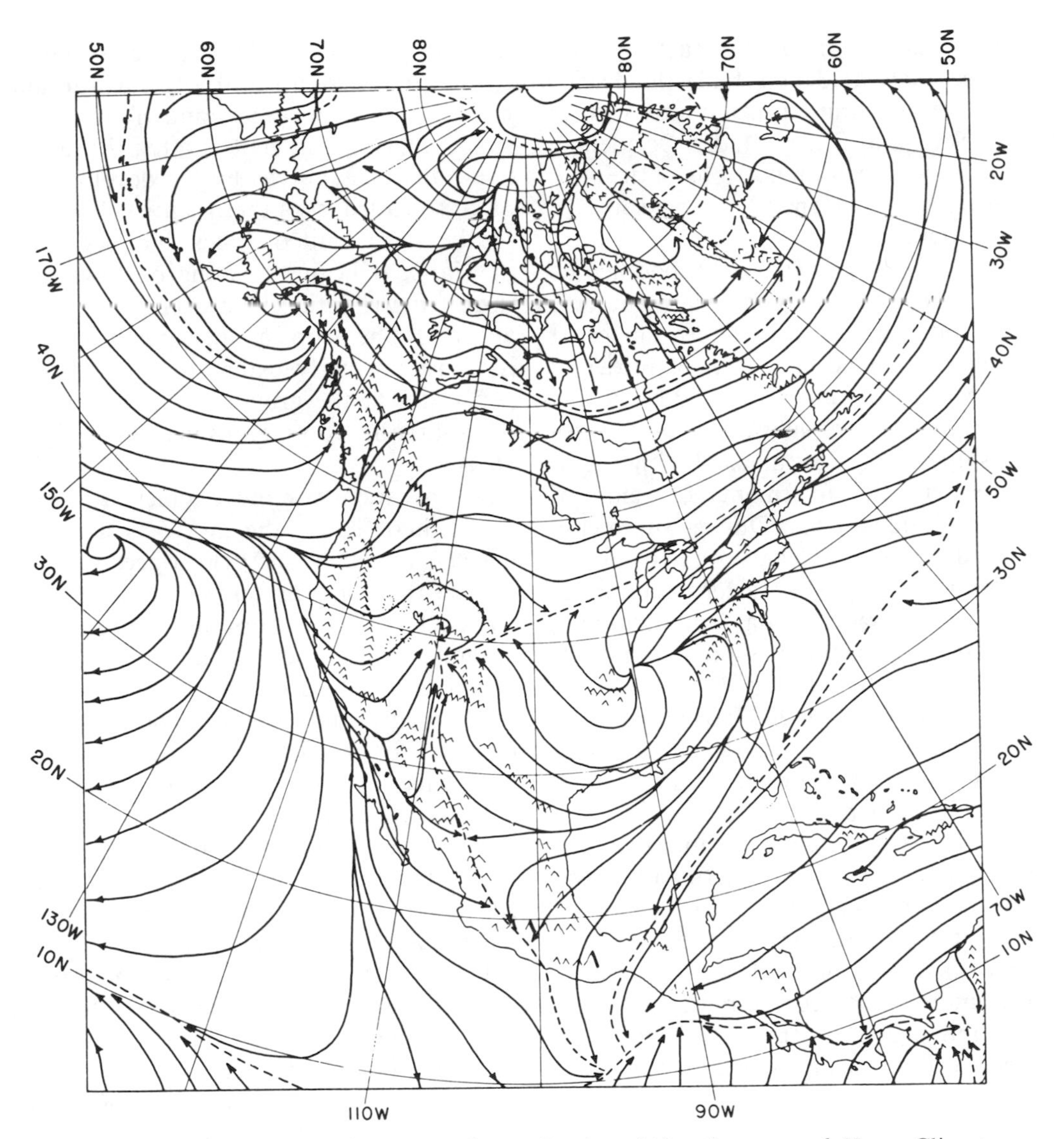

Figure 16.5. Resultant surface streamlines, October. (After Bryson and Hare, Climates of North America, *Elsevier.)*

chinook effect as foehn winds descend the east slopes of the Rockies in Alberta and states farther south. This air flow may be joined at the surface by another Pacific air stream that finds its way eastward through the Columbia River valley between the states of Washington and Oregon, and then up the Snake River valley in southern Idaho, and across breaks in the Rockies in the state of Wyoming.

Also during winter, a southerly Pacific air flow may penetrate the continent through the transverse (west-east) ranges of southern California and the basin-and-range country of Arizona, and the Sonoran Desert of northern Mexico, into the state of Texas. But as can be seen in Figure 16.2, this southern air stream is generally limited to the area west of the southern Rockies and the Sierra Madre Occidental in northern

Mexico. During spring and much of summer (see Figs. 16.3 and 16.4) the Pacific air is generally limited to the area west of the Rockies and Sierra Madre Occidental, except for west-central portions of the prairie provinces of Canada during some of the summer.

The Pacific air is intensely modified as it penetrates inland. It loses a great deal of precipitation on the western slopes of mountains, thus adding large quantities of latent heat to the air, which, added to the adiabatic heating of subsidence on the leeward slopes, produces unusually high temperatures in the intermontane region of western United States. During summer a localized heat low develops in the Colorado and Mohave deserts of eastern California, western Arizona, Nevada, and adjoining states, in a stagnant mass of continental tropical air that is characterized by strong instability in the lower 2000 to 3000 meters, but is capped by a subsidence inversion aloft produced by the eastern extension of the Pacific high pressure cell.

During winter much of this intermontane region, particularly the northern and central portions, is snow-covered; the dry, modified mP air trapped in the basin for days or even weeks takes on cold surface temperatures under the process of radiational heat losses and produces strong, localized high pressure cells in an air mass known as polar basin air. On rare occasions during midwinter when this air surmounts the Rockies and floods the plains eastward, a respite of mild, exceedingly clear weather may last for several days. This is usually the best weather the Great Plains and Middle West receive during winter.

Tropical Stream. The tropical air stream crosses the Gulf Coast of the United States and moves northward up the Mississippi Valley to flood much of the country east of the Rocky Mountains. It originates in the western end of the subtropical high in the Atlantic after having crossed much of the subtropical North Atlantic, where it picked up a great deal of warmth and moisture and became conditionally unstable. During winter it is often limited to the Gulf coastal areas by the development of a so-called "southern anticyclone" in southeastern United States, but beginning in March it begins pushing northward as the southern anticyclone declines and disappears. The tropical air stream reaches its most northerly extent in July when it penetrates at the ground surface as far as the southern tip of Hudson's Bay (Figure 16.4).

As this air stream moves poleward across the United States, the air flows along *isentropic* (constant energy) surfaces that rise toward the north as temperatures become cooler. Hence, the tropical air stream has a tendency to rise off the ground as it proceeds northward, and is thus cooled adiabatically and provides the source of moisture for practically all the precipitation that falls on eastern United States and southeastern Canada. As it moves northward, it curves eastward around the western end of the Atlantic high pressure cell. This curvature usually takes place through the Great Lakes region in summer and early fall, and farther south, in the vicinity of the Ohio River valley, during winter and early spring.

At all times of year a southern branch of the tropical air stream continues westward across the Caribbean to dominate much of Central America and eastern Mexico east of the Sierra Madre Oriental.

Arctic Stream. The Arctic air stream originates in the North Polar area and moves southward across the Canadian Archipelago into the interior of the North American continent. By late November it has usually filled up the Mackenzie River valley in northwestern Canada and has produced a persistent high pressure cell that, during the rest of the winter, sends subsidiary cells southward into interior United States as cold outbreaks of continental polar air. Such cells move southeastward and then recurve northeastward as they get caught up in the upper air streams that are moving across the Atlantic around the southern periphery of a well-established Icelandic subpolar low at this time of year.

During winter the Arctic air overlies a frozen, snow-covered surface in Canada and adjacent seas, and it undergoes strong radiational heat losses that produce very cold surface air temperatures and surface temperature inversions, which induce extreme stability to heights of perhaps 3–4 kilometers. The cold air has little capacity to hold moisture, and therefore its absolute humidity is very low. During mid-winter, the Arctic air stream is often limited to the area north of the 50th parallel by the Pacific air stream, which is strongest in January and floods the northern US–southern Canadian area clear across the continent to the Atlantic coast of New England and the Maritime Provinces. But individual high pressure cells moving southeastward out of the Mackenzie River valley can penetrate clear to the Gulf of Mexico, on occasion, and bring very cold temperatures throughout the interior plains of the United States.

The Arctic air stream reaches its most southerly extent in April when, on average, it penetrates to approximately the 40th parallel in central United States (Figure 16.3). During this time of year the Pacific air stream is generally limited to the area west of the Rockies, and the Arctic air stream meets the tropical air stream head-on along segments of the Polar Front in central and eastern United States. Thus, during spring the Polar Front is accentuated by the most extreme differences in air masses, the most vigorous air flow, and the most violent weather, including the dreaded tornadoes. As the Arctic air stream moves southward it tends to descend along isentropic surfaces and therefore to remain on the ground to act as a cold wedge above which the northward-moving tropical air stream rises across the surface of the Polar Front.

Upper Air Flows

During winter in the upper troposphere, a strong zonal flow covers the entire North American continent (Figure 16.6). On average, a weak ridge occupies the eastern Pacific, and the deepest trough in the standing wave pattern of the earth occupies much of the North American continent, particularly east of the mountains. A jet stream core occupies much of the eastern United States, centered at about 35° latitude. But remember that on a daily basis during winter, this usually consists of two jets, a polar one and a subtropical one, which tend to merge when averaged over a lengthy period.

During summer the zonal flow in the upper troposphere is much weaker, and a westerly extension of the subtropical high pressure cell in the Atlantic extends upward to the tropopause and dominates much of the southern United States and Mexico (Figure 16.7). In middle and late summer a subsidiary high pressure cell tends to develop over the Texas area, which pulls in warm, moist, Gulf air along its southern periphery around the southern end of the Rockies in New Mexico and northwestward into the Colorado Plateau region to bring considerable thundershower activity to this intermontane area during July and August. A weaker jet stream core now has shifted northward into the Great Lakes–New England area.

The magnitude and position of the ridge and trough system in the upper troposphere have profound effects on weather sequences in North America. Remember, as explained in Chapters 6 and 10, that generally opposite types of weather occur on opposite limbs of standing waves of the upper atmosphere. The deep trough over eastern North America usually is centered some-

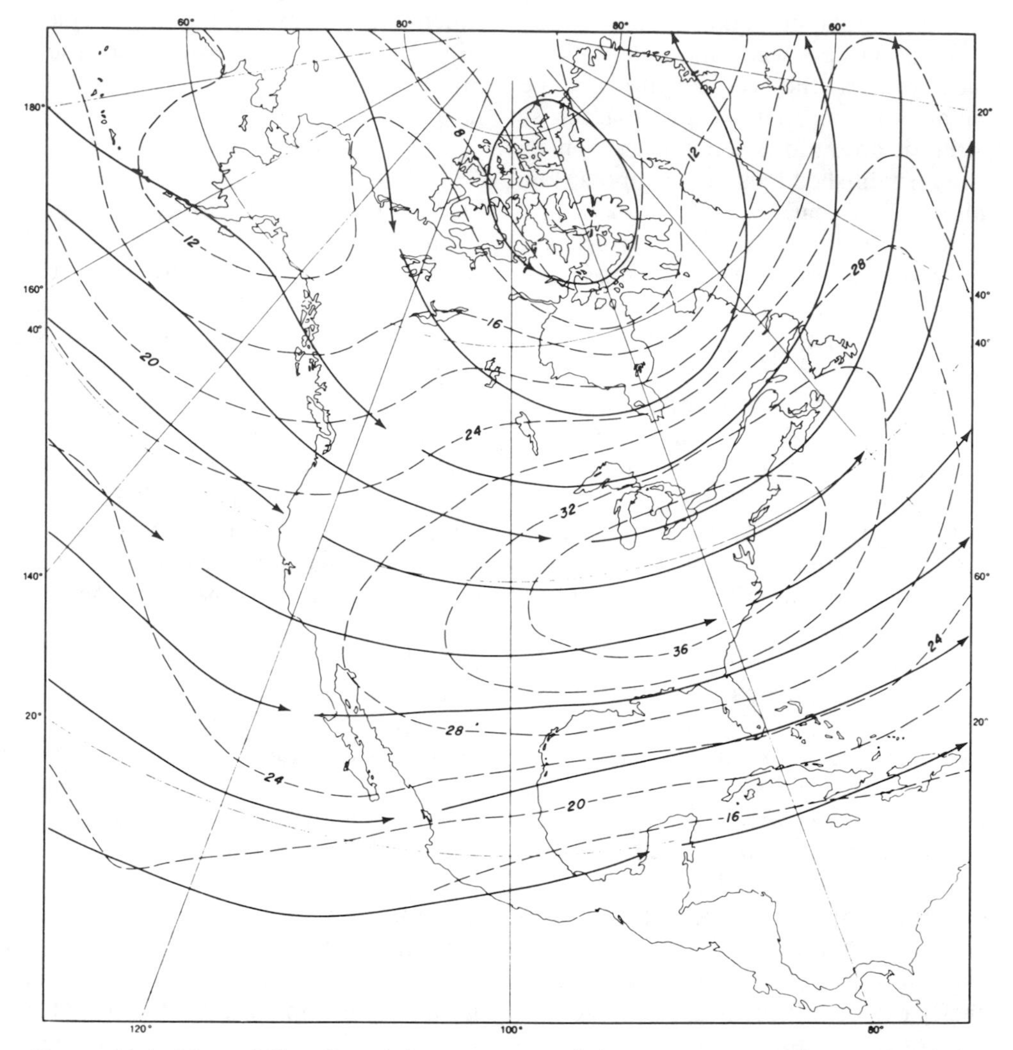

Figure 16.6. Mean 300 mb. winds, January. Solid arrows are resultant streamlines; dashed lines are isotachs in m/s. (After Bryson and Hare, Climates of North America, *Elsevier.)*

where over the midwestern states. If it shifts eastward a few hundred kilometers, the Midwest finds itself along the western limb where upper-level convergence and subsidence tend to suppress storm development and result in a relatively dry period. When the trough shifts westward, the Midwest finds itself on the eastern limb where southwest-northeast flow produces uplift and cyclonic development, resulting in an abnormally wet period. Although these shifts take place relatively slowly, they are still difficult to predict, and hence weather forecasting in this region can be quite tricky. Likewise, the subdued ridge over the eastern Pacific can occasionally become quite pronounced and occupy the west coast region to produce abnormally warm, dry conditions. On occasion these waves become very pronounced and convoluted and trap warm pools of air in the Alaskan–northwest Canadian area, while cold pools are isolated

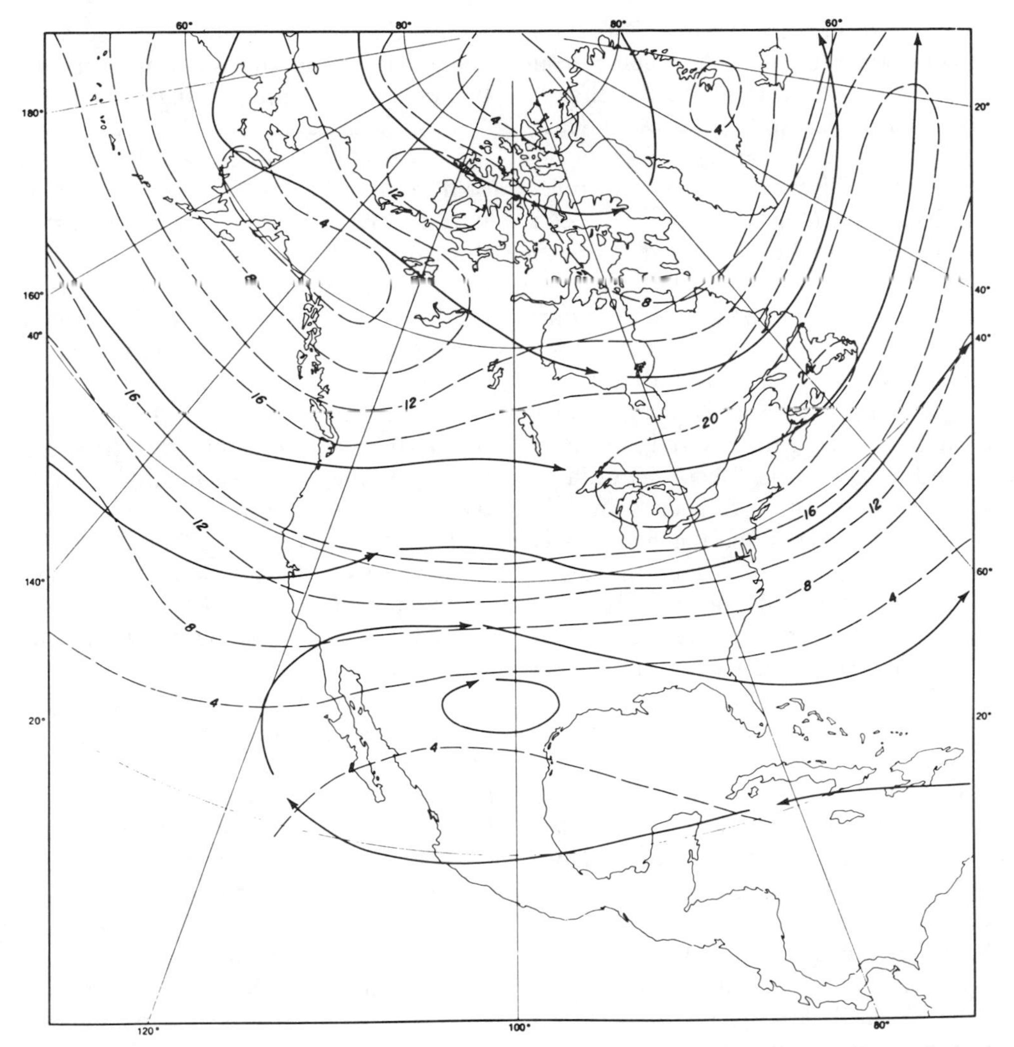

Figure 16.7. Mean 300 mb. winds, July. Solid arrows are resultant streamlines; dashed lines are isotachs in m/s. (After Bryson and Hare, Climates of North America, *Elsevier.)*

in southeastern United States. Under such conditions during winter, Anchorage, Alaska, may have warmer temperatures than St. Louis, Missouri.

Cyclogenesis and Cyclone Paths

Various areas on the North American continent and surrounding seas are preferred localities for the formation of cyclonic storms (cyclogenesis) along portions of the Polar and Arctic fronts. Once formed, the storms tend to follow certain routes, as was shown for the Northern Hemisphere in Figures 10.7 and 10.8. During winter many cyclonic storms move eastward out of the Aleutian Low and strike the North American west coast. After rising over the western mountains, they may regenerate in the intermontane area or the western Great Plains and move on across the continent.

In January, during periods when the zonal (west-east) circulation is particularly strong, a common route crosses the entire cordilleran system in British Columbia, after which the storms often regenerate in the province of Alberta along the eastern foot of the Rockies and move swiftly eastward along the US–Canadian border toward the Great Lakes and the St. Lawrence Valley. Because of their rapid forward speed, these storms have been dubbed "Alberta clippers."

Farther south, cyclones and anticyclones alternate frequently in the Great Basin country of Nevada and surrounding states. These cyclones may have been regenerated from cyclones that earlier entered the Pacific coast and crossed the Sierra Nevada or Cascade mountains. After crossing the Rockies these storms often regenerate once again in the Denver area. It is difficult to trace a single storm all the way from the west coast, across the western mountains, and into the Great Plains because of the profound topographic effects on their forms and delays of their forward speeds, which cause them to appear to dissipate and then to reform on the eastern sides of the mountains. Detection is complicated by the fact that many lows originate in the Denver area that are not regenerated preexisting lows.

Once the Denver Lows leave the Denver area they usually move east-northeastward into the Great Lakes region, particularly during winter when the Great Lakes' heat source tends to attract and strengthen low pressure centers and guide them down the chain of Great Lakes and the St. Lawrence Valley. If an upper-air trough is located over the Great Plains, the Denver Lows first move southeastward into the panhandles of Oklahoma and Texas, and then recurve sharply northward to enter the western Great Lakes region. Such storms are locally dubbed "Panhandle hooks." Since they move almost straight north into Minnesota or Wisconsin, southerly flow in their leading edges typically brings far northerly penetration of warm, moist air from the Gulf of Mexico into the upper Midwest, which during midwinter produces widespread advection and warm frontal fog and brings on the "January thaw" singularity.

East Texas and Gulf coast cyclogenesis usually occurs as open wave formations on cold fronts, which trail southwestward from parent lows that are moving eastward through the Great Lakes area. These secondary waves move northeastward along the front as the entire frontal system progresses toward the east coast. One after another may form in the south and slowly occlude in its northeastward course. These southerly lows are rather rare in summer. In winter they pull warm, moist air from the Gulf of Mexico into their warm sectors and cause deep penetration of maritime tropical air northward, setting the stage for heavy snowfall over large portions of eastern United States and southeastern Canada.

In a similar way, cyclogenesis takes place on the Atlantic Coast along a trailing cold front after it has moved off the east coast of the United States and its parent storm has moved out of the St. Lawrence Valley into the North Atlantic. These storms are particularly active during winter, when cold cP air behind the cold front meets the anomalously warm air over the Gulf Stream just offshore (remember that this region during winter has the highest evaporation rates on earth). Thus, great amounts of heat and moisture are injected into the air as it moves out to sea, and cyclogenesis is often so rapid that a weak open wave on the Polar Front may be transformed into a major cyclone within 12–24 hours. The Cape Hatteras to New England coastal area in winter is proverbially stormy.

Whatever the routes, the storms tend to focus on the New England–Maritime Province region and then move into the North Atlantic, the bulk of them going northeastward into the Icelandic Low, while a few are shunted northward by the southern tip of Greenland along its west coast into the Davis Strait. Because of the great storminess in winter, the top of Mt. Washington, New Hampshire, records the highest average annual wind speed in the United States, 16 meters per second (35 mph). It apparently holds the world's record five-minute wind speed, 84 m/s (188 mph), and the world's highest recorded surface peak gust, 103 m/s (231 mph). In central and northern Canada weak waves may form along the Arctic Front during winter, but they result in only meager snowfall, since the air is too cold to hold much moisture.

During summer, the average position of the Polar Front shifts north of the US–Canadian border, and most of the cyclonic storms are limited to Canada and northeastern United States, although southerly intrusions can certainly occur any time of year (see Figure 10.8).

Regional Characteristics

Western Cordillera

The Coastal Fringe. The entire west coast of North America is dominated by maritime polar air that keeps temperatures consistently mild—cool in summer and relatively warm in winter—for the latitude. Average January temperatures remain above freezing as far north as 57° latitude in the Alaskan Panhandle, so the climate of much of the coast classifies as either temperate oceanic (Do), from the Alaskan Panhandle southward to northern California, or subtropical dry summer (Cs) throughout the rest of western California and the northwestern portion of Baja California (see Figure 15.3). North of 57°N latitude, January temperatures average a little below freezing, and less than four months average above 10°C (50°F), so even the coastal area has a boreal (E) climate. Both extremes of temperature are moderated by the ocean, however, and the outstanding characteristic of the climate along the entire coast is its low temperature range for such high latitude. Yakutat, Alaska, at almost 60°N, averages −2.6°C in January and 12.3°C in July (30, Yakutat).* The cool growing season, not the severity of the winters, sets vegetative limits here.

Two major precipitation controls profoundly affect the west coast and western slopes of mountain ranges and intervening valleys. South of about 40°N, the subtropical high pressure cell dominates the area year-round southward to the Tropic of Cancer, around the southern tip of Baja California. During summer it expands northward to fill up much of the Gulf of Alaska and to bring relatively dry summers to the coastal area as far north as the Canadian border (Figure 16.8). Vancouver receives only 26 millimeters of precipitation during July, as opposed to 164 during December. Nowhere else on earth do such dry summers extend so far poleward. Occasional cyclonic storms bring just enough rain to the Oregon-Washington-British Columbia coasts to keep the summers moist enough to sustain a magnificent forest growth along the coastal mountains, which are watered primarily during winter. Large Ponderosa pines, Douglas fir, and other species comprise some of the best softwood forests in the world. Farther south along the coast of Oregon and northern California, huge redwoods join the forest group (Figure 16.9).

South of the Klamath Mountain knot in southern Oregon and northern California, the summers become truly dry and the forests give way rapidly to a brushy growth of drought-resistant plants, some deciduous, some evergreen, which generally grow no more than 1–3 meters in height. This great variety of species of low plant growth is known collectively in California as chaparral. This kind of vegetation has no commercial value but is very valuable for holding the soil in place on steep slopes that would otherwise undergo extreme erosion during winter rains. This kind of vegetation

*Numbers and place names in parentheses refer to data tables at ends of chapters.

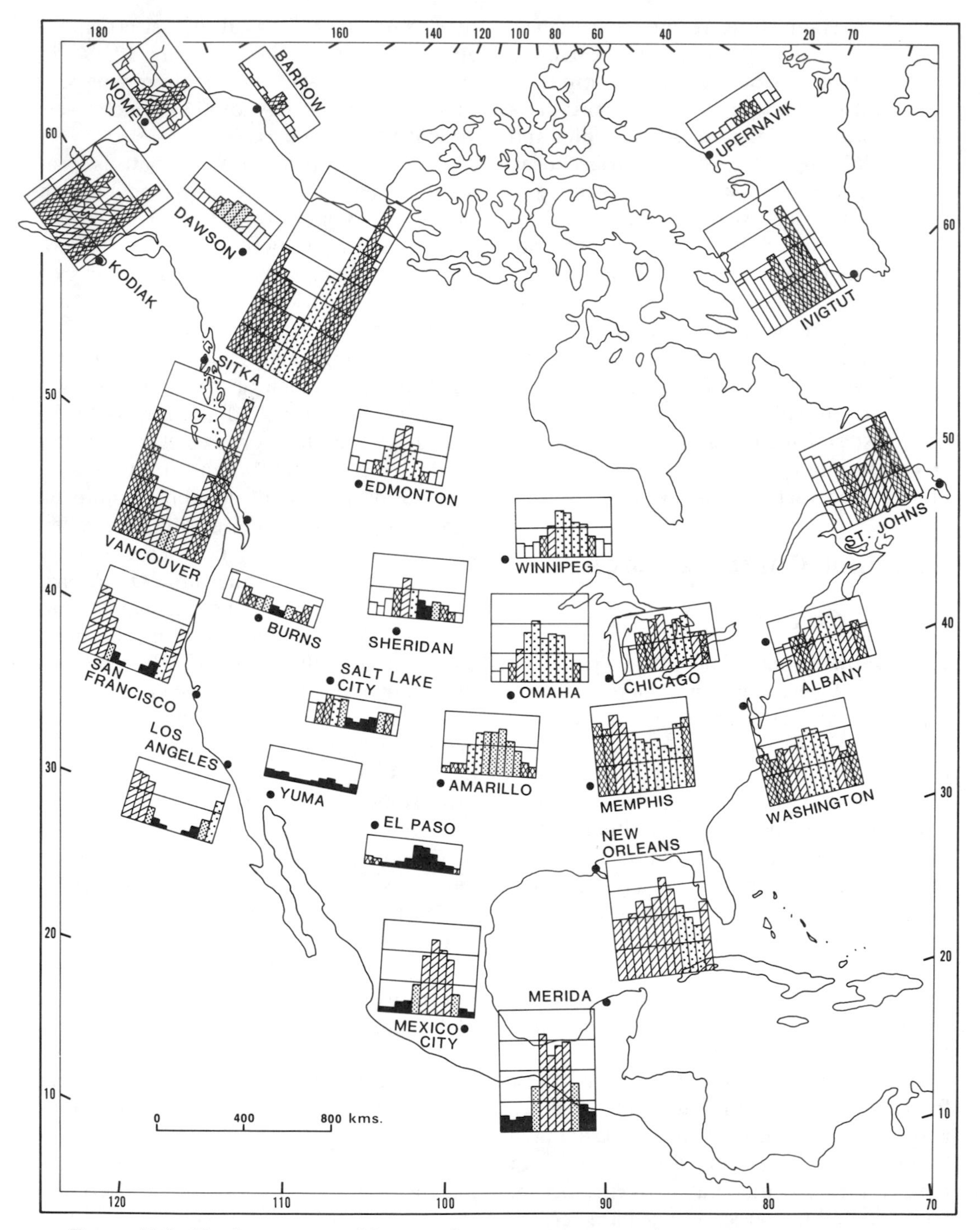

Figure 16.8. Yearly regimes of heat and moisture resources for vegetative growth. (From Agroclimatic Atlas of the World.*)* *See opposite page for key.*

covers much of the coastal ranges as well as some of the interior valleys. But other large areas are also occupied primarily by grass, dotted here and there by live oak trees that give a beautiful parklike appearance to the landscape during summer, when the trees are dark green and the grass is a tawny brown (Figure 16.10).

During winter the subtropical high shrinks southward, and the Aleutian Low engulfs the northern part of the Pacific. Cyclonic storms of great frequency and intensity move eastward to strike the coast of the Alaskan Panhandle and British Columbia, where they drop heavy rainfall in lower elevations and large amounts of wet snow as the air is forced to rise up the steep mountain slopes. Throughout this part of the coast, mountains rise abruptly from sea level, forcing an abrupt initial rise of precipitating air so that some of the highest precipitation totals occur in relatively low elevations. In general along the coast, annual precipitation averages from about 1300 to 3500 millimeters, but it varies greatly over short distances because of the rough topography, and in some places it is considerably greater. The greatest average yearly precipitation in North America occurs at Henderson Lake on Vancouver Island, only 4 meters above sea level, where it amounts to 6650 mm. (262 inches) per year. Precipitation occurs on more than 200 days per year, and overcast skies predominate much of the time. Even so, there is a noticeable dry spell in late spring and summer.

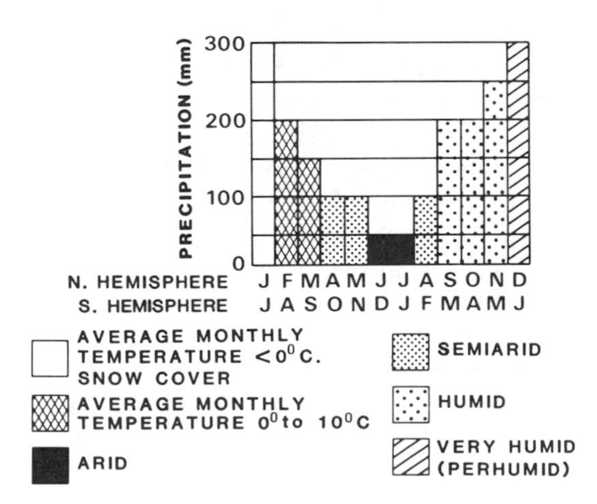

Figure 16.9. Redwoods north of San Francisco.

Such heavy snow packs build up in the mountains that they cannot all melt during summer seasons, and valley glaciers form that in places extend all the way down to sea level, despite the fact that temperatures there generally are above freezing. Although the winter storms become more infrequent southward in the United States, snowfall may still be copious on some of the western slopes of the Cascade and Sierra Nevada ranges in Washington, Oregon, and California (Figure 16.11). The heaviest seasonal snowfall ever recorded in North America, 28,500 mm. (1122 inches), occurred during the 1971–1972 winter at Paradise Ranger

Figure 16.10. Dry grass and live oak trees on the east side of the California Coast Ranges.

Station at an elevation of 1700 meters on the south slope of Mt. Rainier, near Seattle. During one storm 4797 mm. (189 inches) of snow fell on Mt. Shasta ski bowl in northern California. The greatest depth of snow ever recorded on the ground in the United States was 11,447 mm. (451 inches) at Tamarack, California, on the western slope of the Sierra Nevada.

The Polar Front with its cyclonic storms migrates slowly southward along the coast during the winter season. Maximum precipitation occurs during October at Yakutat (30), during December at Seattle (27) and Portland, Oregon, during January at San Francisco (25), and during February at Los Angeles (12) and San Diego (24). Usually the front does not lie as far south as southern California, and this region often receives only the southerly portion of cyclonic storms that are moving inland farther north. But on occasion when the front does lie this far south in the eastern Pacific, perhaps three or four times a winter, one wave after another may form along it and follow each other inland across the California coast, bringing a prolonged period of five days or more of rather continuous rain, punctuated by heavier downpours. In the lowlands, the rain is definitely of cyclonic character, with overcast stratus clouds and little wind or turbulence. Thunder is seldom heard along the Pacific coast. But as the air is forced to rise up the mountains conditional instability may be released, and strong thunderstorms then occur in the mountain areas.

From San Francisco southward the subtropical high pressure cell is dominant even during winter, and rainy spells occur only on those few occasions when the Polar Front invades the area. Thus, although southern California receives all its rain during the winter half-year, even that half-

Figure 16.11. Snow is drifted nearly four times the height of cars near the crest of the Cascade Mountains in the state of Washington during April. Clouds (fog) shroud the mountain slopes.

year is comprised primarily of rainless weather. Total rainfall in the southern half of California is rather meager in the lowlands, diminishing from 529 mm. at San Francisco (25) to 373 mm. in Los Angeles (12) and 264 mm. in San Diego (24). But in the adjacent mountains it may amount to four or five times that amount, mainly in the form of snow. After the passage of a winter cyclonic storm has swept away the pollution of the Los Angeles Basin, the inhabitants of that region are treated to a spectacular view of snow-capped peaks in the San Gabriel and San Bernadino mountains to the north and east, an hour's drive from downtown Los Angeles.

The stability of the surface air along the California coast is enhanced by the cold California Current off-shore. This is the southerly extension of the Japanese Current, which comes across the Pacific and strikes the American west coast at about the US–Canadian border, where it splits, part of it to go north as the relatively warm Alaskan Current, and the bulk of it to continue southward as the cold California Current. The southward-moving current is deflected offshore by the Coriolis Force, which induces upwelling of very cold water from below. This is particularly true during summer when the surface air flow along the eastern edge of the Pacific High most nearly parallels the coast. During July the entire California coast experiences surface air temperatures 10°–11°C below normal for the latitude, which is by far the coldest temperature anomaly over water anywhere on earth (see Figure 14.4). The strength of the upwelling varies with the configuration of the shoreline and the proximity of the open ocean circulation. It is generally strongest just off San Francisco Bay, where the surface water temperatures during summer are usually colder than they are either to the south or to the north. In fact, this area frequently has the coldest surface air temperatures anywhere along the west coast of North America during summer. The rolling motion of the upwelling is very noticeable to ships approaching the Golden Gate. The upwelling subsides during autumn as the atmosphere undergoes its seasonal shift to a winter circulation that no longer so closely parallels the coast. This allows surface air temperatures in San Francisco to rise to their September-October maximum (25).

Sea breezes are exceedingly well developed along this coast during summer, and as they approach the shore they cross the upwelled water, which lowers the surface air temperature below the dew point and produces thick and persistent advection fog that may be carried inland. Much of the west coast experiences at least 60 days per year with dense fog, which is the foggiest part of the entire North American continent. The San Francisco Bay region is particularly noted for its summer sea fogs. The Golden Gate leading into the bay represents the only significant break in the coast ranges that allows an exchange of air between the cold sea to the west, with temperatures hovering in the 10°–15°C (50°–60°F) range during summer, and the heated Central Valley to the east, which on summer afternoons almost always experiences temperatures above 38°C (100°F). This temperature differential sets up a powerful sea breeze regime that generally blows across the hilly San Francisco peninsula much of the day, crosses the bay, and rises up the Berkeley Hills in mid-afternoon. The breeze carries with it the sea fog that has formed over the cold ocean offshore and then cascaded like an avalanche down the leeward slopes of the steep hills of San Francisco and Marin counties to flood the bay area during late afternoon and evening.

Although the oceanic upwelling weakens south of San Francisco, subsidence around the southeastern edge of the subtropical high continues to strengthen to produce the strongest temperature inversions in the Los Angeles–San Diego area. During summer it is not uncommon to have an inversion differential of 15°C or more between base and top of inversion. Typically, the inversion base is a few hundred meters above sea level, with a so-called "marine layer" underneath with near-saturation humidity and near-adiabatic lapse rate. The height of the inversion varies aperiodically with different air flows and diurnally as the sea-land breeze regime oscillates the marine layer back and forth across the coast. During winter the inversion is usually higher and weaker and may be destroyed occasionally by cyclonic passages when they reach this far equatorward.

Active mixing in the marine layer carries moisture and pollutants upward to concentrate just under the base of the inversion. This usually produces a thin layer of stratocumulus clouds above sea level, accompanied by the worst pollution conditions. Since the inversion usually maintains about the same height, or slopes slightly upward inland more slowly than the land surface itself, the inversion usually intersects the ground in the wealthier foothill communities along the northern fringe of the Los Angeles Basin. Hence, some of the better residential areas, such as Pasadena, experience some of the worst air pollution. Since the inversion layer forms a very effective convective lid over the Los Angeles region, and mountains to the north and east block horizontal dispersion of the air, the large metropolitan area of Los Angeles and adjoining cities finds itself in the worst atmospheric pollution conditions in the world. This results not because the cities in this region cause more pollution than cities elsewhere, but because this is the largest urban agglomeration in the world to be built in these meteorological conditions.

During winter the atmospheric circulation does not parallel the coast as closely, so the subsidence and formation of a temperature inversion is somewhat mitigated. Frequently in fall and early winter, an eastern extension of the Pacific High may bulge in over the continent through the states of Washington and Oregon so that the air circulates clockwise across Nevada and Utah and recurves southwestward across California back out to sea. On entering the continent in the north, the air is forced to rise over a number of mountain ranges, which causes precipitation that liberates latent heat into the atmosphere. As the air descends the southern slopes of the San Bernardino and San Gabriel ranges in southern California, it warms adiabatically and extreme temperatures may be reached, considerably higher than anything during mid-summer. Typically in July, even in downtown Los Angeles 24 kilometers inland, daily maximum temperatures do not exceed 27°C (80°F) because by noon the sea breeze has moved in and lopped off the temperature curve during the afternoon. But in October, or even later, the air circulating from northeast to southwest and descending the mountain slopes into the Los Angeles Basin may reach temperatures of 38°C (100°F) or more, and some of the coastal cities, such as Long Beach, may suddenly become the hottest spots in the United States.

This is the famous Santa Ana wind of southern California, a foehn wind, which in the open Los Angeles Basin is not much of a wind at all but a condition of very high temperatures and very low relative humidities. In the mountains, however, and particularly in some of the mountain passes where the air is funneled, winds of 45 meters per second (100 mph) are not uncommon. These blow through Cajon Pass, between the San Gabriel Mountains on the west and the San Bernardino Mountains on the east, carrying air from the Mohave Desert in the north to the alluvial plain in the south. South of the pass a high-speed air stream often blows across the city of Fontana, 85 kilometers east of Los Angeles, and raises dust 3000 meters or more into the air (Figure 16.12). The low humidities and high winds cause extreme fire hazards that may destroy the brush vegetative cover of the southern mountain slopes and foothills and expose the surface to erosion, mud slides, and flooding from the coming winter rains. It is not uncommon in autumn to have a number of forest fires raging simultaneously in various sections of California. So great a hazard are these fires and their consequent flood conditions that the Los Angeles Weather Bureau has a special subsection to forecast only fire conditions during the fall period.

The effects of the temperature inversion are not all bad. In winter it may be very useful for protection of tender crops from frost. Since much of the settled parts of the southern basins are located on alluvial fans with slopes of 5° or more, the temperature inversion may keep the intermediate and upper slopes above freezing, even on cool winter nights when freezing temperatures may be experienced in the bottoms of the basins. Or, if there is danger of frost even on the upper slopes, such devices as smudge pots or windmills that mix the warm inversion air downward may keep surface temperatures warm enough to avoid damage to the fruit. Thus, southern California's high-cost, frost-sensitive agriculture is located in a pattern that very closely reflects microclimatic conditions. Citrus groves are planted on the upper rocky portions of alluvial fans, grapes in the sandy mid-sections, and frost-hardy annual crops on the basin floors.

South of the US–Mexican border the climate rapidly becomes desert on the peninsula of Baja California and the facing Sonoran coast across the Gulf of California. This condition extends southward to the Tropic of Cancer, south of which southern Mexico and Central America are subject to the northward migration of the inter-

Figure 16.12. Dust is raised high into the air by strong foehn winds blowing southward through Cajon Pass. View from top of Mt. San Jacinto looking west.

tropical front during summer, so that the climate changes to tropical wet-and-dry (Aw). Weak tropical storms, usually of less than hurricane strength, infrequently form along this western Mexican coast and progress northwestward along the coast of Baja California. In rare years they may penetrate northward into southeastern California or southwestern Arizona and bring freak summer general rains to these desert areas.

The Western Ranges and Basins. In southern Alaska and British Columbia, the Coast Ranges and Rockies are in close juxtaposition to one another with only narrow valleys and plateaus in between, but south of the US–Canadian border the mountain systems expand west-east to include many intermontane features. Within the Pacific fringe through the states of Washington, Oregon, and California extend three distinct topographic systems:

(a) The Coast Ranges and their intervening valleys run the entire length of the US west coast and cover a width averaging 75–100 kilometers.

(b) Broad structural basins to the east of the Coast Ranges include Puget Sound in the state of Washington and the Willamette Valley in Oregon, separated by the Columbia River along the interstate border and, south of the broad Klamath Mountain uplands in southern Oregon and northern California, the expansive Central Valley of California, which stretches approximately 750 kilometers from Redding in the north to south of Bakersfield in the south.

(c) Rimming the basins on the east are the high Cascade Mountains in central

Washington and Oregon and the Sierra Nevada running through the northern two-thirds of eastern California. Southern California consists of a complex of intersecting northwest-southeast (Peninsular) ranges and west-east (Transverse) ranges, with intervening structural basins that grade eastward into the Basin and Range country of Arizona and Nevada.

Although the Coast Ranges in most cases are not very high, generally less than 500 meters but occasionally rising to 2000 or more meters, they are an effective barrier to the marine layer of air affecting the Pacific Coast. Thus, there is usually a rapid transition in surface temperatures, relative humidities, and cloud conditions across the ranges. Even within the Coast Ranges themselves, such broad valleys as the Santa Clara Valley south of San Francisco Bay and the Salinas Valley south of Monterey Bay have much sunnier, warmer, and drier climates than the coast. But the big contrast occurs between the structural basins on the east side of the Coast Ranges and the coast itself, particularly in California. As was stated earlier, during July when San Francisco is experiencing afternoon temperatures around 20°C (68°F), Sacramento in the Central Valley 100 kilometers inland experiences maximum temperatures over 38°C (100°F) practically every day. Temperatures and aridity increase southward in the Central Valley, much of the southern half of the valley being too dry to classify as subtropical dry-summer climate. The climate of most of the southern half of the valley classifies as steppe, and that of the lowest portion of the valley around Tulare Lake classifies as desert. Extreme surface heating produces strong convection that thoroughly mixes dust throughout the air below the subsidence inversion that is a constant feature during summer a thousand meters or more above the valley floor (Figure 16.13).

The lowlands in Washington and Oregon are significantly more humid and cooler than those in California, but their climate is still characterized by a much greater degree of continentality than their coastal counterparts. Although not absolutely dry, summers here are by far the driest part of the year. Annual precipitation in the Willamette Valley of Oregon varies from about 750 mm. (30 inches) in the south to 1000 mm. (40 inches) in the north. Portland, in the northern portion of the valley, receives 1075 mm. of precipitation per year—only 10 mm. in July but 188 in December. Both summer and winter are relatively mild, July averaging 20°C (68°F), and January averaging 5°C (41°F). But on occasion the weather can go to extremes. The absolute maximum temperature ever recorded at Portland was 42°C (108°F), and the minimum was −16°C (3°F).

The Puget Sound area of Washington displays wide variations of climate over short distances because of the complex topography. Much of the lowland is in the rain shadow of mountainous Vancouver Island and the Olympic Mountains to the south across the Strait of Juan de Fuca. While these mountains may receive well over 2500 mm. (100 inches) of rain per year, portions of the Puget Sound lowland may receive from less than 500 mm. (20 inches) to 1000 mm. (40 inches) per year. At Seattle (27), although 151 days per year record rainfall, the annual total amounts to only 866 mm., and this is highly concentrated during winter. December receives 138 mm., while July receives only 16. During December, on the average, more than eight-tenths of the sky is covered by clouds, whereas in July less than five-tenths is. Temperatures are mild year-round. July averages a cool 19°C (66°F), and January averages 5°C (41°F). Extremes range from a maximum of 38°C (100°F) to a minimum of −12°C (10°F).

The structural basins end abruptly on their eastern flanks with the rugged and continuous Cascade Mountains in central Washington and Oregon and Sierra Nevada in eastern California. These mountain ranges

Figure 16.13. Dust in the air below the subsidence inversion over the Central Valley of California during summer, looking west from Sequoia National Park on the western slope of the Sierra Nevada. Foreground: *coniferous forests at 2000 meters elevation.*

form very sharp climatic divides between well-watered western slopes in the rising Pacific air and very dry eastern slopes in the descending warming air. The heavy winter snowfall on the western slopes of the Cascades and Sierra Nevadas has already been mentioned. The deep snow pack acts as a natural reservoir that holds the precious water for irrigation in the basins to the west until the melting season during summer, when it is most needed by crops. The eastern slopes of the mountains are much drier.

The Intermontane Region. South of the US–Canadian border the Rocky Mountain system bulges eastward and allows for a wide expanse of intermontane basins, ranges, and plateaus between the Rockies in Wyoming, Colorado, and New Mexico on the east and the Cascade–Sierra Nevada ranges on the west. Although almost everywhere the elevations in the intermontane region are lower than those in the mountains on either side, there are great differences in elevation within the intermontane region, and here and there mountain ranges limited in extent rise to 3000–4000 meters or even higher. In addition there are great variations in land form, ranging from lowlands interspersed by small mountain ranges, to high-plateau uplands incised deeply by canyons. Therefore, the climate varies greatly from place to place throughout this region.

If there is any unifying factor in the climate of this area it is probably its rather consistent aridity, except on the higher elevations, where precipitation is still not exceptional. Mean annual atmospheric moisture content in the intermontane region is the lowest anywhere in the United States (Figure 16.14). Forests generally exist only

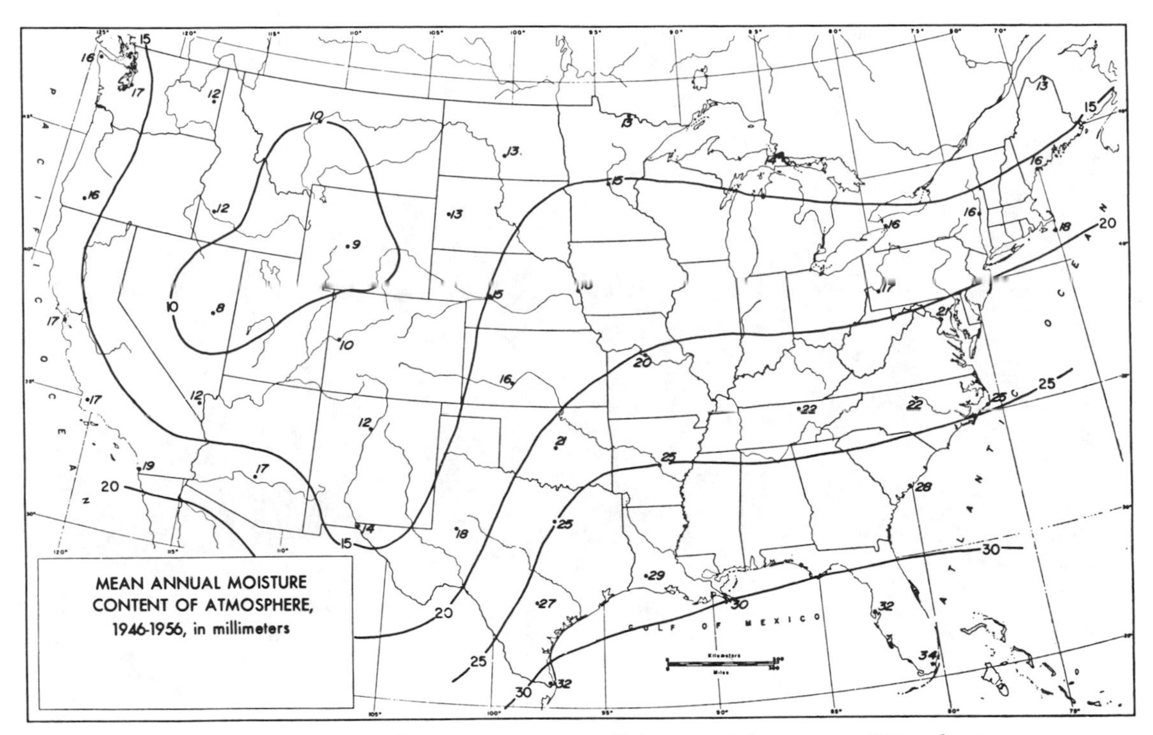

Figure 16.14. Mean annual moisture content of the atmosphere up to 325 mb., in mm. (After Bryson and Hare, Climates of North America, *Elsevier.)*

above 1500–2000 meters, and they are limited again above about 3500 meters because of cold. Throughout western United States there are generally two tree lines, a lower one fixed by drought and an upper one fixed by cold. Hence, forests are limited to intermediate slopes, below which are the chaparral of subtropical California, the cactus of the deserts of southeastern California and Arizona, the sagebrush of Nevada (Figure 16.15), or the bunch grass lands of the better-watered portions of the southern Colorado Plateau of northeastern Arizona and adjacent New Mexico (Figure 16.16) or the Palouse country of eastern Washington. Above the forests lie alpine meadows and tundra vegetation and, in portions of the northern Rockies in Montana and adjacent Canada, perpetual snowfields and mountain glaciers.

Much of the intermontane region is also characterized by extremes in temperature—usually hot in summer and rather cold in winter—except for the very southern portions. In other words, both the meagerness of the precipitation and the ranges of temperatures indicate a region of high continentality, cut off from moisture sources by high mountains on both the west and the east. Only occasionally does moist air filter into this region from either side to bring a little precipitation, but even then the air has already been wrung partially dry by territory traversed before it gets to the region.

The Basin and Range. The hottest, driest part of the United States lies in the basin and range country of Nevada, adjacent western Utah, southeastern California, southwestern Arizona, and adjacent parts of Mexico and New Mexico. In this region the basins are at rather low elevations, particularly in California and Arizona, and

Figure 16.15. Alluvial fans, sagebrush, and desert gravel in the Owens Valley of eastern California.

the mountain ranges are small and widely spaced. In most cases the mountains rise no more than one or two thousand meters above the levels of the basins. In Nevada and Utah the mountain ranges are generally oriented north-south, but in the southern portion of the region they become more haphazard. The entire region is quite hot in summer and is consistently occupied by a heat low at the surface that begins to form in April and usually lasts through September. The hottest parts are generally the lowest in elevation, which lie in a series of down-faulted *grabens* east of the Sierra Nevada in southeastern California. These culminate in Death Valley where, at 84 meters (276 feet) below sea level, a maximum temperature of 54°C (129°F) has been observed, the second-hottest recorded temperature on earth after Azzizia, Libya; (see Figure 16.17). The July average at Death Valley, day and night, is 37°C (98°F). The rest of the region is not quite as bad but is still very hot during summer. For instance, Phoenix, Arizona (20), averages 33°C in July, and Yuma, Arizona, near sea level on the delta of the Colorado River, averages 35°C.

The great surface heating and formation of the heat low during summer are conducive to much air turbulence in the lower troposphere, which is evidenced by frequent dust devils (Figure 16.18), but this is generally capped 2000–3000 meters above the surface by a subsidence inversion in the eastern extension aloft of the Pacific high pressure cell. Since the surface air is so dry, the convection in the first few thousand meters is not enough to lift the air to its condensation level, so this region is essentially cloud-free during much of the summer and receives only infrequent showers. These usually occur in July and August when some moist air may seep into the region

Figure 16.16. Bunch grass in the steppe climate on the elevated Colorado Plateau near Winslow, in northeastern Arizona.

either from the southeast, from the Gulf of Mexico, or from the southwest, from the Pacific Ocean and Gulf of California.

Typically, around 1 July a singularity takes place in the air flow over this region that allows the precipitation to increase suddenly. For instance, at Phoenix (20) the precipitation suddenly increases from a minimum of 2 mm. in June to 20 mm. in July and 28 mm. in August, after which it decreases again during the autumn months. Scattered cumulonimbus clouds then bring short bursts of showers to this parched area. They are much more prevalent at higher elevations over the Colorado Plateau in northeastern Arizona and adjacent states.

Eastern California and Arizona generally receive maximum precipitation during the winter months from remnants of Pacific storms that have moved across western California into this intermontane region. These bring to the area widespread and prolonged light rains, which are more beneficial to the rangelands and reservoir systems than the brief summer downpours. Phoenix in December receives 22 mm. of precipitation, in January 19, in February 22, and in March 17. This is followed by the driest period of the year during April, May, and June. Total precipitation at Phoenix is only 184 mm. (7.3 inches), but the bimodal regime of precipitation with its two very different controls in summer and winter is so characteristic of this area that this type of regime has been dubbed the "Arizona-type rainfall regime." Even less precipitation falls in the delta of the Colorado River at the head of the Gulf of California. Yuma receives only 77 mm. per year, and across the border in Mexico precipitation diminishes to only 30 mm.

Farther north in Nevada and western Utah, elevations are generally higher, sum-

Figure 16.17. Salt accumulation at Bad Water, the lowest spot in Death Valley.

mer temperatures diminish a little, and annual precipitation increases. Elko (8) in northeastern Nevada at an elevation of 1,547 meters has a July average temperature of only 21°C (70°F), with an absolute maximum of 40°C (104°F). The annual precipitation is 231 mm (9.2 inches). Here winter cyclonic storms have become the primary precipitation mechanism, with only a faint semblance of a secondary maximum composed of shower precipitation in May. July, August, and September are the driest months. Winters here are definitely cold, as opposed to the basins in southern Arizona and California. Temperatures at Elko average only −5°C (23°F) in January and have dipped to −42°C (−44°F). Freezing temperatures have occurred at night even in mid-summer. By contrast, Phoenix averages 10.4°C in January and has experienced absolute minimum temperatures of only −8°C. Elko has experienced snow all months except July and August, and its average annual snowfall amounts to 920 mm. (37 inches). A significant snow cover generally occupies the surface from about 1 December through 1 April, whereas in Phoenix only traces of snowfall occur and never remain on the ground.

The cold, snow-covered surface in the so-called "Great Basin" in Nevada and western Utah, between the Sierra Nevada in eastern California and the Wasatch Mountains of central Utah, is conducive to strong radiational cooling of the surface air, which forms the so-called "polar basin" air masses of cold, stable high pressure systems (mentioned in Chapter 9). This air interacts with Pacific air that has come in across the mountains from the west to spawn and rejuvenate cyclones during winter. Both cyclogenesis and anticyclogenesis show high frequencies in this region during winter, in continuously alternating cycles. These open-wave cyclones generally become stalled and difficult to identify over

Figure 16.18. Dust devil during hot afternoon in Nevada. Cumulus of fair weather indicate that surface convection has reached the condensation level, but that it is limited above by a subsidence inversion.

the Rocky Mountains as they progress eastward, but eventually they lower onto the Great Plains and rejuvenate to bring some of the strongest winter storms to the Great Plains, Middle West, and eastern portions of the United States.

Vegetation in the basin and range country varies with elevation and latitude. The low-lying basins in the Mojave, Colorado, and Sonoran deserts of southeastern California, southwestern Arizona, and northwestern Mexico generally have a sparse vegetative cover of various species of cactus and other highly *xerophytic* (drought-resistant) forms of vegetation. Many of the lower-lying areas are occupied by intermittent lakes after rains, and over the years have accumulated salt flats that are barren of any vegetation (Figure 16.19). After the winter rains during spring, there is typically a short period that sports a rapid growth of ephemeral plants that bloom profusely for a week or two and then reseed and die as the summer heat comes on. At this time of year city newspapers in southern California publish maps showing areas that will be blooming during weekends so the urban dwellers can drive into the deserts to see the flowers, much as easterners do in fall to see the autumn foliage. In the somewhat cooler and moister basins of Nevada and western Utah, the vegetative cover runs to sage and other forms of brush. Here some of the mountains rise above 3500 meters, ample elevation to provide a humid climate conducive to forest growth on their intermediate slopes. But in the southern deserts of California, Arizona, and Mexico, the mountains are usually low enough to be dry all the way to their summits.

Large salt flats, such as the Great Salt Lake Desert that stretches for 150 kilo-

Figure 16.19. Playa (intermittent) lake bed in Nevada. Between rains the dry dust and fine sand are picked up by the wind, which deflates the surface below the hillocks held in place by sparse growths of brush.

meters west of present Great Salt Lake, and interconnecting valleys around Carson Sink and Pyramid Lake in western Nevada, are remnants of the extensive interior drainage basins that existed during the so-called "pluvial [rainy] period" of the Pleistocene, when glaciers occupied much of North America to the north and east (Figure 16.20). At that time many of the basins filled up with fresh water that overflowed either to the north, through the Snake-Columbia river system, or to the south, through the Colorado River, to the sea. The largest of these proglacial lakes were Bonneville in northwestern Utah and Lahontan in western Nevada. Today's lakes are much shrunken and have accumulated salts as the whole area developed a negative moisture balance. There is no longer any exterior drainage to the sea throughout this basin and range country except in the lower Colorado River area between California and Arizona; here the stream is an exotic one that loses moisture to evaporation and seepage as its flow, which was gathered in the highlands of the Colorado Plateau and Rocky Mountains of eastern Utah and western Colorado, continues its course all the way to the Gulf of California.

The Colorado Plateau. The Colorado Plateau is an extensive upland in northeastern Arizona, northwestern New Mexico, southwestern Colorado, and eastern and southern Utah. Whereas the basin and range country was low-lying flatlands interspersed with small mountain ranges, the topography of the Colorado Plateau is one of a high, flattish upland, deeply incised here and there by canyons, the ultimate of which is the Grand Canyon of the Colorado River. The plateau surface slopes upward toward the north from elevations of 1000–2000

Figure 16.20. Great Salt Lake Desert, western Utah.

meters in central Arizona to approximately 3000 meters on the north rim of the Grand Canyon. Here and there small mountain groups, often of volcanic origin, rise to as much as 3000–4000 meters.

Although similar atmospheric circulations affect both the basin and range country and the Colorado Plateau, because of the generally much higher elevation in the Colorado Plateau the climate is much different. Most of the plateau's surface and the surmounting mountain groups are relatively cool in summer and receive more showery precipitation (Figure 16.21). Even during mid-summer, nighttime temperatures can dip well below the freezing point. The plateau surface can be positively cold during winter, and receives heavy snowfalls with passing cyclonic storms. The city of Flagstaff, Arizona, at an elevation of 2136 meters (7006 feet) has average temperatures of 19°C (66°F) in July and −2°C (28°F) in January. It has experienced temperatures as low as −30°C (−22°F) in January and −4°C (24°F) in August. Precipitation averages 507 mm. (19.3 inches) per year, with a sharp maximum in July and August, when each month averages 16 thunderstorm days, and a broad secondary maximum in winter from cyclonic storms. In January 1949, 2760 mm. (105 inches) of snow fell.

Much of the plateau's surface reflects a better moisture balance than that in the basin and range country. The typical vegetation on the plateau is bunch grass, which covers large areas in northeastern Arizona and adjacent northwestern New Mexico (see Figure 16.16). As the plateau surface rises toward the Grand Canyon in north-central Arizona at elevations of 1500–2000 meters, the vegetative cover becomes green with scattered piñon trees growing to heights of 3 meters or more, and at the canyon rim at elevations around 2500 meters there are thick stands of pine forests (Figure 16.22). Heavy stands of forests cover the sur-

Figure 16.21. A thunderstorm dots the expansive surface of the Colorado Plateau in northeastern Arizona.

mounting mountains, such as the San Francisco Peaks just north of Flagstaff, which rise to 3900 meters (12,794 feet). In fact, the summits of these peaks rise above the tree line and are snow-covered much of the year. Lower parts of the plateau surface are vegetated with sagebrush and other more xerophytic forms, and the deep canyons are very hot and dry in summer and support only desert forms of vegetation. Much of the plateau surface would classify as a cold steppe climate (BSk), as opposed to the desert climate in lower portions of the basin and range country, hot (BWh) in the south and cold (BWk) in the north.

The Columbia Plateau. The Columbia Plateau is primarily a volcanic upland that is drained by the Columbia River and its major tributary, the Snake River, which occupies the northern third of the intermontane region of the United States in the eastern halves of Washington and Oregon and the Snake River plain of southern Idaho. Elevations on various plateau surfaces vary from 700 to 2000 meters. In northeastern Oregon and again in northeastern Washington, the plateau is surmounted by mountain ranges that rise to 3000 meters or more above sea level, and along the Oregon-Idaho boundary the plateau is cut by the canyon of the Snake River more than 1.5 kilometers deep.

Again, the varied topography makes for a varied climatic situation. Much of the intermediate elevations of the various plateau surfaces would classify as cold steppe climate (BSk), as is shown in Figure 15.3. Spokane, along the northern edge of the wheat-growing Palouse country of eastern Washington, at an elevation of 721 meters receives 437 mm. of precipitation per year, with a maximum from winter cyclonic storms and a minimum during mid-summer. July temperatures average 21°C, and

Figure 16.22. Piñon trees around small volcanic cone south of the Grand Canyon, northern Arizona.

January temperatures −3°C—cold, but not nearly as cold as the Great Plains to the east of the Rocky Mountains at the same latitude.

Boise (3) on the Snake River plain in southwestern Idaho at an elevation of 866 meters receives only 290 mm. of precipitation per year. July and August, the driest months, receive only 4–5 mm. each. Temperature averages range from a warm 24°C in July to a cool −2°C in January. The wide absolute range, from a maximum of 44°C to a minimum of −27°C, attests to the extreme continentality of this area. Climatically, Boise is borderline between a cold steppe (BSk) and a cold desert (BWk). Some of the lower portions of southeastern Oregon would definitely classify as desert. On the other hand, some of the higher mountains, such as the Blue Mountains of northeastern Oregon, are quite humid. Small tributary valleys of the Columbia River on the eastern slopes of the Cascade Mountains, such as the Wenatchee and Yakima valleys, are famous for Washington State apples in every market of the United States.

The Rocky Mountain System

The Rocky Mountains stretch all the way from the Yukon Territory of northwestern Canada southeastward to central New Mexico. In the north the mountains are generally low, and the boreal forest sweeps across them from central Canada into the southern half of Alaska. But south of the mid-portion of British Columbia and continuing through southwestern Alberta, western Montana, and adjacent parts of Idaho, they form a high, continuous mountain massif that is generally humid enough to support forest on intermediate slopes, and alpine meadows, snowfields, and mountain glaciers on the higher peaks. In Wyoming the moun-

Figure 16.23. One of the "parks" in Rocky Mountain National Park, Colorado, in June. The floor of the basin is covered with meadow vegetation, the lower mountain slopes by forests, and the mountain summits by alpine tundra and snowfields.

tains become broken and discontinuous, although many individual ranges still reach elevations of 3000–4250 meters. Yet broad basins in between at elevations of 1500–2000 meters make for an interfingering of original grasslands under semiarid conditions and forests on the intermediate slopes of the more humid mountains. Outliers, such as the Black Hills of western South Dakota and many smaller groups in central Montana, carry islands of forest cover eastward into the Great Plains. A major gap in central Wyoming allows for penetration of modified Pacific air eastward into the plains.

In southeastern Wyoming two or three main ranges extend southward through the entire mid-section of the state of Colorado and eventually play out in north-central New Mexico. Many peaks in this area rise above 4250 meters (14,000 feet). The individual ranges are separated by broad basins called "parks" in Colorado that generally lie at elevations around 2500–3000 meters. The mountain slopes are usually forested from around 2250 meters to about 3350 meters, above which alpine meadows and snowfields predominate. The park floors are primarily meadowland (Figures 16.23 and 16.24).

The basins become lower and drier southward through the San Luis Valley of south-central Colorado and into the Rio Grande Valley of central New Mexico. Albuquerque (1) in central New Mexico at an elevation of 1620 meters receives only 206 mm. of precipitation per year. Unlike the intermontane region to the west, the precipitation at Albuquerque is associated primarily with the atmospheric circulation east of the Rockies. Summer thundershowers account for three times as much rain as winter cyclonic storms do. Central and eastern New Mexico are often sites of the so-called "dew point front" oriented north-

Figure 16.24. Above the tree line on the Continental Divide, 3350 meters, Rocky Mountain National Park, in June. Cirque in the center-distance formed by the head of a previous mountain glacier indicates that the climate was once colder.

south between the moist Gulf of Mexico air, moving northwestward across Texas into the region, and the very dry air, moving northeastward out of the Arizona desert. Although the two air masses are of about the same temperature, their humidity contents, and hence dew points, are quite different. This makes for a convectively unstable situation that leads to numerous severe thunderstorms during middle and late summer. In fact, northeastern New Mexico and southeastern Colorado show a secondary maximum of thunderstorms in the United States, second only to the very southeastern portion of the country. In the upper Rio Grande and Pecos River valleys, thunderstorms occur more than 60 days per year, most of them in middle and late summer (see Figure 10.12).

Even in the southern basins winter can sometimes be quite cold. At Albuquerque, January temperatures average only 2°C (36°F). Snow has fallen there during all but the three summer months and annually amounts to an average of 270 mm. (11 inches).

North America East of the Rockies

The huge expanse of plain east of the Rocky Mountains is wide open to intrusions of air from various directions with little topographical obstruction, except perhaps for certain sections of the Appalachian Mountains, which at their highest rise to only approximately 2000 meters. Their influences are more local than general, affecting local temperatures, rainfall amounts, cloudiness, and fogginess, rather than causing major deflections in the general air flow across eastern United States.

Weather Variability. Eastern North America is the only continental area where the Polar Front is active on land practically all the time. This is particularly true during winter. By contrast, the Polar Front in eastern Asia during winter is almost always well off to sea southeast of the Japanese islands, and the mainland is flooded by continental polar air all the time. Hence, eastern North America experiences the most direct meeting of the two most contrasting air masses, mT and cP, across the Polar Front. Therefore, great interdiurnal variability is normal as these very contrasting air masses are interchanged in the same place over short periods of time. Also, because the Pacific air is blocked by the mountains of the west, the coldest parts of the continent are centered east-west rather than way off to the east, as they are in Eurasia. The greatest continentality, then, occurs in central Canada and north-central United States.

Although day-to-day variability is great throughout much of interior North America, it is probably greater in Alberta, just east of the Rocky Mountains during winter, than anywhere else in the world because of the frequency and intensity of chinook (foehn) winds at that time of year. The Calgary chinooks are world famous. During winter when the zonal flow is strong across the British Columbia and Alberta section of the Rockies, Pacific air may be transported aloft directly across the mountains and then descend upon the plains below. The famous "Calgary arch" is the first indication of the onset of a foehn. This is a high, arched lens of cirrus clouds that are produced over Calgary in the lee wave of the upper air flow over the Rockies. Shortly after it forms, the warm air usually descends to ground surface and temperatures may jump 15° to 25°C within the course of a few minutes. Often there are alternately rapid rises and falls of temperature with the onset of foehns, as the warm descending air momentarily fails to reach the surface and shoots eastward over cold surface air, allowing cold air from the east to reoccupy the foothill areas, and then descends once more to ground surface to cause another temperature jump. For instance, Rapid City, South Dakota, once had three temperature rises and two falls of 22°C (40°F) or more during a period of 3 hours and 10 minutes.

During winter when surface temperatures suddenly rise from way below freezing to above freezing, the snow cover may evaporate in a few hours, subliming directly from solid to gas. This keeps valuable pastures open during the winter for grazing at a latitude that ordinarily would be consistently snow-covered. On the other hand, when cold Arctic air returns to an area after a chinook ceases, temperature falls can be devastating. Browning, Montana, once experienced a temperature plunge of 56°C (100°F) in 24 hours. Such a change is often accompanied by snow and blowing snow, a true blizzard, that blinds and freezes everything in its wake. Chinooks are common during winter along the east front of the Rockies from Alberta southward into Colorado. They may extend eastward well into the Great Plains, where mountain outliers such as the Black Hills of western South Dakota are influential.

Not only is there great interdiurnal variability of temperatures in Alberta, but also tremendous interannual variability. At Edmonton during a period of three years, the January average temperature was as high as −6°C (22°F) and as low as −25°C (−13°F). One November had an average temperature of −18°C (0°F), while four years later it was 3°C (37°F). Havre, Montana, has recorded a January mean as low as −25°C (−13°F) and as high as 1°C (34°F). These great interannual variations are due to varying positions of the mean tracks of storm centers coming over the mountains during different years. If disturbances follow a southerly track, the Alberta region will experience cold Arctic air much of the time, and chinooks will be largely absent. But if the disturbances follow a northerly track, chinooks will be common along the southern peripheries of cyclonic storms.

These storm tracks, of course, are determined by the positions and magnitudes of the waves in the middle troposphere. Shifts in wave patterns also cause great year-to-year fluctuations in precipitation across the Great Plains. While the humid-dry boundary on average runs north-south through the central Great Plains approximately along the one-hundredth meridian, during individual years it can shift westward to the front of the Rockies, leaving the entire Great Plains with a humid climate, or eastward well into the Midwest, thus putting the entire Great Plains and western portions of the Midwest into semiarid or even arid conditions. Standing waves of the upper atmosphere and routes of cyclonic storms tend to lock into place for days or weeks on end, and once the pattern becomes established there may be prolonged periods of either wet weather or dry weather within areas that are in close juxtaposition to one another. Long-term averages tend to show rather simple general patterns of moisture availability, but in any particular growing season this pattern can be quite complex. This is not as true of winter, when more of the precipitation is associated with cyclonic storms and therefore generally occurs in broader, more widespread patterns. But, of course, it is not the winter precipitation that is critical to crop growth. Since throughout large parts of interior North America maximum precipitation is associated with summer thunderstorms, which individually are small in area, precipitation amounts can be very spotty, and even within areas of abundant rainfall there can be localized areas of drought.

The Moisture Pattern. The primary moisture source for practically all of the United States east of the Rockies and southeastern Canada as far north as Hudson's Bay and Labrador appears to be the tropical air stream around the western end of the subtropical high in the Atlantic. This air picks up much moisture and warmth in its long marine trajectory across the southern part of the North Atlantic Ocean around the southern side of the subtropical high, and then curves northward through the Gulf of Mexico, picking up more moisture, and moving on into the heart of the interior plains of the United States. Eventually it curves eastward in the Great Lakes area and returns to the Atlantic across northeastern United States and southeastern Canada. The mean annual moisture content of the atmosphere illustrates this flow nicely (see Figure 16.14). Mean annual precipitation amounts also reflect this moisture source (see Figure 13.9).

Little moisture enters the continent directly from the Atlantic Ocean along the east coast because of the general westerly circulation, except for the peninsula of Florida and the immediate Gulf Coast during summer, when trade winds and easterly waves may penetrate that far poleward. But most of the Atlantic Coast shows a net movement of moisture from west to east every month of the year. The Pacific air is effectively wrung dry by the mountains of the west before it reaches the Great Plains, and when it does break over the mountains it usually appears in central North America as a mild, dry air mass. The Arctic air stream from the north, being constantly cold, carries little moisture, and generally acts primarily as the lifting mechanism for the tropical air stream from the south, interacting with it along the Polar Front.

The overwhelming amount of precipitation in North America east of the Rockies is of cyclonic origin, involved primarily with the Polar Front. Air mass thunderstorms set off by convective activity alone do occur, almost exclusively in the summer, mainly in southeastern United States, particularly in the Florida Peninsula and along the Gulf Coast to the west in the conditionally unstable mT air; but even there many of the thunderstorms are associated with weak frontal situations. Therefore, amounts and seasonal regimes of precipitation are very intimately connected with the configurations of the Polar Front and upper tropospheric features, such as standing waves and jet streams. The seasonal shiftings of these atmospheric circulations bring about a surprising diversity of annual and diurnal precipitation regimes in this broad plains area, which one might assume would be conducive to monotonous uniformity.

Annual Precipitation Regimes. As was mentioned in Chapter 13, interiors of large continents in middle latitudes tend to have pronounced summer maxima of precipitation, the greater influx of moisture and greater convective activity at that time of year outweighing the greater cyclonic activity during winter (see Figure 16.8). And over a large part of interior North America this is true. But the summer maximum is generally not as pronounced as in some other middle-latitude continents, and in portions of North America, particularly the interior southeast and also the extreme northeast, it is not true at all. So the southeastern and northeastern portions of North America show strikingly anomalous rainfall regimes.

By far the most expansive area is covered by a regime with a summer maximum. This stretches from the Rocky Mountain front eastward to east-central Texas in the south, eastern Michigan in the Great Lakes region, and almost the entire width of the continent through central Canada, from eastern British Columbia in the west to central Labrador in the east (see Figure 16.8). Within this area there are important deviations from a simple mid-summer precipitation maximum.

The northern Great Plains to the eastern borders of the Dakotas, Nebraska, and much of Kansas, with a very peaked maximum in June, most closely fits the classic concept of an early summer precipitation maximum in the interior of a continent at latitudes around 35°–50° (28, Williston). But even here there is a hint of a secondary maximum in September, and this becomes much more pronounced eastward through Minnesota, Wisconsin, and Michigan and extending southwestward through Iowa, much of Missouri, central Oklahoma, and the western two-thirds of Texas. Typically, in this region there is a primary maximum in June and a secondary maximum in September, although in places the secondary maximum may occur in August (19, Omaha). And in many cases the early and late maxima are almost equal in magnitude. In Texas the primary maximum may occur as early as May (6, Dallas). The outstanding peculiarity of this regime is the mid-summer dip in precipitation, usually in July and August. This feature continues southward into the western fringes of the Caribbean, eastern Mexico, and portions of Central America.

It appears that during middle and late summer a high pressure ridge associated with a dry tongue of air extending northeast-southwest through this mid-section of the continent develops in the middle troposphere, which damps precipitation processes. This feature seems to be unique to central North America, and is recognized by farmers throughout the American Midwest, who expect to experience some drought during the height of the growing season almost every summer. Since this comes during critical stages of crop development, such as the ear (cob) development of corn (maize), the magnitude of this moisture deficiency often is one of the most determining factors in the yield of crops in the American Midwest. Wherever possible, farmers apply supplemental irrigation to see the crop through this water-deficit period, but over most of interior North America, agriculture must be carried on without this benefit.

From the eastern Great Plains to the Atlantic Coast in northeastern United States and southeastern Canada, the annual precipitation increases by 25 percent or more, and this is due entirely to an increase in the winter precipitation, no doubt because of the convergence of increased cyclonic activity on the St. Lawrence lowland at this time of year. No matter whether the cyclones originate in Alberta, eastern Colorado, eastern Texas, or the Atlantic Coast, they almost all end up in the New England–Maritime Province region of North America. The increased winter precipitation in this northeastern part of the United States and adjacent Canada transforms an annual precipitation regime from one of pronounced early summer maximum in the eastern Great Plains to one that is monotonously even throughout the year on the east coast.

Thus, Omaha, Nebraska (19), which averages 700 mm. of precipitation a year, with 115 mm. in June and only 20 in December, contrasts with New York City (18), which averages 1076 mm. per year with 84 mm. in June and 83 mm. in December. Actually, August and March are the wettest months in New York, but there is not much difference among months. The entire Ohio Valley, from southern Illinois northeastward through Pennsylvania, has rather nondescript precipitation regimes, showing a number of ups and downs among the monthly averages, with none of the months outstandingly wet or dry. The most consistent feature among stations in this area is that the month of October is usually the driest month of the year, reflecting the annually recurring singularity known as Indian Summer: clear, mild, hazy weather due to the stagnation of a high pressure cell over the Appalachian region at this time of year.

The most anomalous precipitation regime in all North America is probably the so-called "Tennessee rainfall regime" that occupies the interior part of southeastern United States, stretching from eastern Texas northeastward to include almost the entire state of Tennessee and the southern two-thirds of Kentucky. This entire region has more precipitation during the winter half-year than during the summer half-year (13, Memphis). The outer portions of this area really have a spring maximum, March commonly being the wettest month (see Figure 16.8). But in a core area extending from central Mississippi into northeastern Tennessee, precipitation during the winter quarter-year exceeds that during the spring quarter by more than 20 percent. One of the peculiarities of this regime is that it does not extend to the Gulf Coast but is limited to an area 75 kilometers or more inland from the Louisiana and east Texas coast (17, New Orleans).

The reason for the wet winters and springs in this area seems to be the concentration of jet stream positions in this region at this time of year and consequent passages of cyclonic storms, which originate in the east Texas area and follow a northeasterly course along cold or stationary fronts that are anchored by parent lows in the New England area. Unlike the Polar Front along the Pacific Coast, which shifts gradually southward from autumn to late winter and then suddenly appears again in the north in spring and summer, the average positions of jet streams in eastern North America show a sudden jump southward from the Great Lakes area in October to southeastern United States in November, where it stays concentrated in a belt about 450 kilometers wide for the next half-year, shifting slowly southward to its most southerly position in January just north of the Gulf Coast, and then shifting back northward to the November position in March, after which it shifts more rapidly but continuously northward to its summer position in the Great Lakes area by June (Figure 16.25). Thus, the positioning of the jet stream in southeastern US during winter and the northeasterly route of cyclonic storms underneath it bring to this region a cloudy, rainy, and sometimes snowy winter. The

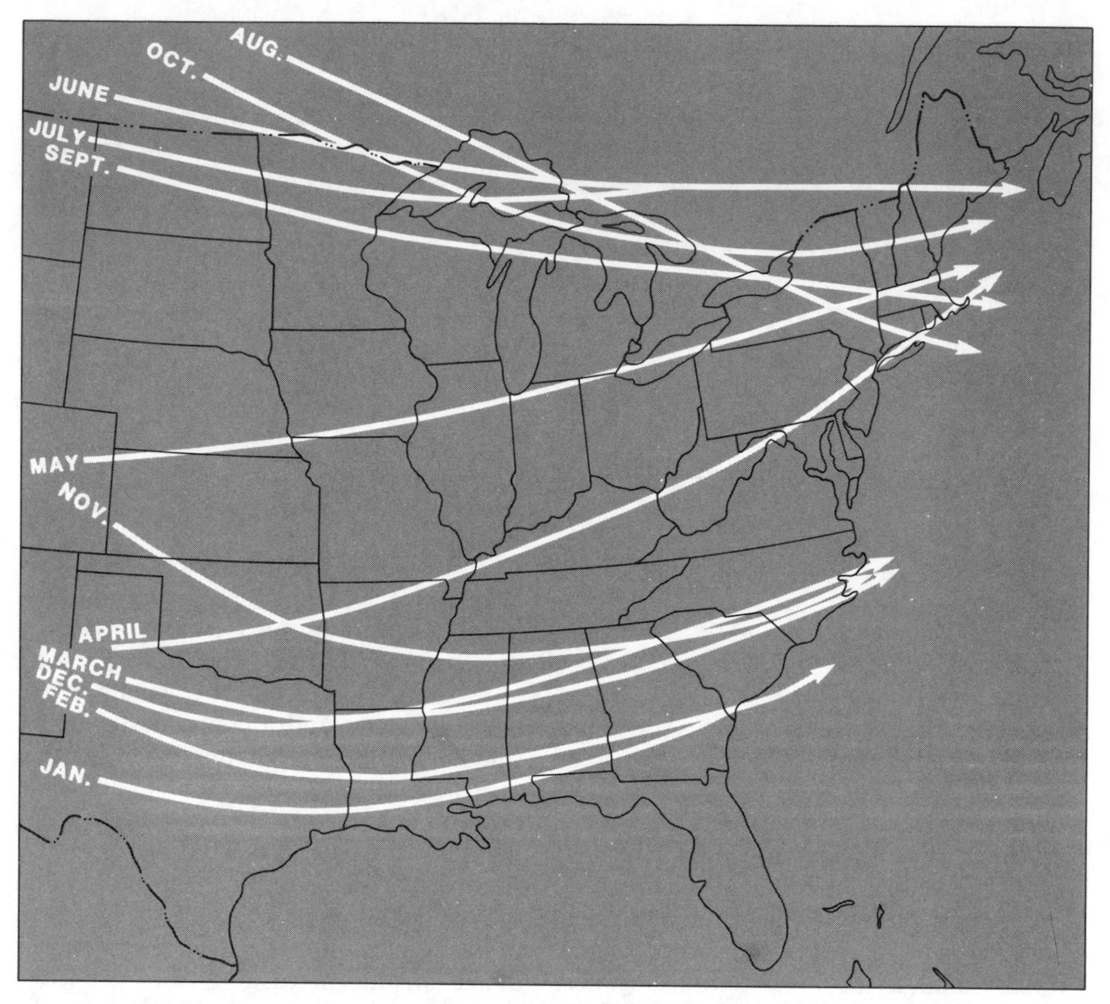

Figure 16.25. Average monthly position of axis of highest frequency of upper air wind speeds at any level in excess of 35 m/s. (After H.E. Landsberg, "Some Patterns of Rainfall in the North-Central U.S.A.," Archiv fur meteorologie, geophysik, und bioklimatologie *[1960]:165–74.)*

area is particularly prone to freezing rain and late-season, heavy, wet snow, which wreak havoc with transportation, although they do not remain on the ground very long.

The winter maximum in this area is due only partially to its abundance of winter precipitation. The summer precipitation here is somewhat depressed from the Gulf Coast inland. During summer while the very coastal area and the peninsula of Florida are receiving abundant rainfall in a moist tongue of air that extends northeastward in the middle troposphere, the area farther to the north appears to lie on the southeastern flank of a dry tongue of air in the middle troposphere extending southwestward across the mid-continent.

The southeastern Gulf and Atlantic coasts north to New Jersey show a strong midsummer maximum of precipitation due to numerous thunderstorms, but in most cases there is still a semblance of the March peak from the interior region due to spring's cyclonic activity. In the immediate coastal area, extending as much as 300 kilometers inland in Georgia and the Carolinas, thunderstorm activity in the warm, moist, unstable air circulating around the western end of the Atlantic High is not damped by stability aloft, as it is farther inland. Particularly over central Florida, mid-summer convective storms are very frequent and often bring heavy rain.

In central Florida the four warmest months, June–September, account for at least two-thirds of the annual rainfall. During the three months June–August, this region receives, on the average, 50 thunderstorm days, more than one every other day, which is as high a frequency during this time of year as that in the world's maximum thunderstorm area near Lake Victoria in equatorial east Africa. The African region, however, maintains this frequency throughout much of the year, averaging 242 thunderstorm days annually, while central Florida averages less than 100. Incidentally, during June–August, the northern New Mexico region averages more than 40 thunderstorm days. The very high frequency of thunderstorms in central Florida during summer, even more than along the coasts of the peninsula, seems to be related to a convergence line running north-south through the center of the peninsula between double sea breezes, which blow into the peninsula from both sides. Typically, in mid-summer these sea breezes meet in the middle of the afternoon, which is the time of maximum occurrence of thunderstorms in this area.

The September rainfall in the southeast coastal area may be partially due to hurricanes, which generally reach their peak frequency at this time of year. In fact, in very exposed coastal locations, such as Miami (15), autumn becomes the season of maximum rainfall. At Miami the precipitation, after a dip in July–August, jumps to 241 mm. in September and 209 mm. in October, after which it plummets to 72 mm. in November. Along the immediate Gulf and Atlantic coasts, tropical storms account for about 15 percent of the total precipitation during the five-month hurricane season, June–November. The amount diminishes inland, but even as far inland as a line extending from central Texas northeastward through central Pennsylvania and New York state and into southern Quebec and the Maritime Provinces of Canada as much as 5 percent of the June–October precipitation is attributed to these storms. As was explained in Chapter 11, although hurricanes depend for their kinetic energy upon the latent heat of condensation, and therefore winds diminish rapidly as they penetrate a mainland, the low pressure, cloudiness, and precipitation may be carried far inland. Often a tropical storm that has penetrated the Gulf Coast and moved northward up the Mississippi Valley may meet up and combine with an extratropical cyclone moving up the Ohio Valley, thereby stalling, intensifying, and bringing copious rainfall to the entire northeastern United States and adjacent Canada. On rare occasions when this happens, a wide area of eastern United States may experience 200–300 mm. (8–12 inches) of precipitation over a period of three or four days.

Severe Thunderstorms and Tornadoes. Although the greatest frequency of thunderstorms in North America occurs in Florida, they are by no means the most violent there. Storms resulting from extreme convective instability in portions of the Great Plains and along the "dew line" of central New Mexico exhibit much more turbulence and extreme thunder and lightning. This is evidenced by the distribution of hail, which of course requires excessive updrafts of air in order to cause large hailstones. The number of days with hail in the United States is greatest in extreme southeastern Wyoming and adjacent parts of eastern Colorado, and the maximum incidence of

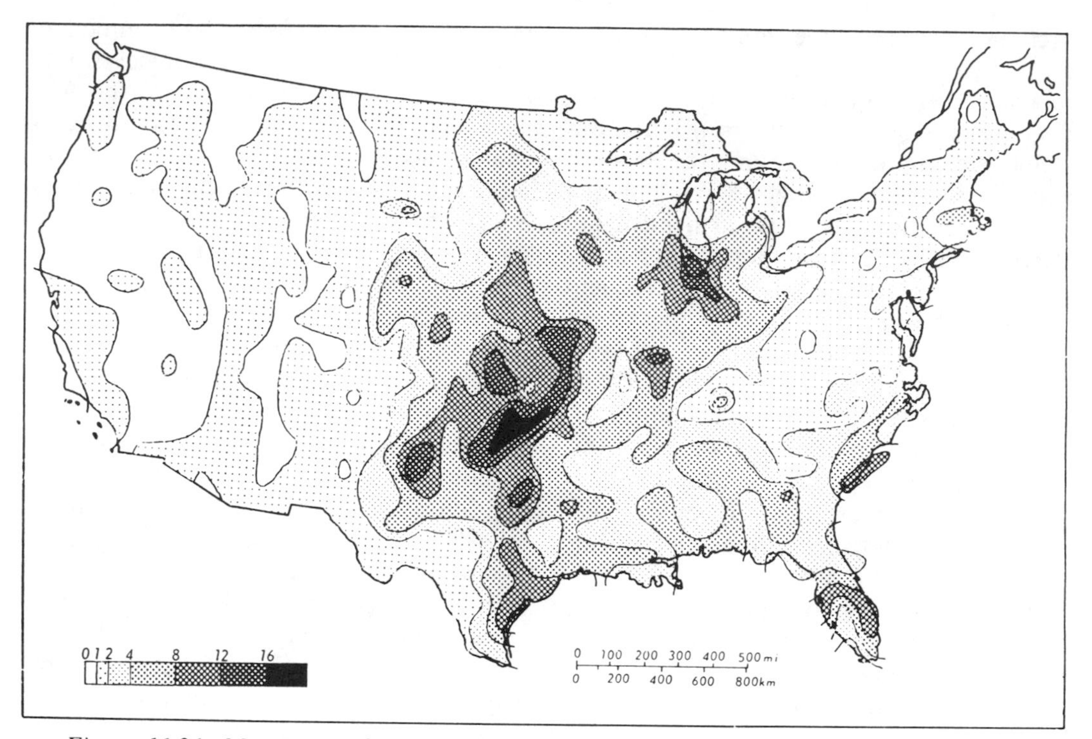

Figure 16.26. Mean annual incidence of tornadoes per 26,000 km², 1955–1967. (After Bryson and Hare, Climates of North America, *Elsevier.)*

large hail, more than 19 mm. in diameter, is in central Oklahoma and Kansas, a distribution that is almost identical with the incidence of tornadoes (Figure 16.26).

As was mentioned in Chapter 10, by far the greatest number of the world's tornadoes are found in the United States east of the Rocky Mountains. They are always associated with severe thunderstorms, frequently in squall lines within the apexes of warm sectors in cyclonic storms. They most likely develop near stalled sections of the Polar Front during periods of most intense air mass contrasts, which usually occur during spring when the Arctic air intrudes to its most southerly position in the United States and the mT air from the Gulf is already very warm, moist, and conditionally unstable. In the states of maximum occurrence, Oklahoma and Kansas, three-fourths of the tornadoes take place during the months of April, May, and June. Most occur during late afternoon and evening during the period of maximum diurnal convective activity. Farther north, maximum occurrence may be delayed until June or July, and they sometimes occur at night, as do many of the thunderstorms in that area. In southeastern United States, particularly along the Gulf Coast, multiple tornadoes are commonly associated with hurricanes during late summer and autumn. Over the Gulf they may appear as water spouts. Tornadoes can occur any time of year, particularly in southeastern United States, on occasions when synoptic conditions are right.

One of the most unusual characteristics of thunderstorms in interior North America is the tendency for a maximum occurrence during nighttime hours over large parts of the interior. In an area stretching from north Texas in the south into the Canadian prairies in the north, and from eastern Montana, western Dakotas, and western Nebraska, Kansas, and Oklahoma in the west to western Wisconsin and central Illinois in the east, more thunderstorms occur between 6 PM and 6 AM than between 6 AM and 6 PM. Over the large central portion of this region the hour of maximum occurrence is 2 AM. Southern Nebraska and northern Kansas is the core area of nighttime thunderstorms. Over almost all the rest of the United States, thunderstorms occur more during daytime hours, except in southern Arizona. The peculiarities of the topography of interior North America, with the high Rockies to the west and the long, tapering slope of the Great Plains eastward, appear to set up a diurnal oscillation of convergence and divergence with occasional formations of low-level jets, which may explain this unusual diurnal occurrence of thunderstorms during the period of negative surface heating.

Lake Effects

Large water bodies significantly affect the climate of surrounding territories. The deep Great Lakes, for example, generally store up enough heat during summer to prevent them from freezing in winter except along immediate shorelines. Only rarely, perhaps once in 40 years, does any of them—except shallow Lake Erie—form an essentially complete ice cover. Hence, during winter the unfrozen surfaces are usually much warmer than the air above, particularly during outbreaks of continental polar air when the surface water may be 30°C or more warmer than the air above. Under such conditions, great quantities of heat and moisture are injected into the atmosphere as it moves across the lakes, which produces strong turbulence in the surface air and dense steam fogs over the lakes, which build into cumulus clouds that drift over lee shores to produce heavy, sharp snow showers.

Muskegon, Michigan, on the eastern shore of Lake Michigan sometimes receives as much as 0.5–1.0 meter of snow during a day, while Milwaukee directly across the lake on the windward western shore is experiencing only clear, cold weather. The so-called "snow belts" around the southern and eastern shores of the Great Lakes are legendary. Buffalo, New York, on the Niagara River between Lake Erie and Lake Ontario, averages 2630 mm. (105 inches) of snow per year, and some years may receive several times that. Many areas on the leeward sides of lakes receive between 2500 and 3750 mm. (100 and 150 inches) of snow per year, while cities on the windward sides of the lakes generally receive only 1000–1250 mm. (40–50 inches). Occasionally, atmospheric circulation across the area will be such that easterly winds bring similar snow showers to western shores. Milwaukee, Wisconsin, in particular, experiences such weather during winter when a high pressure cell moves eastward across southern Canada and sends northeastern winds across almost the entire length of Lake Michigan.

During summer, of course, the lake surfaces are cooler than the surrounding land. This depresses convective activity over the lakes and generally reduces summer rainfall by about 5 percent, compared to adjacent shorelines. Whereas in winter lakes are often enshrouded in cumulus clouds and steam fog while the intervening land is perfectly clear, during summer the lakes usually mark clear atmospheric areas in the midst of cumulus formations over the land (Figure 16.27). On hot summer days with weak synoptic air flow, lake breezes may develop and cool adjacent shorelines as much as 5°–10°C several kilometers inland.

Agriculturally, the Great Lakes have profoundly influenced crop selections along their leeward shores. The "fruit belts" of southern Michigan and upper New York State are well known in the United States. Plantings of various fruits, especially apples, berries, and grapes, are determined primarily by climatic conditions during early spring when the lakes temper spells of extreme rises in air temperature to levels

*Figure 16.27. During summer the air remains clear over cool Lake Michigan (*lower right*), while cumulus clouds build up over the heated land of lower Michigan (*upper left*).*

of 20°–30°C (68°–84°F). Without the lakes' influence the warmth would cause early budding and blooming of fruits and expose them to kill by late spring frosts.

Large water bodies in Canada have similar effects, but they eventually freeze over during winter, so their influences are limited primarily to the first half of winter. Large Hudson's Bay remains unfrozen until about January, and until then the land to the east of it is distinctly less frigid than the land to the west. But after early January the situation is reversed, and the land to the east of the bay becomes colder than that to the west.

Climatic Distribution

If one ignores the many interesting facets of climate just discussed, a climatic distribution based on moisture balance and temperature characteristics forms a simple pattern in eastern North America, since topographic effects are not great. In eastern United States, moisture supply generally decreases steadily westward from the Atlantic Coast to the western Great Plains, and average temperatures decrease steadily northward from the Gulf Coast to Canada. The humid-dry boundary runs essentially north-south through the central Great Plains near the 100th meridian as far north as southern Saskatchewan and Alberta, but north of about 52° latitude the air becomes so consistently cool that the climate in the western plains of Canada no longer is dry, and the cool, humid boreal forest and tundra climates form broad, unbroken west-east zones all the way from western Alaska to eastern Canada (see Figure 15.3).

East of the humid-dry boundary, the southeastern US south of about the southern borders of Kansas, Missouri, Kentucky, and Virginia has 8–12 months with tem-

Figure 16.28. Grazing cattle on the bunch grass steppe of western South Dakota.

peratures averaging above 10°C and humid conditions year-round (see Figure 16.8). Thus, the whole southeastern part of the country classifies as a humid subtropical climate (Cfa). But the western portions of this region may be subject to frequent droughts. The very southern tip of Florida experiences no frosts, and thus classifies as a tropical wet climate (Ar).

Central and northern US and adjacent portions of Canada have four to seven months with temperatures averaging above 10°C and continuously humid conditions. This broad area thus classifies as a temperate continental climate (Dc). Again, its western portions can be subject to severe droughts in individual years, and spotty occurrences of drought occur in most years. The region is often subdivided into two portions according to summer temperatures. If the warmest month averages above 22°C (72°F), the climate is classified as temperate continental, hot summer (Dca). If the warmest month average is below that, it is a cool summer (Dcb). This subcategory boundary runs from central North Dakota southeastward through Minneapolis, Minnesota; Milwaukee, Wisconsin; Detroit, Michigan, and eastward along the Pennsylvania–New York border and through New England to the Atlantic Coast.

The temperate continental, hot summer (Dca) climate to the south of this line is often called the "corn belt" climate, while the temperate continental, cool summer (Dcb) climate to the north is often called the "spring wheat" climate. This northern region might also be dubbed the "dairy belt," since the agriculture in this region largely supports the dairy cow, the basis for cash income for farmers throughout much of Minnesota, Wisconsin, Michigan, and New York State, as well as southern Ontario. Thus, although this is only a subregional boundary, agriculturally it is quite significant: it essentially marks the poleward limits of cash agriculture based on corn and soybeans, and the equatorward

Figure 16.29. A wind break west of Mitchell, South Dakota.

limits of dairying in its more humid eastern parts based on hay crops, oats, and pasture land, and the cash agriculture of spring wheat growing in its drier, western portions.

The dry Great Plains area west of the 100th meridian has a semiarid steppe climate with widely varying temperature extremes throughout much of its north-south extent (BS). Originally, it was vegetated primarily by bunch grass and portions of it still is, which is utilized for grazing (Figure 16.28). But many areas have been plowed under to wheat, or have been irrigated for such crops as alfalfa and sugar beets if river water is available. As is true throughout all subhumid-semiarid regions of the world, soils here are excellent, and their fertility entices farmers into a climatically hazardous region. As was pointed out in Chapter 13, more years experience precipitation amounts below normal than above normal, so the development of agriculture in such regions must be done very cautiously. Wind erosion is a constant threat. Once the native grass cover has been plowed under, the finely textured soil is free to blow. After the dust bowl years of the 1930s many shelter belts were planted to combat wind erosion (Figure 16.29).

Since little moisture is derived from westerly air flow, ordinarily the Great Plains are driest right next to the eastern foothills of the Rocky Mountains, and precipitation increases eastward as the effects of atmospheric circulation from the Gulf of Mexico become more prevalent. But in southeastern Alberta and southwestern Saskatchewan in an area known as "Palliser's Triangle" (the driest part of Canada) annual precipitation is reduced to only about 70 percent of what it is farther west or in any other direction surrounding this region. Cyclone routes tend to go either north or south of the region leaving an intermediate area which, on the average, receives less precipitation than either side, and which experiences higher annual variability of amount.

Throughout the eastern Great Plains and

Middle West, from eastern Nebraska to central Indiana the original vegetation consisted of an intricate fingering of deciduous broadleaf forests and grassland. The forests generally projected westward along stream valleys where soil moisture was consistently available, and the intervening uplands were occupied by grasses that could grow rapidly and reseed themselves, thereby surviving periods of drought. This pattern of native vegetation greatly influenced the settlement of this region. Since most of the early settlers came from forested western Europe, they followed the forested stream valleys westward. They needed the wood to build their log cabins and to heat their homes, and they needed the streams for transport. Thus, long fingers of settlement extended far into the prairies along the wooded stream banks before any of the uplands were settled. Only after population pressure in the valleys became great did new settlers begin to occupy the grassy uplands and only then was it realized that the latecomers generally had come into possession of the better agricultural soils.

Table 16.1 Temperature and Precipitation Data for North America

	Jan	Feb	Mar	Apr	May	June	July	Aug	Sept	Oct	Nov	Dec	Yearly average
1. Albuquerque, New Mexico (35°03′N, 106°37′W, 1620 m)													
T.(°C)	2	4	8	13	18	24	26	25	21	15	7	3	14
P.(mm)	10	10	12	12	19	14	30	34	24	19	10	12	206
2. Barrow, Alaska (71°18′N, 156°47′W, 7 m)													
T.(°C)	−27	−28	−26	−18	−8	1	4	3	−1	−9	−18	−24	−12
P.(mm)	5	4	3	3	3	9	20	23	16	13	6	4	110
3. Boise, Idaho (43°34′N, 116°13′W, 866 m)													
T.(°C)	−2	1	5	10	14	18	24	22	17	11	4	0	10
P.(mm)	34	34	34	29	33	23	5	4	10	21	30	34	290
4. Calgary, Alberta (51°06′N, 114°01′W, 329 m)													
T.(°C)	−10	−9	−4	4	10	13	17	15	11	5	−2	−7	4
P.(mm)	17	20	26	35	52	88	58	59	35	23	16	15	444
5. Churchill, Manitoba (58°45′N, 94°04′W, 11 m)													
T.(°C)	−28	−26	−20	−11	−2	6	12	12	6	−1	−12	−22	−7
P.(mm)	13	14	17	26	30	41	52	61	53	38	39	23	407
6. Dallas, Texas (32°51′N, 96°51′W, 146 m)													
T.(°C)	8	10	13	18	23	27	29	29	26	20	13	9	19
P.(mm)	59	65	72	102	123	82	49	49	72	69	69	68	879
7. Denver, Colorado (39°46′N, 104°53′W, 1610 m)													
T.(°C)	−1	0	3	9	14	19	23	22	18	11	4	1	10
P.(mm)	14	18	31	54	69	37	39	33	29	26	18	12	380
8. Elko, Nevada (40°50′N, 115°47′W, 1547 m)													
T.(°C)	−5	−2	2	7	11	16	21	19	14	8	1	−3	7
P.(mm)	29	23	21	21	24	18	10	8	9	19	23	26	231
9. Fairbanks, Alaska (64°49′N, 147°52′W, 133 m)													
T.(°C)	−24	−19	−13	−1	8	15	15	12	6	−3	−16	−22	−3
P.(mm)	23	13	10	6	18	35	47	56	28	22	15	14	287
10. Fresno, California (36°46′N, 119°42′W, 100 m)													
T.(°C)	8	10	12	16	20	23	27	26	23	18	12	8	17
P.(mm)	52	56	50	29	8	2	0	0	3	11	24	50	285
11. Goose, Newfoundland (53°19N, 60°25′W, 13 m)													
T.(°C)	−17	−15	−8	−2	5	12	16	15	10	3	−4	−13	0
P.(mm)	72	63	68	62	56	72	84	91	76	63	67	63	837
12. Los Angeles, California (34°03′N, 118°14′W, 95 m)													
T.(°C)	13	14	15	17	18	20	23	23	22	20	17	15	18
P.(mm)	78	85	57	30	4	2	0	1	6	10	27	73	373
13. Memphis, Tennessee (35°03′N, 89°59′W, 80 m)													
T.(°C)	6	7	11	17	22	26	28	27	24	18	10	6	17
P.(mm)	154	119	129	118	107	93	90	75	72	69	111	125	1262
14. Mexico City (19°24′N, 99°11′W, 2309 m)													
T.(°C)	12	14	16	17	17	17	16	16	16	15	13	12	15
P.(mm)	8	5	10	23	55	118	160	145	129	49	17	6	726
15. Miami, Florida (25°48′N, 80°16′W, 2 m)													
T.(°C)	19	20	21	23	25	27	28	28	27	25	22	20	24
P.(mm)	52	48	58	99	164	187	171	177	241	209	72	42	1520

	Jan	Feb	Mar	Apr	May	June	July	Aug	Sept	Oct	Nov	Dec	Yearly average
16. Milwaukee, Wisconsin (42°57′N, 87°54′W, 205 m)													
T.(°C)	−6	−5	−1	6	12	17	20	20	16	10	2	−4	7
P.(mm)	46	36	59	64	80	92	75	78	69	53	55	41	748
17. New Orleans, Louisiana (29°57′N, 90°04′W, 3 m)													
T.(°C)	13	15	17	21	25	28	28	29	27	23	17	14	21
P.(mm)	121	106	167	138	138	141	180	163	148	93	102	116	1613
18. New York City, N.Y. (40°47′N, 73°58′W, 40 m)													
T.(°C)	1	1	5	11	17	22	25	24	20	15	8	2	13
P.(mm)	84	72	102	87	93	84	94	113	98	80	86	83	1076
19. Omaha, Nebraska (41°18′N, 95°54′W, 298 m)													
T.(°C)	−5	−3	3	11	17	23	26	25	19	13	4	−2	11
P.(mm)	21	24	37	65	88	115	86	101	67	44	32	20	700
20. Phoenix, Arizona (33°26′N, 112°01′W, 340 m)													
T.(°C)	10	13	16	20	25	30	33	32	29	22	15	11	21
P.(mm)	19	22	17	8	3	2	20	28	19	12	12	22	184
21. Sacramento, California (38°31′N, 121°30′W, 13 m)													
T.(°C)	8	10	12	16	19	23	25	25	23	18	12	9	17
P.(mm)	81	76	60	36	15	3	0	1	5	20	37	82	414
22. St. Louis, Missouri (38°38′N, 90°12′W, 142 m)													
T.(°C)	0	2	6	13	19	24	26	25	21	15	7	2	13
P.(mm)	50	52	78	94	95	109	84	77	70	73	65	50	897
23. Salt Lake City, Utah (40°46′N, 111°58′W, 1286 m)													
T.(°C)	−2	1	5	10	15	19	25	24	18	12	3	0	11
P.(mm)	34	30	40	45	36	25	15	22	13	29	33	32	354
24. San Diego, California (32°44′N, 117°10′W, 4 m)													
T.(°C)	13	14	15	16	17	18	21	22	21	19	16	14	17
P.(mm)	51	55	40	20	4	1	0	2	4	12	23	52	264
25. San Francisco, California (37°47′N, 122°25′W, 16 m)													
T.(°C)	10	12	13	13	14	15	15	15	17	16	14	11	14
P.(mm)	116	93	74	37	16	4	0	1	6	23	51	108	529
26. Saskatoon, Saskatchewan (52°08′N, 106°38′W, 157 m)													
T.(°C)	−18	−15	−10	4	11	15	19	18	12	5	−6	−13	2
P.(mm)	15	16	15	21	34	58	60	45	34	19	18	17	352
27. Seattle, Washington (47°36′N, 122°20′W, 4 m)													
T.(°C)	5	6	8	11	14	16	19	18	16	12	8	7	12
P.(mm)	132	99	84	50	40	36	16	19	42	83	127	138	866
28. Williston, North Dakota (48°09′N, 103°37′W, 579 m)													
T.(°C)	−12	−10	−4	6	13	17	22	20	14	8	−2	−8	5
P.(mm)	14	12	18	24	36	84	48	38	28	19	15	13	349
29. Winslow, Arizona (35°01′N, 110°44′W, 1487 m)													
T.(°C)	0	3	8	13	19	24	27	26	22	14	5	1	13
P.(mm)	11	12	10	11	8	7	26	36	23	17	9	13	183
30. Yakutat, Alaska (59°31′N, 139°40′W, 9 m)													
T.(°C)	−3	−2	0	3	7	10	12	12	10	6	1	−2	4
P.(mm)	276	208	221	184	203	129	214	277	420	498	407	312	3348

17 | CLIMATE OF EURASIA

Primary Controls

A discussion of the huge landmass of Eurasia might well be broken into several chapters since the area covers almost 37 percent of the earth's land surface. But it makes no sense climatically to divide Europe and Asia along their traditional boundary—the Ural Mountain–Caspian Sea–Caucasus Mountains—for the Asiatic High, that great primary control of winter weather, disregards political boundaries and stretches right across this boundary from the central portions of Asia westward into much of Europe. And frontal passages and routes of cyclonic storms sweep across all of Europe and the Russian Plain into the West Siberian Lowland. Winter storms in the Mediterranean move eastward into the Middle East and Central Asia all the way to the Indus Valley of Pakistan and northwestern India. It might make more sense to divide Europe and Asia along some line joining the Baltic and Black seas, for the great Russian Plain of eastern Europe with its high degree of continentality stands in stark contrast to the highly marine-controlled climate of the deeply embayed peninsulas and islands of Europe to the west.

If there is any portion of the Eurasian landmass that is somewhat separate climatically, it is the classic monsoon lands of southeast Asia, including the Indian and Indochinese peninsulas and neighboring island groups that are cut off almost entirely from influences in the north by the highest mountains and plateaus on earth. But were it not for the topography to the north the monsoon circulation would not be developed as it is, so monsoon Asia owes its special climatic characteristics to the rest of the landmass.

Even the entirety of Eurasia is difficult to separate climatically, because much of the area is influenced by the fact that Africa, the second-largest continent on earth, lies just to the southwest across the Mediterranean Sea. This blocks any true maritime tropical air from entering much of Europe and western Asia. Indeed, the north coast of Africa is not the best climatic division line, for coastal Africa climatically (and culturally) belongs to the Mediterranean. Even better might be a west-east line somewhere through the northern Sahara separating Mediterranean coastal Africa from tropical sub-Saharan Africa.

The climate of Eurasia is being discussed immediately after that of North America because Eurasia most nearly resembles the North American situation: both continents are located primarily in the middle latitudes of the Northern Hemisphere, both dip into the tropics in their southern extremities, and both extend into the polar regions in their northern extremities. In both the zonal westerlies in the upper troposphere propel circulation features in the lower troposphere from a general westerly direction to a generally easterly one, with the notable exception of monsoon Asia, primarily during summer, and to some extent the Gulf coastal regions of the United States and adjacent Mexico in North America.

But although the variety of climatic types in Eurasia is similar to that in North America, the pattern of distribution is quite different in the two continents because of very significant differences in the shapes of the landmasses, the orientation of major topographic features, and differences in adjacent oceanic circulations. Whereas the west coast of North America extends westward at higher latitudes, the northwest coast of Europe recedes northeastward with latitude. Thus, western and central Europe do not have a large, cold landmass northwest of them to act as a source region for continental polar air during winter as the United States does, but instead the water of the North Atlantic and the North Sea. This difference in continental-oceanic arrangement completely turns around the concepts of air masses in the two continents. The Europeans during winter receive cold air not from the northwest, but from the east, and the western half of Europe entirely escapes the onslaught of exceedingly cold, fresh cP air. The continental air that occasionally reaches western Europe during winter has had a long trajectory around the southern side of the Asiatic High, and therefore has been considerably modified.

The moderating effects of the ocean to the west of Europe have been enhanced by the fortunate (for Europe) accidents of geography an ocean away in the equatorial Atlantic and adjacent South America. The triangular shape of eastern South America, with Cape São Roque forming the easternmost projection 7°–8° latitude south of the equator, coupled with the northern position of the meteorological equator, causes the bulk of the South Equatorial Current in the Atlantic to be shunted northwestward along the northeastern coast of South America into the Caribbean Sea, where it joins the North Equatorial Current to form a very strong flow northwestward between Cuba and Yucatan into the Gulf of Mexico, and then northeastward through the Florida Straits as the Gulf Stream, which continues northeastward parallel to the eastern coast of North America and on across the North Atlantic as the North Atlantic Drift. A major portion of this oceanic flow continues northeastward through the broad opening between Iceland and Britain into the Norwegian Sea, and some of it even eventually finds its way north of Scandinavia, where it curves eastward into the Barents Sea north of European USSR (see Figures 6.10 and 6.11).

The combined atmospheric and oceanic flows from southwest to northeast across the Atlantic produce the largest positive temperature anomalies on earth during winter (see Figure 14.3). Over portions of the Norwegian Sea about 300 kilometers off the west coast of Norway between latitudes 65°–70°N surface air temperatures during January average as much as 28°C (50°F) above latitudinal normals. The sea remains unfrozen during winter clear around the northern tip of Scandinavia and into the western half of the Barents Sea. The warmth and volume of the Gulf Stream and its extension, the North Atlantic Drift, are significantly greater than that of the Kuroshio (Japan) Current in the North Pacific, and it extends to much higher latitudes along Europe than the Japan Current does along the west coast of North America.

Marine air penetrates much farther inland in Europe than in North America because of the different orientation of highlands. There is no continuous north-south mountain barrier in Europe to block the flow of Atlantic air eastward. During summer when a weak, diffuse low pressure forms over much of Eurasia, Atlantic air may penetrate at the surface all the way to central Siberia. During winter it is usually blocked in the interior by the buildup of the cold Asiatic High, which frequently protrudes westward to occupy a great deal of continental Europe. But even then, Atlantic air may be carried far eastward along the Arctic coast by cyclonic storms moving along the Arctic Front along the southern periphery of the Icelandic Low, which protrudes eastward along the Siberian coast as far as the Lena River delta, and in the south by cyclonic storms moving along the Polar Front through the entire length of the Mediterranean Sea into the Near and Middle East as far as the upper Indus Valley of Pakistan and northwest India.

On the other hand, high mountains in the southern part of Eurasia form a fairly continuous belt of rugged topography that largely rules out northerly penetration of very warm, moist, unstable maritime tropical air, except south of the mountains in southeast Asia. The entirety of Europe is robbed of a maritime tropical source by the existence of Africa just to the south across the Mediterranean Sea. Thus, the segments of the Polar Front in Eurasia any time of year do not separate as contrasting air masses as they do in eastern North America. Throughout Europe the Polar Front generally separates one variety of maritime polar air from another variety that has been modified in one way or another. Hardly ever does fresh continental polar air find its way into central and western Europe, and never does really fresh maritime tropical air find its way into this region. Therefore, air mass contrasts, and hence the intensity of fronts, is generally much less than in eastern North America, and weather is less violent and less changeable. This does not mean that the weather picture is simpler in Europe. The many subtle differences in airmasses, and the complexity of the topography throughout much of Europe, induce many local peculiarities.

The essentially perpendicular differences in the orientation of major mountain barriers in North America and Eurasia induce somewhat perpendicular differences in resultant climatic patterns in the two continents, particularly in eastern United States and eastern Europe, as will be demonstrated later. Thus, there are major differences between the two continents in the arrangements and juxtapositions of climatic types.

The Eurasian landmass is so huge that there is hardly any connection between major controls along the Atlantic coast in the west and the Pacific coast in the east, although such intervening entities as the Asiatic High during winter to some extent affect all parts. Also, the extremely high mountains in south-central Asia almost totally preclude surface air flows between north and south. Therefore, the climate of this supercontinent will be discussed within discrete subregions (Figure 17.1).

Peninsular Europe

The marine influence in Europe is strengthened by deep indentations by the sea on all sides. Hardly any point in Europe west of the Soviet Union lies more than 600 kilometers from the sea. In the north the influences of the Atlantic are carried eastward deep into the continent by the North Sea and the Baltic and its extensions, the Gulfs of Bothnia and Finland, and extended still farther northeastward through the Great Lakes of northwest European Russia to the White Sea and the Arctic. In the south the Mediterranean, with its various subsidiary seas, extends eastward from the Strait of Gibraltar more than 2500 kilometers, and marine influences are continued eastward through the Black and Caspian seas.

The whole of Europe is but a peninsula appended to the western side of Asia, broad on its eastern edge and tapering westward. The deep embayments of the sea divide this peninsula into subsidiary peninsulas and islands: in the north, Scandinavia, Jutland, and the British Isles; in the south, the Iberian Peninsula (Spain and Portugal), Italy, and the Balkans, including Greece. The ubiquity and warmth of the sea impart to Europe a remarkably favorable climate with mild winters, temperate summers,

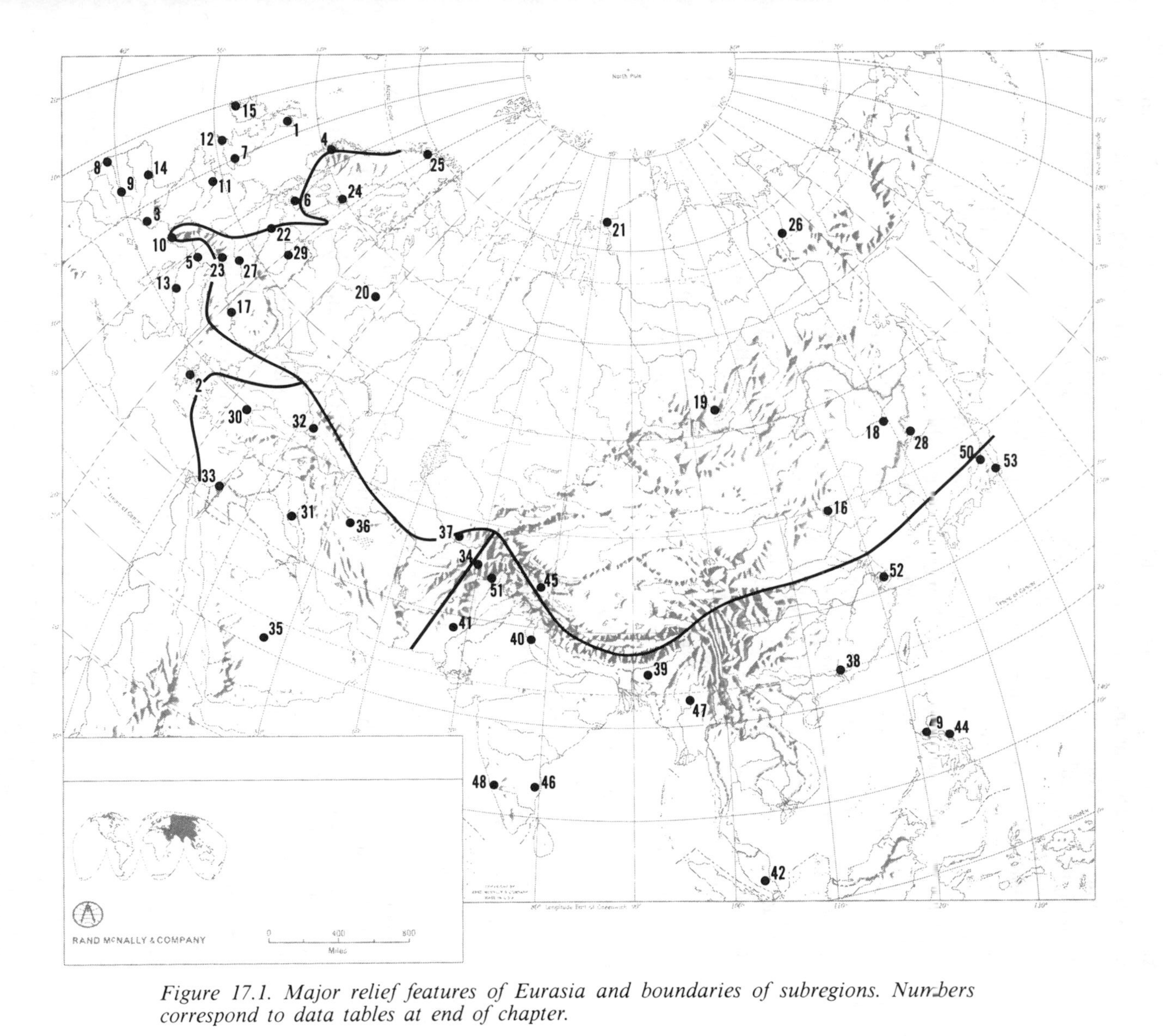

Figure 17.1. Major relief features of Eurasia and boundaries of subregions. Numbers correspond to data tables at end of chapter.

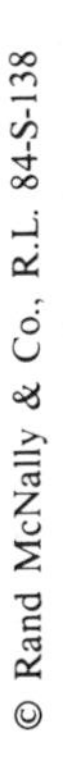

small temperature ranges for the latitude, and generally sufficient and well-distributed precipitation. Europe has a far larger percentage of its total area than any other continent with climate that is suitable for the highest human development. Many crops flourish at considerably higher latitudes in Europe than anywhere else. Barley is grown profitably beyond 70°N in Norway, wheat and vineyards have their farthest poleward extensions in Europe, and subtropical fruits, such as oranges and lemons, grow even beyond 44°N latitude. Date palms are grown in southeastern Spain. Nowhere else do these fruits ripen so far from the equator.

The above characteristics apply only to the most peninsular parts of Europe. Eastward, the plain widens and the climate becomes much more continental. Therefore, the discussion of peninsular Europe will be limited to that very marine-controlled western portion of the landmass where cold month temperatures generally average above freezing and there usually is no winter snow cover (Figure 17.2). This includes territory to the southwest of a line running from the Arctic Circle southward along the western coast of Norway and eastward into the western Baltic along the south shore of Sweden. It then curves southwestward through central Germany and eastern France, keeping the "cold island" of the Alps on the interior side, then east through northern Italy and southeastward into the Balkan Peninsula, keeping the broad coastal areas in milder peninsular Europe, but keeping the interior of the Balkan Peninsula, the Hungarian and Romanian plains, and the Carpathian Mountain countries in the more continental interior. The 0°C isotherm for January enters the mid-section of the west coast of the Black Sea and curves northeastward through the southern part of the Crimean Peninsula. Thus, air temperatures over the shallow northern portions of the Black Sea average below 0°C during January, and this portion of the sea freezes over for part of the winter, while the deeper southern portion remains open.

The 20°C line for the mean annual temperature range might also suffice as the eastern border of peninsular Europe, for it is probably the restricted annual ranges of temperature for such high latitudes that characterize this marine-controlled area better than any other single factor. But this line would include within the maritime fringe such areas as much of the western slopes of the Scandinavian highlands and Sweden south of Stockholm, much of which are covered with a significant snow cover for an unbroken period during winter, and much of the Baltic Sea, the eastern and northern parts of which are frozen over during late winter. In the southeast it runs across the northern edge of the Aegean Sea, which has quite mild winters, the annual range being accounted for by the summer heat in this region.

Although peninsular Europe thus defined is everywhere dominated by maritime polar air practically all the time, and there is no high continuous mountain range blocking the entry of maritime polar air into the interior, some mountains of various orientations and discontinuous natures significantly affect air flow and resultant climate. Therefore, although the limits of climatic variation in peninsular Europe are restricted, within these limits there is great complexity induced by both topography and atmospheric circulation systems.

First, though much of lowland western Europe is characterized by only modest, or even meager, annual rainfall, ranging from around 600 mm. in London (7) and Paris (11) to less than 450 mm. in portions of southern Spain and Greece, windward slopes of some of the highlands, even though not of great altitude, receive more than 3000 mm. per year. Mt. Snowdon in northern Wales, at an elevation of 713 meters, receives 4360 mm. (172 inches) of precipitation, including heavy winter snows. Ben Nevis in the western highlands of Scotland

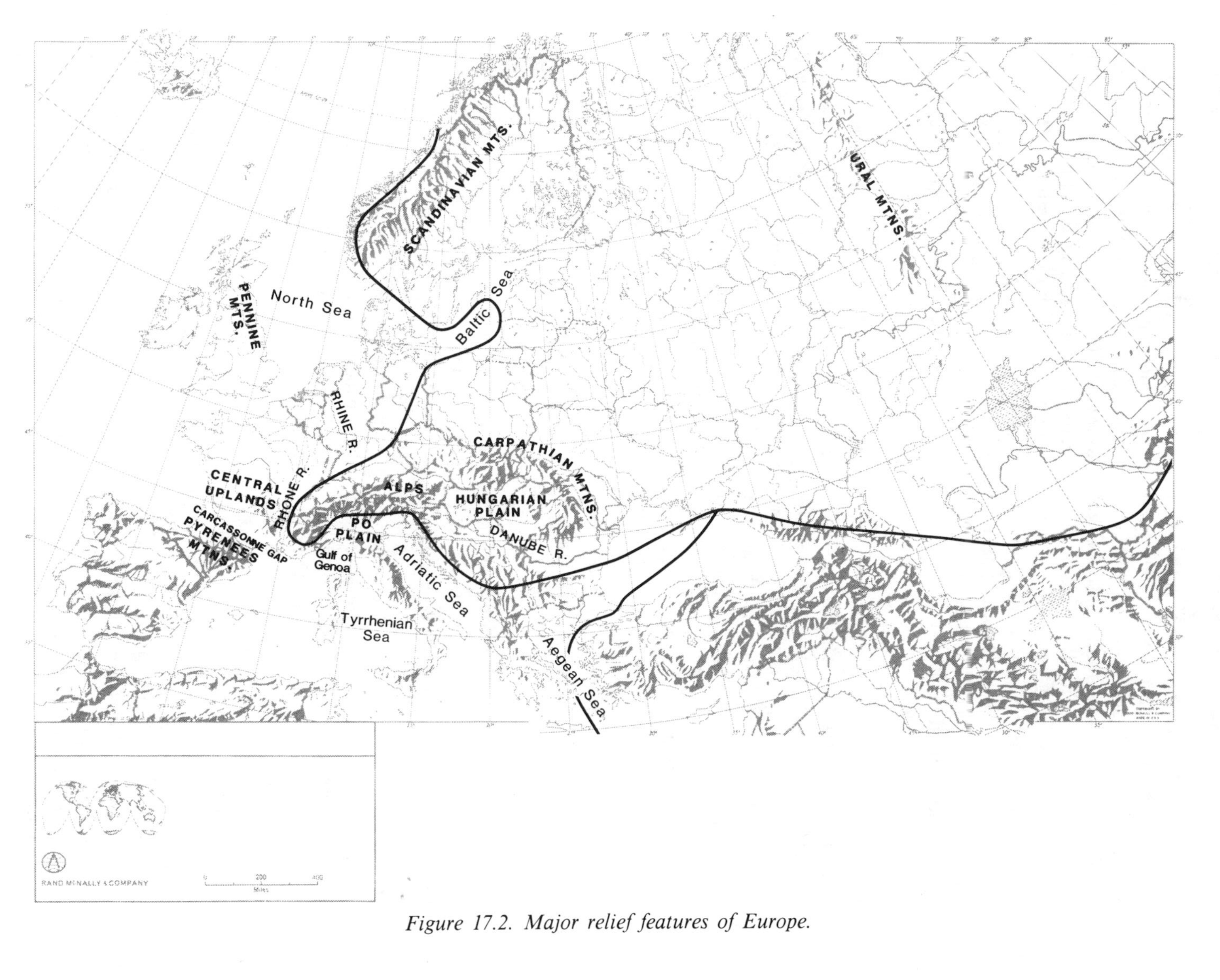

Figure 17.2. Major relief features of Europe.

receives about 3800 mm. at its summit of 1343 meters (4406 feet), but it has been estimated that at the heads of sea lochs facing westward there may be spots that exceed 5000 mm. annually. Precipitation of as much as 4200 mm. has been measured in certain years in small areas of the Central Lake District of England. More than 3000 mm. fall on southwestern slopes of many of the peaks in the Dinaric Alps in western Yugoslavia, portions of the Alps in eastern France and western Switzerland, and many of the coastal highlands of Norway.

In general, the southwestern-facing slopes, and particularly heads of valleys, cause the air to converge as it moves upward and receive heavy precipitation, for much of the precipitation in these areas is caused by the leading northeastern edges of cyclonic storms, which produce long periods of overcast skies, fog, and almost constant precipitation on windward mountain slopes.

Many of the small uplands of Germany may receive as much as 1680 millimeters of precipitation per year, but downwind on their eastern sides precipitation may drop to less than 500 mm. Therefore, the highlands are generally forested, and the lowlands are in cultivation. Throughout history the forest vegetation, more than the upland topography, has impressed the German people, as shown by German folklore and placenames (such as the Black Forest).

In the mountainous peninsulas along the northern fringe of the Mediterranean Sea, lowlands and highland slopes alike are generally wetter and milder during winter on west sides than on east sides. Thus, in northern Italy, Genoa on the protected southwestern side of the Appenine Mountains averages 8°C in January and receives 1295 mm. of precipitation per year, whereas Bologna (5) in the Po Plain northeast of the Appenines averages only 2°C in January and receives 589 mm. of precipitation per year. Annual precipitation amounts in similar topographical locations generally decrease southward toward the Mediterranean, because cyclonic storms that are active year-round in northwestern Europe are usually absent in the Mediterranean during summer, when the Azores High spreads northward and eastward. This, of course, alters the precipitation regime as well.

Precipitation Regimes

Europe can be divided into three main regions according to precipitation regime (Figure 17.3). Much of coastal northwestern Europe from Norway to northern Spain receives adequate precipitation throughout the year, with the maximum occurring in fall and the minimum in spring. This includes much of Great Britain, except for the London Basin, which, being more interior, tends toward a slight summer maximum, and the extreme northwest and southwest coastal areas, which tend toward a winter maximum, as does western Ireland. This regime relates primarily to sea-surface temperatures, which are warmest in fall and coolest in spring, although cyclonic storms in this area are also most frequent in fall. A major portion of the precipitation associated with the passage of cyclonic storms comes in the form of showery weather from cumulus clouds formed by convection in cold air masses behind cold fronts as they sweep southeastward across warmer sea. In some areas this kind of precipitation might even exceed that caused by widespread stratiform clouds ahead of warm and occluded fronts.

Much of the British Isles and the immediate coastal areas of mainland Europe have been noted for their steep temperature lapse rates in the lower portions of the troposphere. The tree line is unusually low in the Pennine chain of England and the uplands of Scotland and Wales, so the windswept heath and moor occupy exceptionally large areas. This steep lapse rate apparently is due to the surface warming that takes place in the marine air when it approaches Europe across the warm water

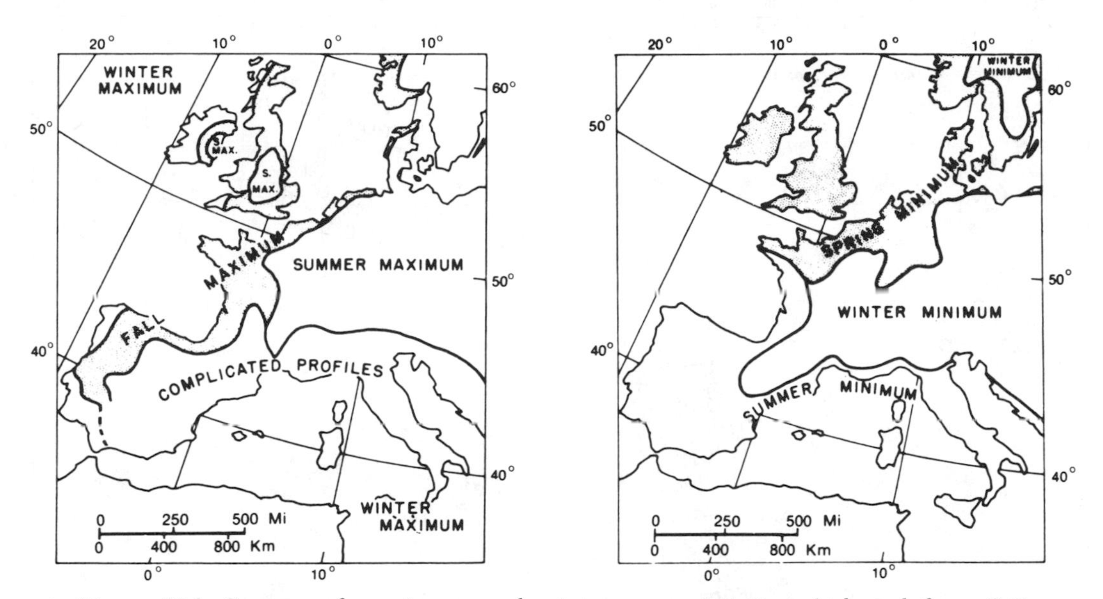

Figure 17.3. Seasons of maximum and minimum precipitation. (Adapted from G.T. Trewartha, The Earth's Problem Climates, *2nd ed. [University of Wisconsin Press, 1981], pp. 248–49. © University of Wisconsin.)*

of the North Atlantic Drift. Such evidence further corroborates the notion that convection is very active in cold, fresh marine air following cold fronts. Throughout much of Britain, about half of the precipitation is associated with warm fronts and warm sectors of cyclonic storms, 25 percent to 50 percent with instability in cold air behind cold fronts, and only about 3 percent with thunderstorms.

Inland from continental Europe's west coast, the precipitation generally remains adequate throughout the year, but the regime quickly turns to a summer maximum and a winter minimum in eastern France, Germany, and Scandinavia. Such a regime is common of continental interiors where summer convective activity produces greater amounts of rainfall than do winter cyclonic storms. Nevertheless, winter generally has the most frequent precipitation, although of a prolonged, light-intensity nature, and is by far the cloudiest period of the year.

South of a line running through northern Spain, southern France, northern Italy, northern Yugoslavia, and southern Bulgaria, summers become distinctly drier, with inadequate precipitation throughout the year (Figure 17.4). This, of course, is a consequence of the northward expansion of the Azores High, which during the winter has been lying over the Sahara to the south. The entire Mediterranean borderlands, including the European peninsulas on the north, the African coast on the south, and the Near East on the east, receive more precipitation during the winter half-year than during the summer. Yet regimes here are generally complicated. Only the African and Near Eastern coastal areas have continuous stretches of land that receive more rain during the winter quarter than during any other quarter of the year. Much of southern Europe has varied and complicated regimes, some with a single fall maximum, some with a fall-winter maximum, some with a winter-spring maximum, and a broad zone running from north-central Spain through northern Italy into northern Yugoslavia and northeastward that has double-equinoxial maxima in spring and fall. This zone appears to lie in a transition

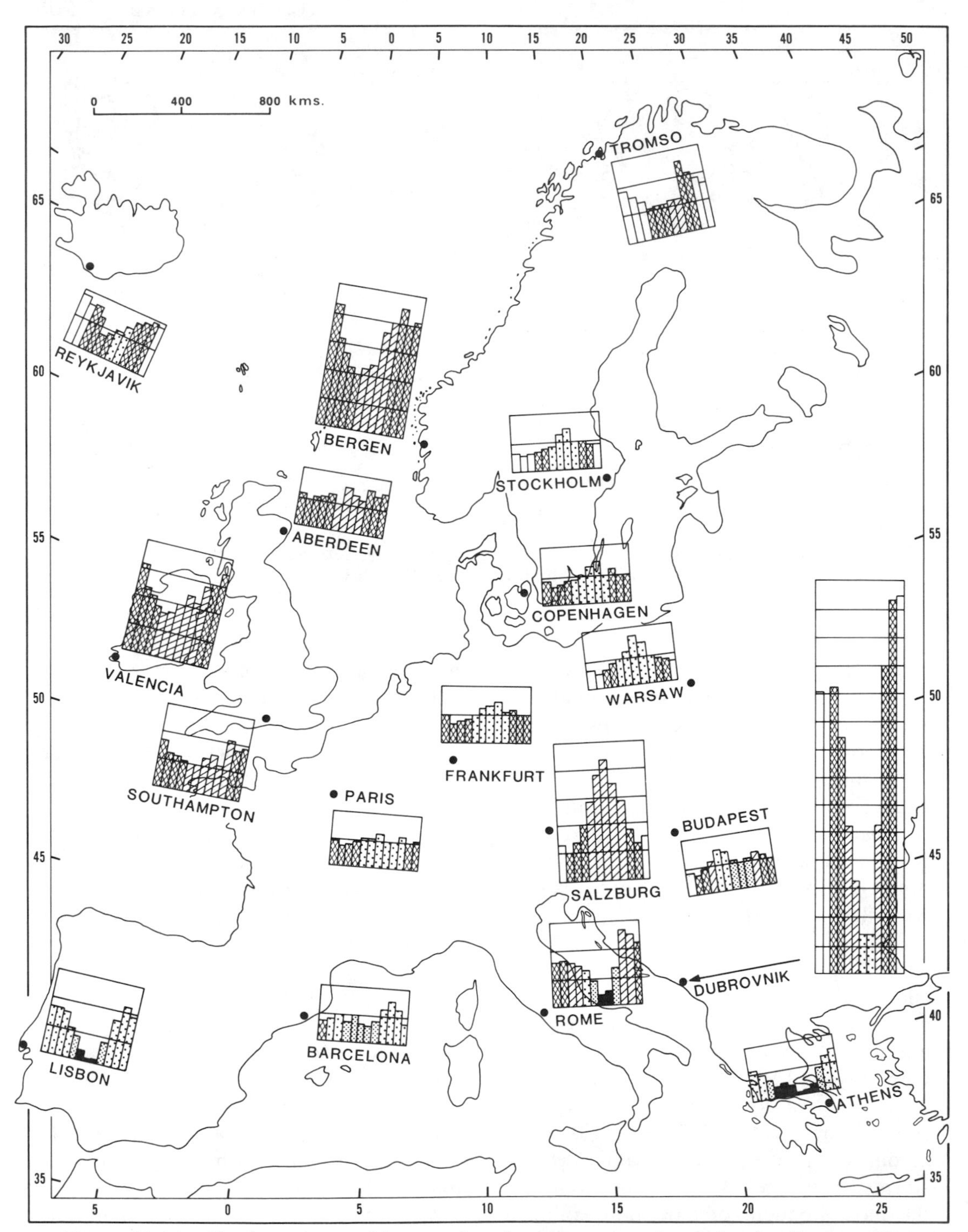

Figure 17.4A. Yearly regimes of heat and moisture resources in peninsular Europe. (From Agroclimatic Atlas of the World.*) Key on opposite page.* *See page 258 for figure 17.4B; see page 260 for Figure 17.4C.*

belt between the summer Azores High in the Mediterranean and the winter Asiatic High north of the mountains of Europe. Since these two high pressure systems are best developed during their respective solstice seasons, these periods are the driest in this transitional zone.

Almost the only consistent feature of the precipitation regimes in the entire Mediterranean borderlands is that practically everywhere the summer quarter is the driest. But summers are not absolutely dry here, as they are in southern California. Only the Atlantic fringes of Iberia and northwest Africa, under the direct influence of the cold Canaries ocean current, approximate the dryness of the California summers. Lisbon (8) receives only 3 millimeters of rainfall in July and 4 in August, and Casablanca receives no measurable rain at all in July and only 1 mm. in August. But since the Mediterranean Sea is unusually warm year-round, cyclonic storms form there even in summer. The Mediterranean is always warmer than the Atlantic at the same latitude, so it unstabilizes all arriving air masses except hot air from the Sahara. By far the bulk of the summer precipitation that falls in the mountainous peninsulas of southern Europe is produced by convective activity when the air becomes conditionally unstable. Thus, although at Rome (13) July and August are the driest months of the year, July still receives 14 mm. of rainfall, and August, 22.

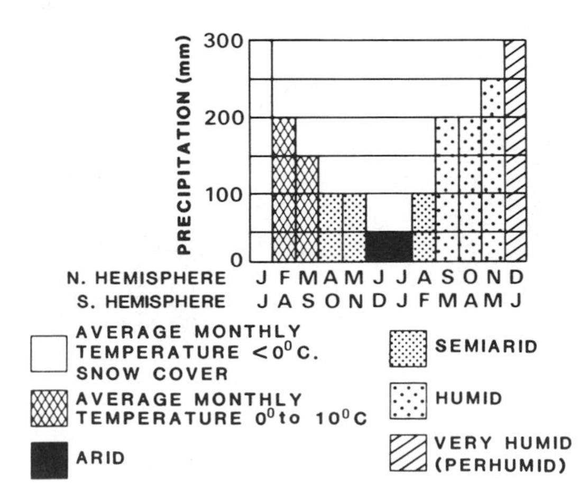

The borderlands of the Mediterranean Sea and eastward through the moister portions of the Near and Middle East are the greatest areal extent on earth of subtropical, dry-summer climate, and for this reason this climatic type has often been called the "Mediterranean climate." Yet this area has the least classical form of subtropical, dry-summer climate; southern California, middle Chile, and the southern tips of Africa and Australia have the near-rainless summers, overcast stratus, and fog which are typical of this climatic type. The Mediterranean region generally has much sunnier weather throughout the year, but particularly in summer, much warmer seaside temperatures, less atmospheric stability, and an appreciable amount of precipitation during the summer half-year, much of which comes in the form of thundershowers.

Cyclogenesis and Cyclone Routes

During winter the Icelandic Low builds up over the North Atlantic and stretches eastward along the Arctic coast of Scandinavia and the Soviet Union as far east as the central Siberian coast (see Figure 6.4). Cyclones form in this area and move eastward and southeastward along two main tracks (see Figure 10.7). One goes directly east along the fringes of the Arctic, and the other goes far to the south and traverses the Mediterranean and Middle East. Sometimes cyclones follow a subsidiary route through the length of the Baltic and join the Arctic route in the Barents or Kara seas. Central Europe is generally less frequented by cyclones during winter because the Asiatic High often extends westward to fill up much of the European landmass. Nevertheless, fronts and secondary waves are common throughout all of Europe during winter.

The great majority of disturbances in

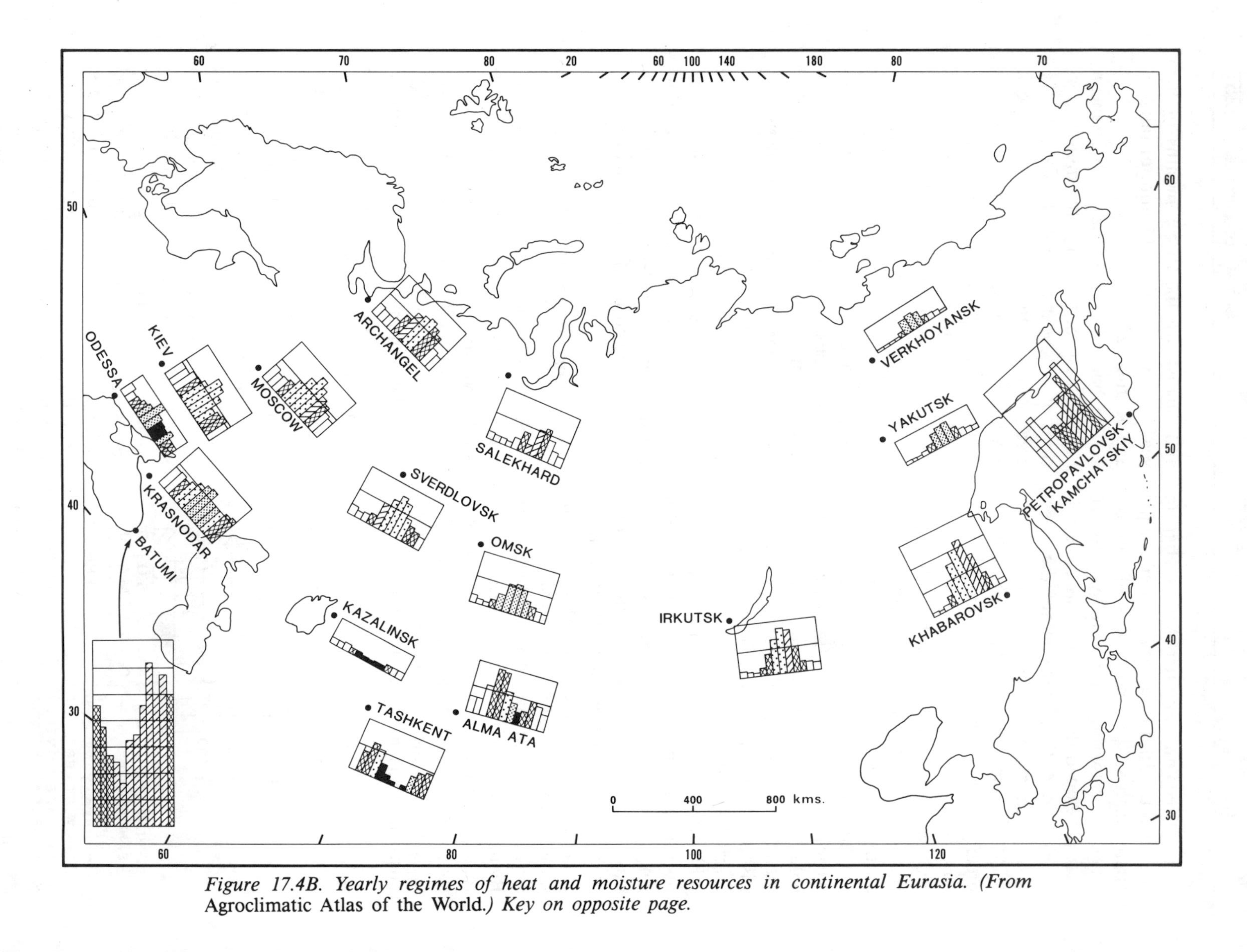

Figure 17.4B. Yearly regimes of heat and moisture resources in continental Eurasia. (From Agroclimatic Atlas of the World.*) Key on opposite page.*

the Mediterranean develop as secondary depressions when primary disturbances lie farther north over western and central Europe. But some enter the Mediterranean as primary systems moving south or southeastward across France, or eastward across Iberia, or through Gibraltar Strait. Other depressions form independently over the Mediterranean itself.

Two preferred areas of cyclogenesis within the Mediterranean region are the Gulf of Genoa along northwestern Italy and the lee (south) side of the Atlas Mountains in North Africa. Other important areas of cyclogenesis are the head of the Adriatic Sea and the eastern Mediterranean around the islands of Crete and Cyprus. The Genoa and Adriatic storms generally move southeastward along their respective coastlines and then curve east into the Near and Middle East or northeastward through the Black Sea and Caucasus areas. The south Atlas storms generally enter the eastern half of the Mediterranean off the Tunisian coast and continue to follow the warm water eastward into the Middle East.

Much of the cyclogenesis in the Gulf of Genoa occurs when outbreaks of cold air through the Rhone and Carcassonne gaps in southern France interact with warm air over the sea to set up counterclockwise circulations. And the continuation southward of the cold air over the warmer sea causes a great deal of convective shower activity in cold air behind cold fronts. Many climatologists claim that this type of showery activity in the cold air brings a greater amount of rainfall to the Mediterranean region than warm frontal rain does. Similar effects are felt in the eastern Mediterranean when cold air surges through the Marmar Gap between the Black and Aegean seas.

During winter, localized high pressure cells tend to form over the colder peninsulas of Iberia, Italy, and the Balkans, while the intervening warm seas tend to generate low pressure centers. Thus, there are often many local circulations superimposed on the general circulation of the area.

During summer, the Icelandic Low is very much weakened, and cyclonic activity in general is far weaker. Most of the cyclones follow a northerly route through Scandinavia and northern portions of the Soviet Union (see Figure 10.8). The Asiatic High no longer exists, and fronts and secondary waves penetrate the European landmass easily and, in conjunction with added convective activity, bring a maximum of precipitation to the interior. The Mediterranean is relatively free of cyclonic passages during summer, but northwesterly air passing through the Carcassonne Gap in southern France may set up cyclogenesis in the Gulf of Lion.

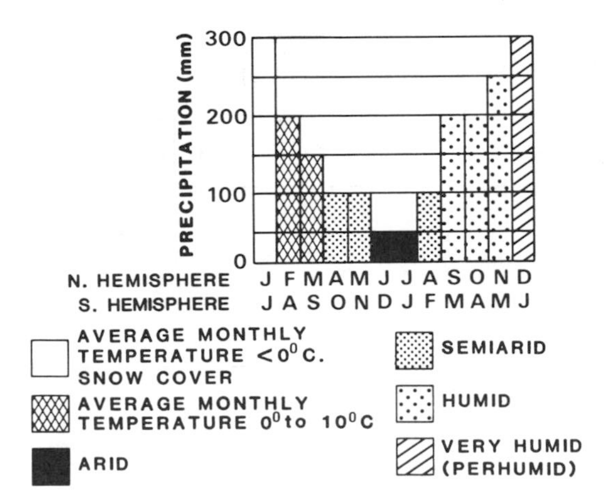

Blocking

Frequently the normal westerly flow at the 500-millibar level is interrupted in the eastern North Atlantic somewhere in the vicinity of 5°W to 15°E by the formation of a quasi-stationary high, which splits the jet stream into two branches, one flowing northward around 75°–80°N, and the other southward into the Mediterranean around 36°N. This accounts for the two major tracks of cyclonic storms. This kind of blocking occurs from less than 20 percent of the time during summer to more than

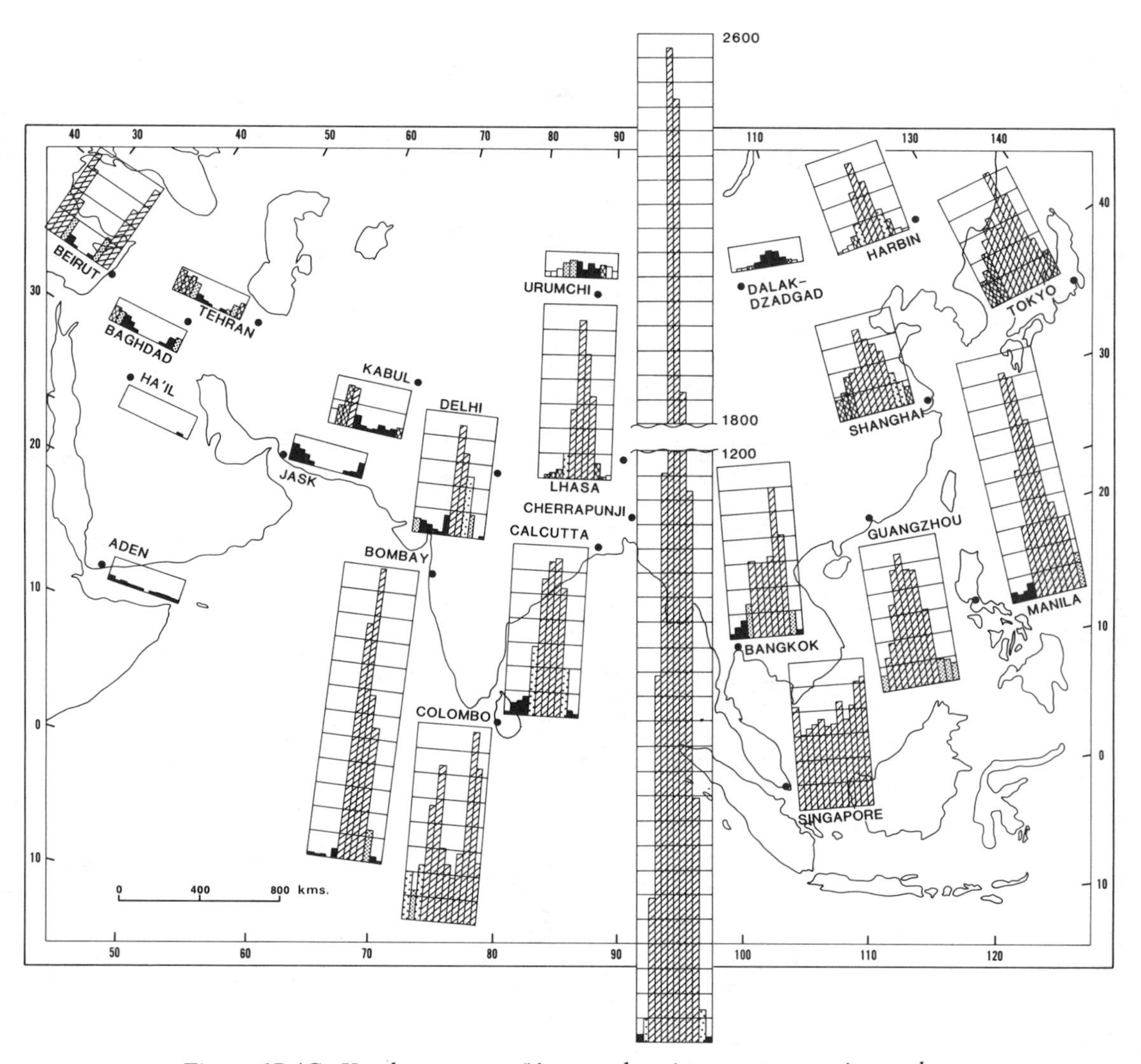

Figure 17.4C. Yearly regimes of heat and moisture resources in southwest and monsoon Asia. (From Agroclimatic Atlas of the World.*) Key on opposite page.*

40 percent during April, the month of maximum occurrence. On such occasions large north-south movements of air bring very contrasting temperatures and weather to adjacent longitudes of western Europe. Typically, it becomes abnormally warm over the eastern Atlantic, the British Isles, and Scandinavia, and abnormally cold in central and southwestern Europe. When the localized blocking high forms over snow-covered Scandinavia during winter, much of western Europe can experience its coldest weather from the northeasterly flow of Arctic air along the southeastern periphery of the high pressure cell.

Local Winds

Because of the intricate topography and interdigitation of land and water, particularly in southern Europe, local wind systems abound, many of which have become famous. The foehns of the northern Alps are world-renowned for the formation of thermal belts on the alp terraces above the deep, U-shaped, glaciated valleys. Here the foehn phenomenon was first identified and the German word for "fall" or "descent" applied to it. Foehns in this region occur primarily with southerly winds associated with the leading edges of eastward-moving cyclonic storms across northern Italy. During these times, strong temperature inversions are formed on alp surfaces between the descending air from above and the stagnant cold, humid, cloud-and-fog-enshrouded glacial valleys below. Under such conditions air pollution becomes a major problem in heavily populated and industrialized valleys a little farther north. Foehns along the northern slopes of the Alps are particularly prevalent during autumn and spring, and they frequently cause disastrous spring avalanches because the snow on the upper slopes is melting so rapidly.

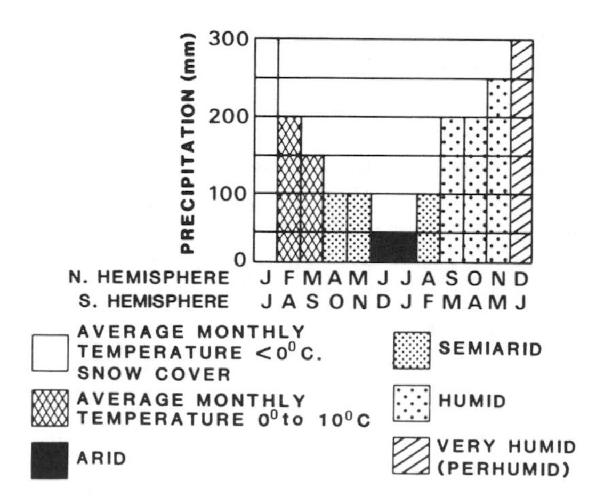

Foehns also occur on the southern slopes of the Alps to the rear of southerly cyclonic storms, but during winter the warm descending air generally remains above the cold, humid surface air in the Po Plain of northern Italy, and the Italians have no such term as foehn. The cold surface air occupying the Po Valley during winter is also frequently overridden by cyclonic systems crossing from west to east. Precipitation from these upper systems frequently saturates the underlying air and produces fogs that persist for 10–20 days per month during winter.

Foehns occur in many other parts of mountainous southeastern Europe, but generally in less spectacular form than along the northern Alps. They are also common down the western slopes of the Scandinavian mountains, where again they may form strong temperature inversions between the warm descending air and the cold air trapped in fiords below. Sometimes the descending air in this area is so cold initially that it arrives at sea level colder than sea-surface temperatures along the Norwegian coast. Then it is known as a *bora* wind. Under such conditions steam fog forms.

During winter, cold air frequently builds up in a localized high pressure cell over the Swiss Plateau north of the Alps and eventually spills around either the western or eastern end of the Alps to descend as a cold bora wind. Although the descending air warms adiabatically, it is originally so cold that it is still abnormally cold for the Mediterranean coast when it arrives at sea level. On the western end of the Alps it funnels down the narrow, down-faulted valley of the Rhone River between the Jura Alps and the Central Uplands of France. It may reach the Riviera with speeds of 20–30 meters per second and cause a great deal of consternation to winter vacationers. In this region it is known as the *mistral.* Similar occurrences take place at the head of the Adriatic Sea when cold air breaks southwestward through the Carnic Gate between the Julian Alps to the north and the Dinaric Alps to the south in western Yugoslavia.

Speeds of more than 35 meters per second (78 mph) are a frequent occurrence with bora winds in the Gulf of Trieste, where the word "bora" was originated. Gusts reach speeds of 60 meters per second (135 mph). Waves on the Adriatic may build up to 2.5 meters in height, which temporarily brings all shipping to a standstill.

A succession of wind systems is commonly associated with eastward-moving cyclones in the Mediterranean. On the leading edge of the storm, southerly winds out of Africa are relatively hot, dry, and dusty. These are known as the *sirocco.* They may pick up considerable moisture as they move northward across the Mediterranean and arrive in northern Italy and southern France as unpleasantly humid, dusty air. As the cyclone passes eastward the northerly air on the rear side pulls in the bora winds from the north. A similar sequence of events may take place along the Adriatic coast of Yugoslavia, where the *jugo* (south) wind precedes the storm and the bora follows it. Since the jugo is not gusty, as is the bora, but rather blows hard and continuously, waves created on the Adriatic may be even higher than those created by the bora, more than 3 meters in height.

Throughout the eastern Mediterranean and continuing eastward through the Near and Middle East, the predominant winds during summer are off the land from the north. In most cases they have subsided and warmed adiabatically, causing clear, dry conditions. These winds are known as the *etesians.* Since they occur in more classic form in the Middle East, further discussion of them will be delayed until that region is under consideration.

Climatic Types

Almost all of peninsular Europe, except the highest mountains, falls into either temperate oceanic (Do) or subtropical dry-summer (Cs) climate categories (see Figure 15.3). These reflect the high degree of control by maritime air, which largely accounts for the mild winters at such high latitudes and the two primary atmospheric circulation systems: the prevailing westerlies with their cyclonic storms in the north-central area, and the Azores subtropical high pressure cell in the south, which shifts northward to occupy the Mediterranean region during summer. The other circulation system which influences western Europe, the Asiatic High in winter, only peripherally affects the climate of peninsular Europe, mainly through alteration of the winter westerlies and their cyclonic systems. (It will be dealt with in much more detail in the next section on continental Eurasia.)

The temperate oceanic climate of northwestern Europe is characterized by its consistently cool, cloudy, moist air and prolonged light rain. Many areas experience measurable precipitation on more than half the days of the year. This, coupled with low evapotranspiration rates, generally makes the moisture balance quite adequate despite only modest rainfall totals at low elevation. These conditions have been conducive to the growth of mixed forests, predominantly coniferous in the north and deciduous broadleaf in the south (Figure 17.5). Of course, in this area of long-time human habitation, much of the land has been under cultivation for centuries, and generally the forests are limited to the rougher topography of the low mountains

Figure 17.5. The lush, green vegetation of the British Isles under stratocumulus overcast.

(Figure 17.6). Although the growing season is very long in much of northwestern Europe, it is generally cool, particularly in Britain and the coastal areas of the mainland, so that crops are limited to, for example, hay, small grains, and root vegetables, plus vineyards in the thermal belts of France, southern Germany, and northern Italy.

In stark contrast, the subtropical dry-summer climate of the Mediterranean borderlands is noted for its sunny, stable air conditions, particularly during summer when for two or three months there is only minimal precipitation. During winter, cyclonic storms frequent the area and bring rainfall in the lowlands, snow in the highlands, and locally strong winds. But even during winter, fair weather dominates the region and storms occur only occasionally. The long droughty summer is conducive to the growth of drought-resistant brushy plants known as *maquis,* the counterpart of chaparral in the Cs climatic region of California. Since humans have inhabited this portion of Europe for thousands of years intensive cultivation of high-value crops has been introduced wherever water is sufficient. Some of the grains can be dry-farmed utilizing the winter rains, but the widespread growth of citrus, vineyards, and other fruits and vegetables is achieved by irrigation. As in all the Cs areas of the world, the topography of this region is mountainous, so adequate amounts of winter precipitation, much in the form of snow, are usually available in the highlands for the irrigation of the lowlands the following summer.

Continental Eurasia

Definition of Area

In direct opposition to the exceedingly temperate marine conditions of peninsular Europe is the vast expanse of land extending eastward all the way to the Pacific

Figure 17.6. Deciduous forests in low mountains of the Odenwald, in southern Germany.

north of the high southern mountains, herein designated as "continental Eurasia" (see Figure 17.1). Winter isotherms in Europe generally parallel the coastlines in the west and are oriented north-south through much of the East European Plain, thereby signifying greater influences of sea-land contrasts than latitudinal controls on temperature. The result is that whereas much of Europe is abnormally warm for its latitude during winter, southeast European Russia and all of Asia north of the high mountains is abnormally cold for its latitude. The zero isanomaly of temperature during January runs from the Ob Gulf in western Siberia southwestward across the Urals, just south of Moscow, and then southward to the eastern end of the Black Sea (Figure 14.3). Everything west of that line is warm for its latitude in winter, and everything to the east is cold. The negative temperature anomalies increase eastward into the northern part of the Soviet Far East, where the greatest negative temperature anomalies on earth are found.

The western border of this region in central Europe is difficult to define since the topography is varied, and although the area to the west is abnormally warm for its latitude, the latitude is high, projecting well above the Arctic Circle in the north, and therefore the winters are relatively severe. In the north European plain of Germany and Poland, the maritime climate of the west changes gradually over a broad zone into the continental climate of the east, since there is no defining topographic barrier. Thus, an imaginary boundary between peninsular Europe and continental Eurasia approximates the western border of at least one month of consistent snow cover during winter (Figure 17.1).

In the south the border has been drawn along the northern slopes of the high mountains of the Middle East and southeastward across the high mountain and plateau area of the Pamir Knot on the border of the USSR, India, and China, and on southeastward along the northern slope of the Himalayas, keeping the high Tibet Plateau and the low interior basins to the north with their exceedingly cold winters within continental Eurasia. In the Far East, where the mountains are lower and less oriented, it is difficult to draw a boundary. It has simply been extended northeastward across north-central China somewhat in line with the low Dabie Shan (Tsinling Mountains) and then eastward across southern Korea and northern Honshu Island in Japan, much as the C-D boundary in Figure 15.3. This boundary somewhat delineates the region to the north, which is overwhelmingly dominated by the Asiatic High during winter, and the region to the south, which is more strongly affected by a monsoonal circulation during summer, although this transitional zone is quite broad in the Far East. Continental Eurasia is considered to extend clear to the Pacific coast in northern Asia and even to include peninsulas and islands offshore, because they are so influenced by winds blowing off the continent during winter that their winter temperatures are abnormally cold. The zero isanomaly of temperature in the Pacific during January runs south-southwest–north-northeast through the Kuril Islands, the eastern coast of Kamchatka, and the very northeastern corner of the Soviet Union west of Bering Strait (Figure 14.3).

In such a broad expanse of land, of course there are great differences in climate. Average January temperatures vary from −3°C at Warsaw (29) in the west to −50°C at Oymyakon in the northeast, and annual precipitation varies from less than 100 mm. in portions of Soviet Central Asia to more than 1300 mm. on Kamchatka Peninsula in the Soviet Far East. But generally these variations are gradual over broad zonations, affected only by different sides of the same atmospheric circulation features, whereas along the southern and western borders rapid transitions occur in climatic characteristics and the circulation features controlling them. The southern fringe, with its high mountains and intermontane basins, shows a great complexity of climatic features. This rugged mountain belt itself might be considered as a climatic region separating the plains and subdued uplands to the north from the monsoon lands of southeast Asia. It is difficult to draw a boundary anywhere, and the boundary that has been drawn leaves some peripheral areas in question. To the south of the boundary many high mountains have below-freezing temperatures much of the year. In general, however, this line runs along major topographic breaks that emphasize divisions between major atmospheric circulation systems.

Primary Controls

The overwhelming control on the climate over most of this region is continentality. This can hardly be overemphasized. The greatest marine influence for most of the area comes from the west, and the Atlantic is a long distance away. The western margins of this region, particularly portions of Scandinavia and the fringes of the eastern Baltic, show considerable maritime influence, but these are high-latitude areas with cold, snowy winters that stand in stark contrast to coastal margins farther west. The Atlantic air penetrates the continent most easily during summer when a diffuse surface low pressure occupies much of the landmass. At this time of year precipitation derived from the Atlantic moisture source may penetrate eastward into central Siberia and occasionally all the way to the Pacific. Net moisture flux across the Pacific coast is from west-to-east throughout the year. But during winter the cold Asiatic High occupies much of the landmass, and Atlantic air has greater difficulty penetrating into Asia.

Some maritime air from the Arctic penetrates the landmass, but this has little moderating effect except in the western Barents Sea, which remains ice-free through the year. Eastward in winter the frozen Arctic seas climatically act more as snow-covered frozen land surfaces than sea surfaces. The primary effect of the Arctic along the Siberian coast is to maintain strong temperature gradients (which reverse themselves seasonally) that produce consistently strong winds across the coastal area. During winter, even though the Arctic water is frozen to an average depth of about 3 meters right up to the coast, significant amounts of heat are conducted upward through the ice from the unfrozen water underneath. As a result surface air temperatures over the Arctic ice are about 15°–20°C higher than over the landmass to the south, which induces strong winds to blow from south to north—from the colder snow-covered landmass onto the frozen Arctic ice—and causes extreme wind chills along the coast. Even though actual temperatures there during January average 8°–12°C warmer than in the interior of Siberia, and absolute minimum temperatures are as much as 20°C warmer than in the interior, the wind chill temperatures along the coast may plunge below −150°C, while the coldest in northeastern Siberia are around −86°C, and in the very calm core area of the Asiatic High in southern Siberia and neighboring Mongolia they reach only about −60°C. During summer the temperature gradient is reversed across the Arctic coast: the surface water temperature remains around freezing as ice floes melt, while temperatures inland warm rapidly under continuous sunlight during June and early July. Strong, cool sea breezes blow from north to south and carry the moist Arctic air inland, where under rapid surface heating it produces sharp, brief convective showers.

Marine air from the Pacific generally affects only the coastal area of eastern Asia and offshore islands, because in most places from Korea northward, rugged mountains come right down to the coast. During middle and late summer a semblance of a southeast monsoon from the Pacific brings heavy precipitation to the coastal slopes of the mountains as far north as the northern coast of the Sea of Okhotsk, and this Pacific air may penetrate inland at the surface several hundred kilometers up the Amur Valley with easterly surface circulations associated with the leading edges of cyclonic storms that move east-northeastward along the Mongolian section of the Polar Front. This surface Pacific air does not bring rain to the area inland from the coastal mountains, but acts simply as the cold air mass over which rises warmer air from the southwest that might have originated as far away as the Bay of Bengal or the South China Sea. (This type of circulation affects southern China much more than northern China and therefore will be taken up in the section on monsoon Asia.) During winter even the offshore islands and peninsulas have abnormally cold temperatures because of the consistency of cold land air blowing off the continent from the northwest.

Along the southern fringe of continental Eurasia, hardly any maritime air is able to penetrate the region from the western end of the Caucasus to central China. Since there is essentially no true maritime tropical air anywhere in the region, the absolute moisture content of the air is rather low. In the cool northern half of the area, the low absolute humidity has no significant consequence because evapotranspiration rates are low, but in the southern half of the region semiarid and arid climates prevail all the way from the southern Ukraine north of the Black Sea, eastward across Soviet Central Asia and Kazakhstan, and into the mountain-enclosed basins of western and northern China and Mongolia. Only the higher mountains of this southern area show evidence of a positive moisture balance.

Thus, air may filter into this landmass

from the west, to some extent from the north, and a little bit from the east, but whatever its origin it generally stagnates over the interior and gradually takes on the characteristics of the land surface. All air masses eventually are modified to what the Soviets call a "temperate continental" air mass, which is locally derived. This would probably be called continental polar at similar latitudes in North America but, as the Soviets point out, this is a misnomer since the air is not derived primarily from the polar region. The word "temperate" is not too accurate, either, since it is used ambiguously in so many ways, but perhaps it is appropriate since "temperate" is generally understood to relate to average conditions and is allowed to include wide-ranging extremes.

Seasonal Temperature and Pressure Reversals. During winter, cold air settles into the interior basins of southeastern Siberia, Mongolia, and northern China. Radiational heat losses during the long, clear winter nights form strong surface temperature inversions that may extend upward 2–3 kilometers into the air and have temperature differences of 20°C or more between base and top. As was explained in Chapter 9, even though the air in the inversion is much warmer than the air below and radiates heat downward, there is still a net loss of heat upward until a very cold equilibrium temperature is reached because of the high efficiency of the radiation of the snow-covered surface that radiates much as a black body across a continuous spectrum of terrestrial wavelengths. Under such conditions, the coldest temperatures on earth at low elevations are produced in mountain-enclosed river basins in northeastern USSR. As has already been mentioned in Chapter 14, Verkhoyansk (26) at an elevation of only 137 meters averages −49°C during January and has experienced an absolute minimum of −68°C. A short distance away at an elevation of 740 meters, Oymyakon has experienced a minimum temperature of −71°C (−96°F).

The upward and downward radiation usually come into balance when the surface temperature drops to about −50°C, after which the surface temperature remains about the same unless it is disturbed by some advection process, which starts the radiational adjustment over again. In northeastern Siberia, equilibrium temperatures may be reached as early as late November, after which temperatures remain about the same for four months or more. These are the "coreless winters" in which temperatures are not arranged symmetrically around some coldest point. Winter becomes the longest season in these northern areas, summer is short and peaked with unusually high maxima temperatures for such high latitudes, and the transitional seasons become very short. Spring, in particular, is almost nonexistent, since the surface air temperature does not rise much above freezing as long as there is snow to melt, ground to thaw, and meltwater to evaporate. By the time all this has been accomplished, the region is well into the June solstice, with its 24-hour daylight, and surface air temperatures shoot up almost instantly.

During winter cool pools of air fill up the intermontane basins and form isolated domes of high pressure underneath inversions, across which transitory circulation systems, such as fronts, may pass without descending to valley floors. The calm surface air within these basins becomes very stagnant and is so cold that even the breath of animals or the combustion of fuel in the stoves of rural huts may saturate the lower air and cause a pall of fog and smoke to hang over settlements much of the winter. The condensed moisture freezes into persistent ice crystal fogs which are continually added to by the downward motion of moisture from warmer inversion air, which has higher vapor pressure. The atmospheric conditions are so calm that one can discern the slight rustling sound of descending ice crystals. In the cold, starry, 24-hour nights of northeastern USSR the native Yakuts refer to this rustle as the "whisper of the stars."

Many of the basin floors in Mongolia and adjacent Siberia and China lie at elevations of 1000 meters or more. When the observed pressures in the cold, dense air in these basins are corrected to sea level using equations based on the standard atmosphere, exaggerated sea-level pressures are obtained, ranging as high as 1075 millibars. Using this procedure, January sea-level pressure maps show an extensive high pressure cell that is usually centered in the intermontane basins of northwestern Mongolia and occupies much of Asia north of the high southern mountains. It often protrudes westward north of the Caspian and Black seas into central Europe (see Figure 6.2), but this is a hypothetical construct rather than a coherent, dynamic circulation system. In the core area many of the basin floors lie 1000–2000 meters above sea level, and the surrounding mountain ranges rise well above that. So the derived sea-level pressures do not accurately represent surface conditions. More representative would be the 850-millibar chart that shows northern Asia and the western Pacific under the influence of an extensive pressure trough stretching from the eastern portions of the Icelandic Low in the Barents Sea, southeastward to the western portions of the Aleutian Low in the Sea of Okhotsk (see Figure 6.17).

This pressure configuration intensifies upward through the troposphere. Thus, during winter northeast Asia and the adjacent Pacific are under the influence of the second most pronounced trough in the standing waves of the upper troposphere, which feeds cold Arctic air southeastward along the western limb of the trough into much of northern and central Asia and eastward across the Pacific coast (see Figure 6.13). This cold air intensifies in the interior basins through radiational exchanges and spills over southward and eastward to dominate much of eastern Asia southward to about the Yangtze Valley of China with cold northwest winds during winter. The winds are particularly strong along the mainland coasts of the Sea of Okhotsk and the northern portion of the Sea of Japan, where coastal mountains allow the cold air to build up behind them and then spill over with great force down to the relatively warmer seas below.

As was the case along the Arctic coast, even though much of the Sea of Okhotsk and the northern portion of the Sea of Japan are frozen over during winter, air temperatures above the ice are much warmer than those over the adjacent landmass. Thus, a strong temperature gradient, pressure gradient, and wind system are common in the vicinity of the coast. The cold northwest winds continue on across the seas and strongly affect Kamchatka Peninsula, the Kuril Islands, Sakhalin Island, Hokkaido Island, and the western coast of Honshu Island in Japan. Where the winds are funneled around mountainous ends of islands and peninsulas, wind chills below −100°C are sometimes experienced.

Abnormally cold temperatures are maintained throughout the winter even in coastal areas. Vladivostok (28), near the southernmost tip of the Soviet Far East, at a latitude of 43°N averages only −15°C in January, while Boston, Massachusetts, at a similar latitudinal and coastal location averages −3°C. Even an interior station in the US at the same latitude, such as Milwaukee, Wisconsin, averages −6°C. At comparable latitudes, then, the east coast of Asia, on the average, is considerably colder during winter than interior portions of North America. Whereas maritime tropical air vies with continental polar air for control of eastern North America, Arctic air reigns supreme in eastern Asia. There is very little penetration of the mainland by Pacific air at any time during winter.

Since the southern two-thirds of the Sea

Figure 17.7. Thaw lake and hummocks over permafrost near Yakutsk, in northeastern USSR. Coniferous forest in background. (Courtesy of Sam Outcalt.)

of Japan generally are open water during winter, the cold northwest winds pick up much warmth and moisture as they cross the sea to produce strong instability showers that dump copious amounts of snowfall on the coastal mountains of northwestern Honshu. But on the mainland of Asia winter snows generally are very light, particularly in the Baykal-Amur-Mongolian region where, often, extensive areas of intermontane basins remain without snow cover much of the winter. Under such conditions the ground freezes to great depths and is subject to frequent hard freezes and thaws during the transitional seasons, which is particularly detrimental to wintering plants. Water pipes in those areas must be buried 4 meters deep. In much of eastern Siberia and the Soviet Far East, as well as in portions of northern Mongolia, the area is underlain by permafrost, permanently frozen subsoil that never thaws out, with only a thin layer of topsoil thawing during the short summer (Figure 17.7).

The same interior and intermontane locations that are conducive to air stagnation and extreme cold in winter are conducive to abnormal warmth for such high latitudes under almost continual sunlight during the brief summer. During July Verkhoyansk (26) averages 15°C and has experienced temperatures up to 35°C (95°F). It thus has a mean annual temperature range amounting to 64°C and an absolute range of 103°C (185°F). This area has by far the greatest temperature range on earth, since such winter cold areas as Antarctica and Greenland do not have warm summers.

Practically all of continental Eurasia, except portions of the Pacific and Arctic fringes, show positive temperature anomalies during summer (see Figure 14.4). In the remote desert basins of western China, anomalies amount to more than +10°C.

The excessive summer heating of the landmass with respect to the surrounding seas causes a weak, diffuse low pressure to form over much of Eurasia, which allows greater penetration of Atlantic air eastward, all the way to central Siberia. This produces a rather pronounced summer precipitation maximum everywhere except southern fringes of the Crimean Peninsula in the Black Sea area and southern portions of Soviet Central Asia (37, Termez), which are influenced by winter cyclones following the Mediterranean–Middle East track (see Figure 17.4). Much of the summer rain comes in the form of showers as the heated land induces surface convective activity.

The summer maximum generally becomes more accentuated eastward, as winter precipitation diminishes to very small amounts under the constant influence of the Asiatic High. Kiev in western USSR receives 34 percent of its annual total of 615 millimeters of precipitation during June–August, whereas Irkutsk (19) in eastern USSR receives 62 percent of its annual precipitation of 458 mm. during June–August.

Since much of continental Eurasia is at fairly high latitude, the summer maximum generally falls during middle and late summer, rather than early summer when it is more needed for the early rapid phase of plant growth. Late summer and autumn rains throughout much of the Soviet Union cause notorious losses during the short harvest season, which may be brought to a premature halt by an early autumn snowstorm. Mid- and late summer rains in the Far Eastern portions of the Soviet Union, China (18, Harbin), and northern Japan are accentuated by a monsoonal inflow from the Pacific at that time of year coupled with the formation of the Mongolian section of the Polar Front that spawns wave cyclones that move east-northeastward through the Amur Valley and surrounding regions. The precipitation maximum lags into September and October in the Japan and Okhotsk Sea areas. Some influences from typhoons may be felt at this time of year, although generally they have recurved well out into the Pacific by the time they reach northern Japan, so that the region under consideration receives only peripheral precipitation and cloudiness from the northwestern portions of these storms.

Cyclones and Other Storm Systems. As was described earlier, cyclonic storms entering from the west during January follow three primary routes: an Arctic route north of Scandinavia and along the Arctic coast of the Soviet Union, a Baltic route, and a Mediterranean-Black Sea-Caspian route. The southern routes swing northeastward through European USSR and western Siberia, so the three routes generally converge in the vicinity of the Ob Gulf along the Arctic coast of western Siberia (see Figure 10.7). The most frequented route is the Arctic route, and the Barents and Kara seas north of European USSR and western Siberia become the stormiest parts of the entire continental Eurasian region during this time of year. At this high latitude all air masses involved in the storms are cold, so precipitation is negligible, but great windiness and changeability of weather are characteristic, in contrast to the calm, clear, cold conditions in the interior of the continent. Since few thaws occur at these latitudes during winter, the snow accumulates on the ground and gives the impression of large amounts of snowfall. The deepest snow packs on the Soviet mainland accumulate on the western slopes of the middle Urals and in the middle Yenisey region of central Siberia, where in both cases mean annual maximum snow depths amount to a little over 80 centimeters (32 inches). Greater snow depths may be found in scattered mountainous areas more exposed to the sea, such as the Kola Peninsula next to Finland and Kamchatka Peninsula in the Far East.

Some of the cyclones along the southern

track work their way through the mountainous territory of the arid Middle East and find their way into the southern portions of Soviet Central Asia to produce a March–April precipitation maximum. On much of the desert floor this amounts to no more than 100–200 millimeters, but in the northern foothills of the high southern mountains 300–400 mm. of rain falls, which is enough for the dry-farming of winter grains. On the middle slopes of the high mountains as much as 1500 mm. (60 inches) of precipitation may fall, most of it as snow, which forms a heavy snow pack, and even mountain glaciers, that produce large amounts of meltwater during summer for irrigation of the desert plains to the north. Hence, there is a similarity to the juxtaposition of the dry Central Valley and snowy Sierra Nevada in central California, where the winter snow pack in the mountains becomes meltwater in streams to irrigate desert plains in mid-summer. And similar to the drier southern portion of the Central Valley of California, Soviet Central Asia has become a prime area for raising irrigated cotton. The very high mountain peaks of the region, extending to heights of 5000–7000 meters and more, loom above most of the winter storm clouds and hence receive very little precipitation. In the high Pamir Knot many areas receive no more than approximately 75 mm. (3 inches) of water equivalent during the year, hence these barren, rocky, windswept surfaces are devoid even of tundra vegetation.

Much of central and eastern Asia north of the high mountains is relatively unaffected by storm systems during winter (Figure 10.7). The Polar Front is present again off the east coast of Asia, but most cyclogenesis takes place southeast of Japan and the Aleutians in the northern Pacific. The cyclones generally move northeastward over a water trajectory and only peripherally influence continental Eurasia.

As in south-central North America, cyclonic activity generally reaches its peak across the East European Plain during spring, but unlike the North American situation the Polar Front in eastern Europe has no unstable maritime tropical air along its southern flank, so spring weather is much less violent in European USSR than in eastern United States. Thunderstorms are much less frequent and less intense in continental Eurasia, even in China and Japan, and tornadoes are almost nonexistent.

During summer, cyclones are very infrequent along the southern route, and are also less frequent and less intense in the north, although they do tend to penetrate directly eastward to northern European Russia and into central Siberia more easily in summer than in winter (see Figure 10.8). The greatest summertime cyclonic activity in continental Eurasia is in the Amur region of the Far East along the Mongolian sector of the Polar Front, which brings a monsoonal late summer rainfall maximum to this region.

Climatic Distribution

The entire huge area of continental Eurasia is known for its cold winters and only modest-to-meager precipitation. Therefore, the climatic types covering it are either moderately humid high-latitude types or dry types which also are classified as cold types, or highland types in the very high Tibet Plateau and its surrounding higher mountains (see Figure 15.3). Along the Arctic coast, stretching the entire length of the continent from northern Scandinavia to the Bering Straits, a fringe of land varying in width from zero to 300 kilometers has tundra climate (Ft) with no month averaging above 10°C (50°F), but with at least one month averaging above freezing (21, Ostrov Dikson). This is a windswept landscape with low forms of vegetation such as mosses, lichens, sedges, and stunted, almost horizontally growing willows and other shrubby trees (Figure 17.8). This type

Figure 17.8. Tundra landscape in northeast European Russia.

of climate and vegetative growth are carried southward on many of the mountain summits, particularly in the Soviet Far East.

South of the tundra stretches a huge belt of boreal climate (E) that occupies all but the very southern portions of Scandinavia, northern European Russia, and most of Siberia and the Soviet Far East. Here one to three months average above 10°C (25, Tromso; 26, Verkhoyansk), which is enough of a sufficient growing season for what the Russians call the *taiga* forest, composed primarily of coniferous trees with admixtures of small-leaved trees such as birch, poplar, and aspen (Figure 17.9). But it is generally too short a growing season for agricultural crops, although marginal forms of agriculture are carried on in some areas, notably around Yakutsk in the Soviet Far East. Also, many intermontane basins along the Trans-Siberian Railroad in the southern fringes, shown as boreal on Figure 15.3, do have fairly highly developed agriculture in small areas of warmer, subhumid climate that do not show on a map of this scale.

Temperate continental climate (Dc) occupies much of the East European Plain (20, Moscow; 22, Potsdam; 24, Stockholm; 27, Vienna; 29, Warsaw). It stretches from the Baltic to the Black Sea in the west and includes practically all of East Germany, Poland, Czechoslovakia, Austria, Hungary, Romania, and the northern two-thirds of Bulgaria, as well as the heart of European USSR (Figure 17.10). It tapers eastward as it stretches across the central Ural Mountains into western Siberia, after which it ends abruptly as it butts up against the Altay Mountains in southern Siberia (Figure 17.11). In this climatic type four to seven months average above 10°C, so the growing season is moderately long, sufficient for a fairly wide variety of crops. Most of the region has a cool summer (Dcb), averaging below 22°C (72°F) during the warmest month, and therefore is more adaptable to growing wheat and other small grains, hay crops, potatoes, flax, and so forth, than to such crops as corn and soybeans. In the southwest, the Hungarian Plain in Hungary, northern Yugoslavia, and adjacent parts of Romania and Bulgaria has a warm-month average above 22°C (Dca), and corn is a major crop.

Another extensive area of temperate continental climate is found in the Far East in the southern portions of the Soviet Union, Manchuria, parts of the North China Plain, much of Korea, and the northern half of Japan. Winters there on the average are colder and drier than they are in the East European Plain, except for the northwestern coast of Japan, which derives considerable warmth and moisture from the Sea of Japan during winter. Harbin, Manchuria (18), 10° of latitude farther south than Moscow, averages 10°C colder than Moscow (20) during January and has only 13 percent as much precipitation. This comparison dramatically illustrates the difference of the primary controls during winter, amelioration from

Figure 17.9. The forest-tundra in north-central Siberia. Sparse stands of spindly coniferous and birch trees occupy the northern fringes of the taiga where it grades into tundra to the north.

the Atlantic Ocean at Moscow, and the severe dry cold of the Asiatic High in Harbin. Yet during July Harbin averages 5.5°C warmer than Moscow and has more than twice as much precipitation.

The humid northern climates are flanked on the south by a broad belt of dry climates stretching all the way from the Black Sea to the Pacific coast in northern China (Figure 17.12–17.16). The North China Plain is the only middle-latitude area in the Northern Hemisphere where dry climate extends to the east coast. This again signifies the extreme drought influence of the Asiatic High in winter and the lack of a direct influx of marine air during summer. Most of the moisture entering this region during summer has had a long land trajectory, either from the south China Sea or all the way from the Bay of Bengal, across the high mountains of southwestern China and adjacent countries.

All this dry area is subject to cold northerly winds during winter, and therefore is classed as a cold dry region. But summers can be very hot. Termez (37) on the Afghan border of Soviet Central Asia averages more than 31°C in July and has experienced temperatures as high as 50°C (122°F). The same station has experienced a winter minimum of −25°C (−13°F). The peripheral fringes of this huge area generally receive 250–400 millimeters of precipitation per year and therefore are semiarid or steppe (BSk), but the large inland basin of Soviet Central Asia containing the Caspian and

Figure 17.10. The deciduous broadleaf forest of the temperate continental climate of eastern Belorussia.

Aral seas and Lake Balkhash is definitely desert (BWk), some portions receiving less than 100 mm. (4 inches) of precipitation per year. The same is true of the Tarim Basin of western China and portions of the Gobi Desert in southeastern Mongolia and adjacent China.

A characteristic feature of the climate of north China in late winter and spring is the great amount of dust blown out of the Gobi and Ordos deserts by the strong northwest winds coming out of the Asiatic High. As a consequence, north China in the great bend of the Yellow River has some of the most extensive and thickest *loess* deposits in the world. To a lesser extent, the same is true of the northern foothills of the high mountains rimming Soviet Central Asia on the south where north winds prevail throughout the year. In May the author, while eating lunch, watched the city of Samarkand fade from view as dust settled over the entire city, a consequence of a wind storm in the desert to the north.

The high mountains and plateaus of the "Roof of the World" in south-central Asia are generally dry and, of course, cold. Peaks of the Tangla Mountains in central Tibet, which rise above 7000 meters, probably receive no more than 50–100 millimeters of precipitation per year, mostly in the form of snow. In the higher northwestern portion of the Tibet Plateau, at elevations around 5000 meters, July temperatures average only about 5°C, and frosts occur every

Figure 17.11. The mixed coniferous and broadleaf forests of the central Ural Mountains. (Courtesy of Ed Glatfelter.)

night. Summer nighttime temperatures have dropped as low as −7°C. The air is thin and clear, and insolation is intense during the long summer days, producing extreme temperature differences between sun and shade. On occasion when the ambient air temperature was only 12°C, the sun temperature measured on a black bulb thermometer was 64°C (147°F)! Intense heating of the barren ground surface during daytime causes continual strong winds, often causing sharp blizzards even in mid-summer, although the snowfall is meager. Because of the scant snowfall, the snowline on the mountains of Tibet is high for the latitude, generally around 6000 meters above sea level.

Although the surface of the Tibet Plateau and surrounding mountains is cold for its *latitude* during summer because of high altitude, the surface is warm for its *altitude* because of the great amount of absorbed heat from the strong insolation. The Tibetan land surface at 4000–6000 meters elevation is considerably warmer during summer than free air temperatures would be at that altitude, and thus this great upland surface acts as a heat source at that height in the atmosphere during summer. This has profound consequences on the upper air flow and resulting climate of much of monsoon Asia to the south. At high altitudes, of course, air pressure becomes a noticeable climatic element, since it has great physiological effects on humans. On the higher portions of the Tibet Plateau air pressure is only about half that at sea level, and on the surrounding mountains it is even less.

The orientation of the humid/dry boundary in continental Eurasia is perpendicular to that in North America east of the Rocky Mountains. Whereas the moisture gradient is directed from east to west in eastern

Figure 17.12. The wooded steppe in the subhumid central Ukraine.

Figure 17.13. Wheat field in the featureless plain of the steppe region of southwestern Siberia.

Figure 17.14. Along the Zeravshan River in the foothills of the mountains of Soviet Central Asia. Late winter and spring cyclonic rains produce a growth of grass vegetation on the hills, which turns brown in the dry, hot summer.

Figure 17.15. Reforestation of young saxaul trees near Bukhara in the desert of Soviet Central Asia. (Courtesy of Edward Furstenberg.)

Figure 17.16. Yurt camp in the Gobi Desert of southern Mongolia. Airplanes land anywhere on the barren gravel surface. The finer materials have been blown southward to form the loess hills of northern China. Cumulus of fair weather signify atmospheric stability, modified by convection due to surface heating. (Courtesy of Frank Messenger.)

North America it is directed from north to south in continental Eurasia. In both continents, of course, the temperature gradient is directed from south to north. In the eastern two-thirds of North America, then, there exist the climatic varieties of hot and humid, hot and dry, cool and humid, cool and dry, and so forth, while in continental Eurasia the only options are cool and humid or warm and dry, since the moisture and temperature gradients are parallel but opposite in direction.

The warm and dry areas in the southern portions of continental Eurasia are not very warm, except in mid-summer. All the steppe and desert regions of Central Asia, China, and Mongolia are classified as cold steppes and deserts, signifying less than eight months averaging above 10°C (see Figure 17.4). The only areas in this region with relatively abundant resources of both heat and moisture are at either end, in the Hungarian Plain and adjacent lowlands in the southwest, and portions of north China, Manchuria, Korea, and Japan in the southeast. In the rest of the region agriculture is squeezed between the cool north and the dry south, with no ideal conditions anywhere. In much of European USSR and western Siberia, and in the Far East, there are extensive areas of fairly tolerable combinations of heat and moisture in the temperate continental regions (Dcb) shown on Figure 15.3; but for 5000 kilometers or more through eastern Siberia and neighboring Mongolia, the boreal climate from the north comes into contact with the dry climate from the south, and no very toler-

Figure 17.17. The dry Khyber Pass between Peshawar and Kabul. (From the American Geographical Society Collection, University of Wisconsin–Milwaukee Library.)

able combination of heat and moisture exists along this stretch except in scattered intermontane basins.

Southwest Asia

Southwest Asia as here defined includes the area in the Near and Middle East and adjacent regions south of the high mountains stretching from the Turkish Straits and eastern Mediterranean in the west, to the Indus Valley of Pakistan in the east. It includes all of the large Arabian Peninsula, much of which could be included instead with the Saharan region of Africa.

With some notable exceptions, this southwest Asian region can be characterized by mild winters, exceptionally hot summers, aridity, and primarily winter precipitation (see Figure 17.4). The eastern extremity has been drawn in the arid Indus Valley in the transition zone between the predominance of meager precipitation from winter cyclonic storms from the west and meager summer monsoon precipitation from the southeast. Although these two seasonal precipitation mechanisms overlap throughout much of the Punjab in northern Pakistan and northwestern India, an approximate boundary has been drawn between the area to the west that receives more rain in winter and the area to the east that receives more rain in summer. The northwestern extremity of the southeastern summer monsoon rains seems to lie somewhere between Peshawar (51) in northwestern Pakistan and Kabul (34) in eastern Afghanistan (Figure 17.17). Whereas Peshawar shows a secondary summer maximum of precipitation, Kabul shows practically no rain at all during the period June–September. Another boundary between winter cyclonic precipitation, to the north, and summer precipitation, to the south, runs through the southern portion of the Arabian Peninsula approximately along the 20th parallel.

Much of the Near and Middle East is mountainous, with broad intervening plateaus and structural basins. Hence, there is considerable climatic variety in spite of general aridity, except in the higher mountains. The highest mountains lie along the northern fringe of the region and, to a considerable extent, protect this southern area from the cold winters of continental Eurasia to the north. In the west the Caucasus, stretching between the Black and Caspian seas, consists of a series of continuously high ranges that reach a maximum elevation of 5633 meters (18,510 feet). The Caspian Sea is fringed on the south by the Elburz Mountains, which reach a maximum elevation of 5604 meters (18,386 feet) in northern Iran and form a very imposing scarp on the north where they plunge to the Caspian Sea 31 meters below sea level. The mountains are continued eastward by the Kopet Dag along the border of the Soviet Union and northeastern Iran, which reach a maximum elevation of 3117 meters (11,908 feet), and these are continued eastward through northern Afghanistan by the Hindu Kush, which rise to 5025 meters (16,872 feet) west of Kabul and then continue rising northeastward to more than 7000 meters as they approach the Pamir Knot. Many lesser ranges are oriented in various directions across Turkey, northern Iraq, Iran, Afghanistan, and western Pakistan that occasionally rise to 3000–4000 meters or more. Highest among these are the volcanic peaks on the Armenian Knot on the border of Turkey, Iran, and the Soviet Union, where Mt. Ararat rises to 5156 meters (16,804 feet).

The Arabian Peninsula, by contrast, is primarily a series of plateau blocks that tilt from elevations around 1000 meters in the west down to sea level along the Persian Gulf in the east. In many places, however, the edges of the peninsula are upturned along major fault zones, and rise to elevations of 3761 meters (12,336 feet) in Yemen in the southwestern portion of the peninsula, 3019 meters (9902 feet) in Oman in the southeastern portion of the peninsula, and 2638 meters (8652 feet) in the Sinai Peninsula in the northwest. Thus, although most of the peninsula is desert, there are a few highlands that receive a modest amount of precipitation. Places in South Yemen receive as much as 500 mm.

Although the local relief in the Palestine region is not great the topography is quite rugged because of the fault-block nature of the region, a northerly offshoot of the great rift system farther south that forms the Red Sea, the Gulf of Aden, and the rift valleys of eastern Africa. The Jordan Valley or "Gor" reaches the lowest elevation on land anywhere on earth, 395 meters (1296 feet) below sea level around the Dead Sea. This stands in contrast to the escarpment to the west, where south of Jerusalem the elevation rises to 1013 meters (3323 feet), and the escarpment to the east in Jordan, which is even higher. Thus, whereas Jerusalem (33) receives 627 mm. of precipitation per year, portions of the Jordan Valley and the Gulf of Aqaba to the south receive as little as 27 mm.

The Transcaucasian mountains, plateaus, and lowlands of the Soviet Union and adjacent eastern Turkey and northwestern Iran have a complexity of rainfall regimes and the greatest amount of rainfall in the entire region. Heaviest rain falls in Soviet Georgia along the eastern end of the Black Sea. Batumi (32) on the coastal hills averages 2504 mm. (100 inches) per year. The precipitation is distributed through the year, coming from cyclonic storms during fall and winter and thunderstorms during summer. Amounts appear to be related to temperatures of the Black Sea. The wettest month, September, when the sea temperatures are warmest, receives approximately four times as much precipitation as May, the driest month, when sea temperatures are coldest. Nevertheless, the region is wet throughout the year. The eastern Black Sea littoral remains quite humid westward into

Turkey, but precipitation amounts diminish to around 700 mm. (28 inches) half-way along the coast of Turkey, and farther west it becomes even drier.

Inland in both the Armenian and Anatolian (Turkish) plateaus the rainfall regime quickly reverses itself, and May in general is the wettest month. Annual totals diminish perceptibly. Yerevan in the middle of the Armenian Plateau receives only 304 mm. of precipitation per year, with 50 mm. in May, the wettest month. August, the driest month, receives only 9 mm. Ankara (30), in the west-central portion of the Anatolian Plateau, receives 359 mm. per year with a May maximum of 50 mm. Again, August is the driest month, with only 8 mm. In this Transcaucasian-Anatolian region, maximum precipitation appears to occur as a combination of the end of the winter-spring cyclonic season and the rapid buildup of late spring thunderstorms before the establishment of the upper-level eastern end of the Azores High during middle and late summer.

Along the west coast of the Caspian Sea the maximum shifts to October–November, while summer remains the driest part of the year and annual amounts generally diminish. They pick up again on the steep northern slopes of the Elburz Mountains along the southern coast of the Caspian, where rain from winter cyclonic storms is enhanced by convective showers as cold, northerly air moves southward across the relatively warm Caspian Sea.

The remainder of the Near and Middle East generally has a rather simple winter maximum–summer minimum rainfall regime, and this is true of the Arabian Peninsula, as well, south to the 20th parallel, where summer monsoon precipitation takes over. Precipitation totals remain low along the southeast coast of Oman in southeastern Arabia because of cool upwelling water in a coastal current moving northeastward along this coast during summer. This effect is most noticeable farther southwest along the Somali coast of Africa, and will be discussed in more detail in the next chapter.

Most inland basins in southwest Asia are quite dry. Already mentioned is the below-sea-level area in the Jordan Valley. Riyadh (35) in east-central Arabia receives only 105 mm. of precipitation per year. Baghdad (31) in the Mesopotamian region of central Iraq receives only 143 mm. Tehran (36) south of the Elburz Mountains in northern Iran, even at an elevation of 1191 meters, receives only 165 mm. of rain per year. Jacobabad (41) in the Thar Dessert of south-central Pakistan receives only 88 mm., and even Kuwait on the Persian Gulf receives only 97 mm.

All these areas are remarkably hot during summer. Jacobabad may have the hottest average summer temperatures on earth. For three months, May–July, daily maxima average more than 40°C (104°F). In June they average 44.3°C (112°F). During both May and June, temperatures have risen to 51°C (124°F). Nighttime temperatures are also warm. During June and July daily minima average more than 29°C (85°F). Riyadh, the capital of Saudia Arabia, has even hotter daily maximum temperatures, averaging above 42°C (108°F) from May through September. During July daily maxima average 44.8°C (113°F). (But Riyadh cools off more at night than Jacobabad.) Daily minima during July average 23.3°C (74°F). A thousand kilometers farther north, Baghdad is still very hot, daily maxima averaging above 41°C (106°F) from June through August. Nearby Abadan, Iran, even though near the coast at the head of the Persian Gulf, is even hotter, maximum temperatures averaging above 42°C from June through September. During the hottest month, August, daily maxima average 44.5°C (112°F). Neighboring Kuwait is similarly hot. The Persian Gulf and southern portions of the Red Sea on either side of the Arabian Peninsula are the hottest water bodies on earth: during summer their surface temperatures average 31°–32°C. On the other hand, during December–February light frosts occur in most of these desert areas.

Why does this area experience such extreme summer heat, with average daily maximum temperatures apparently exceeding anything in the Sahara of northern Africa? Several controls working in the same direction coincide in this area. First, much of the area lies in the latitude that receives the greatest amount of insolation at the surface during summer. Second, the area lies under the eastern extension of the Azores High with its attendant subsidence and adiabatic warming. This is not so evident at the surface, where extreme surface heating during summer produces a thermal low over much of southwest Asia, with core areas in the lower Indus Valley and the Persian Gulf. But aloft this surface heat low is capped by high pressure, which at the 500-millibar level becomes a continuous, subtropical, high pressure zone that stretches across this area half-way around the world in either direction (see Figure 6.14). As a consequence of upper tropospheric high pressure and subsidence, the Middle East is one of the most cloud-free areas on earth during summer (see Figure 13.3). This, of course, allows intense insolation.

One more factor seems to enhance the effects of insolation and high pressure, the descending winds from the high mountains on the north. Summer winds blow constantly from a northerly direction, depending on the topography. In the Mesopotamian region of Iraq they are from the northwest down the Tigris and Euphrates Valley; in Pakistan they are from the northeast down the Indus Valley. But from whatever direction, they have a northerly component and descend from the northern mountains into the basins and plateaus of the Middle East where their foehn effects add to the warming of the surface air. They are so consistently from the north during summer and so noticeably clear, dry, and warm, that they are referred to as "etesian winds" throughout the Middle East. These northerly winds pick up a great deal of dust blowing across the dry plains which gives a murky appearance to the atmosphere. Dust is carried far southward across the dusty plains of Mesopotamia into the Persian Gulf, where the atmosphere typically has a milky-white appearance.

Climatic Types

Practically all of southwest Asia is dry. Only the Mediterranean fringes of the Near East and Turkey, the southern and eastern fringes of the Black Sea, the southern fringe of the Caspian, and the foothill areas of northwestern Iran, northern Iraq and southern Turkey catch enough winter cyclonic precipitation to classify as humid (see Figure 15.3). The arc of Cs climate stretching around the northern fringes of the Arabian Peninsula in the Near and Middle East was the famous "fertile crescent" of biblical times. Some of the highest mountains also have humid climate, but this is limited primarily to the western half of the Great Caucasus Mountains in the Soviet Union, the adjacent Pontic Mountains in northern Turkey, and the higher eastern portion of the Hindu Kush in Afghanistan. Arid climate rises to surprisingly high altitudes in most of the other mountain ranges. The eastern half of the Great Caucasus, even at altitudes of 4000–5000 meters, are generally treeless.

Monsoon Asia

The varied expanse of land and water in southeast Asia and neighboring islands is integrated by the monsoonal aspect of its climate, the seasonal reversal of wind systems, and consequent strong seasonality of precipitation regimes. Much of it is of a tropical nature, and the rest is subtropical or dry. Many rugged mountain areas project through a series of altitudinal climatic zones that range from tropical, at sea level, to tundra and ice cap, in the higher elevations.

The area under consideration stretches

from the Indus Valley of Pakistan in the west through India, Nepal, Bhutan, Bangladesh, Burma, Malaysia, Thailand, Laos, Kampuchea, Vietnam, southeast China, the southern tip of the Korean Peninsula, to central Japan in the east, and from about 38°N latitude on Honshu Island in Japan southward through Taiwan, the Philippines, and Indonesia to more than 10°S latitude just north of Australia. Within this area lives approximately half the world's population. Thus the lives of a large portion of humanity are intimately tied to the rhythm of the monsoon. And, thus, this climatic phenomenon has probably aroused more interest and instigated more scientific investigation than any other single climatic phenomenon. Subsequent investigations have increasingly revealed complexities of atmospheric circulation in this area that cannot be explained by simple heat differentials between land and sea, and they have also revealed much variety within this mountainous, peninsular, and island area.

Primary Air Flows

Winter. Basic surface air flows over the region are shown in Figure 17.18. During January six air streams affect parts of the area. Continental polar air moves out of the Asiatic High to curve anticyclonically southward across eastern China and cyclonically southeastward to Japan. Its leading edge is marked by a segment of the Polar Front, which on the average lies southeast of Japan and trails southwestward through the East and South China seas into northern Vietnam and adjacent countries in southeast Asia. In the western Pacific it meets up with a southeasterly flow of maritime tropical air that moves northward around the western end of the Pacific High. Farther south easterly trades flow westward across the Philippines, the southern half of southeast Asia, Malaysia, and the northern portions of Indonesia into the southern half of the Bay of Bengal, where they converge with cool, dry air descending the southern slopes of the Himalayas and adjoining ranges to curve southwestward across India. These Indian northeasterlies are separated by a segment of the Polar Front in northwestern India and Pakistan from the northwesterlies that blow in from the Middle East and Soviet Central Asia, which are often associated with weak cyclonic storms that bring the meager but critically beneficial rain to this semiarid region without which the winter wheat would fail. The intertropical front lies in the far south near the equator and separates the Pacific easterlies, flowing westward across northern Indonesia, from the equatorial westerlies, flowing eastward across the southern portion of the region. Of course, all these wind systems and frontal zones shift about considerably.

Aloft during winter, the westerlies split around the high mountains surrounding the Tibet Plateau to form two jet streams, the stronger one flowing east-southeastward down the Ganges Plain, and the other curving northward and eastward through north China and Mongolia. These converge on the lee side of Tibet in central China to form a single jet stream across southern Japan into the Pacific, where the highest average jet speeds on earth are observed (see Figure 6.20). In the western Pacific these westerly air streams meet up with the southeast trades across the Polar Front. Easterly winds are limited to the Philippines, southern southeast Asia, and the very southern part of India, Sri Lanka (Ceylon), and the north Indian Ocean. The equatorial westerlies lie across the intertropical convergence zone in southern Indonesia.

Summer. During summer the surface flow becomes predominantly southwesterly across most of peninsular India, southeast Asia, eastern China, Korea, and Japan. This is a deep, humid, relatively unstable extension poleward of the equatorial westerlies. Distorted branches of the intertropical convergence zone form on either side of it and often enclose a large triangular

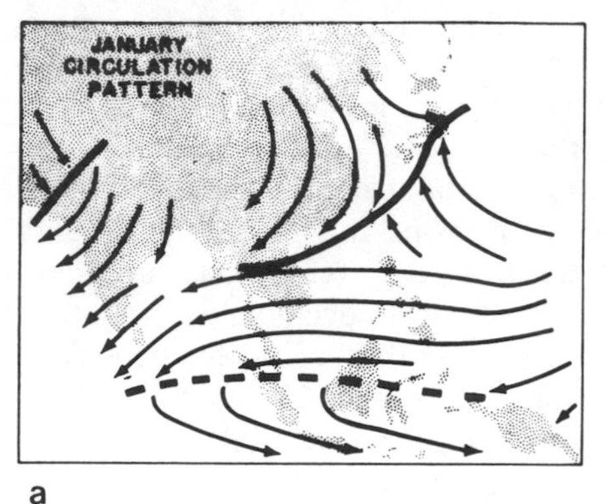

a

b

WESTERLIES
WESTERLIES
TRADES
EQ. WESTERLIES

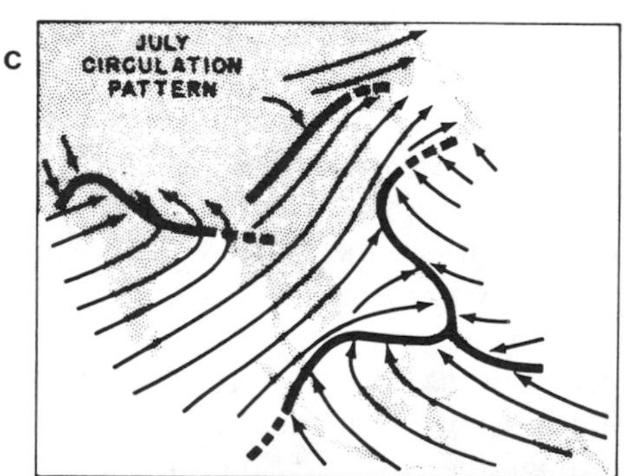

c

Figure 17.18. Schematic air flows over southeastern Asia. (a) Low level, January; (b) upper level, January; (c) low level, July. (After G.T. Trewartha, The Earth's Problem Climates, *2nd ed. [University of Wisconsin Press, 1981], pp. 174–76. © University of Wisconsin.)*

area of doldrums with essentially calm surface conditions, stretching from the Philippines westward to India, particularly during the premonsoon season in April and May. During June–September when the southwesterly flow is strongest, convergence zones take up positions along the Ganges Plain of northern India and among the islands of Indonesia, the Philippines, eastern China, and Japan. The convergence zone in the Ganges Plain passes westward through the heat low in the Thar Desert and southern Persian Gulf and on through southern Arabia and across the southern

Sahara region of Africa. In the Indus Valley during July its average position lies at about 27°N latitude, the most poleward position of the intertropical front anywhere on earth. But it does not produce precipitation, as will be explained later.

The Pacific portion of the intertropical convergence zone during summer lies in a generally north-south orientation in the western north Pacific, where it fluctuates in position and intensity and is sometimes indiscernible except for its bands of cloudiness and precipitation. It separates the southwesterly flow in the west from the deflected trades along the western end of the Pacific High, and in the south, the southeasterlies coming from the Southern Hemisphere out of the Australian winter high pressure cell.

In northern China the southwesterlies are flanked on the northwest by a portion of the Polar Front, the so-called Mongolian Front that is cyclonically active at this time of year. To the north of the front lie the prevailing westerlies of Siberia and adjacent regions. Waves form along this segment of the Polar Front and move northeastward across the Sea of Okhotsk into the Pacific, pulling in Pacific air from the southeast in their counterclockwise circulations underneath warm fronts. As was pointed out earlier, this Pacific influx of air during summer is shallow and usually produces little if any precipitation, except along coastal slopes of mountains in northern Japan, the Soviet Far East, and adjacent portions of Korea and north China. The principal sources of moisture for this entire area are the southwesterlies, which transport moisture all the way from the Indian Ocean and cause precipitation as they ride up the slopes of warm fronts connected with cyclonic circulations along the Polar Front.

Aloft during summer a warm high pressure cell builds up over the Tibet Plateau and surrounding mountains. It is caused in part by the thermal effects of the high-level insolation-absorbing surface, and in part by the tremendous quantities of latent heat released by copious precipitation on the southern slopes of the Himalayas and adjoining mountains and hill ranges to the south. This becomes evident in the middle troposphere and intensifies upward to the tropopause and beyond. It is easily distinguished on the 100-millibar chart for July, where it expands eastward and westward to engulf half the hemisphere in subtropical and lower-middle latitudes (Figure 17.19).

The southerly branch of the winter jet stream along the southern slopes of the Himalayas disappears, and a broad high-velocity easterly jet forms along the southern periphery of the upper level high and stretches all the way from the western Pacific westward through southeast Asia, the Indian Peninsula, the southern half of Arabia, and the southern portions of the Sahara and tropical Africa, into the eastern Atlantic centered at about 15°N latitude. Locally, in the northern portion of the Ganges Valley and foothills of the Himalayas in northern India, it extends downward to the surface and moisture-bearing air flows and disturbances that produce precipitation during summer travel west-northwestward up the Ganges to bring some precipitation to the upper Indus Valley. This air flow gradually dries out along its northwestward course, and precipitation diminishes steadily from copious amounts in northeastern India and Bangladesh to only meager amounts in the Punjab of northwestern India and Pakistan.

This reversal of upper tropospheric flow between winter and summer, and an attendant shift of an upper tropospheric trough westward from a winter-spring position over the western portions of the Bay of Bengal, at approximately longitude 85°E, to the Arabian Sea west of India, apparently has a great deal to do with the peculiarities of the onset of the southwesterly monsoon. The monsoon often occurs with the suddenness of a singularity as it "bursts" around the beginning of June in the south and the end of June in the northwest, accompanied by turbulent weather in the form of thun-

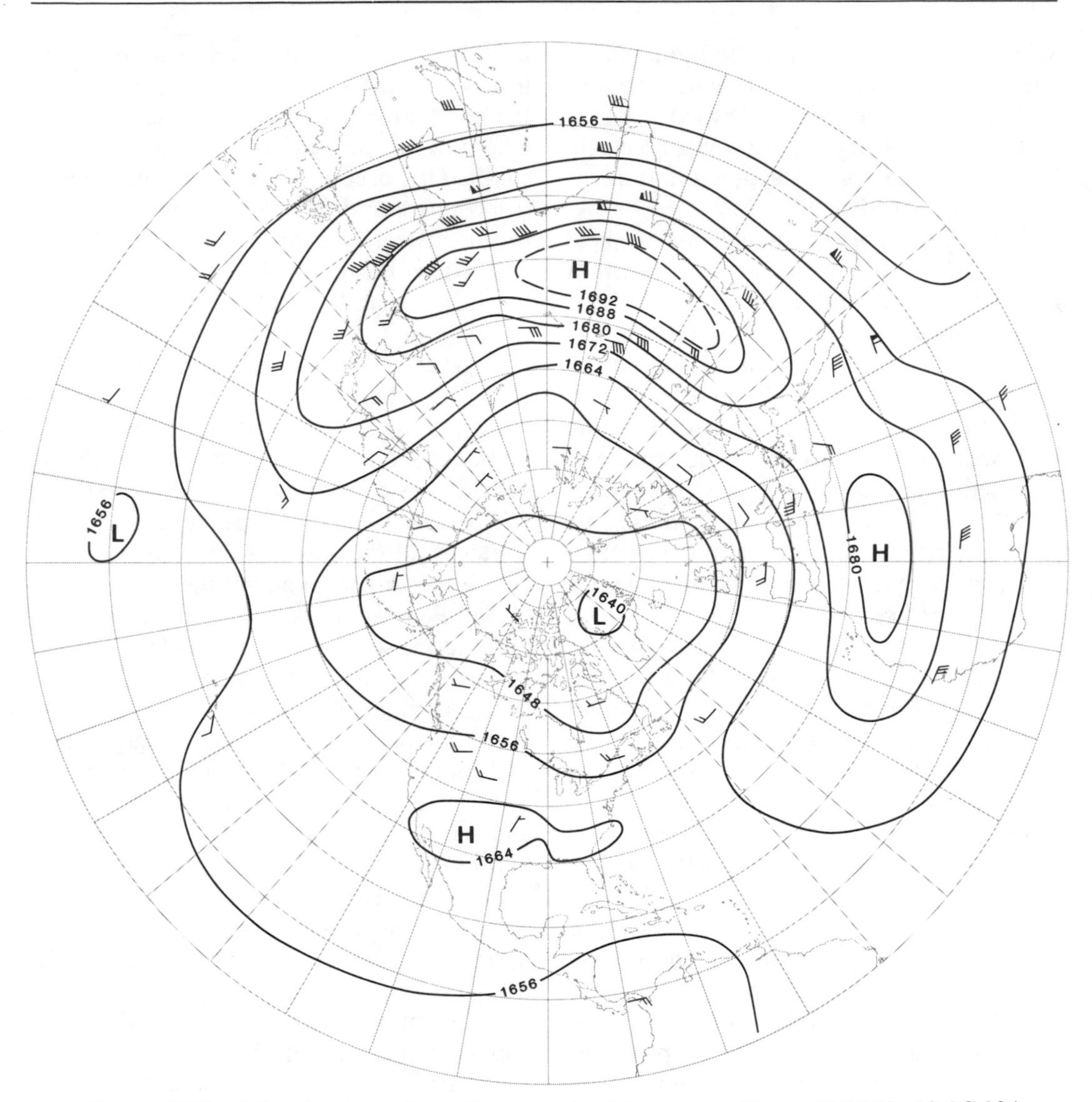

Figure 17.19. 100 mb. chart, July. Contours in dekameters. (From NAVAIR 50-1C-52.)

derstorms and squalls, which generally subside somewhat as the summer circulation becomes established. This phenomenon occurs even earlier along the Burma coast, usually during May, in the southwesterly flow ahead of the upper tropospheric trough before it shifts west of India. Thus, the summer southwesterlies proceed northwestward across the Indian peninsula from late May or early June, in the northwestern part of the Bay of Bengal, to late June and early July, in northwest India and southern Pakistan.

Hence, although the seasonal reversal of the surface atmospheric circulation is undoubtedly related to differential heating between land and sea, the process is strongly modified by the extremely high topographic features of this region, which bring about seasonal shifts in the upper tropospheric and stratospheric flows that greatly affect atmospheric processes at the surface.

Atmospheric Disturbances

The southwesterly surface flow of maritime air during summer provides the environment for most of the precipitation of monsoon Asia, but before precipitation can occur the flow must be disturbed by various kinds of convergence and convectional flows within it. In fact, periods of consistently strong southwesterly flow during summer are coincident with periods of reduced rainfall over much of the region. Since this is primarily a tropical and subtropical region, density fronts play minor roles, and most of the disturbances are of subtle nature, often above surface. Such things as velocity surges in the southwesterly surface flow and easterly waves in the middle troposphere may cause enough convergence and uplift to bring on periods of several days of increased precipitation.

Tropical Cyclones. One of the most significant perturbations of the Indian region is the so-called "monsoon depression," which generally forms somewhere in the Bay of Bengal and moves westward or northwestward with the upper-level easterlies during summer. Locally known as cyclones, they vary in strength from subtle above-surface circulations to tropical storms at the surface, sometimes with hurricane strength. They occur at all times of the year, but they reach their greatest strength during fall and spring. October and November are the months of maximum occurrence of hurricane-strength cyclones, both in the Bay of Bengal and the Arabian Sea, on either side of the Indian peninsula. In the Bay of Bengal the primary maximum in October–November accounts for 46 percent of such storms, and the secondary maximum in April–May accounts for 26 percent. The Bay of Bengal has about four times as many cyclonic storms as the Arabian Sea, which is virtually free of depressions during the summer monsoon season. Many Arabian Sea cyclones originate in the southern Bay of Bengal and move across the narrow southern part of India into the Arabian Sea, but many originate within the Arabian Sea itself. Some of the Arabian Sea storms continue westward to penetrate the southeastern coast of the Arabian Peninsula.

During the autumn season many of the Bengal storms follow a typical parabolic path, moving first northwestward and then curving northeastward. Such paths frequently bring the severest stages of storms into the broad, flat delta of the Ganges and Brahmaputra rivers at the head of the Bay of Bengal, where large tidal waves driven by the winds can engulf this flat lowland to depths of 10 meters or more far inland, inundating this very populous part of India and Bangladesh.

Tropical cyclones are even more numerous in the western North Pacific, where they occur with the greatest frequency and highest wind speeds of any place on earth. The rectangular area in the western Pacific Ocean and China Seas bounded by latitudes 5°–30°N and longitudes 105°–150°E annually averages about 20 storms of hurricane intensity. Here they are generally known as typhoons, although local names are used in different areas. While they can occur any time during the year, 89 percent of them occur during the period June–December. In the Philippines, which lie in the heart of the region, August experiences the greatest frequency, averaging about 3.7 during the month. On daily weather maps one can occasionally observe two or three such storms, spaced 1500 kilometers or so apart, following one another along perturbations on the intertropical convergence zone. They generally form over the ocean somewhere east of Guam where there are few observations, so their origins are obscure. But some of them appear to form when an old cold front moves out of mainland China deep into the tropics to form a "triplepoint" with the intertropical front, around which a wave forms and a cyclonic circulation is set up. Once the winds become strong enough, the air mixes thoroughly throughout the entire storm and the fronts disappear. But not all storms originate this way.

During the early part of the season in the western Pacific, typhoons generally take a southerly course across the southern Philippines, and many cross the China Seas to hit the coast of mainland China. As the season progresses, the routes shift northward and the storms recurve east of Taiwan and follow the Nansei Shoto (Ryukyu Islands) to Japan, wherein September is the month of maximum typhoon occurrence. Many of the storms recurve entirely out to sea and affect little land anywhere. By November and December, the storms are generally following the southern route again.

Thunderstorms. Much of monsoon Asia except for Malaysia has surprisingly few thunderstorms compared to similar locations in southeastern North America. The mean annual number of days with thunderstorms in China reaches a maximum of about 40 around Canton in the south, and fewer than 10 over much of the Japanese Islands. By comparison central Florida, five degrees latitude farther north than Canton, has more than 100 days with thunderstorms. Much of peninsular India has only 20–60 days per year. The reduced thunderstorm activity in the Far East seems to be a function of the shallowness and weakness of instability in the deflected trades around the western end of the Pacific high pressure cell and the lack of strong air mass contrasts over mainland China.

Violent squalls are frequently associated with the northwesterly flow in northern India during May and June. In the drier northwestern portions of India and Pakistan they do not produce much rainfall, but they cause strong winds and well-developed cumulonimbus clouds. These produce huge dust storms that darken the sky about once a week until the beginning of the southeasterly flow up the Ganges around the end of June. In the more humid eastern portions of the Ganges Valley, these squalls are designated as "nor'westers," and they are accompanied by heavy rain showers. Some of the world's largest hailstones fall in the northern foothills of the central Ganges Valley at this time of year, caused by an explosive growth of cumulonimbus clouds as convection in the moist, unstable southerly air from the Bay of Bengal breaks through the overrunning dry air from the northwest.

Polar Front. In China, Korea, and Japan the Polar Front becomes a major consideration. During winter it may swing southward into northern Indochina where it may bring the well-known *crachin* weather to southern China, northern Vietnam, and the Gulf of Tonkin, which is characterized by low stratus, fog, drizzle, and low visibility. Such weather generally occurs in spells of 3 to 5 days for a total of 10 to 15 days per month throughout the winter. This kind of weather generally is associated with the Polar Front as it shifts latitudinally by seasons, and it becomes particularly noticeable when the front stagnates in certain areas. During late spring it brings the *maiu* weather to China and the *baiu* weather to Japan as the front stagnates on its way northward, blocked by a localized high pressure cell that forms over the Sea of Okhotsk during that time of year. This usually accounts for a late spring–early summer secondary maximum of precipitation in these areas, which is followed by a slight reduction of precipitation during mid-summer when the subtropical high in the western Pacific expands westward. An autumn primary maximum occurs during the so-called *shurin* rains, as the front once again halts temporarily on its return southward. The autumn maximum is augmented by typhoon rains.

During winter the Polar Front usually lies off the mainland southeast of Japan.

Cyclogenesis along it is most active in the East China Sea, the Pacific southeast of Honshu Island, and the Sea of Japan. Maximum activity occurs during March. Secondary centers of cyclogenesis are found over the land in the Szechwan Basin in the upper-air convergence zone to the east of the Tibetan highlands and in Manchuria. But many of these systems bring little precipitation to the mainland during winter, since the land is usually occupied by dry Polar and Arctic air moving southward from Mongolia and Siberia. China winters are typically dry, particularly in the north. Harbin (18) averages only 15 mm. of precipitation during the entire three-month period December–February, while in July–August it averages 286 mm. Beijing (Peking) (16) shows a similar seasonality, and even Shanghai (52) farther south receives only about one-fifth as much precipitation during winter as during summer.

Because of the dominance of cold northern air during winter, average temperatures in China are lower than in corresponding latitudes in eastern North America, where a greater exchange of air masses takes place during winter. For instance, Shanghai at approximately 31°N latitude averages only 3.4°C during January, whereas Charleston, South Carolina, at approximately 33°N latitude averages 10.2°C during January. Because of the general southeasterly direction of cold air flow across China, and perhaps due to some shielding effect from the Tsinling and other mountains farther west, the interior of south China during winter has considerably warmer temperatures than the east coast. Perhaps subsidence in the lee convergence zone east of Tibet adds to the warming of this interior region.

Precipitation Distribution

Precipitation amounts and regimes vary greatly throughout monsoon Asia. And since insolation is great year-round throughout the area, local perceptions of seasons and temperature regimes often primarily reflect cloud and precipitation regimes. For instance, throughout much of the Indian peninsula local people recognize three seasons: the hot season (March–May), the wet season (June–September), and the cool season (October–February). Hottest temperatures usually occur during May just before the onset of the summer monsoon rains in June (40, Delhi). This distorted annual regime of temperatures is so peculiar to the Ganges plain of northern India that it is often referred to as the "Ganges-type temperature regime."

Most of the air masses that affect monsoon Asia are warm and moist and have high potentials for precipitation, but sometimes the mechanisms for releasing the precipitation are lacking. Precipitation totals relate strongly to topography—heavy where orographic lifting occurs, and light on leeward sides of mountain ranges.

The Indian peninsula is primarily a plateau with upturned edges. The highest portion, called the "Western Ghats," rises to more than 2600 meters in the southwestern portion of the peninsula. The upturned eastern edge, the Eastern Ghats, is lower and more broken. In the north the peninsula is attached to the Asian mainland by two extensive alluvial valleys, the Ganges-Brahmaputra lowland in the northeast, and the Indus Valley in the northwest.

The southwest summer monsoon rises up the western slopes of the West Ghats and drops 2000–3500 mm. of rain from Bombay southward (48, Mangalore), but on the plateau to the east of the Ghats totals are less than 1000 mm. They increase again toward the northeast as warm, moist Bay-of-Bengal air is drawn into the system, and summer rainfall becomes copious in the lower Ganges-Brahmaputra valleys as marine air funnels into the head of the Bay of Bengal and is forced to rise up the high mountain slopes. The Khasi Hills rise abruptly from this delta plain, in front of the high Himalayas in the background, to an elevation of 1961 meters. Cherrapunji (39) at an elevation of 1313 meters on the southern slope receives 11,419 mm. of rainfall per year and is one of the wettest spots on earth. Most of this falls from May to September. June is the wettest month with 2922 mm., and December is the driest with only 5 mm.

In the high mountains of the Himalaya system, whose name means "abode of snow," the precipitation falls mainly as snow. The western portions of this system also derive considerable snow from the winter cyclonic storms coming from the west. Although good records are lacking, it appears that many places on the southern slopes of the Himalayan range receive at least 15 meters of snow per year.

Summer precipitation decreases northwestward up the Ganges Valley as the southeast winds become more and more remote from the moisture source. Delhi (40) receives only 715 mm. per year, most of it during July–September (see Figure 17.4). Farther northwest, Peshawar (51) in northern Pakistan receives only 331 mm. per year with the maximum during January–April and a secondary maximum in July–September. As was pointed out earlier, the Indus Valley generally is the dividing line between a winter maximum of precipitation to the west and a summer maximum to the east. In much of the Indus Valley, these opposing seasonal and directional precipitation mechanisms overlap.

The driest portion of the entire India-Pakistan area is the lower Indus Valley, where Jacobabad (41) in the Thar Desert receives only 88 mm. of precipitation per year, most of it during July–August. One might wonder why this lowland, open to the sea in the south and penetrated by southwesterly maritime winds during summer, receives so little precipitation, while just to its south precipitation rises to more than 2000 mm. The answer appears to lie in the nature of the intertropical front in the lower Indus Valley. During summer the frontal surface slopes upward to the south over the marine air which is cooler than the hot, dry, etesian flow from the north. What cloud formation takes place occurs because of convective activity in the humid surface air, not because of the uplift of the dry overriding air. Thus the frontal surface with its temperature inversion acts to stifle cloud formation rather than to induce it. For cumulus clouds to build up high enough in the surface layer of marine air to produce showers, the frontal surface must be two or more kilometers above the surface. This does not usually occur until the front extends south of the coast over the northern part of the Arabian Sea. Only occasionally do small showers form over the land underneath the frontal surface. Hence, this is an unusual case of a frontal surface acting as a lid on convective activity below rather than as an incline for cloud formation above.

Some of the Indian peninsula receives maximum rain during autumn and early winter from the northeastern monsoon blowing down the slopes of the Himalayas and crossing the Indian peninsula and portions of the adjacent Bay of Bengal. This is particularly true in the southern portions of the East Ghats and the peninsula interior to them as weak depressions form along the intertropical convergence zone in this region during autumn and move west-northwestward into the interior (46, Madras).

The southeast Asian peninsula is somewhat a replica of the Indian peninsula, except it does not have the Ganges Plain and high mountains to the north. Much of it receives its maximum precipitation during the southwest summer monsoon, which of course drops much more rainfall on the windward western sides of mountains than on eastern sides. The coast of Burma bordering the Bay of Bengal is especially wet, in areas receiving more than 3000 mm. of rainfall per year. But behind the coastal ranges to the east, up the Irrawaddy Valley,

Mandalay (47) receives only 871 mm. per year. This comes in a two-peaked maximum, as classically should be expected around 20° latitude. The primary maximum falls in May–June when the intertropical front is moving northward to the Tropic of Cancer, and the secondary maximum falls during August–October when it is moving back southward again. July, when the intertropical front has moved north of Mandalay, is only about half as wet as June or August. The period December–March is nearly rainless as the cool northeasterly land air settles into the region.

Much of the interior basin-and-plateau country of Burma, Laos, Thailand, and Kampuchea receives only modest precipitation, 1000–2000 mm. But amounts pick up again on the eastern slopes of the mountains in Vietnam, and also southward in the Malay Peninsula, which catch considerable rainfall from the northeasterly monsoon that blows into this region during the cool season. Many of the mountainous slopes in these areas receive more than 3000 mm. per year, but the coastal areas usually receive only about half that amount, except for the east coast of Malaysia, which continues very wet. Northern and central Vietnam have a distinct autumn maximum and spring minimum, but farther south Ho Chi Minh City shows a broader twin-peaked summer maximum, similar to Mandalay, except greater in amount, with a dry late winter and spring. Much farther south, Kuala Lumpur (42) in the western part of the Malay Peninsula only 3°N of the equator is a classic example of an equatorial regime with equinoxial rains and somewhat reduced rainfall during the solstices. Singapore, almost on the equator, has a very even distribution of rainfall throughout the year with a slight maximum during December–January.

Most of the area of the Indonesian Islands receives at least 2000 mm. of precipitation per year, but this varies greatly because of the mountainous character of all the islands. One station in central Java at an altitude of 700 meters has an annual rainfall of more than 7000 mm., while portions of the Palu Valley of Sulawesi (Celebes) Island receive only 574 mm. On some of the higher peaks precipitation appears almost daily during portions of the year. An expedition up one of the high peaks of New Guinea during December–March observed rain and snow showers every day. A snowfall of about 20 centimeters was observed during one night.

Rainfall regimes are varied throughout Indonesia because of different orientations of mountains with respect to prevailing winds during different seasons. Some of the records show a rather typical equatorial regime of equinoxial rains and ample rainfall throughout the year, while others, particularly in the southeast next to Australia, show more distinct wet and dry half-years. In this drier eastern end of the Lesser Sunda Islands, Kupang (43) on Timor Island receives only 1297 mm. of precipitation per year, 391 of which falls during January and 0 in August. In general, Indonesia receives more rainfall during the Northern Hemisphere winter, when the deflected northeasterlies in the Indonesian area blow from the northwest or west, and less rain during the Southern Hemisphere winter, when the counterclockwise circulation out of the winter High over Australia blows from the southeast or east across Indonesia.

Farther north the Philippine Islands, between 6°N and 19°N latitude, are subjected to more simple flows of southwest winds during summer and northeast winds during winter. Since, again, most of these islands are mountainous, precipitation regimes vary from east to west coasts, generally having winter maxima in the east (44, Legaspi) and summer maxima in the west (49, Manila). In general the east coasts are the wettest, some areas receiving more than 4000 mm. per year, whereas maximum amounts along west coasts generally are between 3000 and 3500 mm. Some interior areas behind mountains may receive as little as 1000 mm. or even less. Easterly waves are a major mechanism producing precipitation in this area. Also, during summer the intertropical convergence zone is generally located somewhere in the region. During May–August it usually takes a distorted orientation north-south across the region. As was pointed out earlier, the Philippines lie athwart one of the most active routes of typhoons in the world, the sources of much precipitation.

Since temperatures in the Philippines are warm throughout the year, seasons are identified according to precipitation regimes rather than according to temperatures. And due to the period of Spanish domination, the wet season is known as winter (*invierno*) and the dry season is known as summer (*verano*), as they are in Spain, which, of course, is just the opposite of what it should be in the western portions of the islands.

Precipitation amounts vary considerably in southeastern China because of the mountainous topography, but generally the region receives 1000–2000 mm. per year. On mountainous Taiwan Island to the east, some of the higher elevations may receive as much as 3000 mm. Precipitation diminishes rapidly northward in China, dropping to less than 750 mm. north of the Yangtze Valley. In general southeast China shows a very marked summer maximum of precipitation due to the southwest monsoon (38, Canton; 52, Shanghai). During winter southward surges of Siberian air may be ushered in by sharp blizzard conditions as far south as Shanghai, but snowfall is not great, and it diminishes rapidly southward. There are generally less than five days of snowfall per year south of Shanghai and in the Red Basin in southwestern China. In most of southern China snow lies on the ground less than one day per year. It may lie on the ground 5–25 days in southern Korea.

The Japanese Islands generally are quite humid. Least rainfall occurs in the northern island, Hokkaido, particularly in the northern and eastern portions, where less than 1000 mm. per year may be received. But since the temperatures there are constantly cool, this is quite adequate. On the other islands, lowlands generally receive 1000–1500 mm. of precipitation per year, and mountainous highlands may receive more than 2500 mm. In most places there is a fairly pronounced summer maximum, particularly in the southeastern portions of Honshu, Shikoku, and Kyushu islands, where July may receive more than 300 mm. As was pointed out earlier, much of Japan receives a twin-peaked summer maximum, with mid-summer being a little drier than earlier or later (53, Tokyo). The Shurin rains of late summer–early fall may be enhanced by typhoons, which are notorious in southeastern Japan at this time of year.

As was already mentioned, portions of northwest Honshu may experience heavy snowshowers during winter when cold northwest winds blow off the mainland across the relatively warmer Sea of Japan (50, Niigata). In this area snowfall is often so great that houses are built with roofs overhanging sidewalks so the winter snow can build up over the roof and the sidewalk below can become a tunnel. Strong northwesterly winds are combatted by placing large boulders on the roofs to hold them down.

The driest and sunniest portions of the southern parts of the islands lie in the so-called "Inland Sea" between the southwestern peninsula of Honshu Island, and Shikoku and Kyushu islands. In this region winter, in particular, is dry and sunny. Summer is also notably drier than the exposed Pacific fringes of the islands.

Climatic Types

As can be seen on Figure 15.3, practically all of monsoon Asia is tropical (Ar or Aw) or subtropical (Cfa). But portions of it, particularly northwestern India and adjacent Pakistan, are hot steppes and deserts (BSh and BWh). In the mountains everywhere, climatic zones vary with altitude, to well above the snowline in the higher mountains.

Vegetation consists of broadleaf evergreen rainforests in the wetter areas (Ar) and savanna grasses and brush in Aw areas. Places that receive 2000 mm. or more of precipitation annually but have more than two dry months generally are forested, but many of the trees drop their leaves during the dry season. These are typical monsoon forest, known as semideciduous forests. They are true jungles with much undergrowth and vines that get sufficient sunlight on the forest floor during the leaf-bare season.

Table 17.1 Temperature and Precipitation Data for Eurasia

	Jan	Feb	Mar	Apr	May	June	July	Aug	Sept	Oct	Nov	Dec	Yearly average
PENINSULAR EUROPE													
1. Aberdeen, Scotland (57°12′N, 02°12′W, 58 m)													
T.(°C)	2	3	4	7	9	12	14	14	12	9	6	4	8
P.(mm)	77	54	52	50	62	53	92	73	65	90	91	78	837
2. Athens, Greece (37°58′N, 23°43′E, 107 m)													
T.(°C)	9	10	11	15	20	25	28	27	24	19	15	11	18
P.(mm)	62	36	38	23	23	14	6	7	15	51	56	71	402
3. Barcelona, Spain (41°24′N, 2°9′E, 95 m)													
T.(°C)	9	10	12	15	18	22	24	24	22	18	14	10	16
P.(mm)	33	42	46	47	52	43	29	48	77	80	49	47	594
4. Bergen, Norway (60°24′N, 5°19′E, 44 m)													
T.(°C)	2	1	3	6	10	13	15	15	12	8	6	3	8
P.(mm)	179	139	109	140	83	126	141	167	228	236	207	203	1958
5. Bologna, Italy (44°30′N, 11°21′E, 84 m)													
T.(°C)	2	4	9	13	18	22	25	24	20	14	8	4	14
P.(mm)	41	35	47	54	51	48	32	28	60	74	70	49	589
6. Copenhagen, Denmark (55°41′N, 12°33′E, 9 m)													
T.(°C)	0	0	2	7	12	16	18	17	14	9	5	3	9
P.(mm)	49	39	32	38	42	47	71	66	62	59	48	49	602
7. London, England (51°28′N, 00°19′W, 5 m)													
T.(°C)	4	4	7	9	12	16	18	17	15	11	7	5	11
P.(mm)	53	40	37	38	46	46	56	59	50	57	64	48	594
8. Lisbon, Portugal (38°43′N, 9°09′W, 77 m)													
T.(°C)	11	12	14	16	17	20	22	23	21	18	14	12	17
P.(mm)	111	76	109	54	44	16	3	4	33	62	93	103	708
9. Madrid, Spain (40°25′N, 3°41′W, 667 m)													
T.(°C)	5	7	10	13	16	21	24	24	20	14	9	6	14
P.(mm)	38	34	45	44	44	27	12	14	32	53	47	48	435
10. Marseille, France (43°27′N, 5°13′E, 3 m)													
T.(°C)	6	7	10	13	17	21	23	23	20	15	10	7	14
P.(mm)	43	32	43	42	46	24	11	34	60	76	69	66	546
11. Paris, France (48°58N, 2°27′E, 52 m)													
T.(°C)	3	4	7	10	14	17	19	19	16	11	7	4	11
P.(mm)	54	43	32	38	52	50	55	62	51	49	50	49	585
12. Plymouth, England (50°21′N, 04°07′W, 26 m)													
T.(°C)	6	6	7	9	12	15	16	16	15	12	9	7	11
P.(mm)	105	77	73	55	65	58	71	80	82	94	115	115	990
13. Rome, Italy (41°54′N, 12°29′E, 46 m)													
T.(°C)	7	8	11	14	18	22	25	25	21	16	12	9	16
P.(mm)	76	88	77	72	63	48	14	22	70	128	116	106	881
14. Santander, Spain (43°28′N, 3°49′W, 68 m)													
T.(°C)	9	9	12	12	14	17	19	19	18	15	12	10	14
P.(mm)	119	89	74	82	88	66	59	84	114	134	134	155	1198
15. Valentia, Ireland (51°56′N, 10°15′W, 9 m)													
T.(°C)	7	7	8	9	11	14	15	15	14	12	9	8	11
P.(mm)	164	107	103	74	86	81	107	95	122	140	151	168	1398
CONTINENTAL EURASIA													
16. Beijing (Peking), China (39°57′N, 116°19′E, 52 m)													
T.(°C)	−5	−2	5	14	20	25	26	25	20	13	4	−3	12
P.(mm)	4	5	8	17	35	78	243	141	58	16	11	3	623
17. Belgrade, Yugoslavia (44°48′N, 20°27′E, 132 m)													
T.(°C)	0	2	6	12	17	21	23	22	18	13	7	3	12
P.(mm)	48	46	46	54	75	96	60	55	50	55	61	55	701

	Jan	Feb	Mar	Apr	May	June	July	Aug	Sept	Oct	Nov	Dec	Yearly average
18. Harbin, China (45°45′N, 126°38′E, 143 m)													
T.(°C)	−20	−16	−6	6	14	20	23	22	14	6	−7	−17	3
P.(mm)	4	6	17	23	44	92	167	119	52	36	12	5	577
19. Irkutsk, USSR (52°16′N, 104°79′E, 468 m)													
T.(°C)	−21	−19	−10	1	8	15	18	15	8	0	−11	−19	−1
P.(mm)	12	8	9	15	29	83	102	99	49	20	17	15	458
20. Moscow, USSR (55°45′N, 37°34′E, 156 m)													
T.(°C)	−10	−10	−5	4	12	15	18	16	10	4	−2	−8	4
P.(mm)	31	28	33	35	52	67	74	74	58	51	36	36	575
21. Ostrov Dikson, USSR (73°30′N, 80°14′E, 22 m)													
T.(°C)	−28	−26	−25	−18	−8	−1	4	5	1	−7	−19	−24	−12
P.(mm)	20	13	17	9	11	23	32	46	42	21	14	18	266
22. Potsdam (Berlin), Germany (52°23′N, 13°04′E, 81 m)													
T.(°C)	−1	0	3	8	13	17	18	18	14	9	4	1	9
P.(mm)	44	39	32	42	47	66	71	71	45	47	46	40	590
23. Sonnblick, Austria (47°03′N, 12°57′E, 3,106 m)													
T.(°C)	−13	−13	−11	−8	−4	−1	2	1	−1	−4	−8	−11	−6
P.(mm)	115	108	112	153	136	142	154	134	104	118	108	111	1495
24. Stockholm, Sweden (59°21′N, 18°04′E, 44 m)													
T.(°C)	−3	−3	−1	4	10	15	18	17	12	7	3	0	7
P.(mm)	43	30	26	31	34	45	61	76	60	48	53	48	555
25. Tromsö, Norway (69°39′N, 18°57′E, 114 m)													
T.(°C)	−4	−4	−3	0	4	9	12	11	7	3	0	−2	3
P.(mm)	96	79	91	65	61	59	56	80	109	115	88	95	994
26. Verkhoyansk, USSR (67°33′N, 133°23′E, 137 m)													
T.(°C)	−49	−44	−30	−13	2	12	15	11	3	−14	−36	−46	−16
P.(mm)	7	5	5	4	5	25	33	30	13	11	10	7	155
27. Vienna, Austria (48°15′N, 16°22′E, 203 m)													
T.(°C)	−1	0	5	10	15	18	20	19	16	10	5	1	10
P.(mm)	40	43	45	45	70	67	83	72	41	56	53	45	660
28. Vladivostok, USSR (43°07′N, 131°54′E, 138 m)													
T.(°C)	−15	−11	−4	4	9	13	18	20	16	9	−1	−11	4
P.(mm)	10	13	20	44	69	88	101	145	126	57	31	17	721
29. Warsaw, Poland (52°09′N, 20°59′E, 107 m)													
T.(°C)	−4	−3	1	8	14	18	19	18	14	8	3	−1	8
P.(mm)	23	26	24	36	44	62	79	65	41	35	37	30	502
SOUTHWEST ASIA													
30. Ankara, Turkey (39°57′N, 32°53′E, 902 m)													
T.(°C)	0	1	5	11	16	20	23	23	18	13	8	3	12
P.(mm)	35	38	36	34	50	31	13	8	19	22	28	46	359
31. Baghdad, Iraq (33°20′N, 44°24′E, 34 m)													
T.(°C)	10	12	16	22	28	33	35	34	31	25	17	11	22
P.(mm)	25	24	23	22	7	0	0	0	0	4	17	23	145
32. Batumi, USSR (41°45′N, 41°40′E, 92 m)													
T.(°C)	6	7	8	11	16	20	23	23	20	16	12	9	14
P.(mm)	237	205	136	138	82	165	178	233	315	261	294	260	2504
33. Jerusalem, Israel (31°52′N, 35°13′E, 755 m)													
T.(°C)	8	9	11	15	19	22	23	24	22	20	15	11	17
P.(mm)	143	129	102	25	6	0	0	0	1	9	75	138	627
34. Kabul, Afghanistan (34°30′N, 69°13′E, 1,985 m)													
T.(°C)	−2	1	6	12	18	22	25	24	20	14	8	3	13
P.(mm)	33	38	92	84	22	4	2	2	1	10	15	14	316
35. Riyadh, Saudi Arabia (24°42′N, 46°43′E, 594 m)													
T.(°C)	15	17	22	25	29	33	34	34	31	26	21	16	25
P.(mm)	20	10	17	21	14	0	0	0	0	0	12	12	105

Table 17.1 (continued)

	Jan	Feb	Mar	Apr	May	June	July	Aug	Sept	Oct	Nov	Dec	Yearly average
36. Tehran, Iran (35°41′N, 51°19′E, 1,191 m)													
T.(°C)	5	7	10	15	22	25	30	30	25	18	11	6	17
P.(mm)	47	21	26	16	12	1	0	0	4	15	12	12	165
37. Termez, USSR (37°17′N, 67°19′E, 302 m)													
T.(°C)	3	6	12	19	25	29	31	30	23	17	10	5	17
P.(mm)	21	23	30	19	10	1	0	0	0	3	0	17	133
MONSOON ASIA													
38. Canton, China (23°00′N, 113°13′E, 18 m)													
T.(°C)	14	14	17	22	26	27	29	28	27	24	20	16	22
P.(mm)	27	65	101	185	256	292	264	249	149	49	51	34	1720
39. Cherrapunji, India (25°15′N, 91°44′E, 1,313 m)													
T.(°C)	12	14	17	19	19	20	20	20	21	19	16	13	17
P.(mm)	20	37	179	605	1705	2922	2457	1828	1168	447	47	5	11419
40. Delhi, India (28°35′N, 77°12′E, 216 m)													
T.(°C)	14	17	23	29	34	34	31	30	29	26	20	16	25
P.(mm)	25	22	17	7	8	65	211	173	150	31	1	5	714
41. Jacobabad, Pakistan (28°18′N, 68°28′E, 56 m)													
T.(°C)	15	18	24	30	35	37	35	34	32	28	22	17	27
P.(mm)	7	9	8	2	4	6	27	22	1	0	1	3	88
42. Kuala Lumpur, Malaysia (3°07′N, 101°42′E, 34 m)													
T.(°C)	26	26	27	27	27	27	26	26	26	26	26	26	26
P.(mm)	171	169	237	279	216	126	102	157	188	275	259	230	2409
43. Kupang, Indonesia (10°14′S, 123°37′E)													
T.(°C)	27	27	27	27	27	26	26	26	24	25	28	27	27
P.(mm)	391	263	292	60	41	12	7	0	1	9	56	165	1297
44. Legaspi, Philippines (13°08′N, 123°44′E, 19 m)													
T.(°C)	26	26	27	27	28	28	27	27	27	27	27	26	27
P.(mm)	366	265	218	158	178	194	235	209	252	313	479	503	3371
45. Leh, India (34°09′N, 77°34′E, 3,514 m)													
T.(°C)	−9	−6	0	6	10	14	17	17	13	7	1	−5	6
P.(mm)	12	9	12	7	7	4	16	20	12	7	3	8	115
46. Madras, India (13°00′N, 80°11′E, 16 m)													
T.(°C)	25	26	28	31	33	33	31	30	30	28	26	25	29
P.(mm)	24	7	15	25	52	53	84	124	118	267	309	139	1215
47. Mandalay, Burma (21°59′N, 96°06′E, 77 m)													
T.(°C)	20	23	28	32	31	30	30	29	29	28	25	22	27
P.(mm)	1	5	5	36	150	152	74	102	147	127	64	10	871
48. Mangalore, India (12°52′N, 74°51′E, 22 m)													
T.(°C)	27	27	28	29	29	27	26	26	26	27	27	27	27
P.(mm)	5	2	9	40	233	982	1059	577	267	206	71	18	3467
49. Manila, Philippines (14°35′N, 120°59′E, 16 m)													
T.(°C)	25	26	27	28	29	28	27	27	27	27	26	25	27
P.(mm)	23	11	17	32	128	253	414	437	353	195	138	68	2069
50. Niigata, Japan (37°55′N, 139°03′E, 2 m)													
T.(°C)	2	2	5	10	15	20	24	26	21	16	10	5	13
P.(mm)	194	126	121	104	95	127	193	107	177	165	171	264	1841
51. Peshawar, Pakistan (34°01′N, 71°35′E, 359 m)													
T.(°C)	11	13	17	23	29	33	33	31	29	24	18	13	23
P.(mm)	39	41	65	42	15	7	34	41	14	10	10	15	331
52. Shanghai, China (31°12′N, 121°26′E, 5 m)													
T.(°C)	3	4	8	14	19	23	27	27	23	18	12	6	15
P.(mm)	47	63	85	91	96	177	148	139	132	74	53	38	1143
53. Tokyo, Japan (35°41′N, 139°46′E, 5 m)													
T.(°C)	4	4	8	13	18	21	25	26	23	17	11	6	15
P.(mm)	48	73	101	135	131	182	146	147	217	220	101	61	1563

18 | CLIMATE OF AFRICA

To the southwest of Eurasia across the narrow Red Sea and the somewhat broader Mediterranean lies the continent of Africa, the second-largest landmass on earth. As was pointed out in the last chapter, Eurasia and Africa can hardly be considered separately because their juxtaposition has profound climatic consequences. Some of these influences were seen in Chapter 17 in the case of much of Europe and southwest Asia, which, because of the proximity of Africa, lacks a significant source of maritime tropical air. The same lack also affects much of north Africa to some extent, but probably more profound is the alteration of the climate of east Africa by the lack of a normal trade wind system in the north Indian Ocean.

The presence of the summer heat lows backed up by the high mountains in adjacent southern Asia turns the atmospheric circulation over the north Indian Ocean into a monsoonal one that strongly affects the climate of the eastern Horn of Africa. In fact, it strongly affects much of equatorial Africa, which throughout the year receives much of its surface air flow from the west rather than from the east, as might be expected in near-equatorial latitudes. Hence, in spite of the fact that Africa alone among the continents is situated most symmetrically across the equator, its climatic distribution, particularly in its eastern portion, is not as symmetrically arranged around the equator. For a landmass that is situated primarily within the tropics and so well divided north-south on either side of the equator, Africa contains a surprising number of climatic abnormalities. Most of these are related in some way to the proximity of the Eurasian landmass.

Continental Shape and Form

Unlike the peripheries of Eurasia that are marked by deep penetrations of seas between peninsulas and island groups, the landmass of Africa is quite compact with few indentations of climatic significance (Figure 18.1). In addition, much of the continent consists of uplands at moderate elevations above the sea, often fringed by higher upturned edges near the coasts. This is particularly true of the southern and eastern half of the continent, the Guinea region of northwest Africa, and the Atlas Mountain area in the north facing the Mediterranean and adjacent Atlantic. And much of the northern interior area which has extensive lowlands also has extensive uplands, some of them rising to mountainous proportions. All this greatly influences the types of air masses entering the continent and the processes involved during air mass penetration. Also, the elevation alone in many places lifts the surface climate out of what would be a tropical type into a subtropical or even cooler type because of the altitude.

The Atlas Mountains in the north lie outside the tropics and rise to a maximum elevation of 4165 meters (13,665 feet). During the passage of winter cyclonic storms through the Mediterranean route, while the coastal areas are receiving rainfall, elevations above 300 meters frequently are receiving snow. Whereas Casablanca has never seen snow and Tangier glimpses it only about once in five years, the mountain slopes receive some every year. Between 1000 and 2000 meters elevation, snow remains on the ground for a week or more. In the High Atlas frequent blizzards accumulate snow to more than a meter in depth, some of which remains on northwest slopes 6–12 months of the year. Conditions vary greatly within the Atlas system itself, since it consists of several widely spaced ranges with dry basins in between. The southern side of the eastern end of the ranges acts as a major cyclogenesis region for the formation of lee depressions during winter, which move eastward into the Mediterranean and bring a little rainfall to the dry coastal areas of Libya and Egypt.

At the other end of the continent the upturned southeastern edge of the plateau, known as the Drakensberg (Drake's Mountains), rises to 3657 meters (10,438 feet). In the highest portions snow may occur five to ten times per year. The escarpment forms an imposing climatic divide between the very mild conditions of the southeastern coast and the more continental upland of the interior. The mountains also generate many localized thunderstorms and hail along their higher crests.

Much of eastern Africa consists of plateau surfaces averaging 1000–2000 meters above sea level, with upturned edges of rift valleys and volcanic areas rising much higher. The volcanic upland of Ethiopia rises to a maximum elevation of 4620 meters (15,158 feet), and to the south individual volcanic cones in Kenya and Tanzania rise even higher. The highest, Mt. Kilimanjaro, rises to 5895 meters (19,340 feet). Such cones, of course, show the entire gamut of climatic variations with altitude, including permanent snow caps. But since they are only individual peaks on an otherwise low plain, they have little effect on the climate of the surrounding territory. The short, high fault block of the Ruwenzori Mountains along the western rift zone west of Lake Victoria rises to 5120 meters (16,763 feet). This mountain range lies right on the equator and is constantly enshrouded in clouds and rain.

In the central Sahara, individual, eroded, dead volcanic massifs rise to 3265 meters (11,204 feet) in the Tibesti and to 3003 meters (9,852 feet) in the Ahaggar, the two highest and most extensive of a number of such landforms in the central Sahara. Similar dead volcanoes make up the Cameroon Mountains, a northeast-southwest line of individual peaks that cross the coast of west Africa right at the bend of the continent and continue into the equatorial Atlantic as a short line of low mountainous islands. Mt. Cameroon on the coast rises to 4070 meters (13,353 feet).

Thus, although no extensive main mountain barriers, such as the Cordillera in western North America and the high mountains in southern Eurasia, affect the climate of major sections of the African continent, individual elevations are as high or higher than anything in North America.

Off the southeast coast of Africa the large island of Madagascar stretches from about 12°S to about 25°S latitude. This consists of a large fault-block mountain range, steep on the eastern side, that rises to 2639 meters (8,671 feet) and to a great extent shields the Mozambique Channel and the African mainland to the west.

Primary Atmospheric Circulations

Since the African continent almost exactly straddles the equator, reaching from about 37°N latitude to 35°S latitude, its climate is dominated by the intertropical convergence zone and the subtropical high pres-

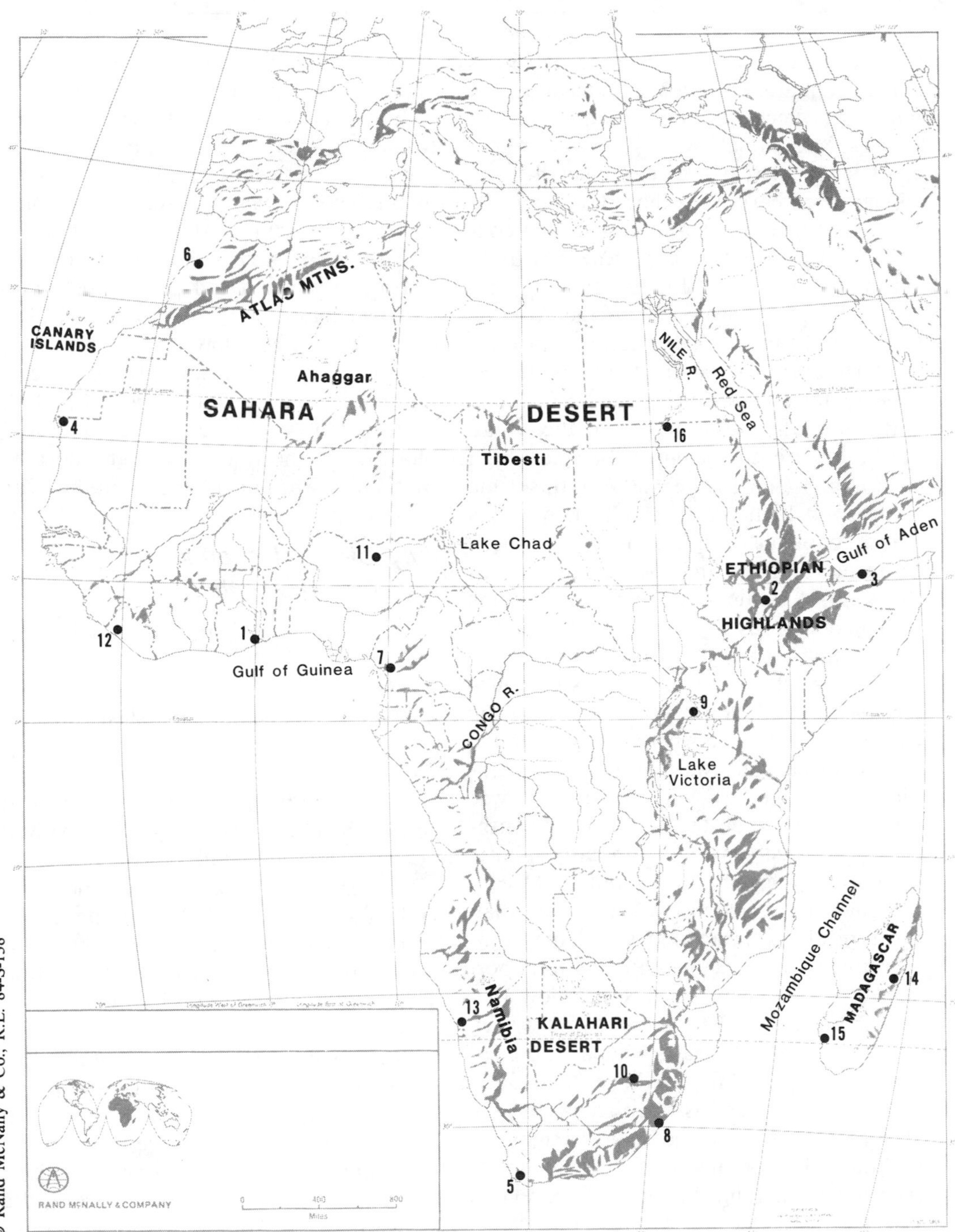

Figure 18.1. Major relief features of Africa. Numbers correspond to data tables at end of chapter.

sure cells on either side, with the westerlies shifting equatorward just far enough during winter to affect the northern and southern fringes. Meridional exchanges of air are not great since there is no major north-south topographic barrier to the zonal flow. No fresh continental polar air reaches the continent. In the south the continent is quite symmetrically located between the subtropical high pressure cell in the Atlantic Ocean and that in the Indian Ocean, particularly during July—the Southern Hemisphere winter—when the Indian Ocean high shifts into the western part of the ocean (and is apparently the only subtropical high to do this) (see Figure 6.3). At this time a localized high pressure cell often forms over the plateau surface of the southern portion of Africa and then tends to join the south Atlantic and south Indian highs into a continuous high pressure zone across the southern part of Africa.

In the north, however, there is no counterpart to the Atlantic Azores High on the east side of Africa, where, of course, the land continues eastward through the Arabian peninsula and southern Asia. As was pointed out earlier, the landmass of southern Asia transforms the entire atmospheric circulation over the north Indian Ocean, turning it into a monsoonal one rather than the typical northeast trades. During summer the Sahara occupies the western portion of an expansive surface heat low that stretches westward from the Indus Valley through the Middle East, Arabia, and clear across northern Africa. Above 2000–3000 meters the Sahara is dominated by the eastern extension of the Azores High.

Thus, the circulation across much of Africa is not arranged in neat zones, related to the zonal belts of pressure, as might be deduced simply from the continent's symmetrical arrangement astride the equator. Striking differences occur between east and west, particularly south of about 10°N latitude, and there are many ideosyncratic features connected with the intertropical convergence zone and adjacent circulation systems throughout much of Africa.

In western and central Africa the intertropical convergence zone at the surface stays well north of the equator year-round, maintained in that position by a consistently strong southwesterly flow off the sea into the Guinea coastland and the Congo Basin to the north and east of the Gulf of Guinea (Figure 18.2). The Africans call this southwest surface flow a monsoon, but it is simply a northern extension of the equatorial westerlies which are well developed in this region, apparently because of the blocking of the trades from the northeast by their disruption in the Indian Ocean.

July Circulation and Associated Weather

During July the southwesterlies generally penetrate to 18°–20°N latitude and eastward to the western edge of the east African highlands. This is the moistest, most unstable air entering the African continent, and it brings much rainfall to the Guinea coastlands and all of central Africa. In the central Sahara it makes contact with the hot, dry etesian winds coming from the north across the Sahara around the eastern end of the Azores High and the western end of the surface heat low. Although the intertropical convergence zone lies at the surface through the middle of the Sahara at this time of year, it is not a rain-bringer to this region, since the frontal surface slopes upward to the south, and the hot desert air from the north overrides the warm, moist maritime tropical air from the southwest. The convective activity that takes place is in the maritime tropical air underneath the front and is largely limited upward by the temperature inversion that exists at the frontal surface. In the southern Sahara this inversion layer is too close to the surface to allow enough convective activity in the marine air to produce showers, but farther south where the marine air is

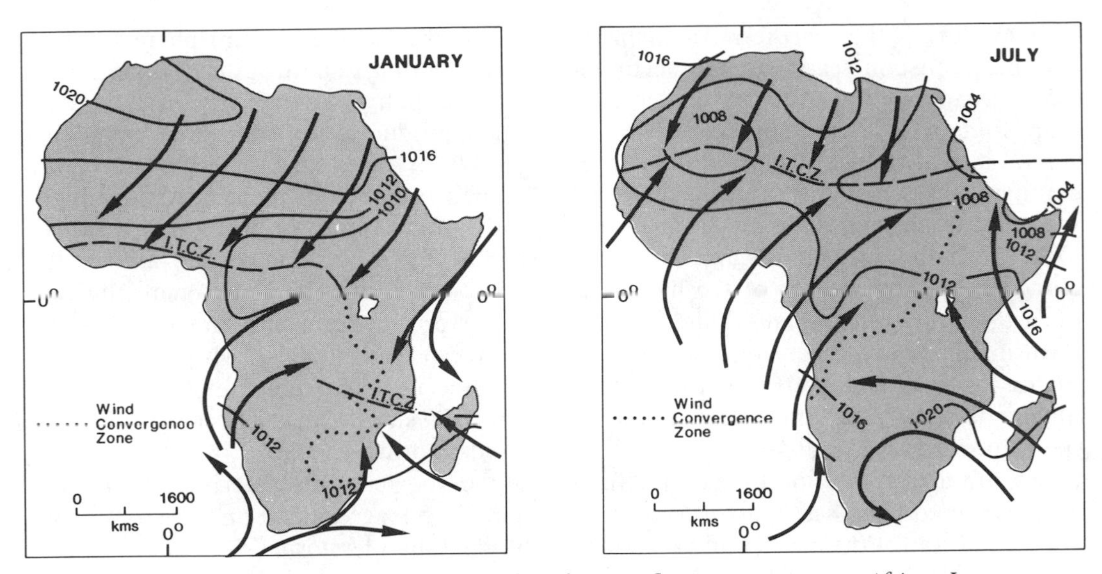

Figure 18.2. Mean sea-level pressure and surface air flow patterns across Africa, January and July. (Modified from J.F. Griffiths, Climates of Africa, *Vol. 10 of* World Survey of Climatology *[Amsterdam: Elsevier, 1972].)*

thicker, cumulus clouds can build up to considerable heights and bring much thunderstorm and rain shower activity to the Guinea and Congo lands.

The frequent thunderstorms imbedded in the maritime tropical air are often organized in disturbance lines, most of which form somewhere east of Lake Chad and move westward with the upper easterlies across the Guinea lands of western Africa. Although they are a product of the moist southwesterly surface flow, their tops build well into the easterly jet that crosses north Africa during summer, and they are carried westward at fairly high speeds (see Figure 17.19). The so-called west African "tornadoes" are squall lines normally oriented north-south in an arc convex toward the west. They are characterized by massive cumulonimbus clouds extending upward into the middle and upper troposphere with well-formed anvils extending 5–10 kilometers downstream to the west. These squall lines often stretch as much as 150 kilometers meridionally and move westward at average speeds of 45 kilometers per hour, with squall gusts up to 25 meters per second. Typically a downpour of rain lasts for about a half hour, and this is followed by prolonged light rain and drizzle falling from much low and middle clouds. According to some west African meteorologists, as much as 90 percent of the rainfall of portions of the Guinea area is accounted for by this mechanism. Easterly waves in the upper troposphere perhaps may initiate formation of the squall lines moving from east to west through the Sudan region. Somewhat similar squall lines occasionally move inland with the southwestern air along the Guinea coast, and the origin of these squall lines is even less understood than that of the ones to the north.

The surface southwesterlies on their southeastern side converge with the southeast trades that dominate the entire width of southernmost Africa, south of about 15°S latitude, during the Southern Hemisphere winter. North of about 10°S latitude these become deflected into southwesterlies that flow essentially parallel to the east coast of the eastern Horn of Africa north of the equator during the Northern Hemisphere summer monsoon season in the north Indian Ocean. As can be seen on the July map, Figure 18.2, the convergence line in central Africa, between the southwesterlies coming from the Atlantic and the deflected southeast trades over eastern Africa coming from the Indian Ocean, lies approximately along the western border of the highlands of east Africa. A great many thunderstorms form in this convergence zone, which fluctuates fairly widely over the east African highlands and brings the summer rain to the Ethiopian highlands in the north. Farther south a large portion of central Africa around the equator has double maxima of equinoxial rains during the two periods when the sun's direct rays cross the equator.

Thunderstorms. Thunderstorms are very active throughout much of equatorial Africa. A broad belt experiencing more than 100 thunderstorm days per year stretches from about 10°N latitude along the Guinea coast of west Africa east-southeastward to the highlands just east of Lake Victoria a few degrees south of the equator (see Figure 10.12). The activity culminates in the vicinity of Lake Victoria where more than 180 days per year experience thunderstorms, sometimes more than one storm a day. On the northern shore of Lake Victoria, Kampala experiences 242 thunderstorm days per year, which may be the highest frequency in the world, although it may be rivaled by certain places in the Indonesian islands where information is lacking. Small hailstones fall more often than 100 days per year in local areas of some of the highlands of western Kenya, which appears to be the greatest frequency of hail on earth, although similar situations may occur in parts of the Peruvian Altiplano in South America (see Figure 13.8).

Obviously, there is a great deal of atmospheric instability in the vicinity of the convergence zone between the southwesterlies and deflected southeasterlies. During the Northern Hemisphere summer often the moist southwesterly surface flow is overlain by a rather dry easterly flow which produces extreme convectional instability as the air rises over some of the localized uplands of the east African highlands. Warm water bodies, such as broad, shallow Lake Victoria sitting astride the equator, feed a great deal of moisture into these storm systems and enhance their activity, especially at night, when land breezes converge into the center of Lake Victoria from all sides. Although most of these thunderstorms probably owe their origin to the moist southwesterly surface flow, there is no agreement on which way individual thunderstorm cells and line squalls move once they are formed. There is such a paucity of reliable observations that no one can be sure that a thunderstorm cell or line of thunderstorms found over one portion of the highlands is the same system that was observed earlier east or west of the current observation.

Precipitation. Despite the high thunderstorm frequency in the east African highlands, rainfall totals are moderate, and only a small percentage of the area receives more than 1500 mm. per year. Most of the area receives between 1000 and 1500 mm. per year, and some enclosed basins and the lowlands on the eastern side of the highlands receive much less, some areas less than 500 mm. Thus, it appears that not much rain falls per storm, which probably indicates the relative shallowness of the southwesterly flow in this eastern part of Africa. Although the temperature and moisture lapse rates are conducive to great instability, the amount of precipitable water in the air is not great. By comparison the west coast Cameroon area where the coast bends from west to south in the Gulf of Guinea receives much more rainfall with somewhat fewer thunderstorms. Debundscha near the base of Cameroon Mountain averages 10,300 mm. (more than 400 inches) of rain annually and is by far the wettest spot in Africa and one of the three or four wettest spots on earth.

Much of interior Africa does not get a great deal of precipitation. Although the Congo Basin does receive 1500–2000 mm. of rainfall per year this is not outstandingly wet in an equatorial region. This central part of Africa appears to be affected by northerly air from the dry Sahara anticyclone during the Northern Hemisphere winter and to some extent by a southeasterly current from the south African anticyclone during the Southern Hemisphere winter. Also, the southwesterly flow that brings much of the moisture to the region generally appears to be rather stable, often with a temperature inversion around 2000 meters above sea level. This may be a remnant of a temperature inversion formed in the stable eastern end of the south Atlantic High, which is further cooled by the Benguela Current along the southwest African coast as the air moves inland. Often a stratus cloud deck is characteristic during night and early morning hours, particularly during the Southern Hemisphere winter season.

January Circulation and Associated Weather

During January the various pressure and wind zones shift southward, but in western and central Africa the intertropical convergence zone still remains well north of the equator, the average position lying between 8° and 10°N latitude (see Figure 18.2). In central Africa it abruptly turns southward and becomes lost in a broad, diffuse heat low that develops over southern Africa during the Southern Hemisphere summer. It can generally be picked up again somewhere along the Mozambique coast of southeast Africa and extends eastward across the island of Madagascar, where it separates the northeast monsoonal winds coming from the north Indian Ocean from the southeasterly trades in the south Indian Ocean. Between the two segments of the ITCZ there is generally an ill-defined convergence line oriented approximately north-south through the southern half of Africa, which is the meeting zone between the westerlies to the west and the easterlies to the east. Much of Africa south of the equator experiences considerable thundershower activity at this time of year, showing a December–February maximum of precipitation, except for the double equinoxial maxima of the equatorial zone to the north and the southwestern tip of Africa, which is dominated by the stable eastern end of the south Atlantic subtropical high during summer.

Extending 5°–10° latitude on either side of the equator, equatorial Africa continues to receive a high degree of thundershower activity in January as the southwesterlies bring warm, moist, unstable air into this region. The reason that this part of Africa probably has more thunderstorms than any other place on earth is that these unstable southwesterlies prevail throughout the year, whereas in other tropical areas they are generally replaced by more stable easterly flows during portions of the year.

In the north the Mediterranean fringe of Africa southward to 30°N latitude or beyond receives general rains from winter cyclonic storms that form along the Polar Front that lies through the Mediterranean and north Africa at this time of year. The northern slopes of the Atlas Mountains and adjacent coastal areas receive 450–650 mm. of rain annually, most of it during the period October–April (6, Casablanca). But eastward along the Libyan and Egyptian coasts rainfall diminishes rapidly. Cairo receives only 24 mm. per year, with 20 of them coming during December–March. As mentioned previously, the High Atlas may experience frequent blizzard conditions from these cyclonic passages and may accumulate snow to a depth of a meter or more on the ground. Rarely, perhaps once in ten years, snow may fall along the Egyptian coast as cold air is drawn in from the east European plain along the back side of a strong cyclonic storm in the eastern Mediterranean. In spring *khamsin* (desert) depressions track farther southward through Egypt. They bring warm, dry, dusty southerly air to the coastal region and, on occasion, more severe sandstorms. They may be accompanied by high-level thunderstorms with bases around 3 kilometers above the surface, which may produce much lightning but only a few large drops of melted hail. Cold fronts associated with these depressions may swing rapidly southward across the Sahara and bring the so-called *harmattan* into the Guinea coastal lands of western Africa. The gusty winds associated with the front pick up dust across the Sahara and moisture as they approach the Guinea coast, and arrive in the Guinea area as muggy, dusty weather with low visibility.

During January the flow in the upper troposphere is from the west across the entire northern half of Africa, the easterlies of summer having been displaced to within 10° of the equator and having become very weak and inconsistent. The weak easterlies extend to the south of the equator to approximately 15°S latitude, south of which much of the continent is under the influence of the eastern nose of the south Atlantic subtropical high pressure cell (Figures 6.13 and 6.15).

Tropical Cyclones. The only portion of the African area affected by tropical cyclones is the region adjacent to the western portions of the south Indian Ocean (see Figure 11.3). These occur infrequently from December through April and describe parabolic paths, moving from the northeast toward Madagascar and then recurving to the southeast. Most of them recurve over the ocean to the east of Madagascar and affect only the eastern side of this mountainous island, but some enter the Mozambique Channel to the west of Madagascar and affect the mainland of Mozambique. On rare occasions when they penetrate the mainland, they may bring heavy rains to the interior portions of Mozambique and Zimbabwe. But by far the heaviest rainfall from these storms affects the small Reunion Islands to the east of Madagascar, where a record of 3858 mm. (152 inches) of rain has fallen during a five-day period.

Ocean Currents

Three cold ocean currents profoundly influence the climate along large sections of the African coast, and two minor warm currents add to the tropical and subtropical nature of a couple of other sections of coasts. The most climatically significant ocean current is the Benguela Current along the southwest coast of Africa, which is the second most climatically significant cold current on earth after the Peru Current along the western coast of South America. The Benguela Current flows almost all the way from the Cape of Good Hope at the southwest tip of Africa northward to the equator, and in fact seems to cross the equator into the Northern Hemisphere during July–September to reduce sea-surface temperatures by as much as 8°–11°C (15°–20°F) in the Accra-Lome region of the Ghana-Togo coast of the Gulf of Guinea. The bulk of the current normally turns westward just south of the equator, where it can be traced half-way across the Atlantic, and it probably adds to the atmospheric stability and aridity of Ascension Island, 8°S latitude in the middle of the Atlantic. The current keeps in close juxtaposition with the Namib coast of southwestern Africa northward to Cape Fria south of the Angolan border, after which the coastline recedes northeastward and loses intimate contact with the current. In the Namibian area the sea water undergoes strong upwelling, which produces very cold water close to shore and further stabilizes the overlying atmosphere, which already has a strong, subsidence-induced temperature in-

Figure 18.3. The escarpment inland from the Namib Coast of southwestern Africa. (Courtesy of Lutz Holzner.)

version a few hundred meters above the surface. As a result, this Namib coast is the driest part of the entire south African desert region. The driest point on the coast, Swakopmund (13), averages about 10 mm. (.4 of an inch) of precipitation per year, whereas inland the driest portions of the Kalahari Desert appear to receive about 150–200 mm. per year.

The cold surface water of the Benguela Current causes much fog and low stratus clouds along the Namib coast underneath the inversion. The low-lying clouds are dark and threatening, but measurable rain hardly ever falls, although a light drizzle, called the *moltreen* (moth rain), often keeps the surface damp and may be equivalent to about 50 mm. of rainfall per year (Figure 18.3). About 20 kilometers inland from the coast, the escarpment along the upturned edge of the interior plateau lifts the surface above the cool, damp marine layer into the dry, sunny, but somewhat rainier, atmosphere above the inversion. Three hundred kilometers inland from Swakopmund, Windhoek at an elevation of 1728 meters averages only two-tenths skycover during the year but receives 370 mm. of precipitation annually, whereas Swakopmund averages eight-tenths sky cover and receives only 10 mm. of precipitation. Mean maximum temperatures during November–January average 30°C at Windhoek and only 20°C at Swakopmund.

The cold Canaries Current flows along northwestern Africa and cools and aridifies the climate from about 33°N latitude to around 20°N latitude, south of which the coast turns southward and recedes from the coldest of the water. The Canaries Current is not quite as climatically effective as the Benguela Current, coming as an offshoot from the North Atlantic Drift in the middle of the Atlantic rather than from the cold sea surrounding Antarctica, as does the Benguela Current. In general, the current is not as voluminous nor as swift as the Benguela Current and causes less upwelling. In its origin and extent it is more comparable to the California Current off the Pacific coast of North America, except it is more climatically significant because the coast of Africa adjacent to the Canaries Current trends southwestward, whereas the coast of North America adjacent to the California Current trends southeastward, and therefore loses close contact.

The driest point along the northwest African coast appears to be at Nouadhibou (Etienne), Mauritania, on Cap Blanc (4), about 21°N latitude at the point where the coast discontinues trending southwestward and abruptly changes to a southerly trend. Here the mean annual precipitation is only 27 mm. Other weather conditions are somewhat similar, although not as pronounced, to those along the Namibian coast of southwest Africa. The coast at Cap Blanc is not backed by an escarpment a short distance inland. Nevertheless, the coast of northwest Africa is drier than the immediate hinterland, although certainly much of the interior is equally or even more dry: some parts of southern Libya, northern Sudan, and adjacent Egypt apparently average no more than 2–3 mm. of precipitation per year.

The most unusual of the cold currents flows northeastward along the coast of Somalia during the Northern Hemisphere summer. First, its location on an east coast at equatorial latitudes makes it unique in the world and, second, the current reverses seasonally according to the reversal of the monsoon winds in the western part of the north Indian Ocean. During the Northern Hemisphere winter, the northeasterly monsoons drive a mild warm current of little climatic significance southwestward along the Somalia coast. But during the Northern Hemisphere summer, the strong southwesterlies produce one of the strongest, most consistent cold currents on earth along the coast from about 7°S latitude to 17°N latitude, in the middle of the Arabian Sea. The coolness of the surface water must be due entirely to upwelling, since it is not coming from higher latitudes. During July sea-surface temperatures at 10°N latitude around the Horn of Africa average 5°C colder than they do 300 kilometers out to sea, and on individual occasions the difference is even greater. Ships have recorded temperature decreases of as much as 9°C (16°F) as they approached the east coast of Somalia, followed by a rise of 14°C (25°F) as they rounded Cape Guardafui into the Gulf of Aden.

During summer the entire Somalia coast is engulfed in fog, mist, or haze 40 percent of the time, while in winter such phenomena are rare. Precipitation is meager, ranging from about 20 mm. annually at Cape Guardafui at 12°N latitude to about 400 mm. on the equator. Mean maxima surface air temperatures during July–September average about 29°C along the east Somalia coast, whereas they average about 42°C along the north coast facing the Gulf of Aden. Surface air temperatures are consistently hot in the Gulf of Aden and adjacent Red Sea coastal areas. Dallol, Ethiopia, 75 meters below sea level in an enclosed portion of the down-dropped rift zone near the shore of the Red Sea, averages 34°C (94°F) throughout the year, possibly the world's highest average annual temperature.

A portion of the South Equatorial Current moving westward through the north Indian Ocean moves southward through the Mozambique Channel as a weak warm current and enhances the subtropical nature of the narrow coastal plain along that part of Africa, which is the main citrus-growing region of southern Africa. In the central Atlantic an Equatorial Counter Current moves eastward north of the equator to warm the waters of the Gulf of Guinea, which helps to bring unusually high precipitation amounts to the southwestern portion of the Guinea coast and the Cameroon region at the bend of the continent in the

Bight of Biafra (12, Monrovia; 7, Douala). In between these two wet areas, however, lies a narrow, unusually dry strip in the Ghana-Togo area (1, Accra). In this area the west-east warm current has a slight offshore set where the coastline trends east-northeastward. Also, as mentioned before, during mid-summer in the Northern Hemisphere the northern extension of the cool Benguela Current largely displaces the warm countercurrent in this area.

Climatic Distribution

As can be seen on Figure 15.3, almost the entire continent of Africa has some kind of tropical or hot, dry climate, which is what might be expected of a landmass that straddles the equator, and all parts of which lie within 37° latitude of the equator. Only the very northern and southern tips of the continent are subtropical. Highland topography throughout much of south-central and eastern Africa lifts the area out of true tropical temperatures, so that on a strictly temperature basis much area would classify as subtropical. For instance, Kimberley in the heart of South Africa at an elevation of 1197 meters averages only 11°C (52°F) during June–July and has experienced temperatures as low as −7°C. Similarly, Nairobi, Kenya, near the equator at an elevation of 1798 meters averages only 15°C during July, and the maximum temperature ever recorded is 30°C (86°F). Thus, these upland tropical areas have consistently comfortable temperatures atypical of the tropics. Yet their weather types and precipitation regimes are definitely of a tropical nature, consisting mainly of frequent thundershower activity and a seasonality that follows the sun. A large area of equatorial Africa exhibits rather classical twin-peaked equinoxial rainfall maxima (9, Entebbe), and farther poleward there is a definite alternation of wet and dry seasons so typical of the Aw climate (Figure 18.4). Thus, upland Africa within the tropics, where not too dry to classify as humid, must be classified in some special way, such as upland savanna, since it is neither truly tropical nor subtropical. Large areas of these uplands are vegetated by tall, coarse savanna grasses interspersed by sparse, widely spaced, stunted, gnarled trees, such as the acacia in the big game country of Kenya and adjacent countries (Figure 18.5).

In western Africa, particularly in the Northern Hemisphere, the latitudinal zonation of climatic types is somewhat as might be expected, but as mentioned before there are many hard-to-explain idiosyncracies. The area of tropical wet (Ar) climate in equatorial Africa is unusually restricted, and much of it is less rainy than might be expected. Much of the area so classified receives only 1500–2000 mm. of rainfall per year, the exception being the Cameroon area to the west which in places rivals the wettest places on earth. Yet, even here the precipitation diminishes rapidly southward along the west coast, and at the equator the climate no longer classifies as Ar, but as Aw. And this equatorial west coastal area is penetrated by moist southwesterly air flow year-round. As mentioned earlier, however, this air is somewhat stabilized by mild subsidence in the northeastern portion of the south Atlantic subtropical high pressure cell and northerly extension of the Benguela Current, so that often a slight temperature inversion exists 1–2 kilometers above the surface. The lowest part of the Congo Basin, the Cuvette Central, is somewhat wetter than the highlands all around it. The only explanation seems to be locally derived additional evaporation from the broad swampy surfaces of this portion of the basin.

A detached area of Ar climate is located along the southwestern Guinea coast which, in places, is much rainier than anything in the Congo Basin. Monrovia (12) on the coast of Liberia averages 4624 mm. (185 inches) of precipitation per year. Yet, to the east, a small portion of the coastal area

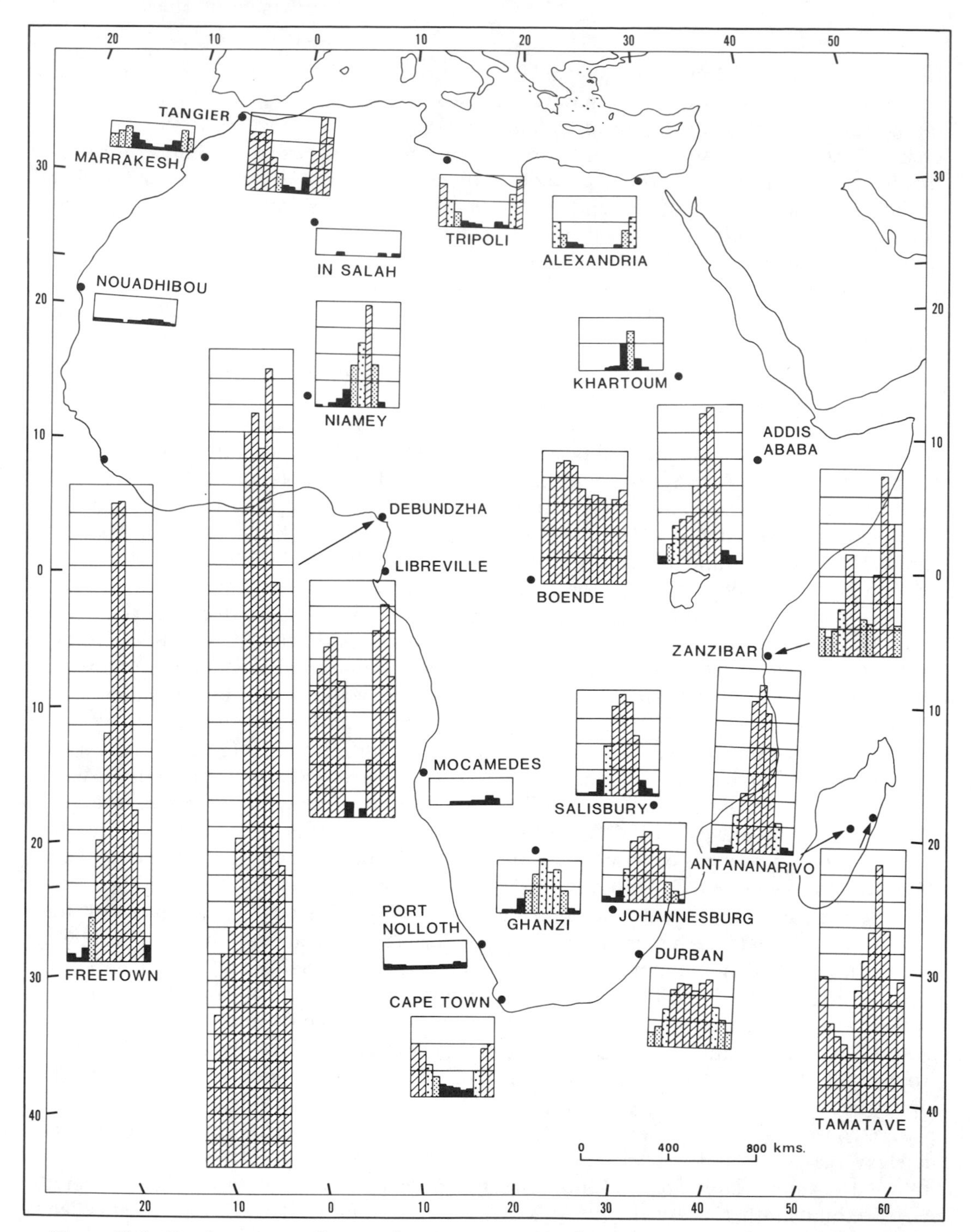

Figure 18.4. Yearly regimes of heat and moisture resources for vegetative growth. (From Agroclimatic Atlas of the World.*)*

between Accra (1) and Lome in Ghana receives less than 750 mm. per year, and perhaps less than 500 mm.

To the north of the tropical rainforest of the Congo Basin and Guinea Coast areas lies a narrow strip of tropical wet-dry climate (Aw) (11, Kano, Nigeria), followed by a narrow strip of low-latitude steppe (BSh), and then the broad Sahara Desert (BWh). The Aw and BS strips are unusually narrow, scrunched equatorward by the unusually broad latitudinal extent of the Sahara. This was explained earlier by the fact that the intertropical convergence zone is not a rain-bringer to this area. Although during mid-summer its surface location lies in the south-central Sahara around 20°N latitude, the slope of the front is upward to the south, with hot dry desert air from the north overriding the maritime tropical air from the southwest. The temperature inversion formed between these two air masses depresses convective activity in the marine air below, and cumulus clouds large enough to produce significant precipitation do not form until the marine layer reaches a great enough thickness, far south of the surface ITCZ.

Figure 18.5. Upland savanna in east Africa. (Courtesy of Lutz Holzner.)

The Sahara is the most extensive dry area on earth, and in fact continues eastward through the Red Sea, the Arabian Peninsula, and into the Middle East and Central Asia, although the causes for the aridity change from subsidence in a subtropical high pressure cell, through north Africa, Arabia, and the Middle East, to interior location behind high mountains in Central Asia. Good records are lacking, but portions of the Sahara (Figure 18.6) may be the most rainless areas on earth, large portions of the interior receiving no more than 2–5 mm. per year (16, Wadi Halfa, Sudan). Also, as expected in a subtropical desert, summers are exceedingly hot. Although higher seasonal and annual temperature averages may be found elsewhere in Africa and the Middle East, the highest individual surface temperature on earth has been recorded at Azzizia in northern Libya, 58°C (136°F). Large diurnal and

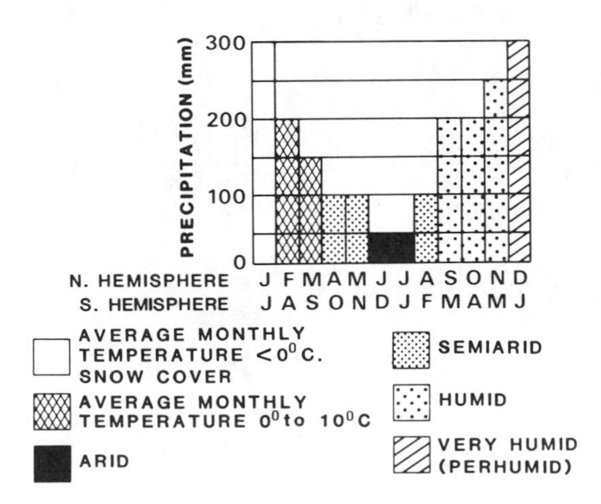

Figure 18.6 The well-watered Nile Valley cuts through the desert near Al Uqsur (Luxor), Egypt. (From the American Geographical Society Collection, University of Wisconsin–Milwaukee Library.)

annual temperature ranges, however, reduce average values in this area.

The northern fringe of the Sahara, particularly in its western portion south of the Atlas Mountains, is ringed by another narrow strip of steppe climate; this area still classifies as a hot steppe, but it differs from the strip of steppe along the southern fringe of the Sahara in its seasonality of rainfall. In the north most of the rain falls from winter cyclonic storms, while in the south it falls during summer thunderstorm activity (see Figure 18.4). The winter cyclonic activity increases on the northern slopes of the Atlas to take the area out of a dry climatic classification and put it into a subtropical, dry-summer (Cs or Mediterranean) type.

South of the tropical rainforest of the Congo Basin lies a very broad strip of tropical wet-dry climate, which is classified as upland savanna over much of the region.

Figure 18.7. The coastal plain south of the escarpment near Cape Town. Settlers have planted elaborate wind breaks against the constant strong winds funneling around the Cape. (Courtesy of Lutz Holzner.)

On the south this merges into a broad area of hot steppe climate, much of which is vegetated by upland thorn forests. In the central and western portions of southern Africa this grades into desert climate in the restricted area of the Kalahari, which, as mentioned before, is most rainless and most extensive along the Namib coast of southwestern Africa.

The very southwestern tip of Africa reaches far enough into the middle latitudes to catch enough rainfall from winter cyclonic storms to put it into a humid climatic category, Cs (Figure 18.7). Cape Town (5) averages 506 mm. of precipitation per year, most of which falls during April–September. Temperatures are constantly mild. January and February average 21°C, and July averages 12°C. The coldest temperature ever recorded is −2°C.

On the southeast coast, temperatures are warmer and precipitation is greater and seasonally reversed. Durban (8) receives 1003 mm. of precipitation per year, the bulk of it falling during November–March. Temperatures average 24°C during January–February and 15°C during June–July. The coldest temperature ever recorded is 4°C.

The eastern Horn of Africa is mostly dry, large portions classifying as upland steppe climate, while the lowlands of Somalia and neighboring Ethiopia and northeastern Kenya classify as desert (3, Berbera, Somalia). As already pointed out, this is a very unusual location for such dry climate. The aridity seems to be due to a combination of monsoon winds that parallel the coast in both seasons, upwelling of cool water during summer and atmospheric divergence around a thermal high over the Ethiopian highlands.

The distribution of climate of the large, mountainous island of Madagascar, which stretches from 12°S to 25°S latitude, is as might be expected in such a location. The northeastern slopes of the mountains are quite wet, receiving upslope winds from the northeast along the western end of the counterclockwise-rotating subtropical high pressure cell in the south Indian Ocean. Rainfall amounts generally average more than 2500 mm. and reach a maximum of 3700 mm. [14, Toamasina (Tamatave)]. They diminish southwestward to only about 340 mm. in a restricted semiarid zone at the southwestern tip of the island [15, Toliary (Tulear)].

Table 18.1 Temperature and Precipitation Data for Africa

	Jan	Feb	Mar	Apr	May	June	July	Aug	Sept	Oct	Nov	Dec	Yearly average
1. Accra, Ghana (5°36′N, 0°12′W, 65 m)													
T.(°C)	27	28	28	28	27	26	25	24	25	26	27	27	27
P.(mm)	16	37	73	82	145	193	49	16	40	80	38	18	787
2. Addis Ababa, Ethiopia (09°02′N, 38°45′E, 2450 m)													
T.(°C)	24	25	26	28	30	29	26	26	28	28	26	25	27
P.(mm)	16	44	70	86	95	136	282	294	192	21	15	6	1256
3. Berbera, Somalia (10°26′N, 45°02′E, 8 m)													
T.(°C)	25	26	27	29	32	37	37	37	34	29	26	26	30
P.(mm)	8	2	5	12	8	1	1	2	1	2	5	5	49
4. Cap Blanc (Etienne), Mauritania (20°56′N, 17°03′W, 4 m)													
T.(°C)	20	20	20	21	22	23	23	24	26	24	22	20	22
P.(mm)	2	1	2	1	1	1	1	3	8	5	5	2	27
5. Cape Town, South Africa (33°54′S, 18°32′E, 17 m)													
T.(°C)	21	21	20	17	15	13	12	13	14	16	18	20	17
P.(mm)	12	8	17	47	84	82	85	71	43	29	17	11	506
6. Casablanca, Morocco (33°34′N, 7°40′W, 49 m)													
T.(°C)	12	13	15	16	18	21	23	23	22	19	16	13	18
P.(mm)	80	68	68	37	22	5	0	1	3	25	77	125	511
7. Douala, Cameroons (4°01′N, 9°43′E, 11 m)													
T.(°C)	27	27	27	27	27	26	25	25	25	26	27	27	26
P.(mm)	57	82	216	243	337	486	725	776	638	388	150	52	4150
8. Durban, South Africa (29°50′S, 31°02′E, 5 m)													
T.(°C)	23	24	23	21	19	16	16	17	18	20	21	23	20
P.(mm)	118	128	113	91	59	36	26	39	63	85	121	124	1003
9. Entebbe, Uganda (0°03′N, 32°27′E, 1146 m)													
T.(°C)	22	22	22	22	22	21	21	21	21	22	22	22	22
P.(mm)	100	86	141	280	257	98	65	91	87	108	146	126	1585
10. Germiston (near Johannesburg), South Africa (26°15′S, 28°09′E, 1665 m)													
T.(°C)	20	20	19	16	13	11	11	13	16	19	19	20	16
P.(mm)	117	101	78	46	25	9	8	6	25	63	110	120	708
11. Kano, Nigeria (12°03′N, 8°32′E, 470 m)													
T.(°C)	21	24	28	31	30	28	26	25	26	27	25	22	26
P.(mm)	0	1	2	8	71	119	209	311	137	14	1	0	873
12. Monrovia, Liberia (6°18′N, 10°45′W, 25 m)													
T.(°C)	26	27	27	27	26	26	25	25	25	26	26	26	26
P.(mm)	51	71	120	154	442	958	797	354	720	598	237	122	4624
13. Swakopmund, Namibia (22°41′S, 14°31′E, 12 m)													
T.(°C)	18	19	18	16	15	16	14	13	13	14	16	17	15
P.(mm)	1	2	2	2	0	0	0	0	0	0	1	0	10
14. Toamasina (Tamatave), Madagascar (18°07′S, 49°24′E, 5 m)													
T.(°C)	27	27	26	25	23	22	21	21	22	23	25	26	24
P.(mm)	420	441	528	404	302	300	257	208	134	87	184	259	3526
15. Toliary (Tulear), Madagascar (23°23′S, 43°44′E, 9 m)													
T.(°C)	28	28	27	26	23	21	21	21	23	24	25	27	25
P.(mm)	71	71	42	6	18	11	4	3	10	14	34	57	341
16. Wadi Halfa, Sudan (21°50′N, 31°15′E, 160 m)													
T.(°C)	15	17	21	26	31	32	32	33	30	28	22	17	25
P.(mm)	0	0	0	0	1	0	1	0	0	1	0	0	3

19 | SOUTH AND CENTRAL AMERICA

Form and Position

Like Africa, South America is primarily a tropical continent. But it is not symmetrically positioned around the equator and is surrounded by ocean, so planetary pressure and wind belts are more as expected than they are around Africa. However, unlike Africa, an unbroken, high mountain chain runs the entire length of the continent near its west coast, which greatly alters the zonal flow of the atmosphere and divides the continent into two almost exclusive climatic parts, the narrow west coast area and the rest of the continent to the east of the mountains. It also produces major atmospheric perturbations that allow the greatest meridional penetration of air masses anywhere in the Southern Hemisphere.

The continent extends much farther poleward than does Africa, to about 56°S latitude, where it comes to within about 7° latitude of the Antarctic Peninsula across Drake Passage. At these high latitudes it is very narrow, having tapered consistently from its broadest point, around 6°S latitude. The equator runs through the northern portion of the continent, of which the northernmost point reaches to a little over 12°N latitude. Central America and the Caribbean islands carry the territory under discussion northward to approximately the Tropic of Cancer.

Topography

The Andes Mountains start in the northwestern part of the continent as three subparallel ranges separated by deep river valleys (Figure 19.1). Elevations range from 4000 to almost 5500 meters (18,022 feet). The ranges converge in Ecuador to form a mountain knot that rises to 6272 meters (20,561 feet). Southward the mountains split again into two major ranges that rise to elevations between 4000 and almost 7000 meters on either side of an intervening high Altiplano, which lies at an elevation around 3500–4000 meters in its broadest portion in Bolivia. The highest peak in the Andes, at 6960 meters (22,831 feet), is Mt. Aconcagua on the Chile-Argentine border at latitude 32°38′S. From there southward the mountains become essentially one range on the border of Chile and Argentina, although to the west there is often an upland vale separated from the Pacific by low coastal mountain ranges. The main range maintains peaks above 5000 meters along the Chile-Argentine boundary to about 35°S latitude, south of which the range lowers and eventually becomes strongly dissected by glacial fiords as the vale to the west dips under the water and the coastal ranges become a labyrinth of offshore islands. However, the main range of the Andes remains an important climatic divide, with a very wet windward side and a much drier leeward side almost to the southern tip of the continent in the island of Tierra Del Fuego across the Strait of Magellan.

The Andes drop off very abruptly on both sides. On the west they often plunge steeply to the Pacific Ocean with little, if any, coastal plain, particularly from Ecuador southward. Through much of Peru and Chile they are somewhat removed from the west coast by much lower coastal ranges and an intervening vale. On the east the Andes plunge to low-lying alluvial basins drained by three major river systems: the Orinoco in the north, primarily in Venezuela; the huge Amazon in the equatorial region, primarily in Brazil; and the Parana-Paraguay system in the south, which drains southern Brazil, Paraguay, northern Argentina, and Uruguay southward into the large estuary of the Rio de la Plata.

The broad alluviated lowlands drained by these river systems separate the Andes on the west from two much more subdued, but broader, uplands on the east, the Guyana Highlands, in the northeast between the Amazon and Orinoco rivers, and the very extensive Brazilian Highlands, in the broad eastern bulge of the continent bordered by the Amazon River on the north and the Paraguay River on the west. Although most of these eastern uplands lie between 500 and 2000 meters above sea level, portions of them rise considerably higher. The highest point in the Guyana Highlands, Mt. Roraima, on the triple point border between Venezuela, Guyana, and Brazil, reaches an elevation of 2772 meters (9,094 feet). Likewise, the Brazilian Highlands reach an elevation of 2890 meters (9,482 feet) at Pico do Bandeiro about 200 kilometers inland from Rio de Janeiro. A continuous, low, interior corridor between the Andes on the west and the Guyana and Brazilian highlands on the east provides easy access for extensive meridional exchanges of air from extreme northern and southern portions through the interior of the continent.

Atmospheric and Oceanic Circulations and Associated Weather

West Coast

The Andes are such a complete climatic barrier that the west coast of South America can be considered as a separate entity from the rest of the continent. The counterclockwise circulation around the eastern end of the subtropical high in the south Pacific Ocean dominates much of the west coast year-round. The shape of the west coast, with its northwestward extension equatorward and the high mountain backdrop inland, forms a near-perfect basin into which the Pacific High can fit snugly and maintain intimate contact with the continent almost to the equator. No other subtropical high pressure cell is so stable in position and so perfectly terminated along a coast.

During the Southern Hemisphere summer, the high is centered at approximately 32°S latitude, and southerly and southeasterly winds along its eastern periphery dominate the west coast of South America from approximately 40°S latitude northward to about 5°S, where the coast turns abruptly northward at the Gulf of Guayaquil. At this point, the southeasterly atmospheric flow diverges from the South American coast, but it continues westward across the Pacific in the vicinity of the equator.

During the Southern Hemisphere winter the center of the high shifts equatorward to approximately 26°S latitude, and southerly and southeasterly winds are a little more limited in latitudinal extent, from approximately 36°S latitude to the equator and beyond. After diverging from the South American coast south of the Gulf of Guayaquil, the southeasterly winds cross the equator and curve to the right to become southwesterlies that affect the southwestern coast of much of Central America during the Northern Hemisphere summer (see Figures 6.8 and 6.9).

The southwesterlies just north of the equator meet with the northeastern trades from the Caribbean that cross Central America to converge with the southwesterlies along the east Pacific section of the intertropical convergence zone, which has an unusually stable location in the Colombian west coastal area. The ITCZ is po-

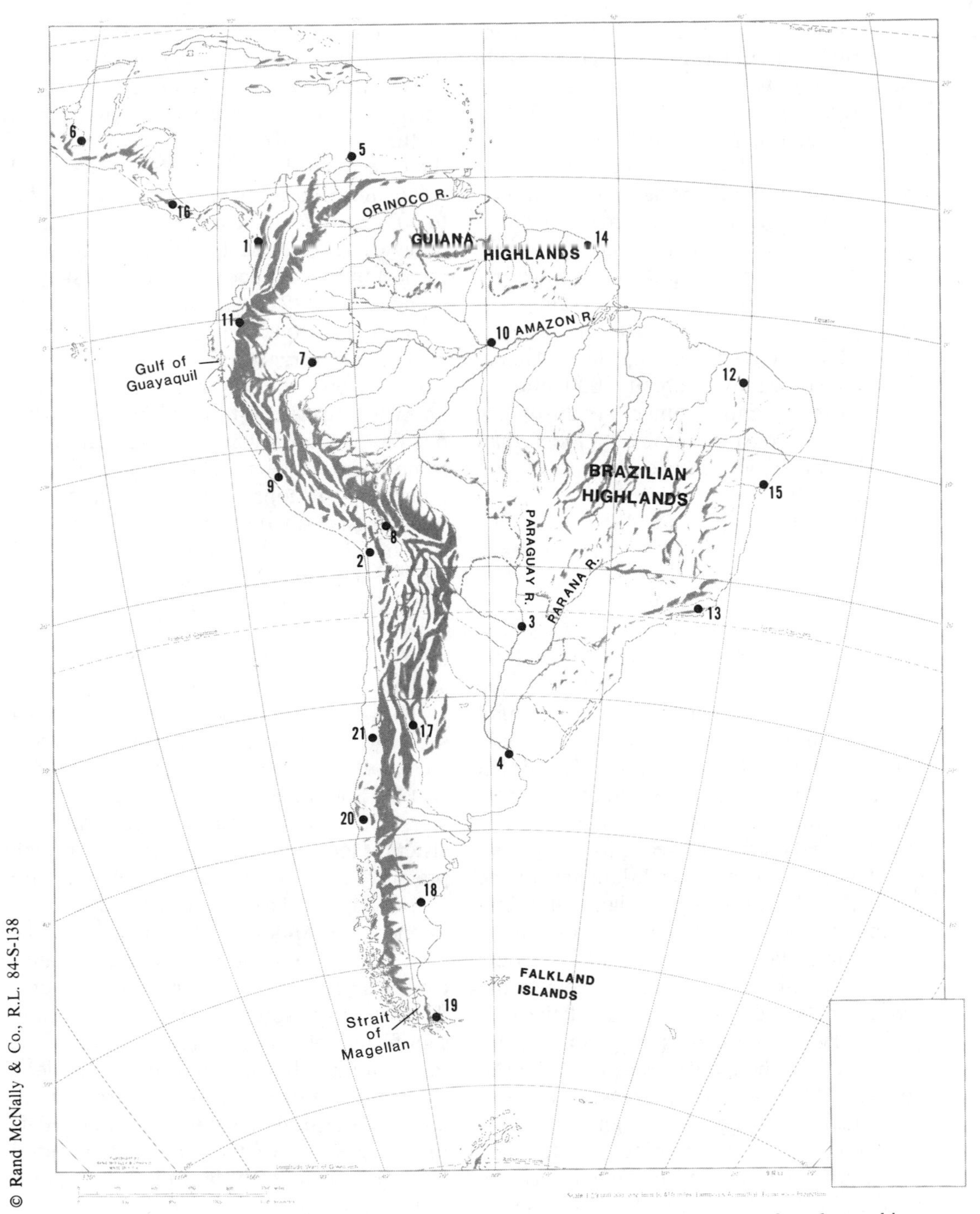

Figure 19.1. Major relief features of South America. Numbers correspond to data tables at end of chapter.

sitioned just north of the equator throughout the year in this area, fluctuating seasonally only about 5° of latitude along the crescent-shaped coast of western Colombia (see Figures 9.3 and 9.4). The consistency of its location and the onshore southwesterly winds bring copious rainfall to this portion of the west coast of South America. Although rainfall records are inadequate, this appears to be one of the four wettest areas of the world, rivaling the rainfall totals of Kauai, Hawaii; Cherrapunji, India; and Cameroon, Africa. Annual amounts range from about 3000 mm. at the northern and southern ends of the Colombian coast to probably more than 10,000 mm. (400 inches) in the central portion.

During the two years 1952–1954, the station of Lloro at 5°31′N latitude averaged 13,473 mm., and during the year 1936, Quibdo at 5°42′N allegedly recorded 19,839 mm. (approximately 800 inches). These stations lie inland from the coast in the hilly portion of the lowland near the base of the Andes, where the belt of heaviest rainfall occurs. In this belt, elevations are only 50–250 meters above sea level. The rainfall decreases up the western slope of the Andes at approximately 5 mm. per meter of elevation, and decreases seaward at approximately 50–100 mm. per kilometer. In the coastal and foothill areas night rains predominate, perhaps as much as 90 percent of the total falling during nighttime hours in the coastal area. Day rains become more common farther inland. Most of the rain falls as heavy downpours from cumulonimbus clouds, although they may be prolonged by several hours of gray overcast and drizzle the next morning. Although the heavy amounts of rainfall seem to be explained simply by the consistency of the location of the ITCZ in this area and by the onshore, unstable southwesterly winds, the site location of the heaviest rainfall on the inland foothills and the nocturnal predominance of rainfall in the coastal area are somewhat of a puzzle, although various theories have been put forth by local weathermen.

Within about 10° of latitude the climate changes from one of the wettest on earth, in the western Colombian area, to one of the driest, south of the Gulf of Guayaquil in northern Peru. From approximately 7°S to 30°S latitude, the coast of Peru and Chile everywhere receives less than 50 mm. (2 inches) of rainfall per year. The driest area, around Arica (2) in northern Chile at the bend of the continent, during the period 1911–1949 averaged 0.7 mm. (0.03 inches) per year, probably the most rainless point on earth. About one-third of that fell during one day in January 1918. Most years there is no measurable rain at all at the three recording stations—Arica, Iquique, and Antofagasta—in this northern coastal portion of Chile stretching from around 18°S to 25°S latitude. These stations lie in the north Chilian province of Atacama, which name is often given to the entire desert coastal strip of the northern half of Chile and the entire coast of Peru.

The extreme aridity along this coast seems to be a function of several reinforcing factors. It has already been pointed out that the South Pacific subtropical high pressure cell is the best-formed and most consistently located of all the subtropical high pressure cells on earth, probably because of the shape of the coastline and the backup of the high Andes Mountains just to the east. The consistent and strong counterclockwise circulation around this high produces the most voluminous and climatically significant cold ocean current on earth, the Peru (Humboldt) Current, which flows along the west coast of South America from about 45°–50°S latitude northward to Point Parinas, just south of the Gulf of Guayaquil at approximately 5°S latitude (see Figures 6.10 and 6.11).

The bulk of the Peruvian Current swings northwestward from that point to continue westward across the Pacific in the vicinity of the equator, but a minor branch, known

Figure 19.2. Upwelling water north of Lima, Peru. Upwelled deep water has lighter color than surface water.

as the Peruvian Coastal Current, continues northward to affect the headlands around the Gulf of Guayaquil all the way to the equator, or even a little beyond. The Peruvian Current draws cold water equatorward from the sea surrounding Antarctica, but a great deal of its coolness is due to very active upwelling along certain portions of the South American coast. Increased upwelling seems to take place wherever a headland juts out into the open-ocean circulation, and it subsides in sheltered embayments, where surface water temperatures are perceptibly warmer. Therefore, its cooling effect varies along the coast, but it is always active to some degree (Figure 19.2).

The juxtaposition of cold ocean water close inshore to a heated tropical and subtropical landmass sets up some of the strongest sea breezes on earth, which in some places may reach almost gale force and blow a great deal of beach sand onto coastal mountain slopes. In some places the sea breeze blows almost 24 hours a day, since the nighttime temperatures over the land remain warmer than those of the ocean surface offshore. Such a consistent temperature and pressure gradient between land and sea in many sectors of the coast sets up a low-level jet in the first 2000 meters or so of the lower troposphere, that blows essentially parallel to the coast from south to north and serves as the core of a helical vertical circulation that carries sensible heat upward from the land surface, out to sea aloft, and down to the sea surface in the subsiding eastern end of the Pacific high pressure cell.* The sea surface remains cold because of upwelling from below, so

*This theory has been proposed by Heinz Lettau.

Figure 19.3. Beach sand driven by strong sea breezes is engulfing the village of Sechura, northern Peru, about 6°S latitude.

the sea breeze at the surface remains strong and the helical circulation carrying heat from land to sea aloft is maintained.

To recapitulate: the high Andes paralleling the west coast of South America abruptly terminate the eastern end of the South Pacific subtropical high pressure cell. This produces maximum subsidence in the coastal area. The consistency of the circulation around this stationary atmospheric cell sets up the largest cold ocean current on earth. This further stabilizes the air in the coastal region by cooling it from below. The shape of the continent, extending northwestward toward the equator, maintains the coast in active contact with the atmospheric and oceanic circulations to lower latitudes than is the case anywhere else on earth. The extreme temperature contrast between the land and the adjacent sea sets up a localized low-level jet with its helical circulation that constantly transfers heat away from the land to the sea. But the sea surface does not warm because of strong upwelling from below. All these reinforcing factors culminate along the north Chilean and Peruvian coasts to produce maximum atmospheric stability and aridity in the immediate coastal area. Precipitation increases inland and, as far as can be determined, out to sea as well (Figures 19.3–19.5).

Special Features of the Desert Coast. Despite its lack of measurable rain, this coastal area is frequented by high surface humidities, dense fogs, and low overcast stratus clouds that often produce a fine mist and keep the streets wet. This is particularly true of the Peruvian sector. Typically, atmospheric subsidence from above and oceanic cooling from below produce a strong temperature inversion a few hundred meters above sea level. In the Peruvian area this

Figure 19.4. An increase in vegetation signifies some increase in precipitation on the hills inland from Tumbes, northern Peru, latitude 3°39′S.

is best developed during winter when the subtropical high has shifted to its equatorward position. At that time, the inversion base at Lima (9) averages about 800 meters above sea level and the top of the inversion about 1500 meters, and the inversion averages about 8°C temperature difference between base and top. Underneath the base of the inversion is a marine layer of cool, moist air carried inland by the sea breeze. Convective activity thoroughly mixes the marine layer and produces a near-adiabatic lapse rate in it. The surface air cools adiabatically as it rises and, being nearly saturated, usually reaches its condensation level before its upward rise is stopped by the base of the inversion. This forms a stratus layer, under the inversion, that averages about 360 meters thick, extending from about 440 meters to 800 meters above sea level. Sometimes the stratus builds down to the ground to produce fog, particularly over the ocean offshore where the base of the inversion is often closer to sea level (Figure 19.6). Fog is also often formed by advection of air from farther seaward that cools as it crosses the cold coastal water on its way inland.

The stratus cloud deck during winter usually persists day and night and absorbs 80 percent or more of incoming solar radiation and almost all the terrestrial radiation. During winter, therefore, diurnal temperature ranges are small. From June through September the mean maximum temperature in Lima is 17°C (63°F) and the mean minimum temperature is 14.4°C (58°F). The average surface air temperature is 15.6°C, which is .6°C warmer than the mean ocean-surface temperature. The mean winter air temperature in Lima is 9°C (16°F) below the latitudinal average.

The meager amount of measurable rainfall on the coast of Peru comes either with

Figure 19.5. Increased precipitation farther inland from Tumbes and higher up fosters the growth of brush and stunted trees.

a mild penetration of tropical air from the north, which may produce widely scattered showers, or an occasional spillover of showers across the Andes from the east, or in far-northward moving Antarctic Fronts coming up from the south. The first two types occur most frequently during the Southern Hemisphere summer, and the latter type during winter. Lima is about the meeting point for these meager precipitation mechanisms. To the north most places have summer maxima of precipitation, while south of Lima a winter maximum prevails. Inland, the lower western slopes of the Andes receive exclusively summer spillover from the east.

In the Lima area, however, measurable rainfall is not the primary concern. During winter, in particular, a stratus cloud deck persists underneath the inversion for weeks on end to produce a fine, penetrating drizzle called *garua.* This mist is most noticeable along the slopes of the coastal hills (*lomas*), where the cloud deck intersects the ground surface at elevations of 150–750 meters. Here winter may receive 100–200 mm. (4 to 8 inches) of drizzle and fog, which is enough to green up the hillsides temporarily and turn them into a prime grazing area. In fact, in the hills north of Lima a mature man-induced eucalyptus grove has maintained itself on this fog drip for more than 30 years. Fog is so persistent along the Pan American Highway overlooking the sea that permanent fog signs have been erected.

During the Southern Hemisphere summer at Lima the atmospheric subsidence is much weaker, since the subtropical high has shifted somewhat poleward, and the summer inversion, which is closer to the surface, shallower, and less strong, is primarily a consequence of the cool ocean. Fog and stratus during this season form offshore over the cool upwelled water and

Figure 19.6. A fog bank hangs over the channel between downtown Lima and mountainous San Lorenzo Island. The inversion limits the top of the fog-stratus buildup below the crest of San Lorenzo Island.

are carried onshore by sea breezes during the morning, but they usually dissipate about noon, when the southerly trade wind flow takes over again, and afternoons are fairly sunny in downtown Lima.

In northern Chile, fog and stratus are not as frequent. In fact, in the coastal portions of the province of Atacama there are less than two days of fog per year. The inversion layer is higher than it is farther north, and the stratus deck, if it does form, is higher, thinner, and usually breaks up by afternoon. Even so, intermediate elevations on the western slopes of the coastal mountains frequently show a vegetation supported by winter fog drip that is similar to the vegetation farther north. The coast ranges here rise abruptly from the sea to heights of 1500–2000 meters. The marine air has a depth of about 900 meters, so it does not override the coast ranges into the Vale of Chile to the east. Hence, this inland area is one of the driest, most cloudless spots on earth. Not even a fog drip moistens the barren landscape. The surface is completely devoid of vegetation, except where a small stream may cross the desert strip on its way from the Andes to the Pacific Coast. Many small streams do this throughout the Peruvian-Chilian desert, and wherever they exist they become a source of water for a thriving strip of irrigation agriculture that interrupts the desert landscape.

El Niño. A few of the tropical thunderstorms that annually reach southward to about Piura in northern Peru, latitude approximately 5°S, in certain years make their way farther southward, sometimes as far as Callao, the port of Lima, around 12°–14°S latitude, to bring brief spells of warm, muggy weather and thunderstorm downpours to this normally desert coast.

The local people call this unusual phenomenon *El Niño* (the child), because it usually occurs in mid-summer around Christmas. (It is actually more frequent in January and February.) The inhabitants, who like to think of natural phenomena occurring in cycles, say that El Niño occurs once every seven years and is particularly severe once every 35 years. What few records are available do not corroborate this regularity, although the frequency is about right.

The phenomenon has attracted much attention because the inhabitants of the area have adapted their lifestyles to almost complete aridity, and when it rains 300–400 mm. in the course of a few weeks it is a catastrophe. The small mountain streams, which normally cross the desert and provide the life-giving water to small strips of irrigated land, become raging torrents and cover fields with coarse rubble, sometimes to a depth of two meters. Villages are wiped out, their adobe huts melted in torrential rain and swept away by floodwaters.

But even more catastrophic on a national scale is what takes place in the ocean offshore. Under normal conditions, upwelling in the cold Peru Current constantly brings nutrients from below to feed some of the richest plankton growth on earth, which supports one of the richest fish populations on earth, which in turn feeds millions of a variety of huge birds that roost on craggy, rocky islands offshore and collectively are known as "guano birds." With the advent of El Niño, the water temperature rises 5°–10°C, much of the plankton dies, and its decomposition releases great quantities of hydrogen sulfide that befouls the air and causes the paint of ships to blacken, a phenomenon known as the "Callao Painter." The fish that normally feed on the plankton either die or migrate southward into colder water, which leaves the guano birds without their normal food supply. Many of them die and litter the beaches, along with dead fish and plankton; others migrate southward and leave their young to die in the nests on the islands offshore.

El Niño has disastrous results on Peru's finances as well as its wildlife and environment. Peru sacks and ships guano (bird droppings) all over the world as fertilizer—which is a nationalized industry that provides much revenue to the national government. Although El Niño usually lasts only a few weeks at worst, it may take the local residents a year or more to recuperate from the calamity. A shortage of guano revenue has even triggered political upheaval.

Many theories have been put forth to explain this unusual phenomenon. The triggering action appears to be the weakening of the southeast trades around the east end of the South Pacific High. This allows the slowing or cessation of upwelling in the ocean and consequent warming of surface water to temperatures slightly above those of the air, which releases the atmospheric convection activity that one would normally expect at these latitudes. Simultaneously, the northeastern trades crossing the Isthmus of Panama into the Pacific may strengthen and push the intertropical convergence zone south of its normal position, but this is difficult to discern since it appears that the entire atmospheric circulation in the eastern Pacific is weak during El Niño, a broad area of doldrums occupying the Ecuadorian and Peruvian coasts during those times.

The Pacific Dry Belt. The aridifying factors along the west coast of South America apparently continue westward along the equator across the eastern and central portions of the Pacific Ocean. (A similar situation occurs with the Benguela Current and the dry Namib coast of Africa, but the Pacific case is a much more pronounced one.) A tapering wedge of dry-to-subhumid climate centered on the equator extends approximately 11,000 kilometers westward from the South American coast to about 160°E longitude. In the South Atlantic west of Africa, the zone is much less extended and is confined to the ocean area south of the equator. In the eastern Pacific, the annual precipitation varies from only

Figure 19.7. Laguna de Vichuguen at about 35°S latitude. Here the Cs climate of central Chile grades into the Do climate of southern Chile. (Courtesy of Clinton Edwards.)

around 50 mm. along the southwest coast of Ecuador, to about 100 mm. in the Galapagos Islands on the equator about 1100 kilometers to the west, to about 700–1000 mm. in the vicinity of Malden and Christmas islands on either side of the equator in the mid-Pacific. The rainfall increases rapidly to the north of this zone. Christmas Island at 1°58′N has an annual rainfall of only 950 mm., whereas Fanning Island at 3°54′N has more than 2000 mm.

Through much of the tropical Pacific there seems to be a double intertropical convergence zone, one on either side of the equator, with a zone of subsidence between them along the equator. Satellite cloud photos tend to corroborate this hypothesis (see Figures 13.5 and 13.6). Throughout the year a continuous narrow cloud band can be seen across the Pacific located about 5°–10°N latitude, and a less-defined one running west-northwestward from the coast of South America to New Guinea and Indonesia south of the equator. A band 5°–10° wide across much of the Pacific along the equator shows a consistent paucity of clouds. Within this dry, relatively cloudless equatorial zone, sea-surface temperatures are consistently a degree or so colder than the surface air temperatures above them, and the atmospheric circulation over this zone diverges, which induces some upwelling in the ocean water below.

Central and Southern Chile. South of about 28°S latitude the rainfall from winter cyclonic storms becomes sufficiently abundant to produce a humid climate. Middle Chile from 28°S to 37°S latitude has been likened to southern California, with its winter rains from cyclonic storms, when the westerlies shift equatorward, and its summer aridity, when the subtropical high shifts poleward (21, Valparaiso) (Figure 19.7).

South of about 37°S latitude the west-

Figure 19.8. Along the Strait of Magellan. (Courtesy of Clinton Edwards.)

erlies prevail year-round and bring cyclonic rain to the southern third of Chile to produce a very cool, humid, cloudy climate similar to the coast of British Columbia and the Alaskan Panhandle in North America. Southern Chile's topography is also very similar to that of the Panhandle of Alaska, with its mountainous terrain deeply cut by glacial fiords into a labyrinth of peninsulas and islands. Thus, orographic uplift is added to the cyclonic uplift to produce copious precipitation in certain areas, primarily rain at lower elevations and snow in the highlands. Valdivia (20) at approximately 40°S latitude receives 3700 mm. of precipitation per year, more than 70 percent of which comes during the winter half-year. It has been estimated that along steep cliffs of coastal mountains between 48°S and 52°S latitude precipitation amounts to more than 6000 mm. per year, with little seasonal variation. On the island of Guarello, at 50°21′S latitude, an annual average of 7330 mm. of precipitation was recorded during a nine-year period. Bahia Felix at the western entrance to the Strait of Magellan during the past 40 years has averaged 320 days of rain annually, accompanied by overcast stratus clouds (Figure 19.8).

Even in this far southern tip of South America, at coastal elevations snow is experienced only 20–30 days per year, but at elevations above 1000 meters the precipitation is almost always in the form of snow. Since temperatures are constantly cool, not all the snow melts, and mountain glaciers accumulate south of 42°S latitude. The great Patagonian ice field, "Hielo Continental," stretches more than 500 kilometers south of 45°30′S latitude and reaches widths of as much as 70 to 80 kilometers. In places along the interior channels of the fiorded Pacific coast, glaciers reach down to sea level.

Even more characteristic of the climate of this southern tip of South America is the incessant wind. The "roaring forties" is an apt name for the latitude south of about 42°S, and it also applies to much of the fifties. The average wind speed at Evangelistas, at latitude 52°24′S, is 12 meters per second (27 mph), with every month recording maxima exceeding 30 meters per second (67 mph). Most of these strong winds are consistently westerly, although they dip southeastward to take on a slight northerly component as they approach the Andes, and occasionally the surface winds may be altered by traveling cyclones, particularly south of the tip of South America. Cyclonic activity and winds are most vigorous during the equinoxial seasons. The persistence of these strong westerlies leads to a very cool marine climate with small annual and diurnal temperature ranges, not only west of the mountains but also across Patagonia to the east.

East of the Andes

The east side of South America south of the equator is also affected by a subtropical high pressure cell. But the landmass lies along the western end of the cell, rather than the eastern end, as is the case along the Pacific coast. Thus, air moves from lower to higher latitudes along the southwest Brazilian coast and drags with it a warm ocean current, the Brazil Current. Therefore, the atmospheric and oceanic flows along eastern South America south of the equator are more akin to those along the coast of southeast North America than they are to the west coast of South America.

The air circulating over eastern Brazil, however, is not nearly as unstable as that circulating across southeastern United States. This may be partially due to the much more easterly position of South America and to the shape of the continent, which bulges far eastward at Cape São Roque, often well within the portions of the high pressure cell where subsidence is significant. The air circulating westward around the northern side of the South Atlantic High reaches the Brazilian coast usually with a weak temperature inversion at an altitude of around 1800–2000 meters. South America tapers poleward, rather than expands, as does North America, so at latitudes comparable to southeast United States the air around the western end of the subtropical high in the South Atlantic has already moved back out to sea.

Also, because of the triangular shape of eastern South America, with the bulge of Cape São Roque south of the equator, the major portion of the South Equatorial Current flowing westward is shunted northwestward along the northeast coast of Brazil, across the equator, and into the Northern Hemisphere, where it passes through the Caribbean to join the Gulf Stream and enhance the North Atlantic Drift, which has such a profound climatic affect on western Europe. The minor portion of the Equatorial Current that strikes the South American coast south of Cape São Roque, and then moves southwestward along the coast of Brazil, is a much weaker current and not very climatically effective, although this coastal area, being largely at tropical latitudes, is quite warm anyway. The northeast-southwest flow of air and water along this coast brings fairly heavy precipitation to most of the immediate coastal area, but is a rather ineffective rain-bringer to the interior.

The northeastern coast of Brazil, the Guyanas, and Venezuela north of the equator are dominated much of the year by the northeastern trades coming around the southern edge of the subtropical high pressure cell in the North Atlantic. These produce heavy rainfall in the coastal area from about 2°S to 9°N latitude, but at either end rainfall decreases rapidly to perhaps less than 500 mm. in portions of eastern Brazil and northern Venezuela. The moist northeasterlies usually extend upward only about 1000 meters, and therefore bring little rainfall to the interior.

The most unstable air, and the greatest rain-bringer to the interior of South America, appears to be a northwest-southeast flow that penetrates the deep interior during spring and summer of the Southern Hemisphere, when a broad thermal low develops over interior South America south of the equator with its center between 20° and 30°S latitude just to the east of the Andes (Figures 19.9 and 6.2). This is probably why the interior of the Amazon Basin is wetter than farther east, where a corridor of tropical wet-dry climate extends almost continuously across the equator to divide the wetter interior from the wet Guyana coastal area. Also, the southeastward penetration of these equatorial westerlies from the north crosses Brazil inland from the Cape São Roque area, which remains quite dry.

Nevertheless, the dryness of the northern Venezuelan coast remains a mystery as far as this northwesterly flow is concerned, and probably must be explained by lack of rain at other times of year, particularly during the Northern Hemisphere summer, when the northeasterly trades enter the Caribbean almost directly from the east parallel to this coast. Frictionally induced divergence between the mountainous coastal area and the water to the north induces subsidence along the coast, and at times appears to induce some upwelling in the water offshore. The north coast of Venezuela has a complex configuration and a complex topography. The eastern branch of the northern Andes, the Cordillera Oriental, swings eastward along the coast in northern Venezuela to produce a mountainous shore area with a coastline that in places is oriented east-west, and in other places north-south. Thus, the atmospheric and oceanic circulations moving from east to west along this region meet with different resistances in different places. Where the coast parallels the atmospheric and oceanic flows the rainfall is meager, and where the coast is perpendicular to the flows rainfall is fairly abundant.

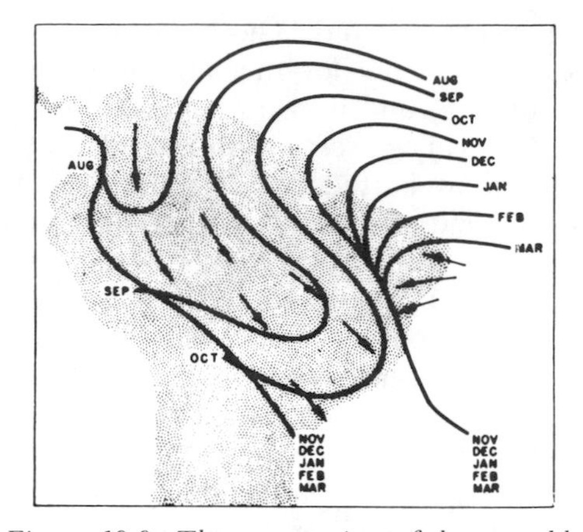

Figure 19.9. The progression of the unstable northwesterlies into interior South America during the Southern Hemisphere summer. (After Coyle, as shown in G.T. Trewartha, The Earth's Problem Climates, *2nd ed. [University of Wisconsin Press, 1981], p. 55. © University of Wisconsin.)*

The other major circulation feature that strongly affects the climate of South America east of the Andes and south of the equator is the succession of cyclonic storms that form along the Polar Front in the south and move northward, either through the interior lowland or northeastward up the coast of Brazil. The most active cyclogenesis region in the Southern Hemisphere appears to be the ocean area to the east of the southern tip of South America and the Antarctic Peninsula, just to the south of the Falkland Islands. These storms are generally too far east to affect significantly the very southern part of South America; but they may penetrate the continent on their northward course in the Argentine Pampas, move up the Paraguay River basin, and continue northward through the topographic break between the Brazilian Highlands and the Andes into the western portions of the Amazon Basin, where cold fronts or *friagems* occasionally move almost to the equator and bring a refreshing spell of cooler air into the upper Amazon during the Southern Hemisphere winter. Many of the storms, however, move northeastward and keep pretty much to sea, affecting only exposed mountainous coastal portions of Brazil, particularly around Rio de Janeiro near the Tropic of Capricorn, or farther north around Salvador at about 13°S latitude, where occurs a curious case of a winter maximum of precipitation in a tropical coastal area, a rare strip of "As" climate.

The Corrientes-Misiones region of northeastern Argentina appears to be an important secondary area of cyclogenesis. After local pressure centers have moved northward in the interior lowland, a secondary wave may form on a cold front trailing through this region and produce a wave cyclone that then moves northeastward and brings more widespread precipitation to southeastern Brazil.

Climatic Distribution

As can be seen in Figure 15.3, highland climates run the full length of the continent along the crest of the Andes and intervening basins and altiplanos. And these separate a narrow north-south string of climates along the west coast from a somewhat different pattern but similar set of climatic types to the east.

West Coast

In the northwest, the west coast of Colombia receives rainfall year-round from the equatorial westerlies in the vicinity of the ITCZ and is one of the rainiest places on earth. The primary recording station in this area, Andagoya (1), has a mean annual precipitation of 6905 mm. (276 inches) but, as was mentioned earlier, some nearby localities may receive much more, particularly a little inland on the foothills of the Andes. Temperatures are consistently warm but not exceedingly hot, averaging 26°–27°C every month of the year, with an extreme maximum of 38.5°C (101°F). These warm temperatures are very sultry, since the average relative humidity is around 80 percent.

The situation changes rapidly southward through Ecuador where Guayaquil receives only 843 millimeters per year with no rainfall at all from July to September. This is definitely a dry form of tropical wet-dry climate (Aw). The climate also varies from the coast to the interior in this region, the coastal area being considerably drier than farther inland (see Figures 19.3 to 19.5). In southern Ecuador west of the Andes, the climate progresses through a very narrow strip of steppe into true desert along the entire coast of Peru and the northern third of Chile. The outstanding characteristics of this coastal desert are its extreme rainlessness, its far equatorward extent, and its relatively cool temperatures for such low latitudes. Lima (9) at a latitude of only 12°S averages only 15°C in July–September and 22°C in February. The maximum temperature ever recorded there was 31°C (88°F), and the minimum, 8°C. The mean annual precipitation is only 10 mm. (.4 inches), but the relative humidity at the surface is consistently high, averaging about 85 percent throughout the year. Vegetation along much of this desert coast is essentially nonexistent and the dune sands constantly shift under the force of the cold sea-enhanced winds (see Figure 19.3).

At about 28°S latitude a narrow strip of steppe climate separates the desert to the north from a subtropical summer-dry climate (Cs) in the central portion of western Chile where typical Mediterranean shrub and grass vegetation prevails. Santiago, at an elevation of 520 meters in the Vale of Chile inland from the coast ranges, receives 356 mm. of precipitation per year, most of it during May–September. Mean monthly temperatures range from 8°C in June and July to 20°C in January. Extremes range from −5°C to +37°C. Hence, this is a consistently mild area, but it can have light frosts from April through October. Summers are dry and sunny, but during winter these conditions are interspersed by brief periods of cyclonic cloudiness and rainfall similar to conditions in southern California.

South of about 37°S latitude the Mediterranean shrub vegetation grades into mixed forest of various varieties of coniferous trees and beeches as annual precipitation becomes much greater and cyclonic storms prevail throughout the year (see Figure 19.7). Portions of the fiorded coastline of western Chile may contain the rainiest spots on earth outside the tropics. Valdivia (20) at approximately 40°S latitude receives 2489 mm. per year with a rather strong winter maximum, but farther south between 48°S and 52°S latitudes portions of the coast appear to receive more than 6000 mm. Also characteristic of this region of southern Chile are the incessant winds and high relative humidities, which make the area feel much colder than it really is. At Ushuaia, Tierra del Fuego (19), monthly mean temperatures range from 2°C during June–August to 9°C during January–February. Extreme temperatures range from a minimum of −19.6°C to a maximum of 29°C. Although this very southern tip of South America has no month averaging above 10°C (Figure 19.10) and hence classifies as tundra climate, in most places it is forested, primarily by conifers. The constantly cool, humid conditions have also fostered a tangle of undergrowth of vines, shrubs, and thick carpets of moss.

Andes

The western countries of South America within the tropics benefit from the height of the Andes, and many of their main settlements are at altitudes above the tropical heat. The only major city at low elevation in western South America between 10°N and 30°S latitude is Lima, which of course, does not have tropical temperatures on the cool ocean coast. The rest of the capital cities and other major centers, including the old Inca settlements, are situated in elevated mountain basins in Venezuela and Colombia and various portions of the Altiplano in Ecuador, Peru, and Bolivia. These tropical highlands experience a type of perpetual spring, mean monthly temperatures showing little variation through the year. But diurnal variations are extreme in this thin, pure air where radiational exchanges of heat are paramount.

Bogota, Colombia, at an elevation of 2556 meters at 4°38′N latitude, has mean monthly temperatures of 12°–13°C throughout the year, but a daily range of temperature averaging 9.3°C. Extreme temperatures range from a minimum of −5°C to a maximum of 25°C. Since all twelve months average above 10°C but occasional frosts do occur, these temperature statistics fall into a subtropical (C) category. Yet the consistency of the temperatures throughout the year do not reflect the typical seasonality of a "C" climate, and the rainfall regime, which has strong twin maxima following the equinoxes, reflects the equatorial position of the station with the double passage of the direct rays of the sun during the year. The high incidences of cloudiness, averaging eight-tenths skycover year-round, and of fog, averaging 167 days per year, also reflect the upland location with upslope

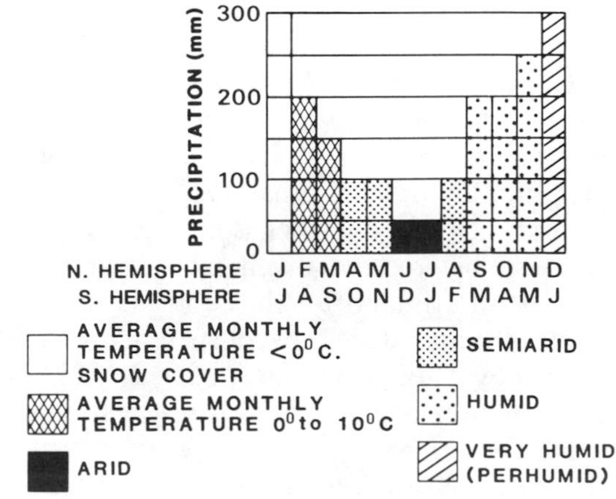

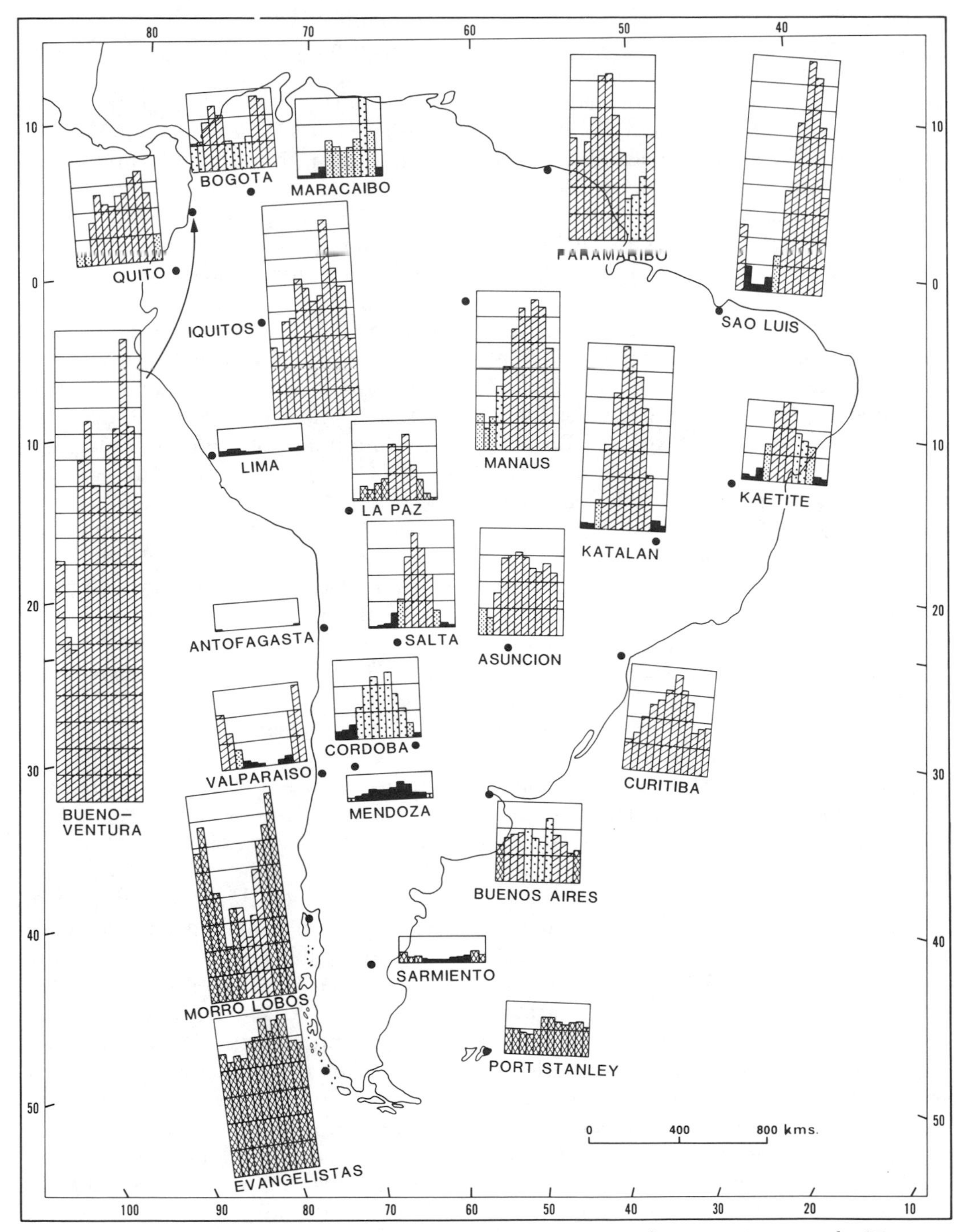

Figure 19.10. Yearly regimes of heat and moisture resources for vegetative growth. (From Agroclimatic Atlas of the World.*) Key on opposite page.*

clouds and fog. Thus, although the climate of these tropical uplands could be classified according to the system used in this book, seasonal and daily regimes and other features of the climate really have no lowland analogs, and so these tropical highlands have a climate of their own. In addition, another climatic element becomes significant for humans: the air pressure, which at Bogota averages only 753 millibars (mb.) during the year (Figure 19.11).

Farther south, Quito, Ecuador (11), almost exactly on the equator at an elevation of 2818 meters (9243 feet), is even higher than Bogota. Its climate reflects the same special characteristics as that of Bogota, except perhaps to a greater extreme. Mean monthly temperatures vary only a fraction of a degree throughout the year, averaging approximately 13°C every month. But because of its lower latitude, it is a little warmer than Bogota, the minimum temperature ever recorded being +2°C. While it would classify as a tropical rainy (Ar) climate, it is really too consistently cool for a tropical climate; its mean monthly daily maximum temperatures range from only 21° to 23°C. However, the annual rainfall regime is an almost perfect example of equatorial equinoxial rains. Mean air pressure at Quito is only 728 mb. (Figure 19.12).

The old Inca center of Cuzco on the Altiplano of Peru, at an elevation of 3312 meters at latitude 13°33′S, has a mean air pressure of only 707 mb. Being farther poleward than Quito or Bogota, it shows somewhat more seasonality, particularly in precipitation, which varies from a mean maximum of 151 mm. during January to only 2 mm. during June. The annual total is only 750 mm., which is usually sufficient in this cool highland for the region's agriculture, much of which is based on grazing of llamas, alpacas, vicunas, and so forth (Figure 19.13). Monthly means of daily maximum temperatures range from 19° to 21°C, and extreme temperatures range from a minimum of −4°C during June–August to a maximum of 27°C during November–December.

La Paz (El Alto), Bolivia (8), at an elevation of 4105 meters (13,464 feet) is the highest of these inland capitals. Air pressure there averages only 625 mb. Mean daily maximum temperatures range from 13°C to 16°C, the seasonal maximum coming during October–November, after the vernal equinox and before the summer clouds and rain. Daily minima average near the freezing point year-round, from −2°C during June–July to +3°C during November–March. Extreme temperatures range from a minimum of −6°C to a maximum of +23°C. Every month of the year has experienced frosts (Figure 19.14).

The inhabitants of these Andean countries are so cognizant of their climatic zones related to elevation that they commonly refer to them, according to their thermal characteristics, as *Tierra Caliente* (hot lands), *Tierra Templada* (temperate lands), *Tierra Fria* (cool lands), and *Tierra Helada* (land of frost). These vertical climatic zones, of course, vary in elevation somewhat with latitude, but near the equator Tierra Caliente extends from sea level to approximately 1000 meters. Mean monthly temperatures in this zone range from 30°C to 24°C. In this hot tropical landscape are grown such crops as bananas, cacao, coconuts, and rubber. The Tierra Templada extends upward from approximately 1000 meters to 2000 meters and has mean monthly temperatures ranging from 24°C to 18°C. This is the zone of coffee, citrus, and sugarcane. The Tierra Fria ranges upward from approximately 2000 meters to 3000 meters, with mean monthly temperatures between 18°C and 12°C. Small grains, such as wheat and barley, and root crops such as potatoes are common in this zone. Toward the upper end of this range, and into the Tierra Helada, are extensive areas of alpine meadowlands on the South American Altiplano known as the Paramos or

Figure 19.11. Sabana (savanna) de Bogota. Fields on the flat surface of an old lake bed at 2600 meters elevation in the upland basin near Bogota, Colombia. (Courtesy of Robert Eidt.)

Figure 19.12. Fields on the Altiplano south of Quito, Ecuador. (Courtesy of Robert Eidt.)

Figure 19.13. Grazing alpacas on the alpine pastures above the tree line on the Altiplano near La Raya, Peru, elevation 4318 meters.

Figure 19.14. The barren, cold Bolivian Altiplano near La Paz. (Courtesy of Robert Eidt.)

Puna. Permanent snowfields and mountain glaciers generally exist above approximately 4800 meters in the South American tropics. But, as was mentioned earlier, the snowline descends to within a few hundred meters above sea level in southern Argentina and Chile, and mountain glaciers may descend to sea level in the fiords of southern Chile.

During summer an upper tropospheric high pressure system develops over the Altiplano of western Bolivia and southern Peru, similar to that over Tibet except on a much smaller areal scale. At the same time, intense convective activity is observed in the form of violent thunderstorms almost every afternoon and evening over the surrounding mountain chains. These *tempestades* are usually accompanied by rain and hail and furious winds, which drive the particles of hail like shot. From about 12°S to perhaps 20°S latitude there are at least 100 days with thunderstorms per year, almost all of them occurring during September–April. Since what few observations exist are generally not in the most mountainous parts of this region, it can be assumed that some places receive even more thunderstorms and perhaps are some of the most thundery places on earth during the summer season.

Of course, the vertical climatic zones are affected by exposure to prevailing winds and precipitation mechanisms as well as to elevation. In the central Andes around the Tropic of Capricorn, the dry climate along the western coast of Peru and northern Chile crosses the mountains in Bolivia, northern Chile, and northern Argentina and continues southward through Argentina east of the Andes. Between latitudes 21°S and 27°S, even the high western range of the Andes and much of the adjacent Altiplano to the east are very dry, with few places receiving as much as 100 mm. of precipitation per year. One recording station near the southern end of this sector gets only 56 mm. per year, all of which falls in a few, brief, widely scattered showers. Many of the high basins in this area have negative moisture balances, and therefore no outflows to the sea. They act as interior drainage basins (*bolsons*) during brief showers, and have accumulated broad *salinas* (salt flats); these are subject to strong wind erosion and produce frequent severe dust storms and blowing salt that descend the eastern slope of the Andes with hot, dry foehn winds into the dry area of the Gran Chaco in northern Argentina and western Paraguay.

North of about 36°S latitude in the western Pampas and Gran Chaco of northwestern Argentina, the rain shadow effect of the Andes is as expected, with the driest part of the territory immediately east of the Andes and precipitation increasing eastward. Much of this area is a hot steppe climate (BSh) grading eastward into subtropical humid (Cf) in the Pampas proper of northeastern Argentina, and adjacent Uruguay, southern Brazil, and southern Paraguay. The western steppe portions of this area receive 300–1000 mm. of precipitation per year with a strong summer maximum, a precipitation regime indicating the moisture source to be primarily to the east along the western edge of the humid tongue of air penetrating the interior of South America from the northwest during the summer. Parts of this region, particularly from about 27°S to 36°S latitude, are similar to the Basin-and-Range country of western United States, with widely scattered fault-block mountains extending north-south 300 kilometers or more through this dry region. Hence, there is considerable variability in rainfall amounts, depending on elevation and exposure to easterly winds. On some of the eastern slopes of mountains, orographic uplift is sufficient to cause the climate to be subhumid to humid, and supports tropical-subtropical jungle vegetation. At a latitude of approximately 32°S, San Juan (17), near the eastern base of the Andes, receives only 87 mm. of precipitation per year, while Cordova, approximately 400 kilometers to the east and on the east side of the easternmost fault block range of mountains, receives 680 mm. Both stations show a marked summer maximum.

South of approximately 36°S latitude, winter precipitation predominates, as it does on the west side of the Andes along the south Chilean coast (18, Sarmiento) (see Figure 19.10). In this region, precipitation decreases eastward through Patagonia. The winter maximum and decrease of rainfall eastward indicate that the primary precipitation mechanisms in Patagonia are the cyclonic storms of the westerlies, which in this lower southern portion of the Andes have some spillover on the east side. This westerly spillover also brings much cloudiness and marine temperatures to Patagonia. Such consistently cool, windy, cloudy, arid conditions on an east coast in middle latitudes is unique in the world. Most stations along the coast of Patagonia record only about 125 mm. of rainfall annually, whereas inland toward the base of the Andes, amounts of approximately 400 mm. are common.

It appears that the Andes induce a midtropospheric trough centered on about 50°W longitude, far east of the Patagonian coast, but intersecting the south Brazilian coast at approximately 30°S latitude. Storms that penetrate Patagonia from the west move across the region in the subsident anticyclonically-curved western limb of the trough and do not experience regeneration until they move northeastward to approximately the Brazilian border. Thus, Patagonia and adjacent parts of the western Pampas receive little rain from these weakened storms, while areas farther northeast receive significant amounts. Of secondary importance might be the stabilizing effect of the cold Falkland Current, which flows northward along the Patagonian coast to approximately 38°S latitude throughout the year (see Figures 6.10 and 6.11). Patagonia gets some winter snow which, in the strong winds that prevail, drifts behind clumps of brown grass 15–20 centimeters high and imparts a look of whitecaps on the sea to this desolate, windswept landscape.

The Pampas of northeastern Argentina, Uruguay, southern Brazil, and southern Paraguay constitute the third large area on earth of subtropical humid climate (Cf), along with southeastern United States and southeastern China and adjacent parts of Japan and Korea. Other Cf areas are located along the coastal strips of southeastern Africa, eastern Australia, and the northern island of New Zealand, but these are considerably smaller in area than the three previously mentioned. The Pampas area is generally somewhat drier and much milder during winter than are the Cf climatic areas of southeastern United States and China. In the Southern Hemisphere continents, outside Antarctica, there is no fresh cP air to bring the cold temperatures that occur in the Northern Hemisphere continents.

Buenos Aires (4), at 34°35′S latitude, during its coldest month, July, averages 10.5°C, and the coldest it has ever recorded is −5°C. Light frosts occur in occasional years, and light snow falls only several times per century. The winters are rather cloudy, averaging about 60 percent skycover, and the surface air is always humid, often 90 percent or more relative humidity, so visitors to the region usually remark about the dreary, chilly winters. Toward the interior, of course, the winters are drier. Summers are warm and humid, as is true of all subtropical humid regions. January averages 24°C, and a maximum temperature of 43°C has been recorded. Annual precipitation in Buenos Aires amounts to 1027 mm. (41 inches), fairly evenly distributed through the year, with winter the driest season. The annual precipitation decreases fairly rapidly west of Buenos Aires to less than 700 mm. (28 inches) about 350 kilometers inland. It increases even more rapidly northeastward to more than 1800 mm. (72 inches) in portions of northeastern Argentina and southern Brazil.

The Pampas, as the name implies, was originally vegetated primarily by middle-latitude grasses, in contrast to the forests of the humid subtropics in southeastern North America and Asia. The reason for the difference appears to be the more infrequent, but heavier, showers in the Pampas region. Based upon the frequency of precipitation alone, much of the Pampas might be classified as steppe. Also, more frequent droughts occur on the Pampas than in the North American and Asian regions. The grass vegetation has developed very fertile black soils, as is usually the case in a subhumid region. In fact, the South American Pampas may well have the best combinations of soil and climate for agriculture anywhere in the world. Nevertheless, large landholdings of haciendas whose economies are based primarily on cattle grazing have slowed the development of this area to its most intensive use for crops. Even so, the Pampas has become the corn and soybean belt of South America.

Figure 19.15. Tropical Atlantic forest along the mainland coast of southern Brazil near Florianopolis. (Courtesy of Augusto Zefferino.)

Rainfall is plentiful along the southeastern coast of Brazil, and much of the coastal area classifies as a tropical rainy (Ar) climate (Figure 19.15). But the rainfall varies greatly over short distances, according to topography and alignment of the coast. Rio de Janeiro (13), almost on the tropic circle, receives a modest 1093 mm. per year, but 300 kilometers to the southwest a spot receives more than 4000 mm. Most of this coastal area, as far north as Cape São Roque, averages between 1200 and 2000 mm. Thus, it is a rainforest climate, but not an exceedingly wet one. Probably the most interesting phenomenon is the reversal of rainfall regime midway up the coast. To the south of 13°S latitude, the area receives about 60 percent of its rainfall during the summer half-year, but to the north of 13°S the area receives 70–80 percent of its annual rainfall during the winter half-year (15, Salvador). This is explained because the northern part of the coast is not directly in line of the summer rains brought by the unstable northwesterly airflow that comes across the interior (Figure 19.9), but the northeastern coast is most prone to cyclonic storms from the southwest during winter.

The Brazilian Highlands inland from the southeastern coast continue to have annual precipitation totals of around 1200–2000 mm., but the seasonality becomes much more marked, and the climate changes to a typical, tropical, wet-dry climate. The long drought period during the low-sun season changes the vegetation from a tropical rainforest along the coast to a scrubby, brushy, thorn forest and tropical grassland known as *caatinga.* Because of the seasonal difference in rainfall and cloudiness, surface temperatures may not reflect the seasonal shifting of the sun. Often the hottest periods occur at the end of the dry season during spring. For instance, at For-

Figure 19.16. The unbroken rainforest of large broadleaf evergreen trees in flat interior Amazon Basin. (Courtesy of Norman Stewart.)

mosa, latitude 15°32′S in the middle of the Brazilian Highlands, September and October are the two hottest months, averaging 23°C, 1°C warmer than any of the summer months. Its elevation of 912 meters above sea level somewhat meliorates the tropical temperatures. The maximum ever recorded is 36°C; the minimum 5°C. Hard frosts occur on some of the higher parts of interior plateaus.

The easternmost portions of the Brazilian Highlands are much drier than the rest of Brazil, as was mentioned earlier. Inland 100 kilometers or so from Cape São Roque, annual precipitation drops below 400 mm. With temperatures averaging 26°–29°C throughout the year in this region at only 6°–8°S latitude, such rainfall is indeed inadequate. A fairly large area classifies as hot steppe climate (BSh), and limited portions may classify as hot desert (BWh).

Much of the Amazon Basin and adjacent river systems, as well as the coastal portions of the Guyana Highlands, classify as tropical rainy climate (Ar) (10, Manaus). But in many cases there is considerable seasonality to the rainfall and, as mentioned previously, areas of tropical wet-dry climate separate a rainier Guyana coastal area from a rainier upper Amazon Basin. Much of the coastal area from the mouth of the Amazon northwestward through the Guyanas receives 2000–4000 mm. of precipitation annually (14, Rochambeau). Some of the higher areas slightly inland receive more than 4000 mm., and some of the windward hills of French Guyana may receive as much as 8000 mm (320 inches). Although there is adequate precipitation year-round in this region, there is usually two to three times as much precipitation during the first half of the year as during the second half. It is difficult to explain

Figure 19.17. Clearing the rainforest in eastern Peru. A small patch of corn is being grown to the left-rear of the hut. The dense forest looms beyond. (Courtesy of Norman Stewart.)

this rather peculiar rainfall seasonality—a maximum during winter–spring and a minimum during summer–fall. In many cases in this region, May–June is the wettest season, and September–October is the driest, with only about one-sixth as much rainfall as May–June.

Precipitation diminishes to about 1500 mm. inland a few hundred kilometers from the mouth of the Amazon and then increases again in the western part of the basin to 2400–2800 mm., and as much as 3200 mm. or more on the Brazilian-Colombian border. Iquitos, Peru (7), on the upper Amazon River, averages 2845 mm. of precipitation per year with a March–April maximum and a July–August minimum. Dense rainforests extend upward to approximately 1800 meters on the eastern slopes of the Andes in this region (see Figures 19.16 and 19.17).

Temperatures in the Amazon Basin, of course, are constantly hot. Daily maximum temperatures at Iquitos average 30°–32°C every month of the year. Maximum temperatures, however, are moderated by the high relative humidity, cloudiness, and precipitation. The maximum temperature ever recorded at Iquitos is only 37°C (98.6°F), and 100°F temperatures have never been observed in the tropical rainforests. Even so, the constantly high temperatures and high humidities, without respite, are very enervating.

To the north of the tropical rainforest of the Amazon, in much of Venezuela and adjacent portions of eastern Colombia, the climate again becomes a tropical wet-dry one, similar to that of the Brazilian Highlands farther south. Poleward, at about 10°N latitude, summer rains diminish, and portions of the coastal strip of northern

Venezuela and Colombia classify as dry climate, as was described earlier. This is one of the more enigmatic dry climatic anomalies on earth, with some east-west sections of coastline that parallel the prevailing winds and ocean drift receiving less than 500 mm. of annual precipitation, and other sections, where the coast is oriented north-south and backed up by low mountains, receiving as much rainfall as 1500–2000 mm.

The aridity continues northwestward from the South American coast into the southwestern portions of the Caribbean. Some of the best records are kept on the Dutch Islands: Aruba, Curaçao (5), and Bonaire just north of the Venezuelan coast. These islands have relatively low relief, and a composite of their stations averages 569 mm. of precipitation per year. They have an unusual rainfall regime as well as unusually low annual totals: 75 percent of the precipitation comes during the fall-winter months of September through January, only 16 percent occurs during February–June, and July–September are also relatively dry. This pattern is particularly unusual in light of the summer rainfall maximum in the Llanos in Venezuela, just to the south. This reversal of seasonality between the interior Llanos and north coastal region has prompted some analysts to theorize that the drier coastal area is under subsidence during summer, which is the compensatory motion for convection over the heated Llanos of the interior.

Central America and the Caribbean

Central America and the islands of the Caribbean all lie within the tropics, from about 8°N to 23°N latitude. They are dominated by the subtropical highs on either side in the North Atlantic and the North Pacific. Their generally southeastward trend toward the equator keeps them in much closer contact with the northeasterly airflow around the western end of the North Atlantic High than with the northwesterly airflow around the eastern end of the North Pacific High. In general, the intertropical convergence zone lies well south of this region, around 5°N latitude, which is an uncommonly low latitude for the convergence zone during the Northern Hemisphere summer. Thus, the Caribbean is relatively free of incursions of any air flows other than the northeasterlies from the Atlantic, except perhaps in late summer and early autumn in its very southwesterly portion along the north coast of Panama. At that time of year, the equatorial southwesterlies from the Pacific alternate with the northeasterlies across much of the western portion of Central America, including the entire Isthmus of Panama (Figure 19.18).

Occasionally during winter, a cold surge of air from the North American continent crosses the Gulf of Mexico and enters the Caribbean as the well-known *nortes* of Mexico, northern Central America, and Cuba and adjacent islands. These and easterly waves imbedded in the trade winds are the principal rain-bringing mechanisms to eastern Central America and the Caribbean during winter. Also, the entire region is affected by hurricanes, primarily from May through November, with September usually the month of maximum occurrence. Most of these form in the middle or eastern parts of the North Atlantic between 10° and 20°N latitude and proceed westward into the Caribbean, where they can hit any of the islands or coastal areas of Central America. However, many of them recurve northeastward into the Atlantic and do not penetrate this area. Tropical storms also form in the eastern Pacific, but these are usually farther north along the southwest coast of Mexico.

Since all of Central America and the Caribbean islands are mountainous, with intervening lowlands and plateaus, the cli-

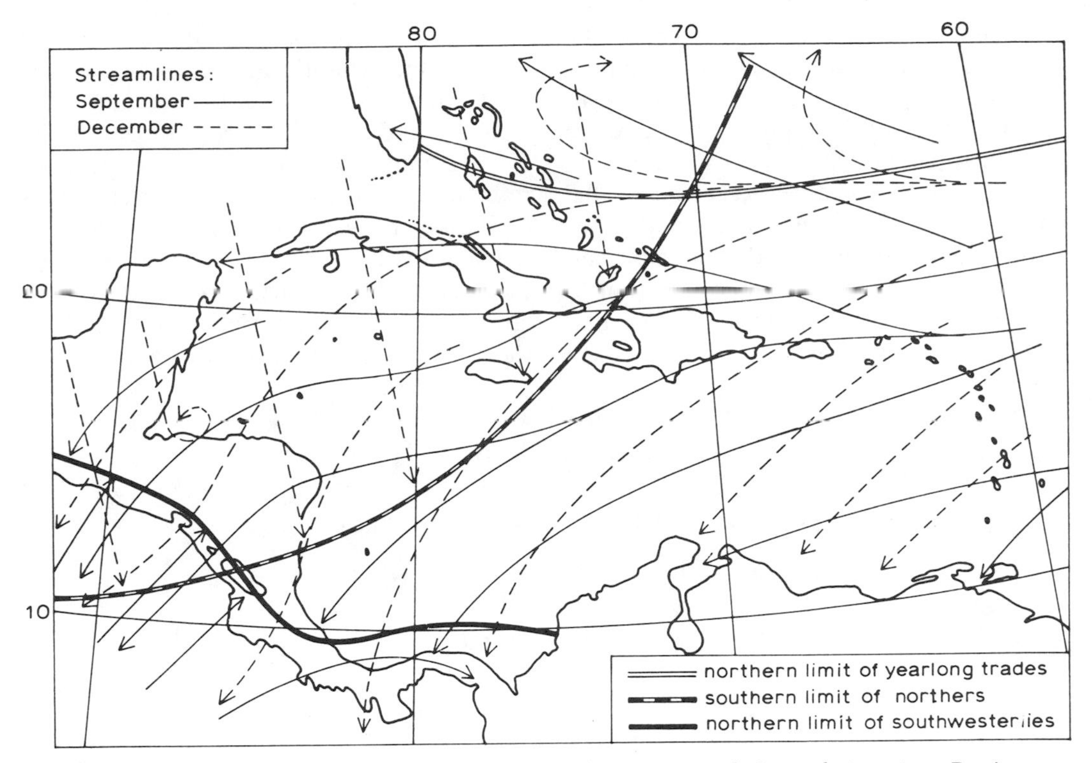

Figure 19.18. Surface air flows over the Caribbean Sea and Central America. During September both the northeast and southwest wind regimes affect the Pacific coast of Central America. (After W. Schwerdtfeger, Climates of Central and South America, *Vol. 12 of* World Survey of Climatology *[Amsterdam: Elsevier, 1976], p. 408.)*

mate is varied from place to place even though the entire region lies within the tropics. This variation applies primarily to amounts and seasonalities of rainfall, but of course temperatures vary with elevation and exposure, as well. The northeastern coasts, which are usually the windward ones, are usually wetter than the southwestern coasts, which are usually the leeward ones, although, as was pointed out earlier, southwestern coasts of Central America are often frequented by the equatorial southwesterlies during late summer and early autumn. Annual rainfall amounts in most places total about 1200–1600 mm., but they range from as little as 800 mm. in west-central Honduras to apparently more than 6000 mm. along the southeast coast of Nicaragua. Similar ranges may occur in some of the larger mountainous islands in the Caribbean. Kingston, on the south coast of Jamaica, receives only 811 mm. per year, and it has been estimated that portions of the south coast of eastern Cuba receive even less. On the other hand, perhaps 6000 mm. or more are received in small mountainous portions of Jamaica, southern Guadeloupe, and northern Martinique. Climatic types thus range from tropical rainy (Ar) through tropical wet-dry (Aw) to hot steppe (BSh). But dry climates are more prevalent farther north in Mexico.

Since the northeasterlies are the most persistent winds in the region, usually there is no marked dry season on the northeastern side of Central America, and most of that area classifies as tropical rainy, whereas the western slopes toward the Pacific generally classify as tropical wet-dry. But because of the complexity of the topography, there are variations in both cases. Almost everywhere, with the notable exception of the northern coast of Honduras, the precipitation maximum comes during summer. But this is often interrupted by a midsummer secondary minimum, the so-called *veranillo,* which appears to be related to a maximum expansion of the subtropical high in the Atlantic at this time of year (6, Guatemala City). Often September or October, and sometimes November, are the wettest months of the year. These autumn rains may be augmented by hurricane precipitation in the Caribbean area, while on the southwestern coast of Central America they relate to the most northeasterly penetration of the equatorial westerlies from the Pacific (16, San Jose, Costa Rica).

Table 19.1 Temperature and Precipitation Data for South and Central America

	Jan	Feb	Mar	Apr	May	June	July	Aug	Sept	Oct	Nov	Dec	Yearly average
1. Andagoya, Colombia (05°06′N, 76°40′W, 65 m)													
T.(°C)	27	27	28	28	27	27	27	27	27	27	27	27	27
P.(mm)	554	519	557	620	655	655	572	574	561	563	563	512	6905
2. Arica, Chile (18°28′S, 70°22W, 29 m)													
T.(°C)	22	22	21	20	18	17	16	16	17	18	19	21	19
P.(mm)	0	0	0	0	0	0	0	0	0	0	0	0	1
3. Asuncion, Paraguay (25°16′S, 57°38′W, 64 m)													
T.(°C)	29	29	27	24	21	19	18	21	22	25	27	29	24
P.(mm)	107	110	160	130	131	87	61	30	87	146	138	133	1302
4. Buenos Aires, Argentina (34°35′S, 60°56′W, 81 m)													
T.(°C)	23	22	20	15	13	10	10	11	13	15	19	22	16
P.(mm)	121	90	111	83	55	59	38	32	56	99	96	101	941
5. Curacao (12°12′N, 68°58′W, 8 m)													
T.(°C)	26	26	27	27	28	28	28	28	29	28	28	27	28
P.(mm)	68	31	14	12	18	26	34	48	31	67	98	85	532
6. Guatemala City, Guatemala (15°29′N, 90°16′W, 1300 m)													
T.(°C)	16	17	18	20	20	19	19	19	18	18	17	16	18
P.(mm)	3	2	7	19	141	265	211	187	257	159	23	7	1281
7. Iquitos, Peru (03°46′S, 73°20′W, 104 m)													
T.(°C)	27	27	27	27	26	26	25	27	27	27	27	27	26
P.(mm)	256	276	349	306	271	199	165	157	191	214	244	217	2845
8. La Paz (El Alto), Bolivia (16°30′S, 68°12′W, 4105 m)													
T.(°C)	8	8	9	8	7	6	6	7	7	9	10	9	8
P.(mm)	116	110	66	26	14	3	6	12	42	29	44	96	564
9. Lima, Peru 12°00′S, 77°07′W, 11 m)													
T.(°C)	23	23	23	21	19	17	16	16	17	18	19	21	19
P.(mm)	1	0	1	1	1	1	2	2	1	0	0	0	10
10. Manaus, Brazil (3°08′S, 60°01′W, 48 m)													
T.(°C)	26	26	26	26	26	27	27	28	28	28	28	27	27
P.(mm)	266	247	269	267	194	100	64	38	60	124	152	216	1996
11. Quito, Ecuador (00°13′S, 78°30′W, 2818 m)													
T.(°C)	15	15	15	15	15	15	15	15	15	15	15	15	15
P.(mm)	94	99	134	171	99	52	24	25	69	139	117	86	1109
12. Quixeramobim, Brazil (5°12′S, 39°18′W, 198 m)													
T.(°C)	29	28	27	27	27	26	26	27	28	28	29	29	28
P.(mm)	67	108	188	169	111	54	26	9	3	2	6	21	763
13. Rio de Janeiro, Brazil (22°54′S, 43°10′W, 31 m)													
T.(°C)	25	26	24	24	22	21	20	21	21	22	23	22	23
P.(mm)	157	125	134	102	63	56	51	40	63	80	92	139	1093
14. Rochambeau (Cayenne), French Guiana (04°50′N, 52°22′W, 8 m)													
T.(°C)	25	25	26	26	26	25	25	26	26	26	26	25	26
P.(mm)	431	423	423	480	590	457	274	144	32	42	122	317	3744
15. Salvador, Brazil (12°55′S, 38°41′W, 45 m)													
T.(°C)	26	26	26	26	25	24	23	23	24	25	25	26	25
P.(mm)	74	116	165	278	296	225	204	116	98	102	116	124	1913
16. San Jose, Costa Rica (9°56′N, 84°08′W, 1120 m)													
T.(°C)	19	19	20	21	21	21	21	21	21	21	20	19	20
P.(mm)	8	5	10	37	244	284	230	233	342	333	172	46	1944
17. San Juan, Argentina (31°36′S, 68°33′W, 630 m)													
T.(°C)	26	24	21	16	12	8	8	11	14	18	22	25	17
P.(mm)	18	12	9	5	1	2	2	2	5	9	12	10	87
18. Sarmiento, Argentina (45°35′S, 69°08′W, 266 m)													
T.(°C)	17	17	14	11	7	4	4	6	8	12	14	16	11
P.(mm)	10	8	11	15	24	16	17	15	10	6	12	9	153
19. Ushuaia (Tierra del Fuego), Argentina (54°48′S, 68°19′W, 6 m)													
T.(°C)	9	9	8	6	3	2	2	2	4	6	7	9	6
P.(mm)	58	50	57	46	48	45	47	49	38	37	50	49	574
20. Valdivia, Chile (39°48′S, 73°14′W, 9 m)													
T.(°C)	17	16	15	12	10	8	8	8	9	12	13	15	12
P.(mm)	65	69	115	212	377	414	374	301	214	119	122	107	2489
21. Valparaiso, Chile (33°01′S, 71°39′W, 41 m)													
T.(°C)	18	18	17	15	14	12	12	12	13	14	16	17	15
P.(mm)	2	2	4	18	97	128	88	67	30	16	7	3	459

20 | AUSTRALIA—NEW ZEALAND

Australia

The Physical Setting

Australia is undoubtedly the simplest continent topographically and climatically. It is located astride the Tropic of Capricorn, well removed from other major landmasses and without barrier mountain ranges (Figure 20.1). Highest elevations lie in the Blue Mountains near the east coast, but these generally are no higher than the Appalachians in eastern United States. The highest peak, Mt. Kosciusko, in what are known as the Australian Alps in the southeastern part of the country, reaches an elevation of only 2230 meters (7316 feet). Small mountain groups scattered about other portions of the country, including the island state of Tasmania in the southeast, generally lie below 1525 meters (5000 feet). Although they exert local influences on climate, particularly where they rise abruptly near coasts, they are not continuous nor extensive enough to alter significantly the broadscale flow of air across the continent.

Primary Air Flows and Precipitation

There are few climatic surprises in Australia. The entire continent is dominated by the subtropical high pressure belt stretching across its mid-section, and the climatic pattern relates to this dominant feature much as one might expect. The driest part of the continent is the central portion, which is dominated by high pressure and subsiding air year-round. The driest spot, southeast of Alice Springs (2), apparently averages about 125 mm. (5 inches) of precipitation per year. Rainfall amounts increase outward in all directions, particularly toward the northeast. Annual regimes reflect the latitudinal shifting of pressure and wind systems with the movement of the sun (Figure 20.2).

During December–February (summer), the belts shift poleward. Subtropical high pressure cells move west-east with their centers well south of the southern coast and allow little rain in this area. At the same time, the north coast experiences its wet season, which is very wet in the northeast. The northeast coast of Queensland at this time of year is crossed almost perpendicularly by the deflected southeast trades, which turn east and northeast as they enter into the clockwise circulation of the Cloncurry Low (4) to the west of the mountains (Figure 20.3). At the same time the eastern half of the north coast westward to Darwin (5) is under the influence of a rainy northwesterly flow of air known in Australia as "the monsoon," which is a Southern Hemisphere deflection of the northeasterlies coming across the western Pacific and Indian Oceans north of the equator during the winter monsoon season. Innisfail (7), at a latitude of 17°32′S along the Queensland coast, during summer is generally located in the confluent zone between these two air flows, and as a result

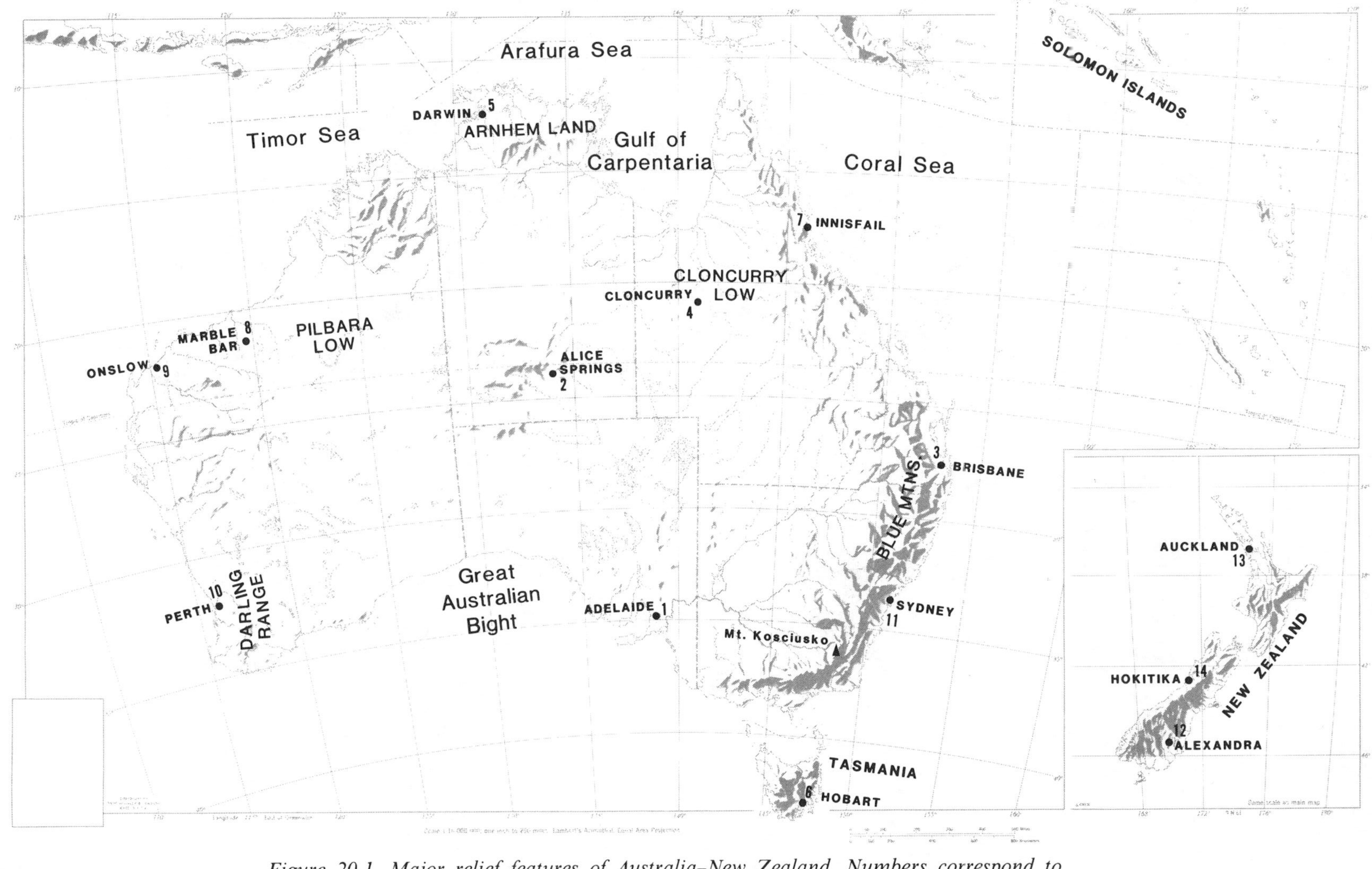

Figure 20.1. Major relief features of Australia–New Zealand. Numbers correspond to data tables at end of chapter.

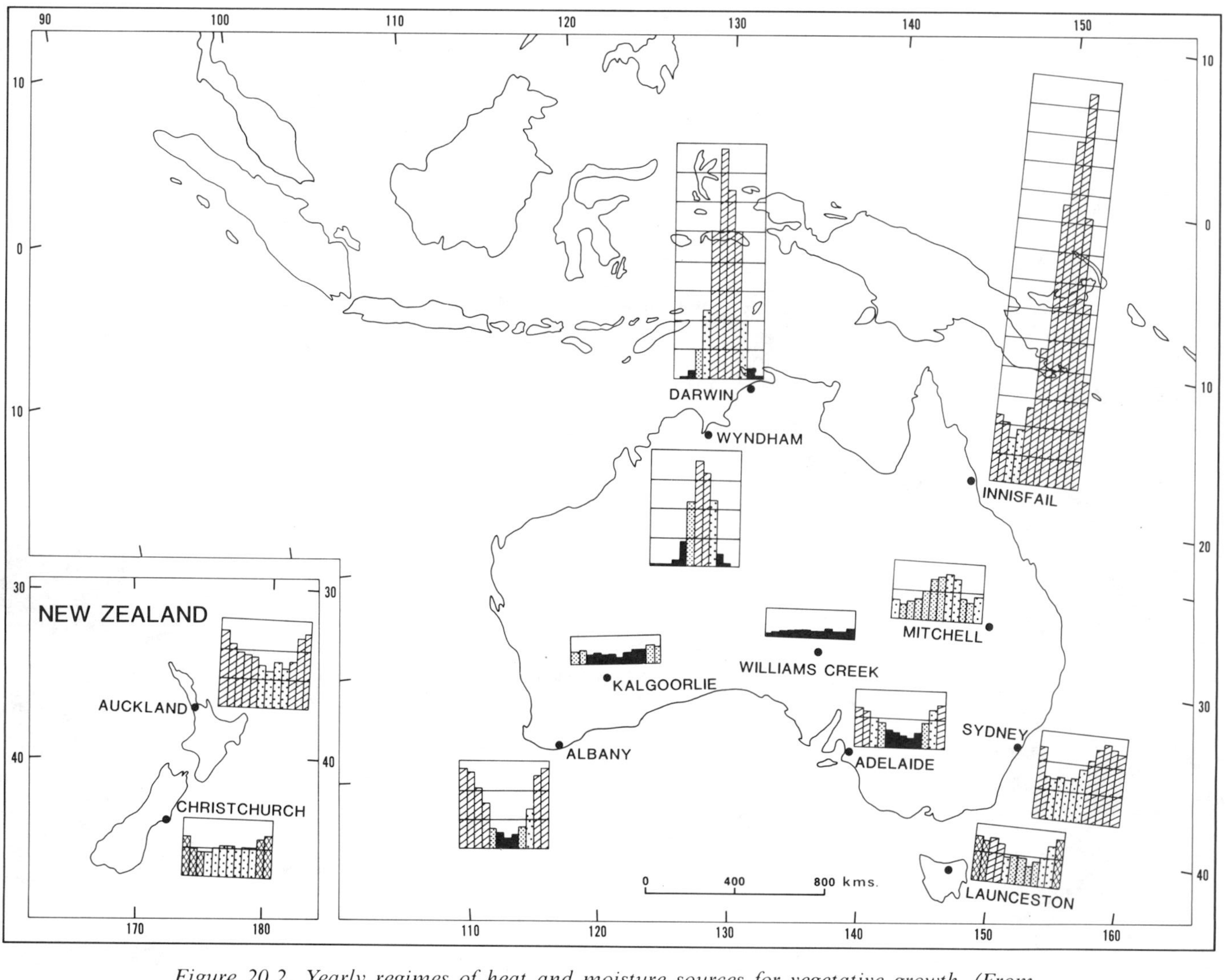

Figure 20.2. Yearly regimes of heat and moisture sources for vegetative growth. (From Agroclimatic Atlas of the World.*)*

records the greatest annual rainfall in Australia, 3535 mm. (139 inches), with a pronounced maximum in December–May. March is the wettest month, with 686 mm. (Figure 20.4).

The western half of the north coast also receives northwesterlies and westerlies, but these have not had long trajectories over the sea. They are extensions of a southerly flow of air that has moved northward along the west coast and then curved inland into the Pilbara heat low that dominates western Australia during summer. These westerlies do not have the moisture content nor the degree of instability of the northwest monsoons farther east. Therefore, the Australians call these short westerly trajectories that enter the Pilbara Low the "pseudo-monsoon."

During June–August (winter) the pressure and wind belts shift northward, and the north coast comes under the influence of northern edges of subtropical high pressure cells, which now have their centers in the southern part of the continent. Along the north coast there is hardly any rain at this time of year, but farther east along the Queensland coast there is still some rain. Cyclonic storms in the southern ocean now migrate west-east along a more northward track, and their northerly extensions of cold or occluded fronts between high pressure cells bring appreciable rain to the two southern tips of Australia. Perth (10), in the southwest, receives 889 mm. of precipitation per year, most of it during May–August; in June it receives 192 mm. and in January only 7.

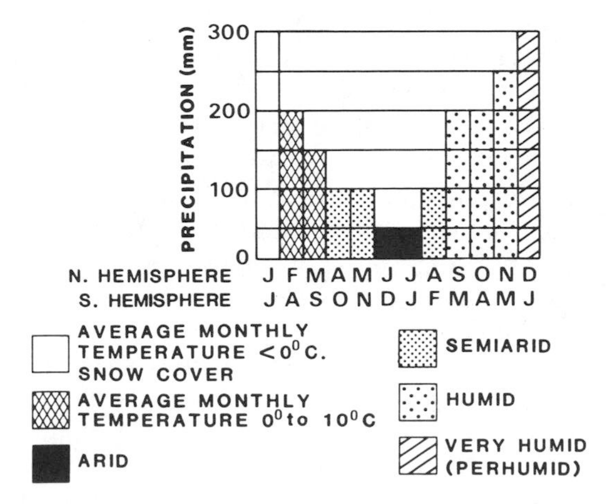

The southeastern part of the country, including the mountainous island of Tasmania, receives some rain both seasons, so there is a fairly even distribution of precipitation through the year. But in the extreme south the winter half-year has significantly more rain than the summer half-year. This reverses around Canberra, north of which summer is definitely the wetter season.

Perhaps the only significant "abnormality" about the air flow across Australia is its normalcy, as compared to other continents in similar latitudes. Whereas the subtropical coasts of western South America, southwest and northwest Africa, and, to some extent, the Baja California coast of North America, situated along the eastern edges of stationary subtropical high pressure cells, are characterized by extreme atmospheric subsidence, cold offshore ocean currents, almost rainless conditions, and much low stratus and fog, the west Australian coast finds itself under a series of moving high pressure cells, with only moderate subsidence and aridity, weak and variable ocean currents, and little stratus and fog. The subtropical highs are far less stationary in the Indian Ocean than in the other oceans of the world, and all year long there is a constant progression from west to east of individual high pressure cells interspersed by weak troughs, which bring some precipitation even to the mid-sections of the western coast and the interior. No place along the western coast of Australia receives less than 225 mm. (9 inches) of precipitation per year. This compares to an almost total absence of measurable rainfall in northern Chile–southern Peru and in the driest parts of the southwest and northwest African coasts. And while those

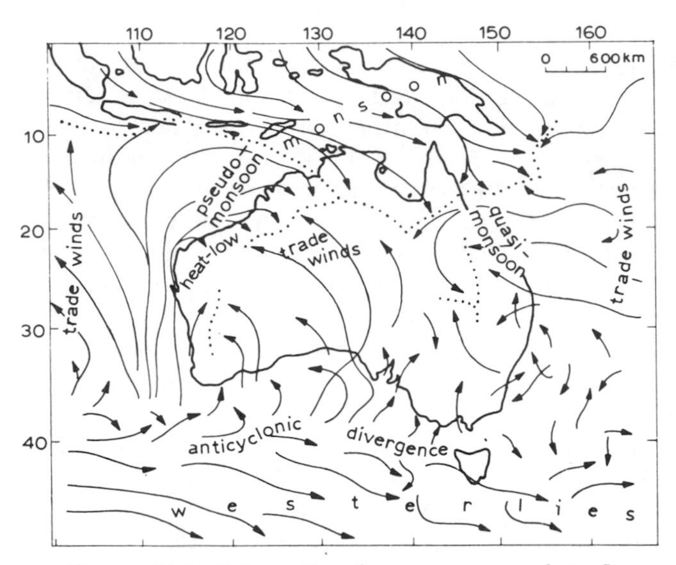

Figure 20.3. Schematic of mean near-surface flow pattern in January. Dotted lines indicate convergence zones, which shift significantly. (After J. Gentilli, Climates of Australia and New Zealand, *Vol. 13 of* World Survey of Climatology *[Amsterdam: Elsevier, 1971], p. 84.)*

coasts are drier than their interiors, the west Australian coast generally receives 75–100 mm. of precipitation more per year than interior parts of the Australian desert, which is not excessively dry, either. Thus, while Australia contains one of the most extensive desert areas on earth, it is by no means one of the driest.

The reasons for the inconstancy of location of subtropical high pressure cells in the South Indian Ocean, and the consequent lack of formation of a cold ocean current and associated weather conditions, appear to be the shape and topography of the western part of Australia, and perhaps the general lack of land at that latitude in the Southern Hemisphere, which allows for a very strong zonal flow in the upper troposphere. The western coast of Australia is convex toward the west, unlike the South American and North American coasts, and is not bordered inland by a continuous high mountain range, although large portions of it are backed up by an escarpment 50–100 kilometers inland.

The subtropical high pressure cells usually move at a rate of about 800 kilometers a day. After crossing the east coast of Australia, they often bump up against a stable high in the western South Pacific and thus tend to pile up on one another. Intervening tails of fronts extending northward from low pressure cells over the southern ocean often induce cyclogenesis in this area, which may account for some of the greater rainfall in southeastern Australia, as well as a considerably different weather pattern in New Zealand farther east. Easterly winds around northern sides of stalled highs centered southeast of Australia, as well as convergence zones between highs often in the vicinity of the southeast Australian coast, bring in warm, moist air off the Coral Sea and produce considerable precipitation along the coast. This is enhanced by orographic uplift along the east-facing escarpment of the upturned edge of the continent, known as the Blue Mountains (Figure 20.5). Farther north the heavy precipitation along the northeast coast of

Figure 20.4. The wet coastal area of Queensland. (From the American Geographical Society Collection, University of Wisconsin-Milwaukee Library.)

Queensland is associated mainly with aperiodic squalls moving in from the sea, but there is no literature about easterly waves in this region.

Except for its southern and northeastern fringes, all of Australia experiences only 20–50 rain days per year. The greatest number lie in the southeast, where portions of the uplands in the Australian Alps and on the island of Tasmania receive as many as 200–250 rain days per year. The very southwestern tip receives 150 days per year, and the northeast coast 100–150 days. The greatest average annual rainfall is located in two spots, Innisfail, in the northeast, and on the higher portions of the mountains on Tasmania, both of which receive more than 3500 mm. per year. In the northeast there are about 80 thunderstorm days per year, but in the southeast only 20–40, and on the west coast 10–20.

Temperature

As might be expected in a subtropical desert, temperatures are quite hot during summer and mild during winter. During summer a large area occupied by the Pilbara heat low in western Australia has a mean daily maximum temperature of more than 37.5°C (100°F), and a small area at its western edge, around Marble Bar (8) inland from Port Hedland, has daily maxima averaging more than 40°C (104°F). Marble Bar holds the record in Australia for length of heat wave: 160 consecutive days with maximum temperatures over 37.5°C (100°F). All of interior Australia is hot during daytime hours in summer, but this western portion of the desert is a few degrees hotter than the rest. This is attributed to heat advection from other portions of the desert by southeast trades along the northern sides

Figure 20.5. The coastal escarpment south of Sydney. From the American Geographical Society Collection, University of Wisconsin–Milwaukee Library.

of subtropical high pressure cells moving eastward with their centers over the ocean to the south of the continent. Also, Marble Bar is situated near the base of the escarpment along northwestern Australia, where the southeasterlies descend a few hundred meters and warm adiabatically a few degrees.

Daily temperature ranges are large, as would also be expected in a desert region. Marble Bar during January averages a daily range of more than 15°C. Minimum nighttime temperatures average about 25°C (77°F). This is about the same as the mean daily maximum temperature during July. Normal winter days at Marble Bar range from about 10°C to about 25°C, making winter very comfortable. The most extreme temperatures ever recorded at Marble Bar are 49.2°C (120.6°F) during daytime in January, and 1.1°C (34°F) during nighttime in June.

Temperatures along the coastal areas are, of course, more temperate than in the interior. Sydney (11) on the southeast coast averages 22°C (72°F) in January and 12°C (54°F) in July, with only 7°–8°C daily temperature ranges throughout the year. On occasion temperatures there have risen to more than 45°C in January and have dropped to 2°C in June and July. Although freezing temperatures have never been recorded at screen (weather shelter) level, at grass level they have dipped to −4.4°C. Thus, Sydney occasionally experiences light ground frosts. In 1836 snow fell on Sydney for about half-an-hour and stayed on the ground for about one hour. At Melbourne there have been five noticeable snowfalls in the last 130 years. Higher up, at an elevation of 1830 meters in the Australian Alps to the south, the lowest temperature ever recorded in Australia is −22°C. In this area snow falls annually, and skiing has become a favorite winter pastime. On Mt. Kosciusko snow falls every year, sometimes as much as a meter a day. As much as 12 meters of hard, dry, crystallized snow has accumulated on the ground there.

The northern portion of Australia, and the coastal areas on the west and east southward to approximately 27°S latitude, are frost-free and hence have a year-round growing season. The central Australian desert averages about 300 frost-free days per year, and the southern lowlands average 200–250 days. The south coast averages about 300 days per year.

Special Weather Phenomena

The Meridional Front. Much of the precipitation in southern and central Australia comes in association with tail ends of cold or occluded fronts that extend northward into the continent from cyclonic centers moving eastward through the southern sea. These fronts occupy troughs between traveling subtropical high pressure cells. They are most prevalent during winter when the pressure belts have shifted equatorward. At that time of year, they move across Australia in rapid succession under strong zonal westerlies in the upper troposphere. Often, two or three move across Australia simultaneously, spaced 1000–1500 kilometers apart. Since most of them are rapid-moving, they bring only meager precipitation to much of Australia. They bring more to southeastern Australia, where they tend to slow down as they approach the western Pacific.

Cool Changes and Southerly Busters. During summer when the traveling high pressure cells are centered south of the continent and the troughs between them are narrow, the weather change between the northeasterly flow on the back side of one high and the southerly flow on the forward edge of the next can be exceedingly abrupt. This is most noticeable along the southern coast, where temperatures may drop 10°–20°C within 24 hours. This cool, southerly flow from the sea is a welcome change during summer from the exceedingly hot, dry, dusty northeasterly flow from the desert. Such changes take place all along the southern coast, and along the western and central portions are in the character of a dry front without significant cloud formation or precipitation. The significant features are gusty winds and temperature drop. Such occurrences are known by the simple term "cool change."

In the western and central portions of the coast, maximum wind gusts are usually no more than 20 to 40 kilometers per hour. In the southeast, however, especially along the coast of New South Wales, the phenomenon occurs with more intensity and is often accompanied by dark roll clouds and some heavy thunderstorms, since in these parts the preceding northeasterly winds have picked up considerable moisture from the western Pacific and thus bring conditionally unstable air onto the southeastern coast. Here the more extreme form of the change is known as the "southerly buster." The change is often very sudden, from a fresh northeastern breeze that ten minutes later becomes a violent gale from the south with gust speeds occasionally exceeding 35 meters per second (78 mph). The busters usually end with thunderstorms and rain.

Southerly busters occur about 30 times a year along the southeastern coast. Before the brickfields were removed from the outskirts of Sydney, the common term for southerly busters in this area was "brickfielder," since the southerly winds brought heat and dust over the city from the brickfields to the south. The strength of the southerly buster in this area may be partially due to the steep scarp that rises to about 1000 meters immediately landward from the sea. At times, a southwesterly component brings hot continental air over the scarp down to the coastal area. But more commonly the coast is occupied by a marine layer of air 1000 meters or more thick which is of tropical origin: warm, moist, and unstable. The front of the southerly buster proceeds up the coast at a rate of 10–20 kilometers per hour, and eventually dies out in Queensland around 30°S latitude.

Tornadoes, Whirlwinds, and Waterspouts. Small vortices—some cyclonic, some anticyclonic—occur frequently in many portions of Australia. These range from nondestructive whirlwinds or dust devils, produced by extreme surface heating during summer, causing superadiabatic lapse rates in the surface air over the hot deserts and steppes of the interior; to more severe tornadic-type storms that occur primarily in humid tropical air in front of stalled or slowly moving fronts, primarily in southeastern Australia during winter and spring. In the southeast many form as waterspouts offshore and move a few kilometers inland with the southeast winds, but most do not move very far into the interior. In the southeastern coastal areas over New South Wales, Victoria, and South Australia, such phenomena may occur as frequently as in portions of southeastern United States, but generally the storms are not as destructive. Over much of the interior deserts in summer, dust devils can be seen every day, and the primary result is short-lived dust storms or "Cobar showers," so named from the small town of Cobar in the dry steppe interior wheatlands of New South Wales. Many a dark cloud on the distant and parched horizon turns out to be a "Cobar shower" of swirling dust, instead of the eagerly awaited rain. Occasionally, a small vortex of greater strength may cause some structural damage to interior settlements.

The terminology for such phenomena is confused and ambiguous. Western Australian press reports use the term "willy willy" to refer to tropical cyclones, tornadoes, and small whirlwinds, but not to coastal whirlwinds associated with winter storms; for these they use the term "cockeye bob." In the subtropical southeast of Australia tornadic-type storms are usually referred to simply as cyclones.

Tropical Cyclones (Hurricanes). Tropical storms affect the entire northern portion of Australia, although they are most common along the northeastern coast of Queensland and the westernmost part of the west coast of the state of Western Australia. The season extends from November to March, with maximum frequency in January. Most of them never reach hurricane wind speeds, but may bring heavy rainfall to extensive areas of northeastern and northwestern Australia and adjacent seas. If the wind speed does not exceed 33 knots (17 meters per second or 38 mph), the low pressure system is called a tropical depression or disturbance. If the wind speed exceeds 33 knots, the storm is called a tropical cyclone. In some portions of Australia, particularly in the southwest, such storms are also referred to as "willy willies."

Tropical storms in the Australian area the western north Pacific or Caribbean regions, or even those in the Bay of Bengal. Storms of tropical cyclone strength average only 3.3 per year in the Queensland–Northern Territory region combined, and 2.1 per year along the coast of western Australia. The highest wind speeds ever recorded in tropical cyclones were 109 knots at Willis Island off the northeast Queensland coast and 125 knots at Onslow (9), near the westernmost part of the coast of Western Australia.

The northeastern storms usually originate in the vicinity of the Solomon Islands and move westward with the trade winds. Some may move almost straight west through the Coral, Arafura, and Timor seas and not affect the mainland of Australia. But most of them migrate poleward and recurve southeastward along the Queensland coast and eventually out to sea in the western part of the South Pacific. A few may penetrate the coast as far south as Brisbane (3) around 27°S latitude. Few of the east coast cyclones penetrate very far inland, although occasionally a weak storm may penetrate interior Queensland from the north out of the Gulf of Carpentaria.

The western storms usually originate in the Timor Sea and proceed southwestward along the northwest coast of Western Australia. Most of them disappear into the south Indian Ocean, but some recurve southeastward onto the continent in the Broome-Carnarvon coastal region and proceed across the southwestern portion of Australia into the Great Australian Bight along the south-central coast. One storm in February–March 1956 made a complete anticyclonic loop through the northwestern half of Australia, proceeded down the west coast again, and crossed the southwestern tip near Perth; then it continued on across the Great Australian Bight to Tasmania, where it joined up with an extratropical storm that was traced all the way to New Zealand. The only time tropical cyclones from the northwest penetrate the interior of the continent is when they happen to enter a col between high pressure cells. When they meet a subtropical high head-on they pull dry air into their systems and undergo cyclolysis (cyclone dying).

Climate Types

The distribution of climatic types in Australia is much as might be expected from the latitudinal location. The large central portion of the continent, which is always dominated by subtropical high pressure cells and is hot in summer and mild in winter, classifies as a hot desert (BWh) (see Figure 15.3). The desert extends all the way to the coast in the west. Around the rest of its periphery it is surrounded by a wide strip of low-latitude steppe (BSh) with a summer maximum of rainfall in the north, a winter maximum in the south, and a fairly even yearly distribution in the east (see Figure 20.2). This is bordered around the peripheries of the continent by climatic types that reflect the alternating influences of the subtropical highs and certain precipitation processes on either side.

In the north and northeast, where temperatures are always tropical and rainfall shows fairly large annual totals and a pronounced summer maximum, the climate classifies as tropical wet-dry (Aw). The poleward migration of the intertropical convergence zone brings heavy rain to this region during summer, while the equatorward shift of the subtropical high pressure belt brings drought to the region during winter. In the two southern extensions of southwestern and interior southeastern Australia, the seasonality is just the opposite: rainfall comes during winter when equatorward extensions of extratropical cyclones bring some rain to the area, whereas in summer the region is dominated by subtropical highs and drought. Since light frosts occur in these southern regions, they no longer classify as tropical, but subtropical dry-summer (Cs).

The east coast and the paralleling mountains south of the Tropic of Capricorn, southward to include the lowland portions of Tasmania, have eight to twelve months that average above 10°C and a fairly even distribution of precipitation through the year from a variety of rain-bringing mechanisms at different seasons. This area classifies as subtropical humid (Cf).

The natural vegetation closely fits the climatic pattern. The large desert area supports mainly sparse shrubs and grasses, with large areas, particularly in the western half, consisting primarily of shifting sands. The surrounding steppe regions, and to some extent the tropical wet-dry region in the north, supports fairly good grass vegetation, which is mainly grazed by sheep and cattle in the northeast and north but may have been plowed up and put into wheat and other drought-resistant crops in the southeast and the southwest. Fairly good mixed forests exist along the southeast coast and along most of the Blue Mountains in the east. These consist primarily of broadleaf trees, some of which are evergreen. Tropical broadleaf evergreens form dense forests along portions of the northeast coast of Queensland and scattered forests in the higher portions of Arnhem Land in the far north. Forests dominated by large eucalyptus trees exist on higher portions of the Darling Range in the extreme southwestern part of the country (Figure 20.6). Much of Tasmania is forested, primarily by conifers.

New Zealand

Climatic controls in New Zealand are significantly different from those in Australia. The islands are generally located farther poleward than Australia and are much more mountainous. This is particularly true of South Island, which stretches from approximately 40°30′S to 46°30′S latitude and has the high range of the "Southern Alps" stretching almost the entire length of the island close to the northwestern shore. The highest peak, Mt. Cook, rises to 3765 meters (12,349 feet). This range, unlike anything in Australia, forms a pronounced

Figure 20.6. Fields and eucalyptus trees near Perth in southwestern Australia. (From the American Geographical Society Collection, University of Wisconsin–Milwaukee Library.)

climatic divide between west and east and causes major perturbations in the upper air flow. North Island is not as high or well aligned, but it is still quite mountainous. The mountains are arranged in various orientations, and the highest peak rises to 2797 meters (9,175 feet).

In addition to enhanced orographic effects on rainfall in the New Zealand region, the islands are also greatly affected by increased cyclonic activity stemming from the cyclogenetic area between Australia and New Zealand. In most places precipitation is abundant or even excessive. Much of North Island annually receives 1000–2500 mm. (40–100 inches). The pattern is quite complex, since the topography is complex and unorganized and rain-bringing airflows can approach the island from almost any direction.

South Island shows much more ordered variation, as it has a more consistent westerly air flow. Higher portions of the western slopes of the Southern Alps receive more than 5000 mm. (200 inches) per year, while the Canterbury Plain on the east receives as little as 335 mm. (13 inches) (12, Alexandra). The greatest average annual rainfall recorded in all of New Zealand is 7094 mm. (279 inches) in the state of Fiordland, along the southern portions of the west coast of South Island. It has been estimated that some of the mountainous areas on the western slopes may have annual precipitation exceeding 8000 mm.

Precipitation is generally of a prolonged, light-intensity nature. The maritime polar air that dominates New Zealand is usually relatively stable, and thunderstorms are not very numerous. The greatest frequency occurs on the western coast of South Island, which has only 15–25 thunder days per year. Clouds and fog are much more prevalent in New Zealand than in Australia, as one might imagine of this cool, marine-controlled area.

Much of the precipitation in the higher mountains falls as snow. The permanent snowline in South Island lies at about 2000 meters, and during winter it descends to about 800 meters above sea level. Mountain glaciers have formed deep fiords along the

Figure 20.7. Governor's Bay near Christchurch on the east coast of South Island, New Zealand. (Courtesy of Paul N. and Joyce Lydolph.)

southwestern coast of South Island and long finger lakes along both slopes of the glacially narrowed divide of the Southern Alps. A small permanent snowfield also exists on North Island above 2500 meters on Mt. Ruapehu. During winter in North Island the snowline descends to about 1000 meters above sea level. Snowfalls are infrequent at low elevations on North Island and in the northwestern part of South Island, but may occur 10 to 12 days a year in many eastern areas of South Island, except immediately adjacent to the coast.

Sea-level temperatures in New Zealand are consistently cool, but never cold. January temperatures average 17°–18°C in North Island and 14°–16°C in South Island. July averages 9°–11°C in North Island and 4°–8°C in South Island. Maximum daytime temperatures during January average only 18°–24°C in North Island and 20°–24°C in South Island. Diurnal ranges average only 8°–14°C throughout the islands.

The lowlands of North Island would generally classify as subtropical humid climate (Cf), and those of South Island as temperate oceanic climate (Do). But both islands have varied topography, so climates vary dramatically with elevation and exposure to moisture-bearing winds and sun. Alexandra (12) in the central part of the southeastern portion of South Island, with a mean annual precipitation of 335 mm. and a mean annual temperature of 10.4°C, classifies as slightly semiarid. And since five months average below 10°C, it classifies as a cool steppe (BSk).

Except for this drier southeastern portion of South Island and higher mountain summits, much of cool, humid New Zealand was originally vegetated by rich forest growth of a mixed deciduous-conifer nature. And much of it still is, interspersed by lush pastures for sheep and dairy cattle and cultivated fields of small grains and hay crops (Figure 20.7).

Table 20.1 Temperature and Precipitation Data for Australia and New Zealand

	Jan	Feb	Mar	Apr	May	June	July	Aug	Sept	Oct	Nov	Dec	Yearly average
1. Adelaide, Australia (34°56′S, 138°35′E, 43 m)													
T.(°C)	23	21	21	17	15	12	11	12	13	16	19	21	17
P.(mm)	23	23	21	50	66	61	61	59	49	47	36	27	523
2. Alice Springs, Australia (23°38′S, 132°35E, 579 m)													
T.(°C)	28	28	25	20	15	12	12	14	18	23	26	27	21
P.(mm)	44	34	28	10	15	13	7	8	7	18	29	39	252
3. Brisbane, Australia (27°28′S, 153°02′E, 42 m)													
T.(°C)	25	25	24	21	18	16	15	16	18	21	23	24	24
P.(mm)	143	183	147	78	57	56	49	30	45	77	92	136	1092
4. Cloncurry, Australia (20°43′S, 140°30′E, 193 m)													
T.(°C)	31	30	29	26	22	19	18	20	24	28	30	31	26
P.(mm)	120	101	47	16	12	20	6	3	4	11	40	48	429
5. Darwin, Australia (12°28′S, 130°51′E, 30 m)													
T.(°C)	29	29	29	29	27	26	25	26	28	29	30	29	28
P.(mm)	411	314	284	78	8	2	0	1	15	49	110	218	1490
6. Hobart, Tasmania, Australia (42°53′S, 147°20′E, 54 m)													
T.(°C)	16	16	15	12	11	8	8	9	11	12	14	15	12
P.(mm)	42	47	52	63	51	66	47	53	53	72	58	64	668
7. Innisfail, Australia (17°32′S, 146°03′E, 7 m)													
T.(°C)	27	26	26	24	22	20	19	20	21	23	25	26	23
P.(mm)	490	602	686	467	323	188	119	106	81	94	134	244	3535
8. Marble Bar, Australia (21°11′S, 119°42′E, 181 m)													
T.(°C)	34	33	32	29	24	20	19	22	25	29	33	34	28
P.(mm)	82	70	58	21	28	19	11	5	2	5	8	31	340
9. Onslow, Australia (21°43′S, 114°57′E, 4 m)													
T.(°C)	30	30	29	26	22	19	18	19	22	24	27	28	25
P.(mm)	21	46	63	18	50	40	20	9	1	1	3	3	274
10. Perth, Australia (31°57′S, 115°51′E, 60 m)													
T.(°C)	23	24	22	19	16	14	13	14	15	16	19	22	18
P.(mm)	7	12	22	52	125	192	183	135	69	54	23	15	889
11. Sydney, Australia (33°51′S, 151°13′E, 42 m)													
T.(°C)	22	22	21	18	15	13	12	13	15	18	20	21	17
P.(mm)	104	125	129	101	115	141	94	83	72	80	77	86	1205
12. Alexandra, New Zealand (45°15′S, 169°24′E, 158 m)													
T.(°C)	17	17	14	11	6	3	2	5	9	12	14	16	10
P.(mm)	46	38	31	33	23	20	18	15	20	30	28	33	335
13. Auckland, New Zealand (36°51′S, 174°46′E, 49 m)													
T.(°C)	19	20	18	16	14	12	11	11	13	14	16	18	15
P.(mm)	84	104	71	109	122	140	140	109	97	107	81	79	1242
14. Hokitika, New Zealand (42°43′S, 170°57′E, 4 m)													
T.(°C)	15	15	14	12	10	7	7	8	9	11	12	14	11
P.(mm)	249	218	213	224	229	203	211	236	211	277	241	251	2764

21 | THE POLAR REGIONS

Because of their latitudes, the two polar areas have much in common. But they also have many differences, because of their topography and the distribution of land and water. At high latitudes, sunlight itself is an important climatic element, analogous to the scarcity of air pressure at high elevations. In much of Antarctica and Greenland both sunlight and air pressure are in short supply. Other unusual climatic phenomena, such as predominance of surface air temperatures by the characteristics of a prevailing temperature inversion, and weird lighting effects due to the diffuse light and a highly reflective surface, also necessitate some reordering of the significance of various climatic factors in the polar regions, compared to those in other parts of the earth. Although practically all polar areas are uninhabited, and therefore local conditions of the various climatic elements have little consequence to humans, the general radiation features and consequent temperatures of the polar areas exert such profound controls on the entire planet's atmospheric circulation that it is important to understand the peculiar processes that operate in these regions.

Figure 21.1 illustrates the unusual light conditions that prevail in high latitudes. The pole, of course, has six months with direct sunlight and six months without. From approximately 21 March to 23 September in the Northern Hemisphere, and from 23 September to 21 March in the Southern Hemisphere, the pole is continuously in the sunlight, although the sun is never very high above the horizon. It reaches its highest angle above the horizon, approximately 23½°, at noon on 21 June in the Northern Hemisphere and 21 December in the Southern Hemisphere. At the equinoxes in March and September the sun circles the horizon all day long. For about six weeks on either side of the continuous daylight period, from about 7 February to 21 March and again from 23 September to 6 November, the sun remains constantly just below the horizon and a twilight condition persists. Just before the spring equinox and just after the autumn equinox, the poles experience twilight 24 hours per day, but the twilight period shortens each day as nighttime darkness lengthens toward winter, February and November in the Northern Hemisphere, or May and August in the Southern Hemisphere. Constant darkness reigns from about 6 November to 7 February at the North Pole, and from about 6 May to 7 August at the South Pole.

Equatorward, the twilight periods lengthen at the expense of constant sunlight or constant darkness, and eventually a significant alternation of day and night takes place. For instance, at 80° latitude, the sun is constantly above the horizon from approximately 17 April to 24 August in the Northern Hemisphere, and from approximately 16 October to 24 February in the Southern Hemisphere. During the periods 5 March–16 April and 26 August–8 October in the Northern Hemisphere, and 5 Feb-

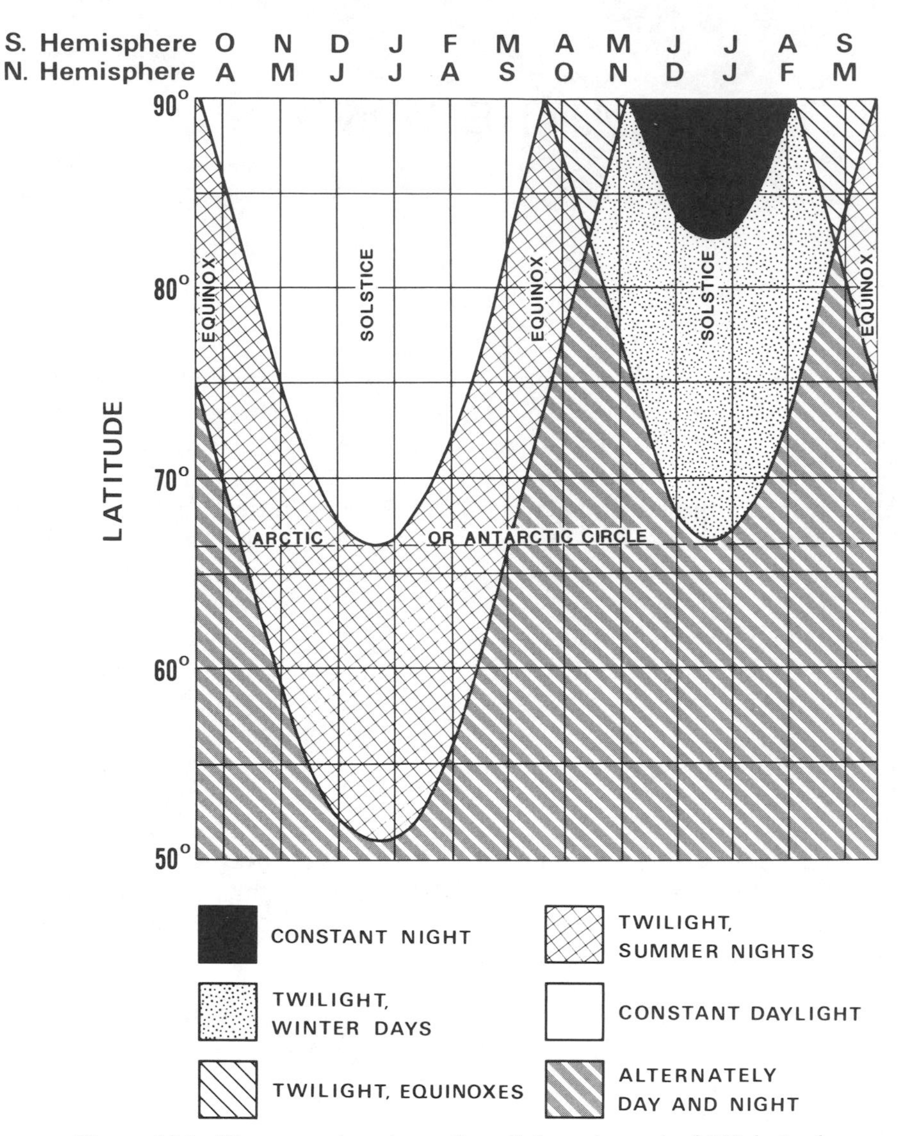

Figure 21.1. The annual regime of sunlight poleward of 50° latitude. (Modified from W. Meinardus.)

ruary–8 April and 4 September–16 October in the Southern Hemisphere, daylight predominates during each 24-hour period, but there is a brief twilight period when the sun dips below the horizon in the middle of the night. A truly dark period begins to occur around midnight on about 4 March and again on 9 October in the Northern Hemisphere, and at about 9 April and 4 September in the Southern Hemisphere, and this period of total darkness increases in length toward the winter solstice in both hemispheres. The sun never rises above the horizon from about 2 October to 20 February in the Northern Hemisphere, and from about 23 April to 21 August in the Southern Hemisphere, but at 80° latitude there is never total darkness throughout a 24-hour period. During daytime hours a twilight condition exists. This twilight period reaches its shortest duration during the winter solstice, when much of the 24-

hour period is in total darkness, but there is a short period of twilight around noon.

Equatorward, the periods of alternating days and nights lengthen. At the Arctic and Antarctic circles there is only one 24-hour period during the year when the sun is constantly above or below the horizon. For more than 4.5 months during the summer season, the nights are characterized by twilight rather than by total darkness, since the sun never dips far below the horizon. The period with nighttime twilight shortens equatorward, but even as far equatorward as 51° latitude there is no complete darkness during the summer solstice. Thus, Moscow, at 55°45′N latitude, during late June and early July experiences no total darkness—only twilight for about six hours during the middle of the night, and broad daylight from about 3:00 AM until about 9:00 PM. This is balanced during winter, of course, by very short daylight periods.

The peculiarities of insolation regimes and the constantly high albedos of snow-covered surfaces at high latitudes impart special characteristics to temperature regimes and related phenomena. As was mentioned in Chapter 14, winter becomes the dominant season, summer is very short, and the transitional seasons are practically nonexistent. Temperatures fall rapidly after the sun sets, and insolation ceases at the pole with the autumn equinox. Rapid radiational heat losses from the earth's surface quickly form a strong surface temperature inversion that stabilizes the lower troposphere and allows the surface air temperature to reach minimum values in mid- or late autumn. At that time an equilibrium is reached between heat loss upward, emitted by nearly black-body radiation from the snow's surface, and heat gain, from atmospheric back radiation from the warmer air in the inversion, from vertical eddy heat flux downward from the inversion, from heat conduction upward from the ground, and from the heat of sublimation of the deposition of atmospheric water vapor as hoarfrost on the surface. From late fall through winter and into early spring there is little change in surface air temperature, except during brief periods of unusual air advection that may affect the structure of the inversion or may alter cloud conditions, which then call for a new balance among heat exchanges. But average monthly temperatures remain about the same from late autumn into early spring, resulting in the so-called "coreless winter" of high latitudes.

The polar areas receive no sunlight during winter and no darkness during summer, so there is little diurnal temperature variation during those seasons. The poles are the only places on earth where there is no diurnal period at any time of year. Also, at high latitudes there is negligible latitudinal heating differences during mid-winter and mid-summer, since the entire areas are dark during mid-winter and light during mid-summer. Therefore, at high latitudes, unlike the middle latitudes, the largest latitudinal temperature gradients, and thus the most vigorous and stormiest atmospheric circulations, occur during the equinoxial periods. (Refer to insolation receipt graphs in Figure 4.3.) The equinoxial storms of the subpolar ocean south of Cape Horn at the southern tip of South America are legendary, as they are along the Arctic coast of Western Siberia and in many other subpolar areas of the world.

Since Antarctica is primarily an ice plateau at high elevation, while the Arctic is primarily a frozen sea surface near sea level, heat exchanges and resultant temperatures and associated climatic characteristics are much different in the two polar regions. Therefore they will be considered separately. Since Antarctica in most respects presents the simpler case, it will be considered first.

Antarctica

Area and Topography

Antarctica consists primarily of a high ice plateau surmounted here and there by higher mountain ranges, particularly in the northwest, and essentially surrounded by a sea-level ice shelf whose width fluctuates greatly with the season. Fifty-five percent of the ice plateau lies at an elevation higher than 2000 meters above sea level, and approximately 25 percent at an elevation of more than 3000 meters. The highest portion rises to more than 4000 meters at the pole of maximum inaccessibility northeast of the geographical South Pole, which is at an elevation of 2800 meters (9,186 feet). The ice plateau is asymmetrically positioned around the South Pole, the larger, higher portion being in the Eastern Hemisphere (Figure 21.2). The ice thickness averages approximately 2440 meters (8000 feet), but in places where the underlying land dips below sea level it is twice as thick. Thus, Antarctica underneath the ice cover is not a continuous landmass, but an archipelago of large and small islands. Climatically it acts as a continuous landmass, however, since the ice is frozen down to the land surface whether it lies above or below sea level.

In the west a discontinuous range of mountains continues the Andes chain of South America southward across Drake Passage, through the entire length of the Antarctic Peninsula, and on southwestward along the coast of Ross Sea. In this general region many individual ranges rise abruptly from sea level to well above the elevations of the ice plateau. The highest peak in Antarctica is Vinson Massif (16, Sentinel Range), near the base of the Antarctic Peninsula, which rises to an elevation of 5140 meters (16,864 feet).

The area of the ice-covered landmass is approximately 14 million square kilometers. The area of the surrounding sea covered by ice varies from about 2.6 million square kilometers in March, when it reaches a minimum after the summer melting, to approximately 18.8 million square kilometers during September, when it reaches its maximum after the winter freezeup (Figure 21.2). Thus, the total ice-covered surface varies from approximately 16.6 million square kilometers in March to about 32.8 million square kilometers in September. The following remarks will apply to this ice-covered surface, although superlative comments will usually refer to the high ice plateau.

Heat Exchanges and Temperature

It is interesting to note that the place where the earth's coldest surface temperatures were recorded is also the place of maximum monthly amounts of solar energy receipt. Both places are at around 80°S latitude on the high ice plateau of Antarctica. During the short summer (December–January) the earth is in its nearest position to the sun, and insolation is very intense in the high, thin, clean, dry atmosphere. During December, 1958 insolation receipt measured slightly more than 30,000 langleys. During the long winter night, of course, no insolation at all is received; an extreme surface temperature inversion forms under the influences of vertical heat exchanges, and very cold surface temperatures are achieved.

The coldest temperature ever recorded at the surface of the earth, −89.2°C (−128.6°F) was at the Soviet station of Vostok (6) at 78°28′S latitude, 106°48′E longitude, elevation 3488 meters (11,441 feet) on 26 July 1983. During August, Vostok averages −68.4°C. During December, the warmest month, it averages −32.7°C. The warmest temperature ever recorded at Vostok was −21°C. Thus, the mean annual range is only 35.7°C, and the absolute range is 68.2°C (123°F). These ranges are not as extreme as those in northeastern Siberia, because summers in Antarctica are not warm. Typical surface temperatures during summer range from 0° to −4°C over the open seas surrounding Antarctica, to −6° to −10°C over the ice shelves, to −20° to −30°C on the ice plateau. During winter these averages range from 0°C at the outer

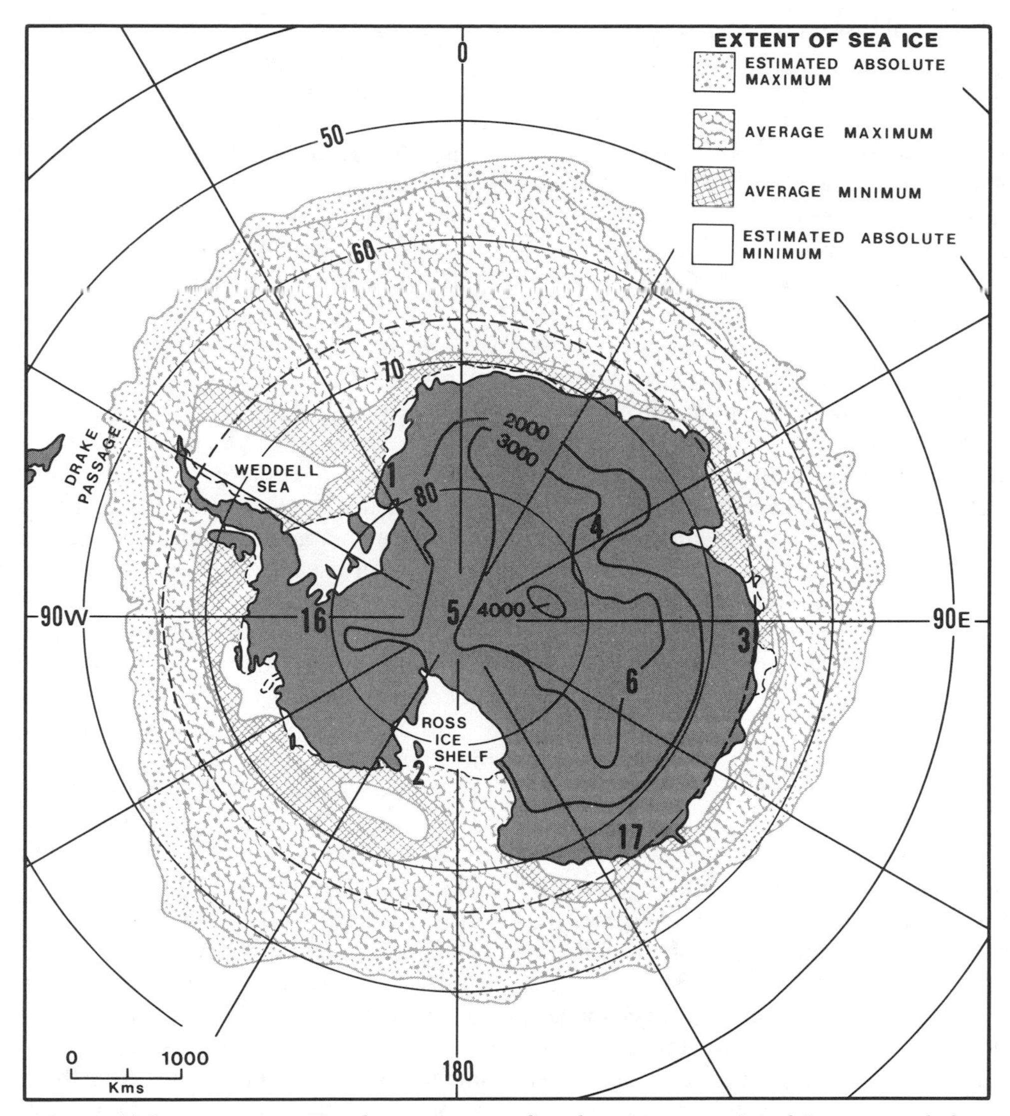

Figure 21.2. Antarctica. Numbers correspond to locations mentioned in text and data tables at end of chapter. (Adapted from Central Intelligence Agency, Polar Regions Atlas *[Washington, 1978], p. 38.)*

edges of the expanded shelf ice, to −30°C over the inner shelf ice, and to −70°C and below over the highest portion of the ice plateau. Only a few stations near sea level north of 70°S latitude have observed absolute minima below −40°C. On the ice shelves the lower limit is close to −60°C, and in the interior, −80°C can be expected once in ten years at the South Pole and more frequently at higher elevations on the central plateau.

The Surface Temperature Inversion. Probably the most conspicuous characteristic of the Antarctic temperature regime, other than the very low temperatures of the interior, is the surface inversion. Except for December and January, when temperatures are essentially isothermal up to a kilometer or more above the surface, the surface inversion is an ever-present feature over the high plateau and a frequent one over the rest of the ice sheet, coastal areas included. Its intensity is greatest during winter (June–August) when the average temperature difference between bottom and top may be as much as 25°C on the high ice plateau and 5°C around the margins of the landmass. Surface temperatures depend much upon the height of the thermometer above the ground, and interdiurnal temperature changes are due more to fluctuations in the structure of the inversion than to anything else. Nevertheless, advection of heat, both sensible and latent, released from deposition (negative sublimation) of atmospheric water vapor as hoarfrost on the ice surface, as well as changing cloud conditions, do bring about aperiodic temperature changes.

Atmospheric Circulation

The lower troposphere circulates clockwise around the margins of the ice plateau in the subpolar low belt, which is generally located north of 65°S latitude during the solstitial months and south of that during the equinoxial months. Surface low pressure cells are imbedded in this belt over the sea and the ice shelves. During January these are most prevalent over the sea to the west of the Antarctic Peninsula, adjacent to the Ross shelf ice, and near the 20°E and 90°E meridians (see Figure 6.6). These core areas tend to deepen during July when the shelf ice expands, except for the area to the west of the Antarctic Peninsula (see Figure 6.7). At any time of year, however, the daily circulation shows these lows to be moving rapidly from west to east.

Of course, one cannot positively determine the conditions at the surface over the high ice plateau. The surface air pressure over wide areas of the eastern half of Antarctica is less than 700 mb., and over the highest parts of the central plateau even less than 600 mb. Therefore, the first standard pressure level that can represent the free atmosphere over the entire continent is the 500-mb. level, shown in Figures 6.15 and 6.16. At this level there is still a semblance of lows over the sea just to the east of the prime meridian and around 90°E. But the main feature is a central low that stretches year-round in an elongated oval from the Ross shelf ice across the narrow part of Antarctica to the Weddell Sea east of the Antarctic Peninsula. Thus, the circumpolar whirl of the Southern Hemisphere is centered almost, but not quite, on the South Pole. Although upper-air data are largely lacking, constant-level balloons released from New Zealand have circled Antarctica several times and reveal a wave pattern similar to that in the upper tropospheric flow of the Northern Hemisphere (see Figure 6.25).

Although it appears that the air flow in the high latitudes of the Southern Hemisphere is rather symmetrically centered around Antarctica and the South Pole, warm and cold advection seems to prevail over different sectors of the continent. Warm air advection and upward vertical motion are found more frequently in west Antarctica, while cold advection and subsidence predominate over east Antarctica.

Surface Winds. Surface winds over the high ice plateau are not strong. At Vostok they vary from a minimum monthly average of 4.4 meters per second (9.8 mph) to maximum monthly averages of 5.4 meters per second in May and 5.5 meters per second in September. Note the equinoxial maxima produced by stronger latitudinal temperature gradients and more vigorous atmospheric circulation at that time of year. Nevertheless, the surface winds over the plateau are very constant and, at the low temperatures that prevail, produce constant wind chills that are difficult to endure. Hence, the winds draw attention to themselves more than their speeds would warrant. Also, an outstanding characteristic of the winds over the plateau is their constancy of direction. This is confirmed not only by sporadic observations, but also by the extraordinary buildup of *sastrugi*, wind-blown firn (crystallized snow) in the shape of longitudinal dunes, the orientations of which reveal the prevailing wind directions.

Average wind speeds are much greater and speeds vary much more along the steep bottom slopes of the ice plateau around the fringes of the continent. Speeds average 10 meters per second or more and occasionally reach extreme strengths. Cape Denison (17) on the southeast coast holds the world's highest annual average wind speed, 19.2 meters per second (43 mph). Gusts of more than 60 meters per second (135 mph) are not uncommon. When D. Mawson revisited his hut at Cape Denison after 20 years he found that planks had been reduced in thickness by more than a half-inch by the snow-blast erosion of the winds. Average wind speeds are reduced by lulls between gusts, since wind speeds on the lower slopes are much less consistent than on the plateau surface.

Near-surface winds in Antarctica seem to be a function of one of two mechanisms, the so-called "inversion" (equilibrium) wind or the "katabatic" (gravity) wind. The inversion over the plateau is maintained at about the same height everywhere over the slightly domed surface. Therefore, there is a gentle slope to the inversion, and at any constant height above sea level there is a temperature, and hence a density, difference through the inversion layer. This induces a gentle air flow several hundred meters above the surface that, essentially free of friction, flows parallel to the contours of the ice surface, with higher elevations to the left. Closer to the surface, friction upsets the equilibrium between pressure gradient and Coriolis Force and causes the wind to cross contour lines at angles of approximately 45° to the left of the fall line (slope of the land). The high angle of deviation in spite of the relatively smooth snow surface is apparently due to the extreme stability of the vertical structure of the lower air.

Where slopes become steeper near the edges of the ice plateau, the gravity winds take over and cause the wind to blow more directly downslope. Gravity winds, however, depend upon a source of cold air (at the top of the slope) that is colder, and hence more dense, than air at the same elevation farther out from the slope. Therefore, katabatic winds soon exhaust their cold air source and cease blowing intermittently while more cold air is formed to tumble down the slopes. That is why the katabatic winds near the base of the plateau are characterized by alternately high speeds and lulls. True blizzard conditions, with great quantities of blowing snow, occur in this basal region and, in general, do not extend more than 1–3 kilometers out to sea. Deviations in the katabatic flow result in weird phenomena such as Mawson's "whirlies"—small vortices with vertical axes similar to dust devils, except they are composed of snow, and vertical walls of clouds along the lower slopes surmounted by revolving cumulus.

The general surface air flow for Antarctica, then, is a diverging one away from the highest part of the ice plateau, semiparallel to the contours in the central portions of the plateau, but becoming more aligned with the slope of the land as slopes become steeper near the edge (Figure 21.3). Since the ice plateau is in the shape of a dome, cold air does not pile up to any extent, but constantly slides downslope. Therefore, the cold air over Antarctica does not move *en masse* as a unified high pressure cell into lower latitudes, as do the cold cP highs in Canada and parts of Eurasia. It therefore lacks the initial vigor to reach as low latitudes as does the cold air in the Northern Hemisphere. Also, of course, the Antarctic air must cross wide expanses of

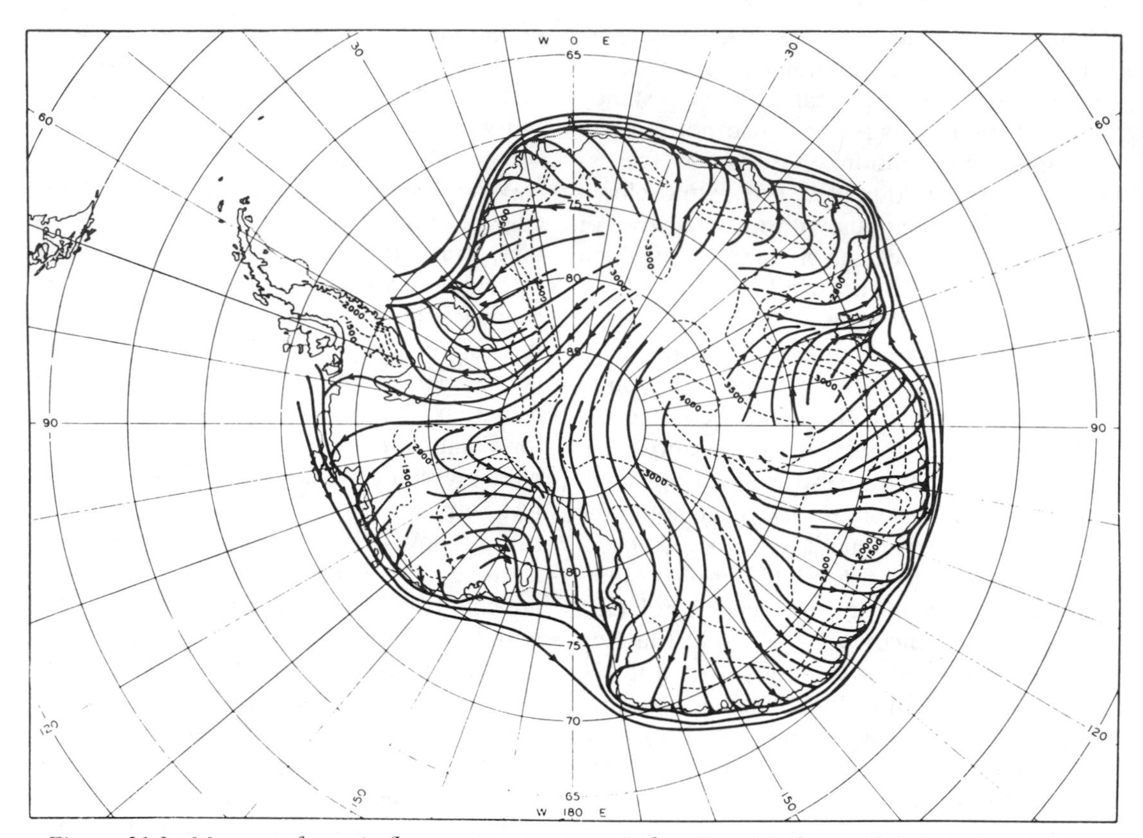

Figure 21.3. Mean surface air flow over Antarctica. (After K.B. Mather and G.S. Miller, "Notes on Topographic Factors Affecting the Surface Wind in Antarctica, with Special Reference to Katabatic Winds," University of Alaska Technical Report UAG-R-189 *[1967]; as shown in S. Orvig,* Climates of the Polar Regions, *Vol. 14 of* World Survey of Climatology *[Amsterdam: Elsevier, 1970], p. 287.)*

unfrozen ocean before it reaches the southern tips of other continents, and is thereby transformed from continental polar air into maritime polar air.

Condensation, Precipitation, and Deposition

The difficulty of measuring all forms of moisture, which is bad enough under optimum conditions, is multiplied many times in the Antarctic. First, no one quite knows what the vapor pressure is in the lower atmosphere. Much of the time the surface layer underneath the inversion is supersaturated with respect to ice, but the unanswered question is "how much?" Second, there is usually a haze all over the sky, which is often so thin as to be hardly noticeable. One then has to decide whether this is a clear condition or a complete overcast; there is no mid-value.

The band of maximum cloudiness is clearly north of the belt of lowest pressure at sea level and south of the belt of strongest westerly winds, but above the ice plateau it is difficult to determine. Complicating the estimates is the reduction of visibility by blowing snow and by lack of light in long, moonless winter nights. Satellite detection of clouds is difficult because the albedo of cloud tops is comparable to that of the snow on the plateau surface at similar heights above sea level, and the infrared radiations of the two are also similar. The best estimates seem to indicate that annual averages of cloud cover reach their maximum around 61°S latitude, where 8.7 tenths of the sky is covered. This diminishes to about 3.6 tenths of sky cover at 79° latitude, the highest latitude for which estimates are available. At 61° latitude a significant seasonal variation occurs: summer is the cloudiest season, averaging 9.3 tenths sky cover, while winter is the clearest season, averaging 7.9 tenths. At 79° latitude there is little seasonal variability, spring and summer being only slightly cloudier than winter and fall. The clouds are usually stratiform, and very thin over the ice plateau. The most commonly used descriptive name is altostratus, although the clouds are frequently based no more than 1000 meters above the ice surface. Fog is a frequent occurrence in coastal areas, caused mainly by the advection of relatively warm maritime air over colder surfaces.

Except for occasional rain in coastal regions, all precipitation is in solid form and almost impossible to measure, since scarcely more than a trace falls on any one day. Over the two-year period 1967–1968, snow was observed to fall on more than 100 days at Plateau Station (4), but there was never enough to measure. Hence, the emphasis is on determining accumulation instead of precipitation.

Adding to the observable snowfalll are deposition and so-called "cloudless precipitation." Since the vapor pressure in the warmer air at the top of the inversion is almost always greater than the saturation vapor pressure of the air underneath the inversion, there is a vapor pressure gradient downward that leads to the deposition of prodigious amounts of hoarfrost on the snow surface. Over the course of a year, hoarfrost growth on a metal tower 33 meters above the surface amounted to an accumulation 35 centimeters (14 inches) in radius. The downward transport of moisture by eddy diffusion also leads to the presence of ice crystals in the lower part of the inversion layer, which gradually settle to earth. Some observers have contended that this cloudless precipitation of ice crystals is a constant phenomenon on the high plateau. This seems to be confirmed by flashlight observations during dark winter nights when it is easy to observe ice crystals floating in the air. They can also be seen in the summer sunshine. During 1967 and 1968, floating ice crystals were observed on more than 300 days per year. During five winter months of 1968 no snowfall was reported, but ice crystals were observed 23 to 29 days per month, and the net accumulation seemed to be about 1.2 centimeters per month (height changes on snow stakes, not converted to water equivalent). Thus, it appears that the accumulation over the ice plateau is the net result of light snow *plus* ice crystal precipitation *plus* a deposition of hoarfrost by negative sublimation *minus* a small amount evaporated (positive sublimation) and the possible effects of snow drift.

It is not too difficult to measure accumulation of these parameters on the plateau surface where storm winds are almost totally lacking. It has been determined that significant snow drifting does not begin until the surface wind reaches a speed of 7–8 meters per second, which is usually twice the speed of the surface wind on the high plateau. But around the plateau edges where winds are very strong, drifting and blowing snow make any determination essentially impossible. At Cape Denison blowing snow is reported 55 percent of all observations during the year, reaching a maximum of 89 percent during autumn and a minimum of 11 percent during summer. Ten percent of all weather observations there report visibility of less than 10 meters, due to intense blowing snow. Since there is usually a rather strong offshore component to the surface winds all along the coast, huge amounts of snow are blown into the southern ocean and become part of the precipitation loss from the continent. Even so, the quantity lost is much less than that lost by icebergs calving into the sea.

Taking all these difficulties into account, the average annual accumulation, reduced to water equivalents, seems to be on the order of 50 mm. (2 inches) over much of the high eastern half of the ice plateau. It increases in all directions toward the coast, where it ranges from 200 to 400 mm. (8–16 inches) along much of the coast and as much as 600 mm. (24 inches) southwest of the Antarctic Peninsula.

The Arctic

Whereas the Antarctic is an ice-covered landmass completely surrounded by sea, the Arctic is an ice-covered ocean almost surrounded by land. Since the ice cover continues southward over much of the landmass of Greenland and large portions of adjacent islands of the Canadian Archipelago, these areas will be included in the consideration of the climate of the Arctic basin (Figure 21.4).

At its maximum extent in March or April, the ice in the North Polar area covers approximately 17 million square kilometers (6,600,000 square miles). At this time the ice covers the entirety of the so-called Central Polar Ocean and its surrounding marginal seas—the Beaufort Sea north of Alaska and northwestern Canada, the East Siberian Sea north of the Soviet Far East, the Laptev Sea north of eastern Siberia, and the Kara Sea north of western Siberia, as well as northern portions of the Barents Sea north of European Soviet Union and Scandinavia and the Norwegian Sea in the North Atlantic, as well as most of Greenland and large portions of the islands and intervening seas of the Canadian Archipelago. This is about the same area as that covered by ice on Antarctica and surrounding seas at the minimum period—March—in the Southern Hemisphere. Remember, in September Antarctica and surrounding seas are covered by an ice sheet with an area of approximately 32,800,000 square kilometers. During summer the Arctic ice cover shrinks to about 11,370,000 square kilometers, 1,657,000 of which are on Greenland.

Since the thick, high ice cap on Greenland and the thin ice cover over the Arctic Ocean have very different climatic effects, the two will be considered separately. Since the climate of Greenland is most similar to that of Antarctica, it will be considered first.

Greenland

Greenland is sort of a miniature Antarctica. Its area, including various islands underneath the ice, is 2,186,000 square kilometers. The main ice cap covers about 1,726,000 square kilometers, or about 80 percent of the island. Peripheral glaciers cover another 76,000 square kilometers, and annually ice-free land covers about 383,000 square kilometers. In most places the landmass rises abruptly from the sea in steep cliffs and deeply cut fiords. Eighty-seven percent of the ice cap lies above 1220 meters. The highest point on the ice reaches 3230 meters, which compares to more than 4000 meters in the highest portion of the ice plateau of Antarctica. The mean thickness of the ice cap is about 1500 meters, about one-third of which extends below sea level at the bottom. The maximum thickness is about 3300 meters. The total volume of the ice is about 2,600,000 cubic kilometers, which is about one-ninth the volume of the ice on Antarctica. If all the ice on Greenland melted it would raise sea level about 6.5 meters, and if it melted on both Antarctica and Greenland it would raise sea level about 65 meters.

Unlike Antarctica, which is almost centered on the South Pole, Greenland is off center from the North Pole and stretches to much lower latitudes, for a distance of

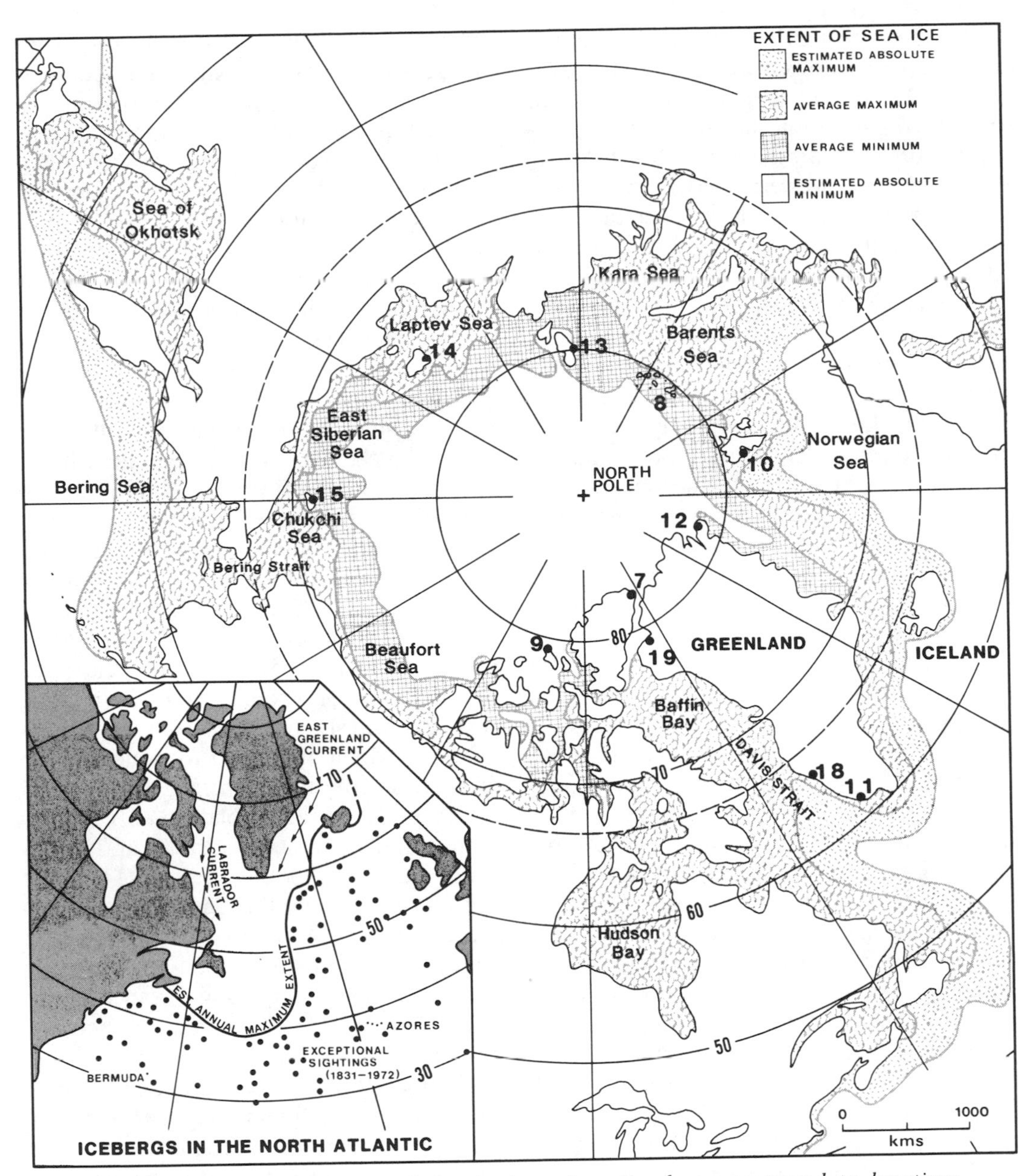

Figure 21.4. Major features of the Arctic region. Numbers correspond to locations mentioned in text and data tables at end of chapter. (Adapted from Polar Regions Atlas, *p. 12.)*

about 2400 kilometers from north of the 83rd parallel to about the 60th parallel. Thus, it pokes southward well into the westerlies of the middle latitudes, and thereby has a much greater effect on the zonal air flow of the Northern Hemisphere than Antarctica does on that of the Southern Hemisphere. Along with the Rocky Mountains to the west, Greenland probably exerts the greatest influence on the west-to-east flow of the Northern Hemisphere and is partially responsible for the development of the deep trough in the standing waves of the upper troposphere that persists over eastern North America. It thus has a great deal to do with the far southward penetration of Arctic air in interior North America along the western periphery of this trough and the far northerly penetration of warm marine air in the North Atlantic along the eastern limb of the trough.

The height of the ice plateau over Greenland retards shallow lows approaching from the west and usually sends them northward along the west coast through Davis Strait and Baffin Bay. They frequently regenerate over the open water of Smith Sound northwest of Thule (19) along the northwest coast of Greenland, which is one of the most peculiar phenomena of the Arctic region in winter. Lows approaching the southern tip of Greenland along a more southerly track often split into two parts, one moving northward along the west coast and the other northeastward into the Icelandic Low. It appears that some lows may occlude orographically as they approach the west coast of Greenland, and their upper portions and frontal disturbances may cross the ice cap. Many of them regenerate on the lee side of Greenland due to the extreme temperature contrasts between cold air coming off the ice sheet from the west and the relatively warm air over the North Atlantic Drift farther out to sea. Between the warm North Atlantic Drift and the Greenland coast, the East Greenland Polar Current flows southward along the east coast of Greenland and keeps the coastal area cool, particularly during summer.

The edges of the Greenland ice sheet have considerably steeper slopes than do the edges of the Antarctic ice sheet, and therefore more annual precipitation is necessary to maintain the Greenland ice cap than to maintain the Antarctic ice cap, about 400 mm. water equivalent, as compared to about 140 mm. in Antarctica. As is the case in Antarctica, amounts of precipitation over the ice plateau, as far as can be determined, vary significantly from one part to another. Maximum amounts of more than 640 mm. water equivalent per year fall in west central Greenland, whereas a study of ice cores along a traverse along the 78th parallel of northern Greenland revealed precipitation of only 120–150 mm. per year. Observations are lacking north of that latitude, but estimates have placed precipitation there at less than 100 mm.

As is the case in Antarctica, a strong temperature inversion persists much of the time over the ice cap, and the air underneath the inversion is often supersaturated with respect to the ice surface. Therefore, hoarfrost and ice crystal precipitation make up significant portions of the total accumulation. All the observable precipitation on the ice plateau falls as snow, but usually not enough to be measured. During a year's observations at Eismitte ("middle of the ice"), snow fell on 56 percent of the days, but hardly ever enough to measure. Fog was reported 15 percent of all days, two-thirds of which amounted to clouds on the ground. During these times a stratus overcast hovers over the ice plateau only 30–100 meters above the snow surface, and sometimes descends to the surface. The light reflection between the snow surface and the cloud base is intense and produces a condition known as "white-out," a term used to describe the disappearance of the horizon when there is no contrast between sky and snow surface.

Temperature inversions are much more pronounced in winter than in summer. During January the inversion base is normally at the surface, with temperatures around −35° or −40°C at a mean height of about 3000 meters above sea level. The top of the inversion is usually about 330 meters above the surface, with a temperature averaging about 10°C higher than that on the surface. Above this often exists an isothermal layer with an average thickness of about 30 meters. During June, on the other hand, the base of the inversion averages about 125 meters above the surface with a temperature around −15°C, which averages about 1°C lower than the surface temperature. The thickness of the inversion layer averages about 230 meters with a temperature increase of only about 2.5°C, above which an isothermal layer extends for another 70 meters or so. These are merely average conditions during the extreme months, and can fluctuate quite widely aperiodically. There is much more meridional movement of air across Greenland than across Antarctica. Particularly during winter, tropopause heights fluctuate drastically, and interdiurnal temperature variations are large, often of the magnitude of about 30°C during winter.

Surface temperatures in the central portion of the ice plateau average about −42°C in January and −13°C in July. The coldest temperature ever recorded on the Greenland ice sheet is −66°C (−87°F) at Northice at 78°4′N, 38°29′W, elevation 2343 meters. The temperature has risen to +2.2°C during August at Site 2 at 76°59′N, 56°4′W, elevation 2128 meters.

On the ice plateau, as was the case over Antarctica, there is always a sensible wind, averaging around 5 meters per second, which is a result of the slope of the temperature inversion. Therefore, there is hardly ever a calm condition on the ice plateau, and even the light inversion winds can be very trying in the prevailing low temperatures.

On the coasts, of course, temperatures are much warmer. The highest summer temperatures occur in the southwest, where they average around 9°C. Godthaab (18) on the southwest coast has experienced a maximum temperature of 23°C during June. In this area temperatures as high as 17°C have been observed in December. Absolute minima in the southwest are around −27°C, and along the northern coast, around −51°C.

Many of the individual record maximum temperatures are associated with foehn winds that can raise temperatures along the coast more than 10°C, particularly during December–March. Foehns are often induced by the passage of low pressure centers during that time of year, which may cause the wind to ascend one slope of a headland, where condensation and precipitation release latent heat, and then descend the other slope of the headland into a fiord where foehn conditions can be quite pronounced. Foehn winds are strongest at heads of fiords and often peter out before they reach the mouths. They may pass over a cold bottom layer of air in fiords. Often fiords during winter experience perfectly calm conditions at sea level, while hurricane-force winds thunder overhead at only 100 meters higher in the air.

The North Polar Basin

Climatic conditions over the frozen Arctic Ocean are much different than they are over the ice plateau of Greenland. As explained earlier, during winter a considerable amount of heat is conducted upward through the ice from the unfrozen water underneath, and this keeps the surface air temperatures in the central Arctic at around −34°C. During summer the melting of snow and a portion of the ice surface maintains surface air temperatures at about the freezing point. Thus, temperature extremes over the Arctic Ocean are considerably less than they are over the Greenland ice cap. During winter, air over the Arctic ice usually is 10°–20°C warmer than over the frozen landmasses surrounding the Arctic, and during summer surface air temperatures over the Arctic ice are generally 10°–20°C cooler than over surrounding landmasses. This leads to a very definite monsoonal reversal of winds between winter and summer, as was described along the Siberian coast in Chapter 17.

The Arctic sea ice fluctuates drastically in extent and thickness by season. Of the 14 million square kilometers of ocean area north of 65°N latitude, ice covers a maximum of about 83 percent from March

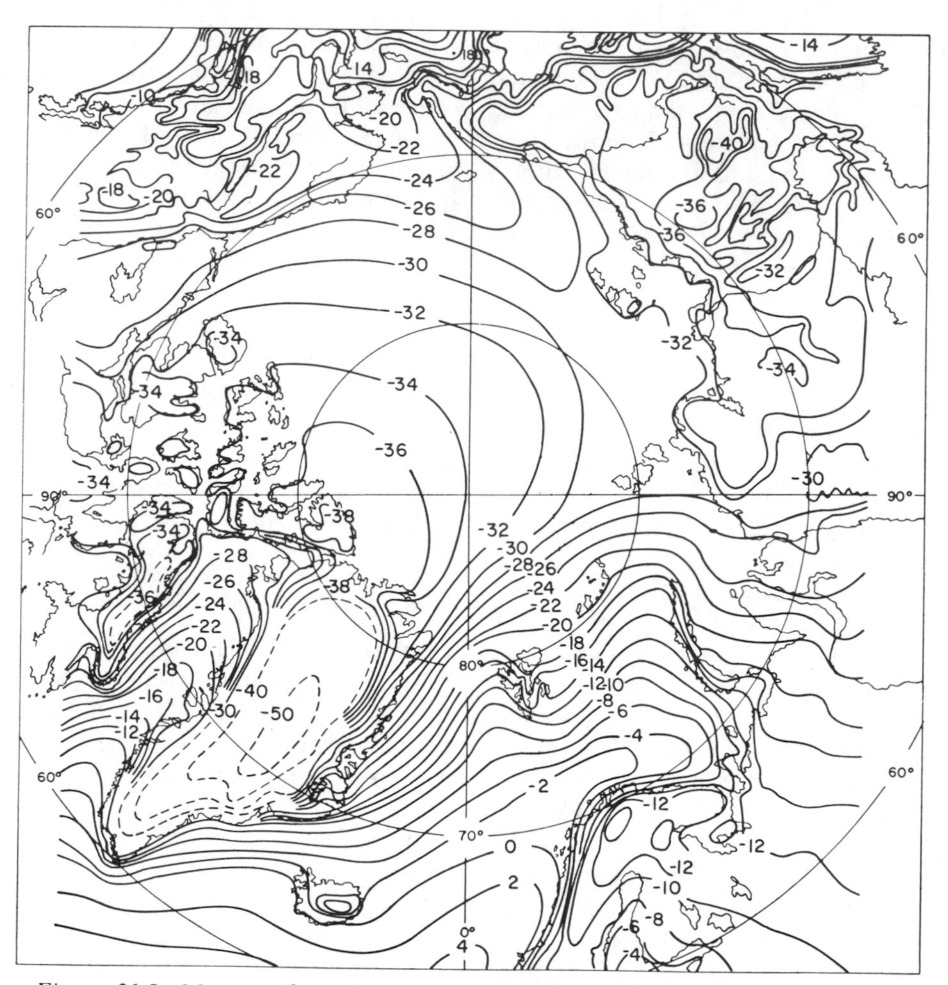

Figure 21.5. Mean surface air temperature in °C, February. (After Z.M. Prik, "Mean Position of Surface Pressure and Temperature Distribution in the Arctic," Tr. Arkticheskogo Nauchn. Issled. Inst., *Vol. 217 [1959]:5–34; as shown in Orvig,* Climates of the Polar Regions.*)*

through May and a minimum of 54 percent from August through September. The average thickness of ice is about 3.7 meters, but this varies greatly from place to place. During the melting period winds may pile up ice floes that refreeze into thick hummocks during the following winter. In these areas ice may pile up to 6–7 meters above the water surface over the open ocean and as high as 13 meters above the surface along coasts. Since only one-tenth of the ice projects above water, such hummocks can have thicknesses as great as 130 meters. This greatly alters the heat exchanges that take place between the overriding air and the underlying water. Since most Arctic observations have been taken at surface stations established for prolonged periods of time on slowly circulating ice floes, most of which have been selected for their potential durability (such as ice thickness), observations may have built-in biases toward more extreme conditions.

The pattern of surface air temperatures and pressures are very asymmetrically located in the Arctic basin with respect to

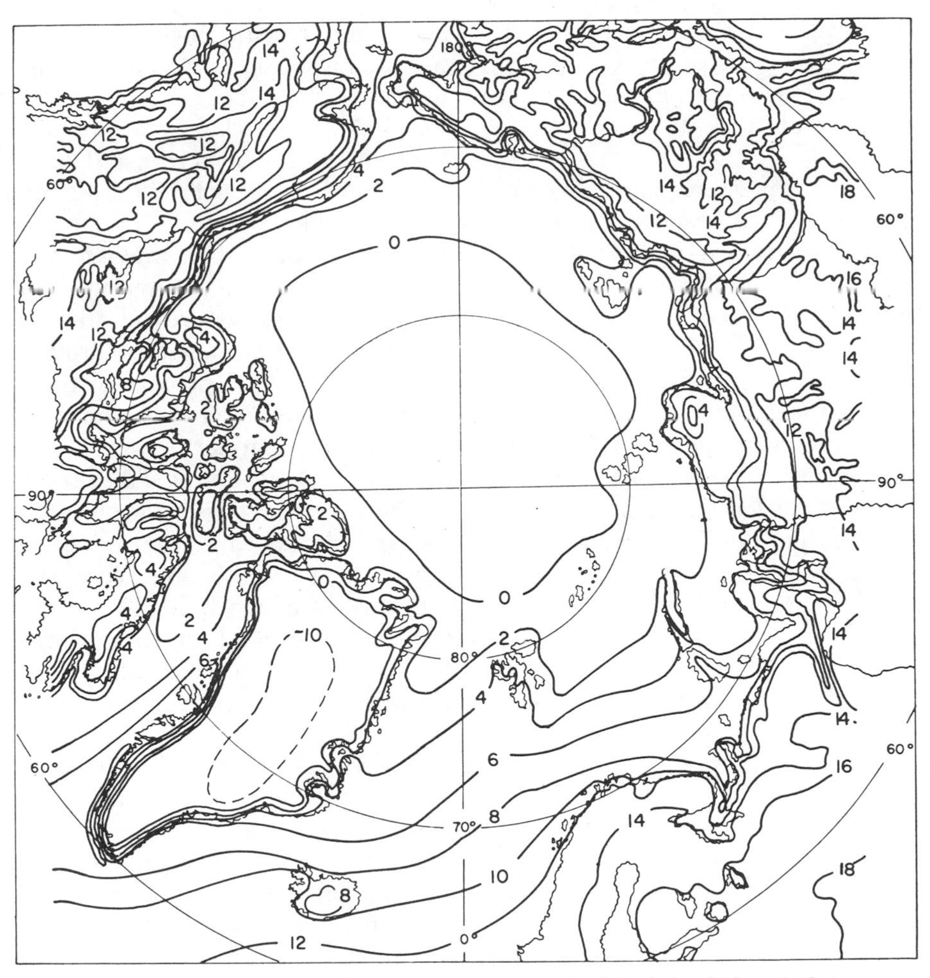

Figure 21.6. Mean surface air temperature in °C, July. (After Prik.)

the North Pole. This stems from the great influence of the North Atlantic Drift (extended Gulf Stream), which reaches nearly to the Pole through the broad gap between Greenland and Scandinavia. The warming effect is particularly great during winter when the water of the North Atlantic Drift around the latitude of Iceland maintains a temperature of 6.7°–8.1°C. On the other hand, the North Pacific Drift has little if any influence on the Arctic, as does the Aleutian Low in the north Pacific, since it is limited to the north by mountain ranges on both the Siberian and Alaskan sides of the narrow Bering Strait.

Surface air temperatures during winter decrease over a distance of about 2000 kilometers from around freezing in the central Norwegian Sea to around −32°C near the northeast coast of Greenland (Figure 21.5). The coldest area over the Arctic at this time is to the north of Greenland and the eastern portion of the Canadian Archipelago. There is some melioration of temperatures to the north of the Bering Strait, but surface temperatures there average around −20°C, as compared to about +2°C at similar latitudes over the Norwegian Sea diametrically across the Arctic. It appears that if Greenland were not present, the North Atlantic Drift and Icelandic low pressure cell in winter would dominate the entire Arctic, and the climatically significant boundary between the Arctic Ocean and the rest of the oceans of the Northern Hemisphere would be at or near the Bering Strait.

The surface air pressure pattern in winter essentially divides the Arctic into two parts, an Atlantic part, from approximately 90°W longitude eastward to 90°E longitude, which is dominated by the northern portions of the Icelandic Low, and the other half of the Arctic between America and Siberia, which is occupied by a ridge of pressure that joins the conspicuous cold high pressure cells that develop over eastern Siberia and northwestern Canada at that time of year. The Aleutian Low in the north Pacific is essentially barred from the Arctic by the mountains across the Bering Strait (see Figure 6.4).

During summer the surface air temperature pattern over the Arctic is much more symmetrical around the North Pole. The latent heat use by the melting of snow and ice holds surface air temperatures at approximately 0°C over much of the ocean area during summer (Figure 21.6). Strong temperature gradients develop in coastal areas, particularly along the Siberian and American coasts. The surface air pressure pattern at this time of year is rather ill defined (see Figure 6.5).

The Temperature Inversion. There are two main areas of semipermanent inversions in the world: the subtropical belt and the polar regions. The one in the subtropical belts is above the surface, and the one in the polar regions is usually at the surface. Gradients are usually much steeper in the polar inversions.

The Arctic inversion is often a composite of two or more inversions, a surface one formed by surface cooling and an upper one formed by subsidence, as well as by warm air advection aloft. These may combine to form an inversion layer that averages about two kilometers in the Arctic region and often reaches higher than that. On occasion when a surface radiation inversion merges with an upper subsidence inversion, the total inversion layer may extend through four kilometers of height and have a difference of 25°C between base and top. Maximum temperatures in the upper inversions are often higher than temperatures at the surface. In the central Arctic the surface inversion is present 80–90 percent of the time from January through April, but only about 20 percent of the time in June. The subsidence inversion is most prevalent in summer, however, when it is present about 40 percent of the time, whereas in January it is present only about 10 percent of the time. During March and April the two inversions combined exist 100 percent of the time. This decreases to only 60 percent of the time during June but rises again to more than 80 percent during late summer, fall, and early winter.

Moisture Conditions. The main precipitation over the polar ocean is frontal in nature. It thus decreases toward the pole, since few fronts cross the central Arctic. Annual precipitation of the central ocean probably amounts to no more than 135 mm. per year, water equivalent, which falls mainly in the form of snow during autumn and early winter. Minimum precipitation occurs during late winter. In the central Arctic, 64 percent of the annual precipitation falls during July–December and only 36 percent during January–June.

The central ocean area is much cloudier during summer and fall than in winter and spring, because of the greater availability of moisture from pools of water on the ice during the melting period. A large area in the central portion of the Arctic basin experiences at least 90 percent cloud cover during July, mostly of a stratus nature, while the same area experiences only 35 percent cloud cover during January. The summer stratus has a mean thickness of about 350 meters and extends as large sheets over wide areas as monotonously uniform cloud. The unfrozen portions of the Norwegian Sea, on the other hand, experience much winter cloudiness in the form of cumulus, caused by instability in cold air to the rear of cyclonic storms as they pass over the relatively warm ocean surface. Thus, large areas of the sea between Iceland and Scandinavia experience 80–90 percent sky cover during much of the year.

The stormiest time of year throughout much of the Arctic is autumn, when sea surface temperatures have reached their highest values and latitudinal temperature gradients are greatest. The stormiest portion of the area extends from the Norwegian Sea eastward through the Barents, Kara, and Laptev seas into the east Siberian region. During winter the Arctic region between northwestern Canada, Alaska, and the Soviet Far East is essentially free of cyclonic storms. In the central polar ocean, mean wind speeds at the surface are not great, about 4–5 meters per second, but in well-developed cyclones along the Eurasian coast winds may frequently reach 25–27 meters per second.

During winter the relative humidity of the air over the polar ocean always remains around 100 percent, perhaps more than 100 percent over ice. Hoarfrost accumulates to as much as 20 mm. during September when much open water is still exposed, and this amounts to about 15 percent of thc annual precipitation in the central ocean. The melting of the ice during summer leads to the formation of persistent fog and low clouds. Usually more than 100 days per year experience fog in polar ocean stations, most frequently during summer. Arctic sea fog also occurs frequently above leads of open water in the ice.

Table 21.1 Temperature and Precipitation Data for the Polar Regions

	Jan	Feb	Mar	Apr	May	June	July	Aug	Sept	Oct	Nov	Dec	Yearly average
ANTARCTIC													
1. Base General Belgrano (77°58′S, 38°48′W, 50 m)													
T.(°C)	−6	−13	−21	−27	−30	−32	−33	−33	−31	−22	−13	−6	−22
2. Little America (78°18′S, 163°00′W, 40 m)													
T.(°C)	−7	−13	−22	−28	−26	−25	−30	−31	−30	−21	−13	−6	−21
3. Mirny (66°33′S, 93°01′E, 30 m)													
T.(°C)	−2	−5	−10	−14	−16	−16	−17	−17	−17	−14	−7	−3	−12
P.(mm)	13	19	51	44	92	67	77	95	52	43	46	26	625
4. Plateau Station (79°15′S, 40°30′E, 3625 m)													
T.(°C)	−34	−44	−57	−66	−66	−69	−68	−71	−65	−60	−44	−32	−56
5. South Pole (90°00′S, 2800 m)													
T.(°C)	−29	−40	−54	−59	−57	−57	−59	−59	−59	−51	−39	−28	−49
6. Vostok (78°28′S, 106;°48′E, 3488 m)													
T.(°C)	−33	−44	−57	−66	−66	−66	−67	−68	−66	−57	−44	−33	−56
ARCTIC													
7. Alert, Greenland (82°30′N, 62°20′W, 63 m)													
T.(°C)	−32	−33	−33	−24	−11	−0	4	1	−10	−20	−26	−30	−18
P.(mm)	6	6	6	6	9	12	15	28	30	16	6	7	146
8. Bukhta Tikhaya, Franz Josef Land (80°19′N, 52°48′E, 6 m)													
T.(°C)	−17	−17	−14	−17	−8	−1	1	1	−2	−8	−13	−17	−9
P.(mm)	5	8	5	5	5	8	18	28	18	8	5	5	117
9. Isachsen, Canada (78°47′N, 103°32′W, 25 m)													
T.(°C)	−35	−37	−35	−24	−12	0	4	1	−8	−19	−28	−32	−19
P.(mm)	2	2	1	4	8	3	22	23	18	10	4	2	98
10. Isfjord Radio, Svalbard (Spitsbergen) (78°04′N, 13°38′E, 9 m)													
T.(°C)	−11	−11	−12	−9	−3	2	5	4	1	−3	−6	−9	−4
P.(mm)	29	30	33	17	20	24	30	38	38	46	39	34	378
11. Ivigtut, Greenland (61°12′N, 48°10′W, 30 m)													
T.(°C)	−5	−5	−3	0	5	9	10	9	6	2	−2	−4	2
P.(mm)	92	129	87	79	89	96	82	97	162	172	146	77	1308
12. Nord, Greenland (81°36′N, 16°40′W, 35 m)													
T.(°C)	−30	−30	−33	−23	−11	0	4	2	−8	−19	−24	−26	−16
P.(mm)	23	20	8	5	3	5	12	19	21	16	35	37	204
13. Ostrov Domashniy, Severnaya Zemlya (79°30′N, 91°08′E, 3 m)													
T.(°C)	−26	−25	−28	−22	−10	−2	1	0	−3	−11	−19	−25	−14
P.(mm)	5	5	3	3	5	8	28	13	15	5	3	5	96
14. Ostrov Kotelny, New Siberian Islands (76°00′N, 137°54′E, 10 m)													
T.(°C)	−29	−30	−29	−21	−9	0	3	2	−1	−10	−21	−26	−14
P.(mm)	3	3	3	1	5	10	31	25	13	5	3	5	105
15. Ostrov Vrangelya (Wrangel Island) (70°58′N, 178°32′W, 3 m)													
T.(°C)	−24	−26	−23	−17	−8	1	3	2	−2	−8	−17	−21	−12
P.(mm)	5	5	5	5	5	10	15	23	13	10	3	5	104

EQUIVALENT UNITS AND CONVERSIONS

Entity	International System(SI)	cgs System	English System	Conversion
Energy(heat)	Joule(J)	Calorie(cal.)		1 cal.=4.1868J
Power	Watt(W)=1 J/sec.			
Radiant Flux Density	W/m^2	Cal/cm^2min Langley(ly)=1 cal/cm^2		1 ly/min=697.5 W/m^2
Temperature	Degrees Kelvin(K)	Degrees Celsius(°C)	Degrees Fahrenheit(°F)	K=C+273 F=9/5C+32 C=5/9(F−32)
Mass	Kilogram(kg)	kg	pound (lb)	1 kg=2.2 lb.
Length		millimeter(mm) centimeter(cm)	inch(″)	1 mm=.0394″ 1 cm=.394″
	Meter(m)	meter(m) kilometer(km)	foot(′) mile(mi)	1 m=3.28′=39.37″ 1 km=.62137 mi.
Force	Neuton(N) 1N=1 $kg.m/sec^2$			
Pressure	Kilopascal (kPa) 1 Pa=1 N/m^2	millibar (mb)	inches of mercury	1 kPa=10 mb 1 inch≃34 mb
Speed	meter/second(m/s)	m/s	miles per hour(mph) knots	1 m/s=2.237 mph 1 m/s=1.943 knots
Area		km^2	$mile^2$	1 km^2=.3861 mi^2

FURTHER READINGS

General Books

Anthes, Richard A., Hans A. Panofsky, John J. Cahir, and Albert Rango. *The Atmosphere.* 2nd ed. Columbus, Ohio: Charles E. Merrill, 1978.

Barrett, E.C. *Climatology from Satellites.* London: Methuen, 1974.

Barry, Roger G., and Richard J. Chorley. *Atmosphere, Weather, and Climate.* 4th ed. New York: Methuen, 1982.

Barry, Roger G., and A.H. Perry. *Synoptic Climatology: Methods and Applications.* London: Methuen, 1973.

Battan, Louis J. *Fundamentals of Meteorology.* Englewood Cliffs, N.J.: Prentice-Hall, 1979.

Boucher, Keith. *Global Climate.* New York: Halsted, 1975.

Chang, Jen-hu. *Atmospheric Circulation Systems and Climates.* Honolulu: Oriental Publishing, 1972.

Cole, Franklyn W. *Introduction to Meteorology.* 2nd ed. New York: John Wiley & Sons, 1975.

Critchfield, Howard J. *General Climatology.* 4th ed. Englewood Cliffs, N.J.: Prentice-Hall, 1983.

Gedzelman, Stanley David. *The Science and Wonders of the Atmosphere.* New York: John Wiley & Sons, 1980.

Griffiths, John F., and Dennis M. Driscoll. *Survey of Climatology.* Columbus, Ohio: Charles E. Merrill, 1982.

Hidore, John Jr. *Introduction to Weather and Climate.* Champaign, Ill.: Park Press, 1985.

Landsberg, Helmut E., ed. *World Survey of Climatology.* Amsterdam: Elsevier Science. 15 volumes, 1969–present. The most comprehensive, up-to-date compendium of information on the climate of the earth.

Lockwood, John G. *World Climatic Systems.* Baltimore: Edward Arnold, 1984.

Mather, John R. *The Climatic Water Budget in Environmental Analysis.* Lexington, Mass.: Lexington Books/D.C. Heath, 1978.

Neiburger, M., J.G. Edinger, and W.D. Bonner. *Understanding Our Atmospheric Environment.* 2nd ed. San Francisco: W.H. Freeman, 1982.

Oliver, John E., and John J. Hidore. *Climatology: An Introduction.* Columbus, Ohio: Charles E. Merrill, 1984.

Petrov, M.P. *Deserts of the World.* New York: John Wiley & Sons, 1977.

Riehl, Herbert. *Introduction to the Atmosphere.* 3rd ed. New York: McGraw-Hill, 1978.

Riley, Dennis, and Lewis Spolton. *World Weather and Climate.* 2nd ed. New York: Cambridge University Press, 1981.

Roberts, Walter Orr, and Henry Landsford. *The Climate Mandate.* San Francisco: W.H. Freeman, 1979.

Rumney, George R. *Climatology and the World's Climates.* New York: Macmillan, 1968.

Saltzman, Barry, ed. *Theory of Climate.* Vol. 25, Advances in Geophysics. Orlando, Fla.: Academic Press, 1983.

Stringer, E.T. *Foundations of Climatology.* San Francisco: W.H. Freeman, 1972.

______. *Techniques of Climatology.* San Francisco: W.H. Freeman, 1972.

Trewartha, Glenn T. *The Earth's Problem Climates.* 2nd ed. Madison: University of Wisconsin Press, 1981.

Trewartha, Glenn T., and Lyle H. Horn. *An Introduction to Climate.* 5th ed. New York: McGraw-Hill, 1980.

Circulation Systems

Dutton, J.A. *The Ceaseless Wind: An Introduction to the Theory of Atmospheric Motion.* New York: McGraw-Hill, 1976.

Gossard, E.E., and W.H. Hooke. *Waves in the Atmosphere.* Amsterdam: Elsevier, 1975

Hoskins, Brian, and Robert Pearce, eds. *Large-Scale Dynamical Processes in the Atmosphere.* Orlando, Fla.: Academic Press, 1983.

Palmen, E., and C.W. Newton. *Atmospheric Circulation Systems: Their Structure and Physical Interpretation.* New York: Academic Press, 1969.

Perry, A.H., and J.M. Walker. *The Ocean-Atmosphere System.* New York: Longman, 1977.

Tropical Climatology

Ayoade, J.O. *Introduction to Climatology for the Tropics.* New York: John Wiley & Sons, 1983.
Newell, Reginald E., John W. Kidson, Dayton G. Vincent, and George J. Boer. *The General Circulation of the Tropical Atmosphere.* Vols. 1 and 2. Cambridge: MIT Press, 1972 and 1974.
Ramage, C.S. *Monsoon Meteorology.* Orlando, Fla.: Academic Press, 1970.
Riehl, Herbert. *Climate and Weather in the Tropics.* Orlando, Fla.: Academic Press, 1979.

Meso-Scale Systems

Atkinson, B.W. *Meso-Scale Atmospheric Circulations.* Orlando, Fla.: Academic Press, 1981.
Barry, Roger G. *Mountain Weather and Climate.* New York: Methuen, 1981.
Dewey, Kenneth F. "An Analytical Study of Lake-Effect Snowfall." Ph.D. dissertation, University of Toronto, 1973.
Eagleman, Joe R. *Severe and Unusual Weather.* New York: Van Nostrand Reinhold, 1983.
Elton, Wallace M. "Forest and Lake Breeze Patterns and the LaPorte, Indiana, Rainfall Anomaly." Ph.D. dissertation, Michigan State University, 1970.

Microclimate

Geiger, Rudolf. *The Climate Near the Ground.* Cambridge: Harvard University Press, 1965.
Lee, R. *Forest Microclimatology.* New York: Columbia University Press, 1978.
Oke, T.R. *Boundary Layer Climates.* New York: John Wiley & Sons, 1978.
Rosenberg, N.J. *Microclimate: The Biological Environment.* New York: John Wiley & Sons, 1974.
Yoshino, M.A. *Climate in a Small Area.* University of Tokyo Press, 1975.

Climatic Variation

Bach, Wilfred, Jurgen Pankrath, and Jill Williams, eds. *Interactions of Energy and Climate.* Boston: Kluwer Academic Publishers, 1980.
Budyko, M.I. *Climatic Changes.* Washington, D.C.: American Geophysical Union, 1977.
———. *The Earth's Climate: Past and Future.* Orlando, Fla.: Academic Press, 1982.
Ford, Michael J. *The Changing Climate: Responses of the Natural Flora and Fauna.* London: George Allen & Unwin, 1982.
Gribbin, John, ed. *Climatic Change.* New York: Cambridge University Press, 1978.
Lamb, Hubert H. *Climate: Present, Past and Future.* Vol. 1, *Fundamentals and Climate Now.* London: Methuen, 1972.
———. *Climate: Present, Past and Future.* Vol. 2, *Climatic History and the Future.* London: Methuen, 1977.
Lockwood, John G. *Causes of Climate.* New York: Halsted, 1979.
McCormac, Billy M., and Thomas A. Seliga, eds. *Solar-Terrestrial Influences on Weather and Climate.* Hingham, Mass.: D. Reidel, 1979.
Pittock, A.B., L.A. Frakes, D. Jenssen, J.A. Peterson, and J.Q. Zellman, eds. *Climatic Change and Variability—A Southern Perspective.* New York: Cambridge University Press, 1978.
Skinner, Brian J., ed. *Climates Past and Present.* Los Altos, Calif.: William Kaufmann, 1981.
Smith, Catherine D., and Martin Parry, eds. *Consequences of Climatic Change.* University of Nottingham Dept. of Geography, 1981.

Climate and Vegetation

Bourliere, F., ed. *Tropical Savannas.* Amsterdam: Elsevier Science, 1983.
Budyko, M.I. *Global Ecology: The Circulation of Energy and of Different Types of Matter Within the Biosphere.* Moscow: Progress Publishers, 1980.
Di Castri, F., D.W. Goodall, and R.L. Specht, eds. *Mediterranean-Type Shrublands.* Amsterdam: Elsevier Science, 1981.
Ovington, J.D., ed. *Temperate Broadleaved Evergreen Forests.* Amsterdam: Elsevier Science, 1983.
Walter, Heinrich. *Vegetation of the Earth in Relation to Climate and Eco-Physiological Conditions.* New York: Springer-Verlag, 1973.
West, N.E., ed. *Temperate Deserts and Semi-Deserts.* Vol. 5 of *Ecosystems of the World.* Amsterdam: Elsevier Science, 1982.

Climate and Man

Amiran, David H.K., and Andrew W. Wilson, eds. *Coastal Deserts: Their Natural and Human Environments.* Tucson: University of Arizona Press, 1973.
Bach, W., J. Pankrath, and W. Kellogg, eds. *Man's Impact on Climate.* Amsterdam: Elsevier Science, 1979.
Breuer, G. *Weather Modification.* New York: Cambridge University Press, 1980.
Bryson, Reid A., and Thomas J. Murray. *Climates of Hunger: Mankind and the World's Changing Weather.* Madison: University of Wisconsin Press, 1977.
Glantz, Michael H., ed. *Desertification: Environmental Degradation in and around Arid Lands.* Boulder: Westview, 1977.
Gribbin, John R. *Future Weather: Carbon Dioxide, Climate, and the Greenhouse Effect.* New York: Penguin Books, 1983.
Griffiths, John F. *Climate and the Environment: The Atmospheric Impact on Man.* Boulder: Westview, 1976.
Hess, Wilmot N., ed. *Weather and Climate Modification.* New York: John Wiley & Sons, 1974.
Kellogg, William W., and Robert Schware. *Climate Change and Society: Consequences of Increasing*

Atmospheric Carbon Dioxide. Boulder: Westview, 1981.
Lamb, Hubert H. *Climate, History and the Modern World.* London: Methuen, 1982.
Liss, P.S., and A.J. Crane. *Man-Made Carbon Dioxide and Climatic Change: A Review of Scientific Problems.* Norwich, England: Geo Books, 1983.
Massachusetts Institute of Technology. *Inadvertent Climate Modification: Report of the Study of Man's Impact on Climate.* Cambridge: MIT Press, 1971.
Norwine, Jim, ed. *Climate and Human Ecology.* Houston: D. Armstrong, 1978.
Rotberg, Robert I., and Theodore K. Rabb. *Climate and History.* Princeton: Princeton University Press, 1981.
Seidel, Stephen, and Dale Keyes. *Can We Delay a Greenhouse Warming?* Washington, D.C.: Environmental Protection Agency, 1983.
Wigley, T.M.L., M.J. Ingram, and G. Farmer. *Climate and History: Studies in Past Climates and Their Impact on Man.* New York: Cambridge University Press, 1981.

Applied Climatology

Griffiths, John F. *Applied Climatology, An Introduction.* 2nd ed. London: Oxford University Press, 1976.
Hobbs, John E. *Applied Climatology: A Study of Atmospheric Resources.* Boulder: Westview, 1980.
Mather, John R. *Climatology: Fundamentals and Applications.* New York: McGraw-Hill, 1974.
Oliver, John E. *Climatology: Selected Applications.* New York: John Wiley & Sons, 1981.
Slater, Lloyd E., and Susan K. Levin, eds. *Climate's Impact on Food Supplies: Strategies and Technologies for Climate-Defensive Food Production.* Boulder: Westview, 1981.
Smith, Keith. *Principles of Applied Climatology.* New York: Halsted, 1975.
Taylor, James A., ed. *Climatic Resources and Economic Activity.* New York: Halsted, 1974.

Agroclimatology

Chang, Jen-hu. *Climate and Agriculture: An Ecological Survey.* Chicago: Aldine, 1968.
Cusack, David F. *Agroclimate Information for Development: Reviving the Green Revolution.* Boulder: Westview, 1982.
Jackson, I.J. *Climate, Water, and Agriculture in the Tropics.* New York: Longman, 1977.
Kovda, Victor A. *Land Aridization and Drought Control.* Boulder: Westview, 1980.
Raper, C. David, Jr., and Paul J. Kramer. *Crop Reactions to Water and Temperature Stresses, in Humid Temperate Climates.* Boulder: Westview, 1982.
Reifsnyder, W.E., and L.P. Smith, eds. *Agricultural Meteorology.* Amsterdam: Elsevier Science, 1983.
Rosenberg, Norman J., ed. *North American Droughts.* Boulder: Westview, 1978.

Urban Climate

Bryson, Reid A., and John E. Ross. "The Climate of the City." In Thomas R. Detwyler and Melvin G. Marcus, *Urbanization and Environment.* Belmont, Calif.: Duxbury, 1972.
Chandler, T.J. *Urban Climatology and Its Relevance to Urban Design.* Technical Note no. 149. Geneva: World Meteorological Organization, 1976.
Goward, Samuel N. "The Physical Nature of the Interface as a Factor in Urban Climate." Ph.D. dissertation, Indiana State University, 1978.
Landsberg, Helmut E. *The Urban Climate.* Orlando, Fla.: Academic Press, 1981.

INDEX